Principles
of
Artificial Lift

Principles
of
Artificial Lift

Niladri Kumar Mitra
Former Director (Offshore)
Oil & Natural Gas Commission, India

Adesh Kumar
Deputy General Manager
Oil & Natural Gas Commission, India

ALLIED PUBLISHERS PVT. LTD.

New Delhi • Mumbai • Kolkata • Lucknow • Chennai
Nagpur • Bangalore • Hyderabad • Ahmedabad

ALLIED PUBLISHERS PRIVATE LIMITED

1/13-14 Asaf Ali Road, **New Delhi**–110002
Ph.: 011-23239001 • E-mail: delhi.books@alliedpublishers.com

47/9 Prag Narain Road, Near Kalyan Bhawan, **Lucknow**–226001
Ph.: 0522-2209942 • E-mail: lko.books@alliedpublishers.com

17 Chittaranjan Avenue, **Kolkata**–700072
Ph.: 033-22129618 • E-mail: cal.books@alliedpublishers.com

15 J.N. Heredia Marg, Ballard Estate, **Mumbai**–400001
Ph.: 022-42126969 • E-mail: mumbai.books@alliedpublishers.com

60 Shiv Sunder Apartments (Ground Floor), Central Bazar Road,
Bajaj Nagar, **Nagpur**–440010
Ph.: 0712-2234210 • E-mail: ngp.books@alliedpublishers.com

F-1 Sun House (First Floor), C.G. Road, Navrangpura,
Ellisbridge P.O., **Ahmedabad**–380006
Ph.: 079-26465916 • E-mail: ahmbd.books@alliedpublishers.com

751 Anna Salai, **Chennai**–600002
Ph.: 044-28523938 • E-mail: chennai.books@alliedpublishers.com

5th Main Road, Gandhinagar, **Bangalore**–560009
Ph.: 080-22262081 • E-mail: bngl.books@alliedpublishers.com

3-2-844/6 & 7 Kachiguda Station Road, **Hyderabad**–500027
Ph.: 040-24619079 • E-mail: hyd.books@alliedpublishers.com

Cover page is designed by Shri Arrnavb Mitraa

Website: www.alliedpublishers.com

ISBN: 978-81-8424-764-0

Published by Sunil Sachdev and printed by Ravi Sachdev at Allied Publishers Pvt. Ltd. (Printing Division), A-104 Mayapuri Phase II, New Delhi-110064

Preface

Computers are being used in petroleum production engineering works much more efficiently than ever before in day to day activities. Advance computing facilitates, incredible network are present where huge information is readily available. However reference text books and printed material had and is always most used and reliable method. During my several years of teaching for petroleum engineering courses for acedemics and industry, I have gathered much more experience and witnessed this technological shift in knowledge sharing. During my student life, I deficiently felt that there is a need for good text and reference books on Artificial Lift. Since then I cherished a strong resolve for bringing out a good book myself in future.

My first venture was, "Fundamentals of Floating Production Systems," written exclusively on offshore Technology, based on my life long experience in teaching, writing technical papers and industry. The present book is my second endeavor. I am fortunate to have Shri Adesh Kumar as Co-Author. This book is written primarily for Production Engineers and Petroleum Engineering college students of senior level as well as graduate level.

Although the purpose of this book is to help as well as teaching artificial lift, it is supposed to be useful as a reference book to the engineers, performing artificial application in Petroleum Industries. We recognize that the topic of 'Principle of Artificial lift' is not complete without a basic understanding of the concept regarding well-inflow performance and multiphase flow in pipes. This inflow performance is being elaborated in most easiest manner at very beginning of the book.

Regarding presentation, this book focuses on presenting and illustrating engineering principles used for designing and analyzing well bore lifting systems, rather than in depth Reservoir engineering theories. Since the material of this book is virtually boundless in depth, knowing what to omit was greatest difficulty with its editing. Many of the industry known basic formula are used instead of deriving the same.

Book encompasses numerous examples on Artificial Lift problems and have been included in the text with solutions for better understanding. For most part, each new concept is accompanied by at least one example. Wherever possible, I have chosen examples from published literature or field experience for reader to get practical understanding over pure theoretical knowledge.

This book is based on numerous documents including technical papers, manuals and articles accumulated through years of work in Industry and Institute. Also it has latest software program developed overtime with most proven computational methods and has been included for easy calculation and monitoring.

For wide understanding we have chosen to use conventional oil field measuring units such as barrels per day, feet and millidarcy. Mixed sets of units (SI & FPS) have been used in line with normal practice in the Petroleum Industry.

Finally I like to convey my gratitude along with Shri Adesh Kumar to our family, friends and colleagues for providing us the needed support and understanding as and when required, even during the weekends and holidays without which completion of this book would have not been possible for with valuable contribution and constant encouragement.

It will be my pleasure to see this book receiving well response from the students as well as working Production engineers throughout the E&P Industry.

Niladri Kumar Mitra

*Ex-Director Oil & Natural
Gas Commission, India*

Contents

CHAPTER 1

Introduction

Artificial lift methods continue to be crucial for optimizing oil production. In fact, according to the Society of Petroleum Engineers, more than 80% of the world's oil wells are on some kind of artificial lift. Eventually most of the wells will need some form of artificial lift, since the technology accelerates and then replaces natural well reservoir flow. The main types of artificial lift are gas lift and pumping methods.

The first lift type described in this book is gas lift, works by injecting natural gas deep in the oil well tubing to decrease the density of the liquid column and increase the flow of fluids into the well bore and then to the surface. A high reliability gas lift valve can stimulate production substantially through reduction of well servicing frequency of the well.

The second type of lift, pumping, occurs when additional energy is added to the well system by either electric or hydraulic means.

The chapter 2 contains the inflow performance of flowing oil and gas wells. Several other forms exist such as sucker rod pumping and plunger lift and still others are in development stage. The industry as a whole does not prefer one type of lift over another. Some of the technological applications are still relatively new, and different conditions call for specific methods. A number of factors play into each type of application: bubble size, gas injection point, flow control, quality of data and the various difficulties associated with the depth, distance, reservoir pressure and expense of oil production etc.

The various types of artificial lift systems which are mostly used:
- Gas lift
- Electrical submersible pump
- Sucker rod pump

- Progressive cavity pump
- Plunger lift
- Jet pump.

Chapter 3 presents the principles of Sucker rod pumping operation and maintenance.

They are large cylinders with both fixed and moveable elements inside. The most important components are: the barrel, valves (travelling and fixed) and the piston. It also has another 18–30 components which are called fitting. The pump is designed to be inserted inside the tubing of a well and its main purpose is to gather fluids from beneath it and lift them to the surface.

Chapter 4 presents the Gas lift method of artificial lift techniques.

It is widely used artificial lift method. As the name denotes, gas is injected in the tubing of an oil well to reduce the weight of the hydrostatic column, thus reducing the back pressure on the formation and allowing the reservoir pressure to push the mixture of produced fluids and gas up to the surface. The gas lift can be deployed in a wide range of well conditions (up to 30000 bpd and down to 15000 ft). They handle very well abrasive elements and sand, and the cost of work over is minimized. The gas lifted wells are equipped with side pocket mandrel and gas lift injection valves. This arrangement allows a deeper gas injection into the tubing. But it has some disadvantages. There has to be a source of gas, some flow assurance problems like hydrates can be triggered by the gas lift.

Chapter 5 presents the design and operation of Electrical Submersible Pumps (ESP).

They consist of a) down hole pump, which is a series of centrifugal pumps, b) a separator or protector, which function is to prevent the produced fluids from entering the electric motor, c) the electrical motor, which transforms the electrical power into kinetic energy to turn the pump and d) an electrical power cable that connects the down hole motor to the surface control panel. ESP is a very versatile artificial lift method and can be found in operating environments all over the world. They can handle a very wide range of flow rates (from 200–9000 bpd) and lift requirements (from virtually zero to 10000 ft (3000 m) of lift). They can be modified to handle contaminants commonly found in oil, aggressive corrosive fluids like H_2S and CO_2 and exceptionally high down hole temperatures.

Increasing water cut has been shown to have no significant detrimental effect on ESP performance. It is possible to locate them in vertical, deviated or horizontal wells, but it is recommended to deploy them in a straight section of casing for optimum run life performance. Although latest developments are aimed to enhance ESP capabilities to handle gas and sand, they still need more technological development to avoid gas locked and internal erosion. Until recently, ESPs have come with an often prohibitive price tag due to the cost of deployment.

Chapter 6 addresses the design and operation of Hydraulic jet pumping systems.

They transmit energy to the bottom of the well by means of pressurized power fluid that flows down in the wellbore tubular to a subsurface pump. There are two types of hydraulic subsurface pumps: a) a reciprocating piston pump, where one is powered by the injected fluid while the other pumps the produced fluids to surface and b) jet pump, where the injected fluid passes through a nozzle creating a jet effect pushing the produced fluids to surface.

These systems are versatile and have been used in shallow depths (1000 ft) to deeper wells (18000 ft), low rate of production in the 10 barrels/day to an excess of 10000 barrels/day (160 m³/d). In addition to this, certain fluids can be mixed with the injected fluid to help or control with corrosion, paraffin and emulsion problems. They are also suitable for wells where conventional pumps such as the rod pump are not possible due to crooked or deviated nature of the wells.

These systems also have some disadvantages. They are sensitive to solids and it is the efficient lifting method. While typically the cost of deploying these systems has been very high, new coiled tubing umbilical technologies are in some cases greatly reducing the cost.

Chapter 7 describes design and operation of Progressive Cavity Pump (PCP).

They are also widely applied in the oil industry. It consists of a stator and a rotor. The rotor is rotated using either a top side motor or a bottom hole motor. The rotation creates sequential cavities and the produced fluids are pushed to surface. The PCP is a flexible system with a wide range of applications in terms of rate (up to 5000 lbs/day and 6000 ft). It offers outstanding resistance to abrasives and solids but they are restricted to setting depths and temperature. Some components of the produced fluids like aromatics can also deteriorate the stators elastomer.

Although, it was a sophisticated idea that was not widely accepted when this initiative stage intelligence with real time production monitoring and surveillance technology is developed world wide. The deployment of intelligent lifting systems and techniques has increased during the past 5 yrs, which has given complex decision support tools. Applying real time production operations with artificial lift intelligence and advanced control application has shown values of optimizing production, minimizing downtime and reduced costs of producing wells and fields. In a huge offshore pipeline network need to have both pressure and flow control of injection gas quantities. Where gas has to be distributed to many other offshore plat-forms. These processes can be optimized by digital distributed control system with fiber optic network and SCADA system.

APPENDIX D

Artificial Lift Feature Analysis

The table adopts many arguments originally authored by S.G. Gibbs and published by Brown (1982). Other arguments, reflecting the experience of major operating companies were presented by Neely *et al.* (1981). These chapters are useful for comparative analysis of various artificial lift methods.

Table 1.1: Features of Various Artificial Lift Methods

Sucker Rod Pumps	
Positive Features	*Negative Features*
Reliable—long time between failures.	Crooked holes present friction problem.
Relatively simple—rugged mechanical system.	High Solids Production is troublesome.
Possible to vary rate and easy to match well capacity.	Gassy wells usually lower volumetric efficiency.
Units easily changed to other wells with minimum cost.	Depth limited, primarily due to rod capability.
Efficient, simple, and easy for field people to operate.	Obtrusive in urban locations.
Applicable to slim holes and multiple completions.	Heavy and bulky in offshore operations.
Can pump a well down to very low pressure.	Susceptible to paraffin problems.
Easy to determine pumping problems.	Tubing cannot be internally coated for corrosion protection.
Can lift high-temperature and viscous oils.	H_2S limits depth at which a large-volume pump can be set.
Can use gas or electricity as power source.	Limitation of downhole pump design in small-diameter casing.
Corrosion and scale treatments easy to perform	
Applicable to pump-off control if electrified.	
Available in different sizes.	

(Contd...)

(Table contd...)

Can pump large volumes from great depth.	Large power oil storage.
Crooked holes present minimal problems.	Relatively high failure rate.
Unobtrusive in urban locations.	Repaired by specially trained mechanics.
Power source can be remotely located.	High solids production is troublesome.
Easy to vary rate and match to well capacity.	Operating costs are sometimes higher.
Can use this for electricity as power source.	Usually susceptible to gas interference, usually not vented.
Downhole pumps can be circulated out in free systems.	Vented installations are more expensive because of extra tubing required.
Can pump a well down to fairly low pressure.	Treating for scale below packer is difficult.
Applicable to multiple completions.	Difficult to obtain valid well tests in low volume wells.
Applicable offshore operation	Requires two strings of tubing for some installations.
Closed system will combat corrosion.	Problems in treating power water where used.
Easy to pump in cycle by time clock.	Safety problem for high-surface-pressure power oil.

Electric Submersible Pumps	
Positive Features	*Negative Features*
Can lift extremely high volumes.	High failure rate—short time between failures.
Unobtrusive in urban locations.	Long downtime for repairs—requires tubing retrieval
Simple to operate.	Not applicable to multiple completions.
Easy to install downhole pressure sensor for telemetering pressure to surface via cable.	Only applicable with electrical power.
Crooked holes present no problem.	High voltages (1000V) are necessary.
Applicable offshore.	Impractical in shallow, low—volume wells.
Corrosion and scale treatment easy to perform.	Difficult to match to well capability.
Available in different size.	Expensive to change equipment to match declining well capability.
Lifting cost for high volumes generally very low.	Cable causes problems in handling tubulars.
	Cables deteriorate in high temperatures.
	System is depth limited (10,000 ft) due to cable cost and inability to install enough power downhole (depends on casing size).
	Gas and solids production are troublesome.
	Casing size is a limitation.
	Difficult for well problems analysis.

Jet Pumps	
Positive Features	*Negative Features*
Simple to vary production rate.	Relatively inefficient lift method.
Retrievable without pulling tubing.	Requires at least 20% submergence to approach best lift efficiency.
Has no moving parts.	Design of system is more complex.
No problems in deviated or crooked holes.	Pump may cavitate under certain conditions.
Unobtrusive in urban locations.	Very sensitive to any change in backpressure.
Applicable offshore	The producing of free gas through the pump causes reduction in ability to handle liquids.
Can use water as a power source.	Power—oil systems are fire hazard.
Power fluid does not have to be so clean as for hydraulic piston pumping.	High-surface power fluid pressures are required.
Corrosion—scale-emulsion treatment easy to perform.	
Power source can be remotely located.	

Table 1.2

Gas Lift	
Positive Features	*Negative Features*
Can produce high rates from high-productivity wells.	High initial investment.
Flexible, easy to change rate.	Limited reservoir pressure drawdown.
Can handle large volume of solids with minor problems.	Lift gas is not always available.
Unobtrusive in urban locations.	Not efficient in lifting small fields or one-well leases.
Power source can be remotely located.	Difficult to lift emulsions and viscous crudes.
Easy to obtain downhole pressures and gradients.	Not efficient for small fields or one-well leases if compression equipment is required.
Lifting gassy wells is no problem.	Gas freezing and hydrate problems.
Sometimes serviceable with wireline unit.	Problems with dirty surface lines.
Crooked holes present no problem.	Some difficulty in analyzing properly without engineering supervision.
Corrosion is not usually as adverse.	Cannot effectively produce deep wells to abandonment.
Applicable offshore—platforms and subsea completions.	Casing must withstand lift pressure.
	Safety problem with high-pressure gas.

Courtesy: S.G. Gibbs; modified by Brown (1982), Tables 1.1 and 1.2.

CHAPTER 2

Inflow Performance

2.1 Introduction

In the extraordinary process of formation of oil and gas deep under the earth crust, followed by their migration and accumulation as oil and gas reserve, a great amount of energy is stored in them. This energy is in the form of dissolved gas in oil, pressure of free gas, water and overburden pressure. When a well is drilled to tap the oil and gas to the surface, it is a general phenomenon that oil and gas comes to the surface vigorously by virtue of the energy stored in them. Over the years/months of production, the decline of energy takes place and at one point of time, the existing energy is found insufficient to lift the adequate quantity of oil to the surface. From that time onwards, man-made effort is required and this is what is known as artificial lift. In other words artificial lift is a supplement to natural energy for lifting well fluid to the surface.

Therefore, the flow of oil from the reservoir to the surface can be fundamentally termed as self flow period and artificial lift period.

2.2 Definition of Artificial Lift

When a self flowing oil well ceases to flow or is not able to deliver the required quantity to the surface, the additional energy is supplemented either by mechanical means or by injecting compressed gas.

Let us consider a well which can deliver the required quantity of oil to a certain height in the well, say 500 meters from the surface, subsequently artificial lift methods/equipments help to lift the required quantity from 500 meters up to the surface.

2.3 Purpose of Artificial Lift

The purpose of artificial lift is to create a steady low pressure or reduced back pressure in the well bore against the sand face, so as to allow the well fluid to come into the well bore continuously. In this process, a steady stream of production to surface would result. In other words, maintaining a required and steady low pressure against the sand face, which we call steady flowing bottom hole pressure, is the fundamental basis for the design of any artificial lift installation.

2.4 Path-Sectors Influencing the Design of Artificial Lift System

Broadly four main sectors influence the design and analysis of artificial lift system. The first and second is the reservoir component from the periphery of drainage area to around the wellbore and then from around the well bore to the wellbore which represent the well's ability to give up fluids into the well bore. The third component of flow path is the entire tubing in the vertical/inclined/horizontal path which include all systems like, down-hole artificial lift equipment, sub-surface safety valves, non return valves etc. The fourth component includes the surface flow path which consists of length and diameter of flow-line, valves, bends, wellhead, chokes, manifold, separator etc.

Any change in the relevant parameters in any of the four sectors, influences the parameters of other sectors. The required changes of parameters should be made till the flow gets steady. The individual sectors of flow-path area have been discussed as under:

2.5 Flow through Porous Medium Around the Well Bore

No definite shape of flow conduit can be conceptualized in this sector of flow through porous medium. So, it is largely an area of concern for determining the flow parameters. In order to understand this, the fundamental concept of Reservoir engineering which includes reservoir drive mechanism and P.I. (Productivity Index) of individual wells are dealt. The productivity index is the measure of the ability of well to produce fluid into the wellbore. Mathematically, it can be expressed as:

$$Q \, \alpha \, (P_r - P_{wf})$$

Where, Q = Total quantity of fluid

$\qquad P_r$ = Reservoir pressure

$\qquad P_{wf}$ = Flowing bottom hole pressure at sand face.

Therefore, Q = Constant $\times$ $(P_r - P_{wf})$. This constant is the productivity index (PI) of the well and is generally abbreviated as "J". In other words,

$$J \, = \, PI \, = \, Q/(P_r - P_{wf}) \hspace{4cm} ... \, (2.1)$$

In fact, *J* is not a constant value but it varies with the type of reservoir, type of drive mechanism, production rate, time of production, cumulative production, perforation density, skin, sand bridging, gas coning, infill wells on production etc.

In order to define PI more correctly, the concept of inflow performance relationship (IPR) is introduced to define the liquid inflow in the wellbore. It is basically a straight line or a curve drawn in the two-dimensional plane, where X-axis is Q, the flow rate and Y-axis is P_{wf}, flowing bottom hole pressure. Therefore, the concept that *J* is always a constant is not correct. *PI* here can be described as just a point on IPR curve. The following are some of the typical IPRs being mainly influenced by different reservoir drive mechanisms.

2.5.1 *IPR in Case of Active Water Drive*

Out of all types of reservoir drives, water drive is regarded as the strongest. However, the intensity differs in different types of water drive reservoirs. Some are moderately weak and some are strong. Such as edge water drive is weaker than bottom water drive. In bottom water drive, when the oil pool is underlain with a large aquifer of dynamic source, reservoir pressure is generally not mellowed at all with the advancing years of production—that is, the reservoir pressure practically remains constant and is not influenced by cumulative production. In this case, the IPR curve will simply be a straight line i.e. the IPR curve will provide only one value of PI.

2.5.2 *IPR in Case of Solution Gas Drive*

This type of drive is also called as internal gas drive or depletion drive. This is the least effective drive mechanism. If excessive draw-down is created, it results in increase of permeability to gas and correspondingly decrease of permeability to liquid, thereby, ability of well to deliver liquids is greatly reduced. Generally, the reservoir pressure for this type of reservoir declines at a very fast rate and accordingly it influences the pattern of IPR curve.

2.5.3 *IPR in Case of Gas Cap Expansion Drive*

This drive mechanism is also called segregation drive because of the state of segregation of oil zone from gas zone, where oil zone is overlain by gas zone called gas cap. Also, as production continues, the gas cap swells and because of this the drive is also known as gas cap expansion drive. This type of reservoir drive mechanism is more effective than solution gas drive and less effective than water drive. Therefore, the profile of IPR curve for gas cap expansion drive lies somewhere in between those for solution gas drive and water drive.

2.5.4 *IPR—When* P$_r$ > *Bubble Point Pressure (Saturation Pressure)*

Up to a point *B* in the profile, *AB* is a straight line representing constant *PI*. At *B*, the gas separation starts in the reservoir. With more draw-down *i.e.* by further dropping-in of bottom-hole pressure, more and more gas will come out and this affects the flow of liquid due to generation of more gas around the wellbore.

2.5.5 *Change of PI with Cumulative Recovery (percentage of original oil in place) with Time*

The pattern of IPR curves, with cumulative recovery that is percentage of oil in place, can be best described when a reservoir is allowed to produce over the years without any pressure maintenance either with the help of water injection or gas injection which results in continuous decrease of reservoir pressure. A series of IPR curves with time are obtained where reservoir pressure indicates a downward trend. The successive IPRs tend to approach the origin (0, 0) of the producing rate - pressure axis. This type of IPR curves trend indicate that the reservoir is attaining fast the state of senescence, as such, reservoir pressure has overbearing effect on the inflow of liquid in the wellbore.

2.6 Vogel's Work on IPR

A publication by Vogel in 1968 offered an extra ordinary solution in determining the Inflow Performance Curve for a solution gas drive reservoir for flow below the bubble point or gas cap drive reservoir or any other types of reservoir having reservoir pressure below bubble point pressure. Vogel's performance curve is generated in the following manner.

From general IPR equation i.e.,

$$J = \frac{q_o}{P_r - P_{wf}} \qquad \qquad \text{... (2.2)}$$

Where P_{wf} is zero, then q_o become maximum and is denoted as q_{max}.

Then,

$$J = \frac{q_{max}}{P_r - 0} \qquad \qquad \text{... (2.3)}$$

or

$$J = \frac{q_{max}}{P_r} \qquad \qquad \text{... (2.4)}$$

Dividing equation (2.3) by (2.4),

$$\frac{J}{J} = \frac{q_o}{P_r - P_{wf}} \times \frac{P_r}{q_{max}}$$

or, $$\frac{q_o}{q_{max}} = \frac{P_r - P_{wf}}{P_r}$$

or, $$\frac{q_o}{q_{max}} = \frac{P_r}{P_r} - \frac{P_{wf}}{P_r}$$

or, $$\frac{q_o}{q_{max}} = 1 - \frac{P_{wf}}{P_r}$$

It is a straight line form of equation.

Since IPR curve below bubble point is not a straight line, he created a parabolic equation from the above.

He distributed $\left\{\frac{P_{wf}}{P_r}\right\}$ in the following manner

20% of $\left\{\frac{P_{wf}}{P_r}\right\}$ & 80% of $\left\{\frac{P_{wf}}{P_r}\right\}^2$

Therefore, the new equation is established as,

$$\frac{q_0}{q_{max}} = 1 - 0.2\left\{\frac{P_{wf}}{P_r}\right\} - 0.8\left\{\frac{P_{wf}}{P_r}\right\}^2$$

This is known as Vogel's equation.

He then plotted dimensionless IPRs in two dimensional plane.

Where X-axis represents $\frac{q_0}{q_{max}}$ and Y-axis represents $\frac{P_{wf}}{P_r}$ (both are dimensionless quantity)

The minimum and maximum values of $\frac{q_0}{q_{max}}$ and $\frac{P_{wf}}{P_r}$ in each case is 0 and 1.0.

When, $\frac{P_{wf}}{P_r} = 1$, $\frac{q_0}{q_{max}} = 0$ and when, $\frac{P_{wf}}{P_r} = 0$, $\frac{q_0}{q_{max}} = 1$

2.7 Standing's Extension of Vogel's IPR for Damaged or Improved Well

While deriving the equation, Vogel assumed that flow efficiency is 1.00 which implies that there was no damage or improvement in the well. Standing extended the Vogel's equation by proposing the comparison chart where he has indicated flow efficiency either more or less than one.

According to him, flow efficiency is defined as,

$$F.E. = \frac{\text{Ideal drawdown}}{\text{Actual drawdown}} = \frac{P_r - P_{wf}^1}{P_r - P_{wf}}$$ (in actual drawdown 'skin' has not been considered)

Where,

$$P_{wf}^1 = P_{wf} + (DP)_{skin}$$

$(DP)_{skin}$ defined by Van Everdingen is as below,

$$(DP)_{skin} = \frac{S_{q\mu}}{2\pi kh}$$

$$DP = \Delta P_S = \frac{162.6 \, q \, B\mu}{kh} \times (0.87 \, s)$$

$$DP = 0.87 \, s = 0.87 \, s \left(\frac{162.6 \, q \, B\mu}{kh} \right)$$

h = Pay thickness
q = Flow rate
μ = Viscosity
k = Permeability (md)
S = Skin factor
S = + indicates damage
S = 0 indicates no damage/no improvement
S = – indicates improvement
B = Formation volume factor (bbl/STB).

Therefore,

$$q_0 / q_{max} = 1 - 2\left(\frac{P_{wf}^1}{P_r} \right) - 0.8 \left(\frac{P_{wf}^1}{P_r} \right)$$

Where, $P_{wf}^1 = P_r - F.E. (P_r - P_{wf})$

[Since, from equation (2.3), $F.E. (P_r - P_{wf})_1 = P_r - P_{wf}^1$ or $P_{wf}^1 = P_r - F.E.(P_r - P_{wf})$

Flow efficiency value has to be either obtained or assumed.]

2.7.1 *Fetkovich IPR Equation*

Fetkovich opined that oil well also behaves like gas wells so that IPR equation being used for gas well will also be applicable for oil wells.

Therefore the equation used for gas wells is also the same as that for oil well.

i.e., $Q_0 = C(P_r^2 - P_{wf}^2)^n$

For determining the value of C, at least one flow test required. Let one flow test.

Then, $C = \dfrac{Q_0}{(P_r^2 - P_{wf}^2)^n}$

2.8 Preparation of Future IPR Curves

For the planning of future requirement of artificial lift and other surface and down-hole infrastructure, it is imperative to know the future production potential of oil wells. Therefore generation of future IPR curves assumes a paramount importance.

Combination of Fetkovich and Vogel procedure for the generation of future IPR curves is being commonly used.

Fetkovich has proposed the future IPR equation by correlating the current reservoir pressure with the productivity indices of the present and future as,

$$Q_{o2} = J_{o1} \frac{P_{r2}}{P_{r1}} (P_{r2}^2 - P_{wf}^2)^n$$

P_{r2} is the future reservoir pressure and P_{r1} is the present reservoir pressure.

Eckmier put forward that the Fetkovich equation of the current and future IPRs for Qmax for both the times can be obtained in the flollowing way,

$$Q_{max} = J_{o1} \left(P_r^2 - P_{wf}^2 \right)^n$$

2.9 Reservoir Inflow Performance

The expression *inflow performance relation* (IPR) customarily is used to define the relation between surface oil rate and well bore flowing pressure. Another expression, *back pressure curve,* is commonly used by engineers dealing with performance of gas wells. *Bottom hole flowing pressure,* p_{wf}, used in the IPR and back pressure equations, is usually expressed at the depth of mid perforations. In this book the term bottom hole flowing pressure is used interchangeably with the term *wellbore flowing pressure.*

Perhaps the simplest and most widely used IPR equation is the straight line IPR, which states that rate is directly proportional to pressure drawdown in the reservoir.

The constant of proportionality is called the productivity index, defined as the ratio of rate to pressure drop in the reservoir. Nowadays, the straight-line IPR is only used for under saturated reservoir, so the equation can be written as,

$$q_o = J(p_r - p_{wf})$$ ... (2.5)

Where p_r is the average pressure in the volume of the reservoir being drained by the well. It is not uncommon to use initial reservoir pressure, P_i, or pressure at the external boundary of the drainage area, P_e, instead of P_r, the difference is inevitably small and can be neglected. Figure 2.1 is a plot of the straight-line IPR. Several important features of the straight-line IPR can be seen in Figure 2.1:

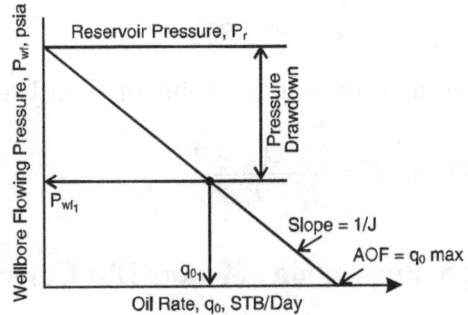

Fig. 2.1

1. By convention, the dependent variable rate defines the x axis and the independent variable, wellbore flowing pressure, defines the y axis.
2. When wellbore flowing pressure equals average reservoir pressure (sometimes referred to as static pressure), rate is zero and no flow enters the wellbore due to the absence of any pressure drawdown.
3. Maximum rate of flow, q_{max} or *absolute open flow*, AOF, corresponds to wellbore flowing pressure equal to zero. Although in practice this may not be a condition at which the well can produce, it is a useful definition and has widespread usage in the petroleum industry, particularly for comparing the performance or potential of different wells in the same field.
4. The slope of the straight line equals the reciprocal of the productivity index (slope = $1/J$).

Example 2.1 illustrates the construction and use of the straight-line IPR.

Example 2.1: Straight Line IPR Calculation

A well was tested for eight hours at a rate of about 38 STB/Day when wellbore flowing pressure was calculated to be 585 psia, based on acoustic liquid level measurements. After shutting the well for 24 hours, the bottom-hole pressure reached a static value of 1125 psia, as per acoustic level readings. The rod pump used on this well is considered undersized, and a larger pump is expected to reduce wellbore flowing pressure to a level near 350 psia (just above the bubble-point pressure). Calculate the following:

1. Productivity index J
2. Absolute open flow based on a constant productivity index
3. Oil rate for a wellbore flowing pressure of 350 psia.
4. Wellbore flowing pressure required to produce 60 STB/Day.

Draw the IPR curve on Cartesian coordinate paper, indicating the calculated quantities.

Solution

1. The productivity index J is calculated from equation (2.5), giving,

   $J = 38/(1125 - 585) = 0.0704$ STB/Day/psi

2. Absolute open flow is merely,

   $q_{max} = JP_r = 0.0704 \,(1125) = 79.2$ STB/Day

3. The rate that can be expected from a flowing well bore pressure of 350 psia is,

   $q = 0.0704 \,(1125 - 350) = 54.6$ STB/Day

4. The wellbore flowing pressure providing a rate of 60 STB/Day is,

   $p_{wf} = 1125 - (60/0.0704) = 273$ psia.

It also indicates the calculated quantities. The straight-line IPR can be derived using Darcy's law and certain simplifying assumptions about rock and fluid properties. Field observation shows conclusively that, for under-saturated oil reservoirs (and water producing wells), equation (2.5) applies with the accuracy needed for well performance calculations. The productivity index as a concept is very useful for describing the relative potential of a well. It combines all rock and fluid properties, as well as geometrical considerations, into a single constant, thus making it unnecessary to consider these properties individually. A constant productivity index states that the ratio of rate to pressure drop is always the same for varying rates:

$$J = q_{o1}/(p_r - p_{wf1}) = q_{o2}/(p_r - p_{wf2}) = \dots = q_o/(p_r - p_{wf})$$

Units of J are *STB/Day/psi*, if the rate is given in *STB/Day* and pressure in psia.

It is important to realize that the concept of stable inflow performance and constant productivity in particular, assumes the condition of *pseudo-steady-state (pss)*. Simply stated, pss represents the condition when the entire drainage volume of a well contributes to production. A certain time is usually required to reach the condition of pseudo steady state. However, in high-permeability formations, pss is reached almost instantaneously. Figure 2.2(a) shows well behavior during a Multi-rate test in a reservoir with high permeability.

In low-permeability reservoir the pss condition may not be reached for years. The time before reaching pss conditions is often referred to as infinite-*acting* flow indicating that the well responds as, if it were in an infinite reservoir. Figure 2.2(b) shows the response of pressure during a Multi-rate flow test for a well producing from a lower-permeability formation (compare with Figure 2.2(a)). The duration of infinite-acting flow may last years in some tight oil and gas wells, in which case the concept of stabilized IPR loses its practical application.

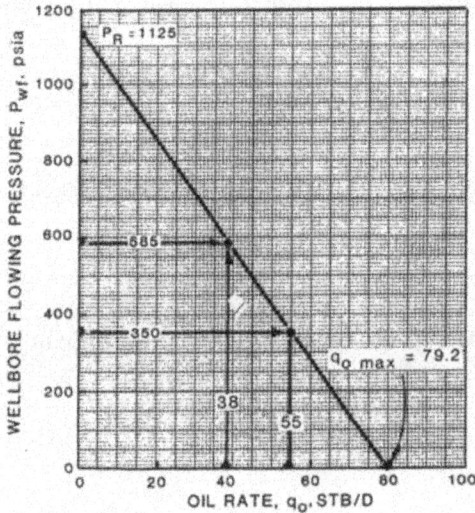

Fig. 2.2(a)

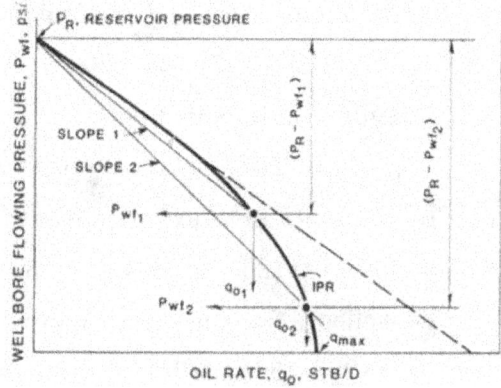

Fig. 2.2(b)

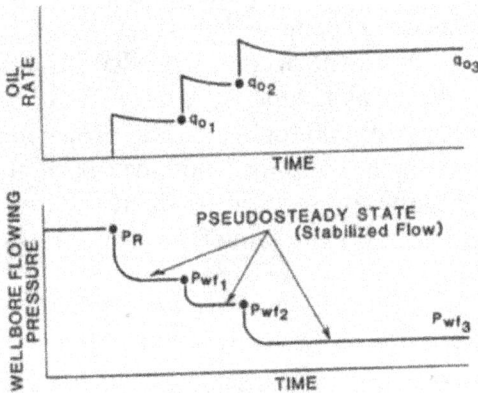

Fig. 2.2(c)

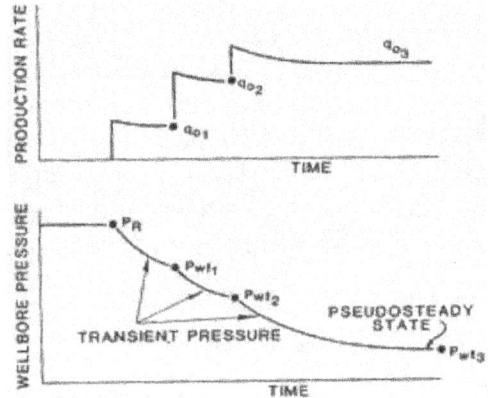

Fig. 2.2(d)

A limitation on the straight-line IPR is the assumption that oil is under-saturated, that is, only slightly compressible. Obviously, this condition does not apply to gases or saturated oil reservoir (which evolve considerable amounts of gas), both of which are highly compressible. The effect of compressible gas and two-phase flow on IPR was observed in the 1920s and 1930s during field testing. Instead of linear rate increase with pressure drawdown, it was observed that larger-than-linear pressure drops were required to increase the rate. The rate pressure relation shows curvature pronounced at higher rates. In terms of productivity index, J decreases with increasing drawdown. Figure 2.3(a) illustrates the continuous variation in J with drawdown. Note that J is not represented by the tangent to the rate pressure curve but is defined as,

$$J = q_0/(p_r - p_{wf})$$

Several equations have been suggested to represent the nonlinear IPR resulting from gas and two-phase flow. The observations of Bureau of Mines Engineers resulted in the simple but accurate relation

$$q = C\ (p_r^2 - p_{wf}^2)^n \qquad \qquad \text{... (2.6)}$$

for both gas and saturated oil wells . The exponent 'n' ranges in value from 0.5 to 1.0. A plot of rate versus $p_r^2 - p_{wf}^2$ (written Δp^2 in shorthand) on log-log paper results in a straight line with slope $1/n$. Note that q is defined along the x-axis and Δp^2 along the y-axis (by convention), as shown in Figure 2.3(b). Equation (2.6) often referred to as the backpressure equation, generally has been accepted for gas wells. It has yet to receive widespread use for oil wells, even though Fetkovich (1973) reconfirmed its general application to oil wells. Example 2.2 shows the use of equation (2.6) for describing the inflow performance of a gas well.

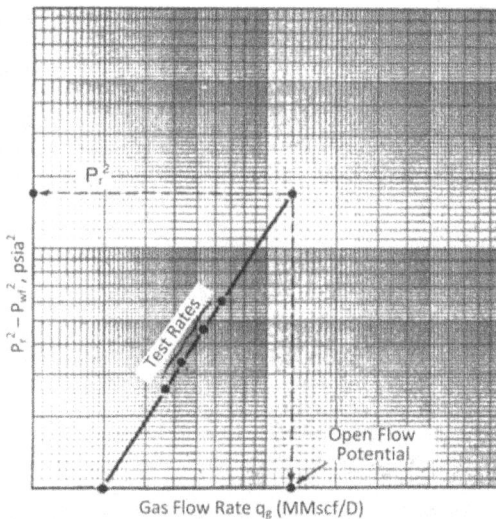

Fig. 2.3(a)

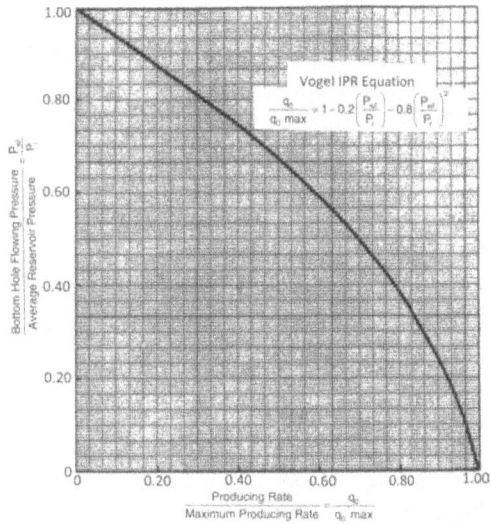

Fig. 2.3(b)

An IPR equation traditionally used to describe oil well performance in saturated oil reservoirs is the Vogel (1968) equation,

$$q_o/q_{omax} = 1 - 0.2(p_{wf}/p_r) - 0.8\,(p_{wf}/p_r)^2 \qquad \qquad \text{... (2.7)}$$

where q_{omax} is the maximum oil rate (AOF) when wellbore flowing pressure, p_{wf} equals zero. Vogel's equation is a best-fit approximation of numerous simulated well performance calculations. No field verification is given by Vogel, although his equation is very simple to use, as explained in example 2.3.

Example 2.2: Multirate gas well test to determine inflow

A gas well produced gas at rates varying from 1.0 to 5.5 MMSCF/Day during a four-point multi-rate isochronal test as given in Table 2.1.

Table 2.1: Multirate Data for Gas Well

Point	p_{wf} (psia)	$p_r^2 - p_{wf}^2$ (psia²)	q_g (MMSCF/D)
S.I.	3355	0.000	0.000
1	3314	0.273×10^6	1.012
2	3208	0.965×10^6	2.248
3	2992	2.304×10^6	3.832
4	2651	4.228×10^6	5.480

1. Calculate the backpressure slope 'n' by plotting rate versus the change in pressure squared $(q_g$ vs. $\Delta p^2)$ on log-log paper.
2. Determine the IPR constant C in equation (2.6) from the log-log plot, and then calculate the absolute open flow (i.e., calculated maximum rate).

Solution

1. The log-log plot of rate versus change in pressure squared is shown in Figure 2.4. The slope of the curve is about 1.61, which gives a value for n of 0.62 $(= 1/1.61)$.

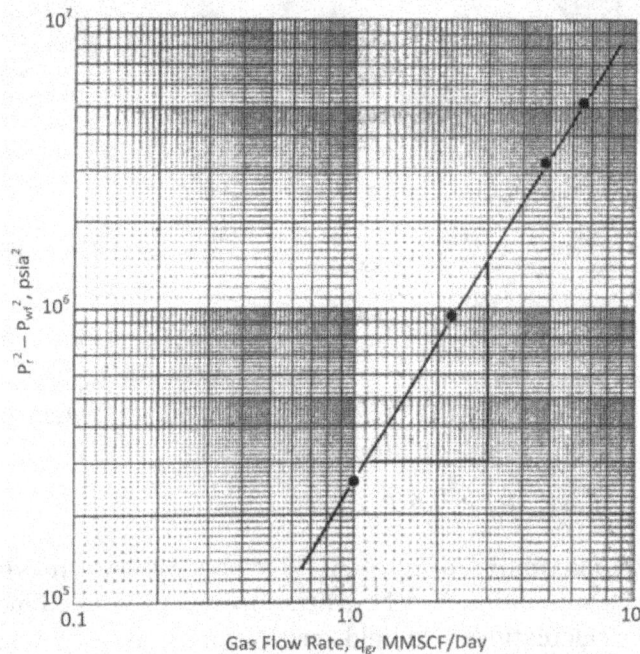

Fig. 2.4

2. Having calculated the value of '*n*', the constant *C* is easily found by choosing an arbitrary point from the straight line (it does not necessarily need to be one of the measured points). Substitute the rate and difference in pressure squared into equation (2.6) and then calculate *C* directly:

$$C = q/(p_r^2 - p_{wf}^2)^n$$
$$= 2.0 \times 10^6/[8 \times 10^6]^{0.62}$$
$$= 438 \text{ scf/D/psi}^{1.24}$$

which gives,

$$q_g = 438 \ (p_r^2 - p_{wf}^2)^{0.62}$$

or in dimensionless form,

$$q_g/q_{gmax} = [1 - (p_{wf}/p_r)^2]^{0.62}$$

where $q_{gmax} = 438 \ (p_r^2)^{0.62} = 438 \ (3355)^{1.24} = 10.3$ MMSCF/D.

Example 2.3: Vogel's IPR equation

An oil well was tested at a rate of 200 STB/Day with a bottom hole flowing pressure of 3220 psia. Bubble point pressure was calculated with a correlation using surface data measured when the well was producing at a low rate. Estimated bubble point of 3980 psia indicates that the well is draining saturated oil, since initial reservoir pressure was measured at 4000 psia. Plot the IPR using Vogel equation.

Solution

$$q_{omax} = 200/[1 - 0.2(3220/4000) - 0.8(3220/4000)^2] = 625 \text{ STB/Day}$$

Now calculate several rates at specified draw-downs to have enough points to plot the IPR. Figure 2.5 shows the plot of the IPR using only the preceding points.

q_o (STB/D)	p_{wf} (psia)
4000	0
3000	250
2000	437
1500	508
1000	562

Perhaps Vogel's principal contribution was the idea of normalizing the IPR equation and including AOF (q_{omax}) as the primary constant to be determined. In fact, if equation (2.6) had been written in the same form as the equation (2.7) in the late 1920s when it was suggested, then it would be obvious that Vogel's equation is substantiated by field observations. We consider the normalized form of equation (2.6)

$$q_o/q_{omax} = [1 - (p_{wf}/p_r)^2]^n \qquad \qquad \text{... (2.8)}$$

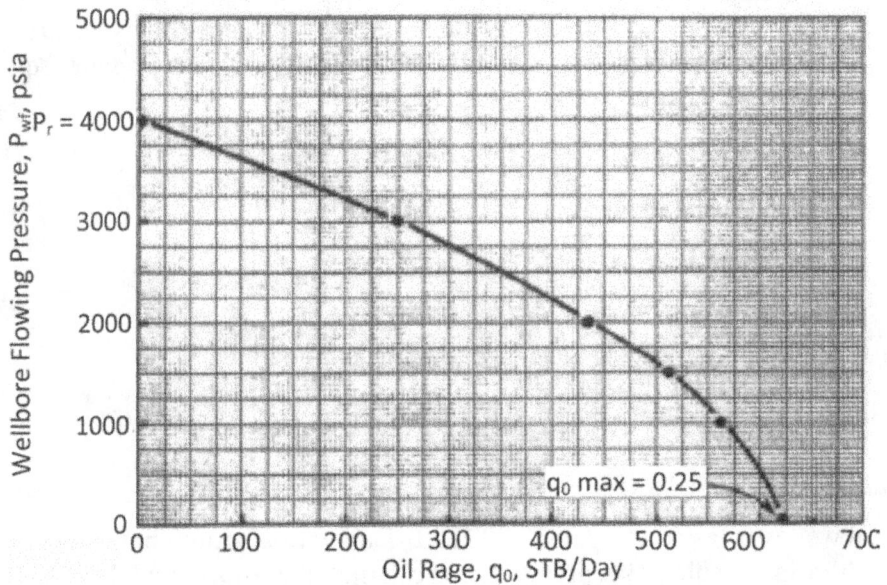

Fig. 2.5

as a better alternative than Vogel's equation for saturated oil reservoir because it is simpler and it considers the effect of high velocity (non-Darcy, turbulent) flow through the inclusion of exponent 'n'.

It is easily shown that equations (2.7) (Vogel) and (2.8) are nearly identical if $n = 1$, as was first noted by Fetkovich (1973). For $n = 1$, equation (2.8) becomes,

$$q_o/q_{omax} = 1 - (p_{wf}/p_r)^2 \qquad \qquad ...\,(2.9)$$

This equation is only slightly different from Vogel's, somewhat more conservative but simpler and easier to use. Most of all, it is based on field observations. In this book, we will consider equation (2.9) as the basic working IPR equation for saturated oil reservoir. If multi-rate data are available, then equation (2.8) is preferred, since it includes the effect of high velocity (turbulent) flow, a factor important for high rate wells. A logarithmic plot of q_o vs $\ddot{A}p^2$ (just as for gas multi-rate tests) will result in a straight line with a slope of $1/n$. Example 2.4 illustrates the analysis of a multi-rate oil well test, using the general IPR equation for a saturated oil reservior (equation 2.8).

Example 2.4: Multi-rate test analysis of a saturated oil well reservoir.

A multi-rate test data for a well are as in Table 2.2.

Table 2.2: Multi-rate Test Data for an Oil Well

Point	p_{wf} (psia)	q_o (STB/Day)	Point	p_{wf} (psia)	q_o (STB/Day)
1	166	2435	9	1194	1474
2	183	2460	10	1267	1260
3	351	2352	11	1342	1045
4	534	2260	12	1470	720
5	787	1965	13	1476	610
6	867	1895	14	1497	565
7	996	1765	15	1558	235
8	1066	1625	16	1600	0

Tasks

1. Calculate the productivity index J, based on the lowest four rates. What is the extrapolated maximum oil rate, based on the straight-line IPR assumption?
2. Plot the data on Cartesian coordinate paper. Use only point 10 (1260 STB/Day, 1267 psia) to determine q_{omax}, with the Vogel equation (equation 2.7). Plot and tabulate the calculated rates corresponding to the bottom hole flowing pressure given in table.
3. Repeat step 2 using the normalized back pressure equation (2.8) with $n = 1$ instead of the Vogel equation.
4. Plot the data on log-log paper to establish if equation (2.6) is valid. What are the constants C and n for this well?
5. Repeat step 2 using equation (2.8), with $n = 0.7$ instead of the Vogel equation. The value $n = 0.7$ is taken from the log-log plot in step 4. What is the calculated AOF?

Solution

1. The lowest rates from Table are: 235, 565, 610 and 720 STB/Day. The calculated values of J for each of these rates are: 5.59, 5.48, 4.92, and 5.54 STB/Day/psi respectively. An average value is 5.4 STB/Day/psi, resulting in a calculated maximum oil rate of 5.4 × 1600 or 8608 STB/Day. Note that the measured rate at the lowest flowing pressure of 166 psia is only 2435 STB/Day, which is much lower than predicted by the straight-line IPR (7743 STB/Day).
2. Figure 2.6(a) is a Cartesian plot of the mulitrate data from Table 2.2. Using the Vogel equation and point 10, the calculated q_{omax} is 3706 STB/D. Table 2.3 lists the calculated rates and compares them with measured values given in Table 2.4. Figure 2.1 plots the calculated IPR and compares it with the measured IPR.
3. Using point 10 and equation (2.8) with $n = 1$ results in a maximum oil rate, q_{omax} of 3379 STB/Day, which is slightly lower as predicted by the Vogel equation.

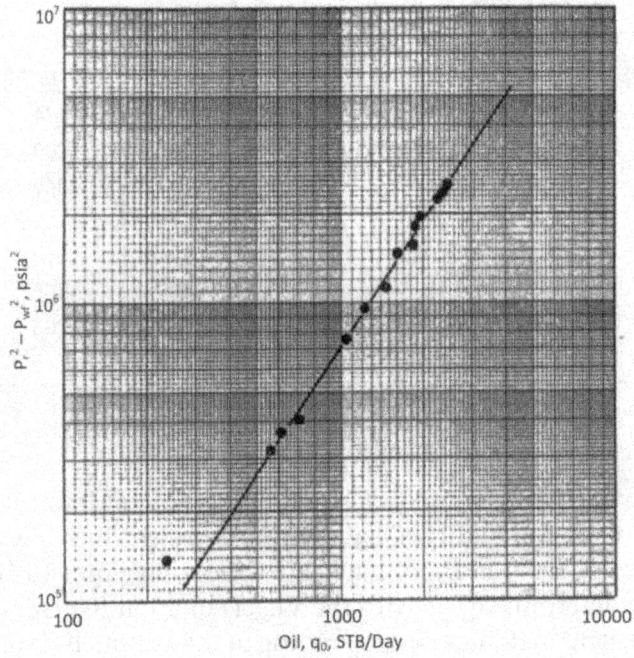

Fig. 2.6(a)

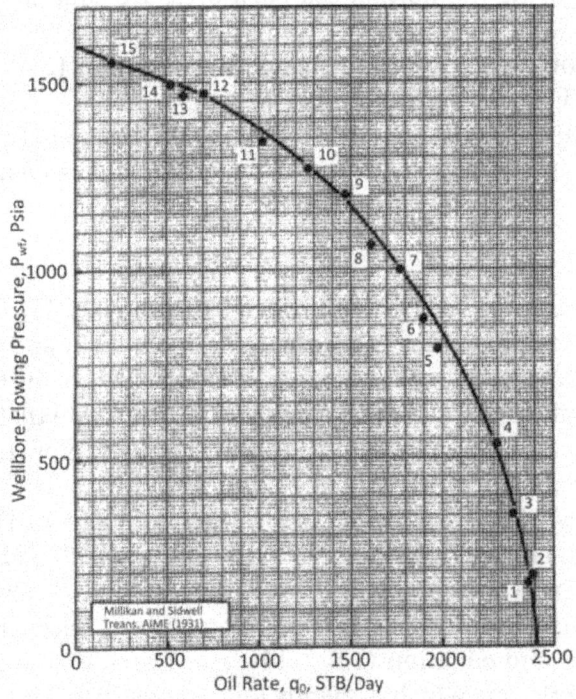

Fig. 2.6(b)

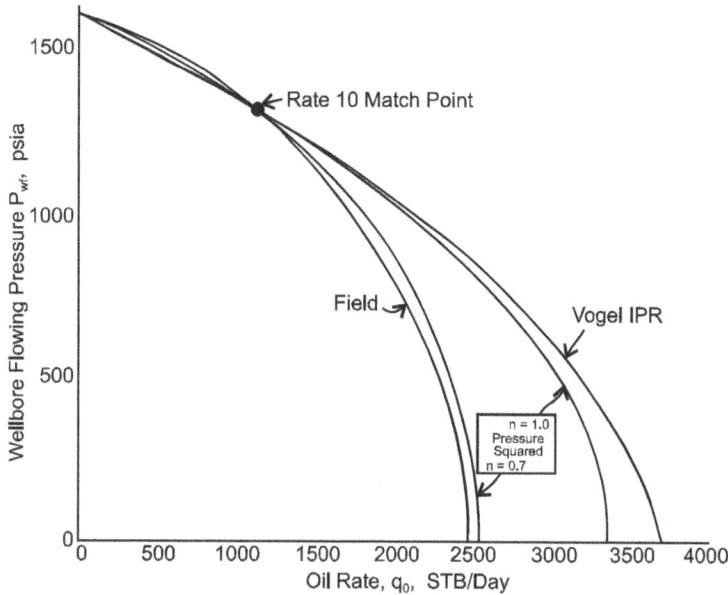

Fig. 2.6(c)

Table 2.3 compares the calculated rates with measured rates at the flowing pressures given in Table 2.2.

Table 2.3: Calculated IPR Results for the Oil Well

Point	p_{wf} (psia)	q_o (STB/Day)			
		Field	Vogel (n = 1)	Pressure	Squared (n = 0.7)
1	166	2435	3597	3343	2494
2	183	2460	3582	3335	2490
3	351	2352	3401	3216	2428
4	534	2260	3128	3003	2314
5	787	1965	2624	2561	2070
6	867	1895	2434	2387	1970
7	996	1765	2096	2070	1783
8	1066	1625	1896	1879	1667
9	1194	1474	1501	1497	1422
10	1267	1260	1260	1260	1260
11	1342	1045	999	1002	1073
12	1470	720	522	527	684
13	1476	610	499	503	663
14	1497	565	417	421	585
15	1558	235	173	175	316
16	1600	0	0	0	0

Note: Rate No. 10 was used to fit the Vogel and pressure-squared (*n* = 1) equations. The pressure-squared equation with *n* = 0.7 was fit using the complete test sequence and a back pressure plot.

4. Figure 2.6(b) is a log-log plot of the multi rate data. The constants in equation (2.6) obtained from the plot are $C = 0.08$ and $n = 0.7$.

5. Using equation (2.8) with $n = 0.7$ results in a maximum oil rate q_{omax} of 2513 STB/Day, which is lower than the rate predicted by either the Vogel equation or equation (2.8) with $n = 1$. Table 2.3 compares the calculated rates ($n = 0.7$) with measured rates at the flowing pressures given in Table 2.1. The calculated IPR is compared with the true IPR in Figure 2.6(c).

From this example we find that the Vogel equation and pressure-squared equation with $n = 1$ are not adequate to describe observed rate-pressure data. The pressure-squared equation with $n < 1$ gives an excellent fit of well test data for saturated oil well, substantiating numerous observations made by Fetkovich (1973) for high-rate oil wells.

Many oil wells produce from reservoirs with pressure above the bubble-point pressure but with well-bore flowing pressure below the bubble point. The IPR for such wells is illustrated in Figure 2.7, showing a straight line at flowing pressures above the bubble point and curvature below. Considering only the straight-line region above the bubble point, for p_b d", p_{wf} d", p_r, the IPR equation is,

$$q_o = J(p_r - p_{wf}) \qquad \qquad \text{... (2.10)}$$

If flowing pressures are below the bubble point, then the following IPR equation can be used:

$$q_o = J(p_r - p_b) + (J/\,2p_b)\,(p_b^2 - p_{wf}^2) \qquad \qquad \text{... (2.11)}$$

where J is the productivity index indicated by production when flowing pressure p_{wf} is above the bubble-point pressure p_b, given by equation (2.10). Example 2.5 illustrates the use of equations (2.10) and (2.11).

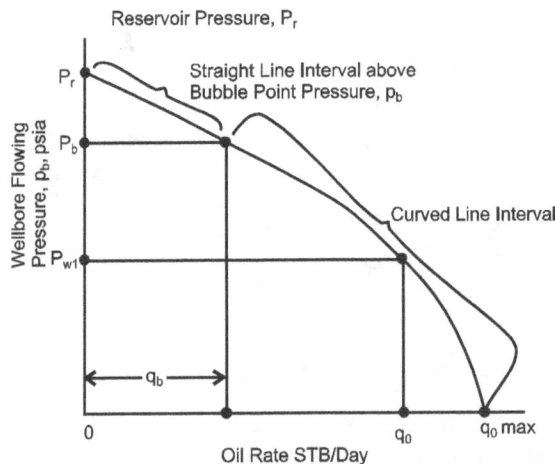

Fig. 2.7

Example 2.5: IPR calculation for an under-saturated oil reservoir/well producing at flowing pressures below the bubble point.

An oil well produced at a rate of 108 STB/Day with a bottom hole flowing pressure of 1980 psia. Surface samples were collected and sent to a PVT laboratory, which determined the bubble-point pressure of the recombined reservoir fluid to be 1825 psia at a temperature of 195°F. An initial reservoir pressure of 3620 psia was recorded during the 48-hour pressure buildup following the flow test.

Tasks

1. Determine the under saturated productivity index J that is valid if flowing bottom hole pressure is greater than the bubble point.
2. Calculate the oil rate if bottom hole flowing pressure is held at the bubble point to avoid gas blockage in the near-well bore region.
3. Using equation (2.11), calculate the maximum oil rate that can be expected from, the well.
4. Calculate and plot the IPR for the entire range of wellbore flowing pressures, both above and below the bubble point.

Solution

1. The productivity index J is simply 108/ (3620–1980), or J = 0.066 STB/Day/psi. If wellbore flowing pressure had been below the bubble point, then equation (2.11) should be solved for J instead of using the definition of productivity index.
2. If the bottom hole flowing pressure equals bubble point pressure, the expected oil rate is,

   q_{ob} = 0.066 (3620–1825) = 118.5 STB/Day

3. The maximum oil rate q_{omax}, expected from the well is found by solving equation (2.11) when, p_{wf} = 0

   q_{omax} = 0.066 (3620–1825) + (0.066/2/1825) (1825^2) = 118.2 + 60.2 = 178.7 STB/Day

4. Figure 2.8 shows the IPR plotted from the equation,

   q_o = 0.066 (3620 – p_{wf}) for p_{wf} > p_b = 1825 psia,
   q_o = 118.5 + 1.81 (10^{-5}) (1825^2 – p_{wf}^2) for p_{wf} < p_b, with several points as tabulated in Table 2.4.

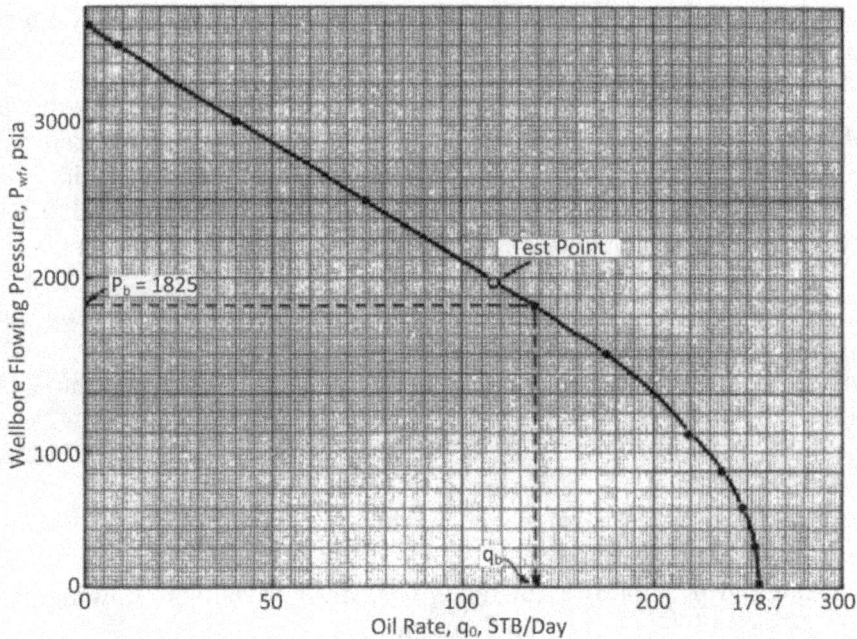

Fig. 2.8

Table 2.4: Calculated IPR for the Oil Well

p_{wf} (psia)	q_o (STB/Day)	p_{wf} (psia)	q_o (STB/Day)
3620	0.00	1500	138.1
3500	7.9	1000	160.7
3000	40.9	750	168.6
2500	73.9	500	174.3
2000	106.9	250	177.7
1825	118.5	0	178.7

2.10 Tubing Performance and Gradient Curves

Pressure drop required to lift a fluid through the production tubing at a given flow rate is one of the main factors determining the deliverability of a well. It therefore appears in most well performance calculations. First, we fix either the wellhead or bottom-hole flowing pressure given the rates of oil, gas, and water. The pressure drop along the production tubing can be calculated by charts or correlations, and the resulting flowing pressure at the other end of the tubing can be determined. For example, if wellhead pressure is specified, then a gradient curve can be used to determine the wellbore flowing pressure at several different oil rates. The resulting relation between bottom-hole flowing pressure and oil rate is called Tubing Performance Relation (TPR) and it is valid only for the specified wellhead pressure.

The pressure drop in tubing due to flow of homogeneous (single-phase) fluid can be calculated by conventional pipe flow equations. Gas and highly undersaturated oil wells come under this category. Just a small quantity of free gas mixed with oil and/or water creates considerably more complicated flow conditions which can be described only by approximately empirical corrections to the conventional pipe flow equations. The tubing performance relationship of wells producing multiphase mixtures is therefore difficult to estimate with any accuracy.

$$q_g = 200,000 \left[\frac{sD^5 (p_{in}^2 - e^S p_{wh}^2)}{} \right]^{0.5}$$ ... (2.12)

where,

q_g = gas flow rate, scf/D
Z = average gas compressibility factor
T = average temperature, °R
f_M = Moody friction factor,
γ_g = gas gravity, air = 1
D = tubing diameter, in
p_{in} = flowing tubing intake pressure, psia
p_{wh} = flowing wellhead pressure, psia
H = vertical depth, ft.
s = 0.0375 γ H/TZ

For dry gas wells there are several methods for calculating pressure loss in vertical or inclined pipe. A simple and accurate equation for vertical flow of gas (Donald Katz *et al.*, 1959, p. 306), which can be solved directly (i.e., without integration or trial and error).

Average temperature is simply the arithmetic average between the temperatures at the wellhead and intake to the tubing (usually reservoir temperature). The average compressibility factor is evaluated at the average temperature and the arithmetic average between the flowing wellhead and intake pressures.

A valid assumption for most gas wells is that flow is turbulent, resulting in an expression for f_M that depends only on the relative roughness of the pipe:

$$f_M = \{2\log [3.71/(\varepsilon/D)]\}^{-2}$$ ... (2.13)

where ε is the absolute pipe roughness and ε = 0.0006 in. for most commercial pipes.

The simplest application of equation (2.12) is for calculating a table of rates vs. flowing intake pressure, given a fixed wellhead flowing pressure and pipe size.

Example 2.6: Illustrate the use of equation (2.12) and (2.13), assuming a fixed wellhead pressure. Higher well head pressures would result in tubing performance curves that shift upward and to the left, while lower wellhead pressures would shift the TPR down and to the right.

Example 2.6: Tubing performance relation for a gas well

A well is to be produced into a high pressure, gas gathering line requiring a minimum wellhead pressure of about 800 psia. Available tubing has $2^{7/8}$-in. nominal diameter (internal dia. is about 2.5 in.). Other relevant data for the well are:

Vertical length of tubing	:	7250 ft
Depth to mid perforations	:	7300 ft
Gas gravity	:	0.75 (air = 1)
Average tubing temperature	:	120 °F
Average gas Z-factor	:	0.78
Pipe roughness	:	0.0006 in.

Use equation (2.12) to calculate the tubing performance relation of this well (up to the rate of 24 MMSCF/Day).

Solution: The first step in calculating the approximate tubing performance relation is to determine s in equation (2.12):

$$s = 0.0375\ (0.75)\ (7300)/(120 + 460)\ (0.78)$$
$$= 0.454$$

The Moody friction factor is estimated from equation (2.13),

$$f_M = \{2\log[3.71/(0.0006/2.5)]\}^{-2}$$
$$= 0.0142$$

The approximate tubing performance relation then becomes,

$$q_g = 9368\ (p\ \ - e^{0.454}\ 800^2)^{0.5}$$
$$= 9368\ (p\ \ - e\ \ \ \ 800^2)^{0.5}$$
$$= 9368\ (p\ \ - 1.0 \times 10^6)^{0.5}$$

A few values of intake flowing pressure are chosen, and Table 2.5 gives the calculated flow rates corresponding to each pressure. Figure 2.9 is a plot of the rate pressure data in Table 2.4.

The approximate TPR equation (2.12) can be used only for dry gas. If water or condensate is produced as an entrained liquid phase (GOR > 7000 scf/STB), then gas velocity generally exceed 18 to 20 ft/sec, if equation (2.12) is used.

Table 2.5: Calculated Tubing Performance Relation for the Gas Well

p_{in} (psia)	$p_r^2 - p_{in}^2$ (psia^2)	q_g (MMSCF/Day)	p_{in} (psia	$p_r^2 - p_{in}^2$ (psia^2)	q_g (MMSCF/Day)
1150	9.9×10^6	5.32	2000	7.3×10^6	16.23
1250	9.7×10^6	7.03	2250	6.2×10^6	18.88
1500	9.0×10^6	10.47	2500	5.0×10^6	21.46
1750	8.2×10^6	13.43	2750	3.7×10^6	24.00

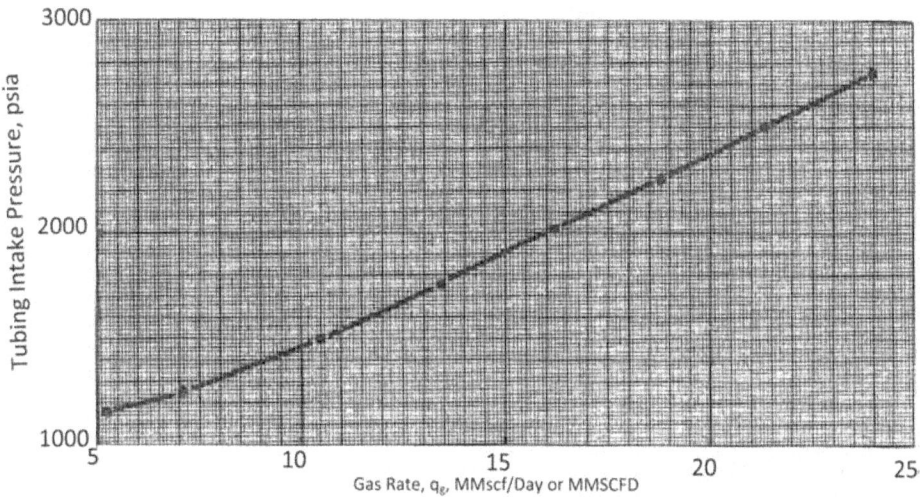

Fig. 2.9

At lower velocities, it has been observed that liquid accumulates, thereby increasing pressure loss considerably above that calculated from equation (2.12). If velocity decreases to 10 to 12 ft/sec, then the well will probably die. Thus equation (2.12) can not be applied to gas-condensate reservoir wells or water producing gas wells with a gas/liquid ratio less than 7000 SCF/STB; gradient curves or multiphase correlations must be used instead.

Pressure elements constituting the total pressure at the bottom of tubing:

1. Back pressure exerted at the surface from choke and wellhead assembly (wellhead pressure).
2. Hydrostatic pressure due to gravity and elevation change between the wellhead and the intake to the tubing.
3. Friction losses, which include irreversible pressure losses due to viscous drag and slippage.

Figure 2.10 illustrates the three components of pressure in a TPR curve for a single-phase liquid, a dry gas, and a two-phase gas/oil mixture. Additional pressure loss due

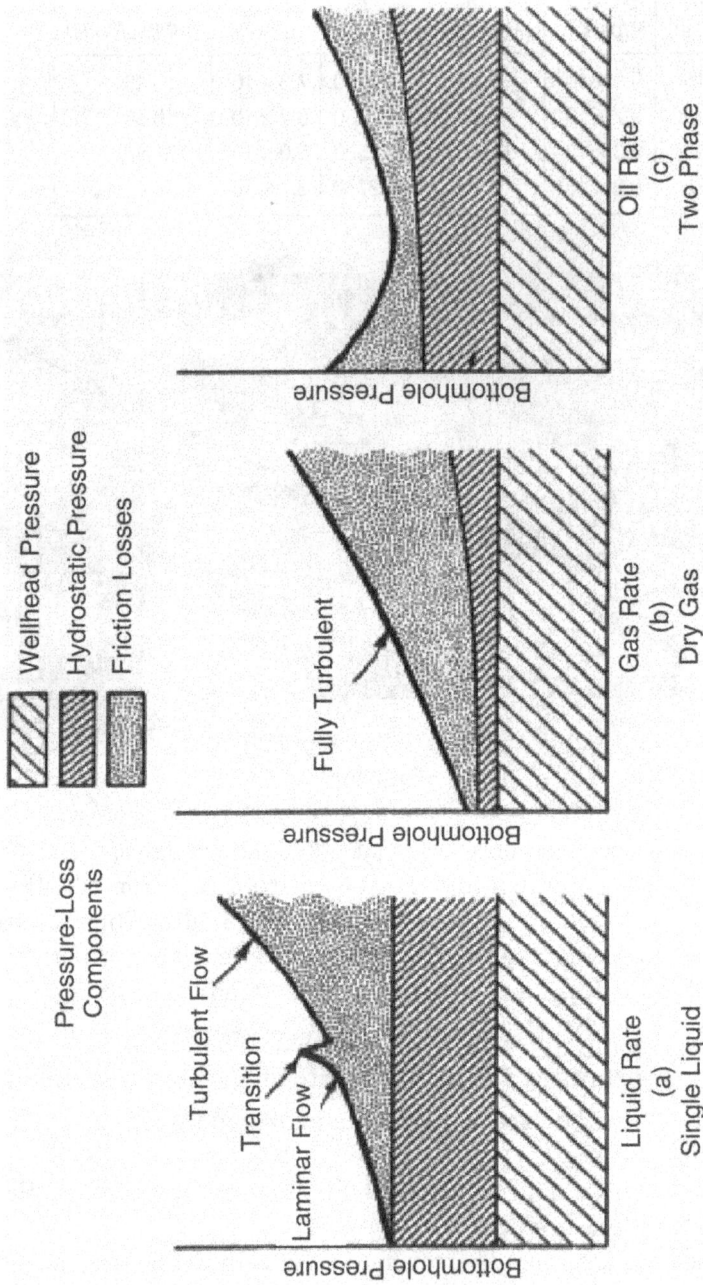

Fig. 2.10

to acceleration of an expanding fluid is usually insignificant when compared with the other losses and therefore neglected in most design calculations.

In the case of single phase liquid (e.g., under-saturated oil or water), density is assumed constant and the hydrostatic pressure gradient (pressure loss per unit length) is a constant. Friction loss, on the other hand, is a rate dependent, characterized by two flow regimes—laminar and turbulent—separated by a transition zone. The rate dependence of friction related pressure loss differs with the flow regime. At low rates the flow is laminar and pressure gradient changes linearly with rate or flow velocity. At high rates, the flow is turbulent and the pressure gradient increases more than linearly with increasing flow rate.

In gas wells, there is interdependence between flow rate, flow velocity, density and pressure. In general, increasing gas rates results in increasing total pressure loss as computed by equation (2.12).

In multiphase mixtures, friction related losses vary with rate in a much more complicated manner than for gas. Increasing rate may change the governing pressure loss mechanism from predominantly gravitational to predominantly frictional. The result of this shift is a change of trend in the TPR curve.

For a given flow rate, wellhead pressure, and tubing size, there is a particular pressure distribution along the tubing, starting its traverse at the wellhead pressure and increasing downward toward the intake to the tubing. The pressure-depth profile is called a pressure traverse, as shown in Figure 2.11.

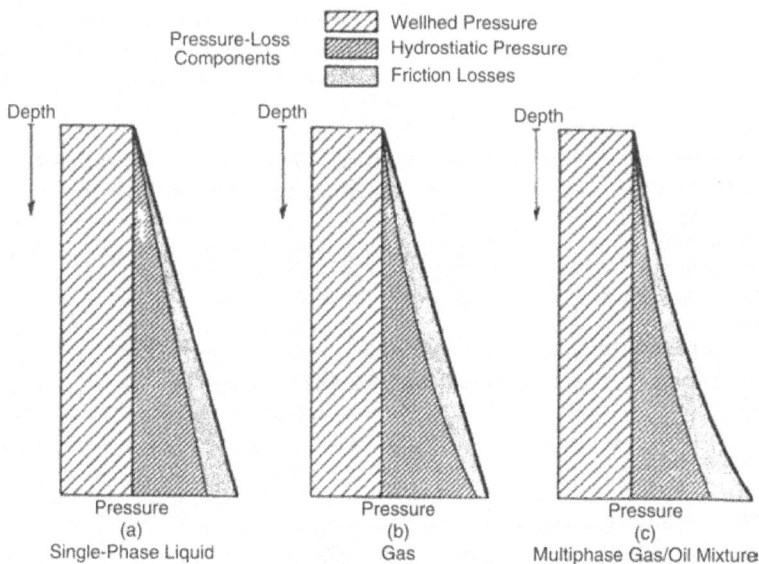

Pressure-Loss Components

Wellhed Pressure
Hydrostiatic Pressure
Friction Losses

(a) Single-Phase Liquid

(b) Gas

(c) Multiphase Gas/Oil Mixture

Fig. 2.11

In single phase liquid both gravitational and friction pressure gradients are constant along the tubing and therefore the pressure traverse is linear with depth. In gas, it is very nearly linear even though the friction and hydrostatic pressure gradients vary significantly with depth.

In multiphase mixtures, there is general trend of increasing pressure gradient with depth. We do not have analytical equations or simple procedures for calculating the pressure traverse of multiphase mixtures. Rather, there are numerous correlations based on field and experimental observations, which take the form of either generalized pressure versus distance curves, called the gradient curves (*e.g.* Gilbert (1954)), or empirical pipe flow equations. A severe drawback of the correlations based on experimental data is that application to producing wells is limited to the conditions of rate, geometry, gas/oil ratio, and fluid properties used in the experimental study.

Multiphase pipeflow correlations can be classified into three broad categories:

1. Gradient curves
2. Homogeneous mixture correlations
3. Flow regime correlations.

Historically, the development of these engineering tools began with homogeneous mixture correlations, followed by the development of gradient curves based on field data, and finally the use of sophisticated correlations based on flow regime concept.

Gradient curves have the salient feature of being graphical, and thus are independent of the need for any computing device. Recently developed gradient curves are based on flow regime correlations, and not on field data as originally proposed by Gilbert. Published gradient curves (Appendix-A) based on field data are generally limited to small and medium oil rates, but they generally cover wide ranges of gas/liquid ratios, these two parameters being the most important in gradient curve application. We have selected only a few gradient curves, to illustrate their use in solving well performance problems at Appendix-A. They are taken from the original work by Gilbert (1954), who developed his correlations using light (25 to 45 °API) crudes produced mainly from California fields, representing tubing sizes of 1.66, 1.90, 2.875 and 3.5 in. for oil flow rates from 50 to 600 STB/Day and gas liquid ratios of 0 and up. We include in Appendix-A only the curves for 2.875 in. and 3.5 in. tubings.

Gilbert noted that main parameters in vertical multiphase pipe flow are diameter, oil rate, and gas/liquid ratio. Other parameters that have an effect on pressure gradient include liquid surface tension, viscosity, densities (oil, gas and water gravities), flowing temperature, gas/liquid solubility, and water cut. Gilbert also noted that his gradient curves would not apply, if an emulsion forms in the tubing. It should be noted that pipe diameters usually reported on gradient curves are given in nominal size, representing the outside diameter and not the actual flow path (inner) diameter. Example 2.7 illustrates the use of gradient curves.

Example 2.7: How to use gradient curves?

The oil well No. 1 produces at a depth of about 5000 ft. The oil is relatively heavy and contains little solution gas, thereby requiring gas lift to produce. Available tubing and gas lift equipment from another lease can be used to test the success of artificial lift without investing in new equipment. The tubing is 3.5 in. nominal diameter. Preliminary calculations indicate that 1000 SCF/STB gas can be injected at an economical cost. Added to the 200 SCF/STB solution gas, this gives a total of 1200 SCF/STB total GOR. Consider production at an oil rate of 200 STB/Day. Assume that the gas is injected at the bottom of the tubing (tubing shoe):

1. Determine the required tubing intake pressure if the wellhead is fixed at 500 psia.
2. Determine the available wellhead pressure when tubing intake pressure (located near the wellbore) is 2000 psia.

Solution: First locate the gradient curve in Appendix-A that corresponds to 3.5 in. nominal tubing diameter and oil rate of 200 STB/D. Figure 2.12 is a reproduction of the gradient curve.

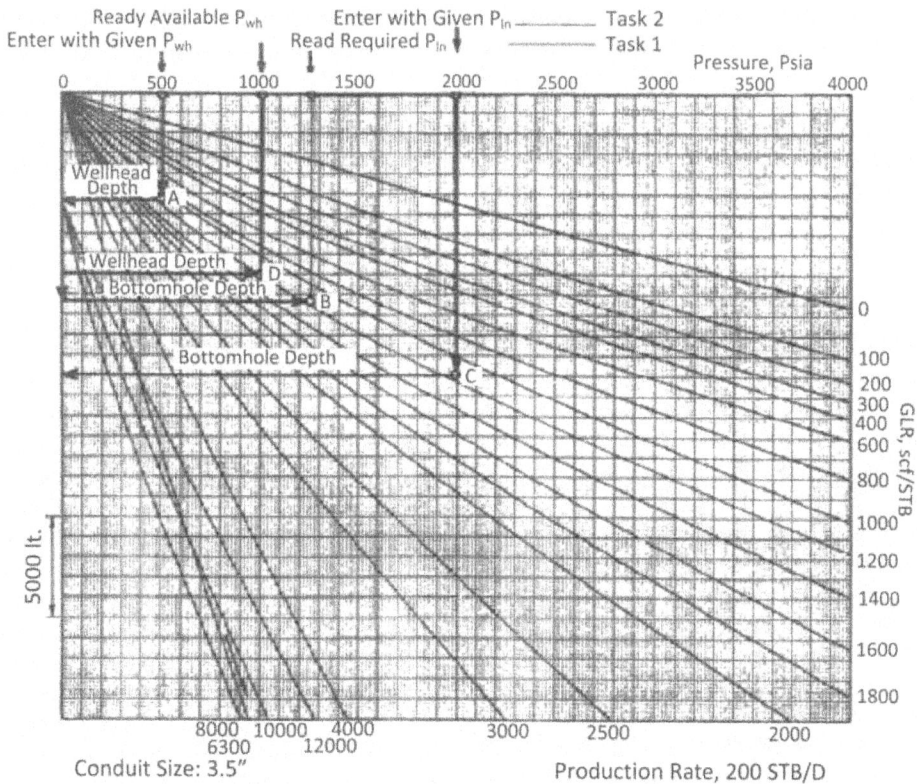

Fig. 2.12

1. Find 500 psia along the axis at the top of gradient curve. Place a straightedge vertically along the 500 psia grid. Move down until the GLR = 1200 SCF/STB curve is found, and label this point A.

   Place the straightedge horizontally along the grid intersecting point A. This represents zero datum. Now add 5000 ft. to the zero datum by moving downward. Position the straightedge horizontally at this depth, which represents the intake to the tubing.

   Locate the GLR = 1200 SCF/STB curve that intersects the horizontal line representing the intake depth. Label this point B and move upward to the x-axis scale that gives the pressure at tubing intake, 1250 psia.
2. Enter the x-axis at 2000 psia, representing the tubing intake pressure. Move down vertically until the GLR = 1200 SCF/STB curve is found. Label this point C.

Move upward a distance of 5000 ft and place the straightedge horizontally along the grid. Move horizontally to the left until the GLR = 1200 SCF/STB curve is located and mark the intersection point *D*.

Move vertically upward to the x-axis to read the pressure corresponding to the wellhead (5000 ft above tubing intake) 1000 psia.

Figure 2.13 shows how to use a set of gradient curves to construct the tubing performance of an oil well producing through tubing with a given diameter and length at a specific gas/liquid ratio and wellhead pressure. The wellhead pressure is specified as a constant. Selecting a gradient curve with the specified GLR, we find the point where pressure equals wellhead pressure. Zero depth corresponds to this point. Moving down vertically a distance equal to the tubing length and then horizontally until the same GLR curve is reached, the bottomhole pressure is read on the x-axis scale. This pressure is the intake flowing pressure for the rate corresponding to the gradient curves chosen. Similarly intake pressure is determined for several other rates. The rate-intake points are then plotted to form the tubing performance curve. Example 2.8 illustrates the procedure for using gradient curves.

Example 2.8: Constructing the tubing performance curve for an oil well.

The well No. 3 produces at a depth of 8000 ft. Solution gas/oil ratio is 600 SCF/STB. Use of 3.5 in. nominal tubing is suggested by the production engineer, who claims there will be a need for gas lift after only one to two years of production. Construct the present tubing performance curve, assuming a wellhead pressure of 200 psia.

Solution: The 3.5 in. gradient curves are found in Appendix A. The procedure given for part-1 of example 2.7 should be followed for each flow rate and a constant

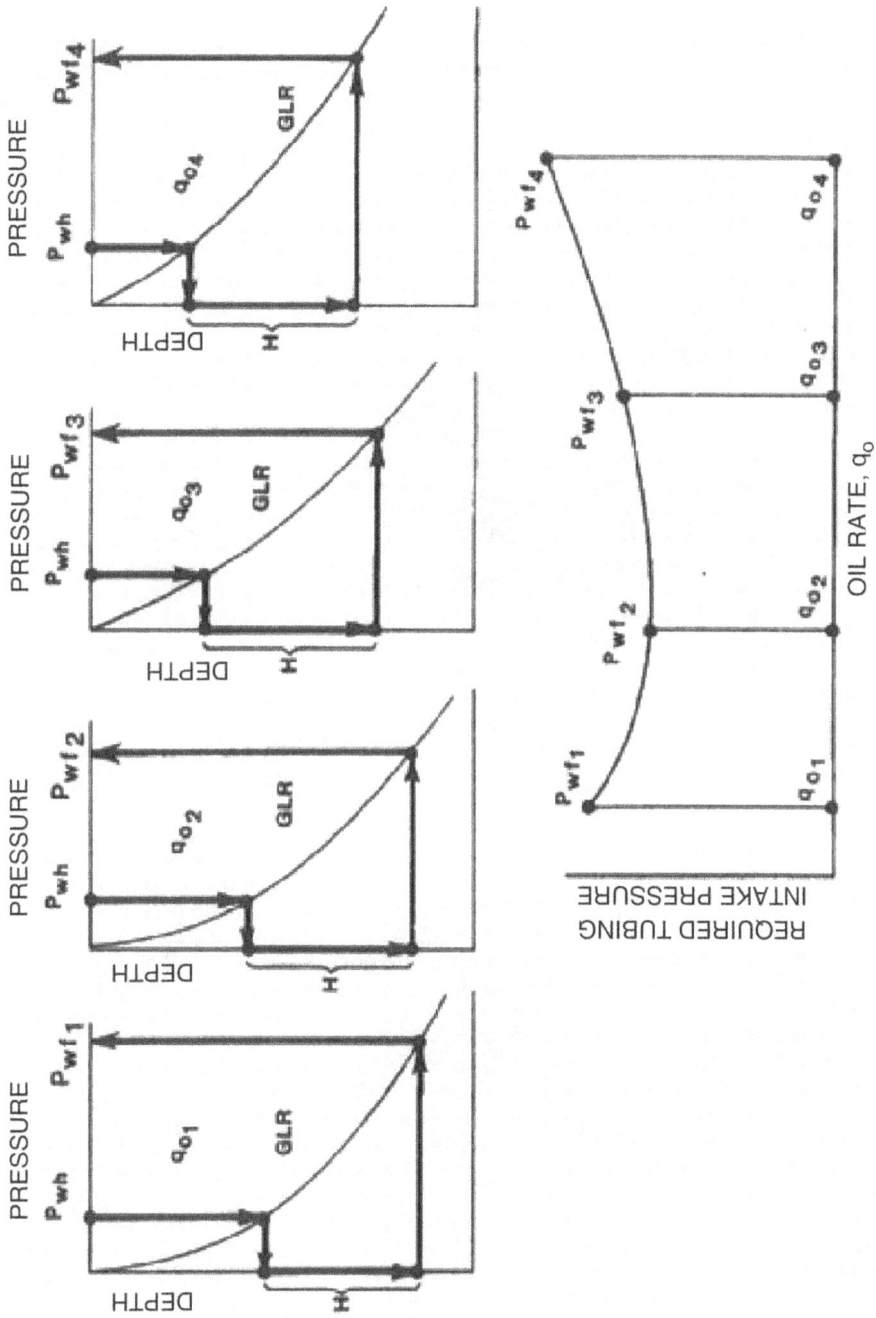

Fig. 2.13

wellhead pressure of 200 psia. The depth from wellhead to tubing intake is 8000 ft and equal the total vertical depth of formation.

Results are tabulated as below and plotted in Figure 2.14(a)&(b). Note the curvature and minimum at a rate of about 350 STB/Day and an intake pressure of 1600 psia.

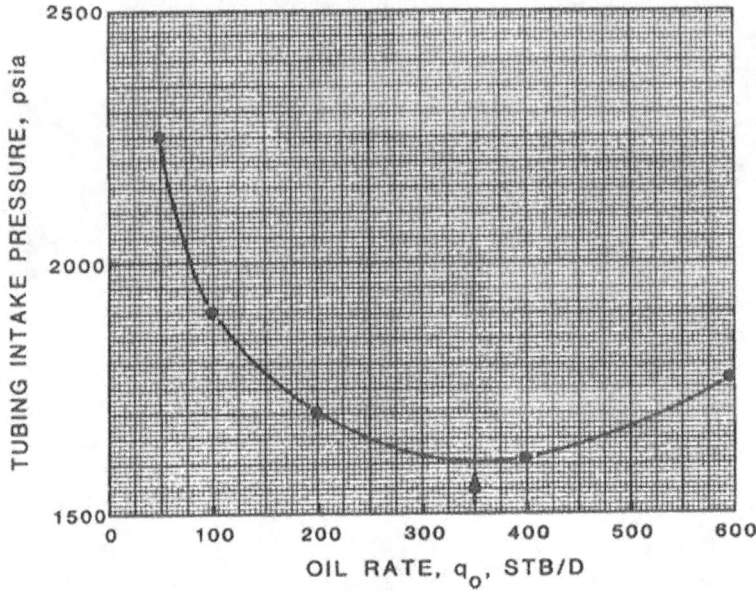

Fig. 2.14(a)

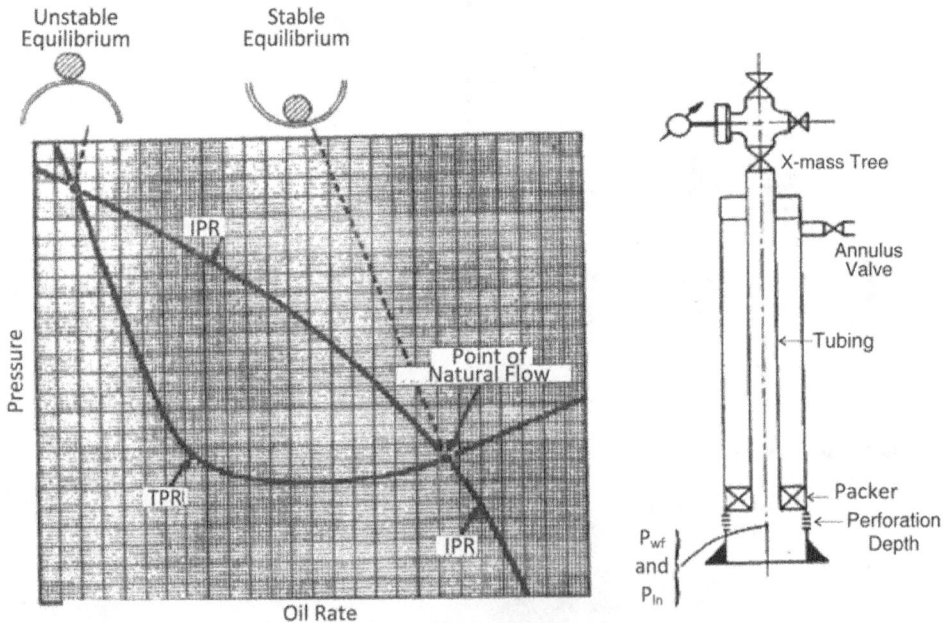

Fig. 2.14(b)

The TPR is valid only for a specific set of well data. Changing well head pressure, gas/liquid ratio, or tubing dimensions will change the tubing performance curve and will require construction of a new curve.

q_o (STB/Day)	p_{in} (psia)
50	2250
100	1900
200	1700
400	1610
600	1760

Natural Flow

Based on only a few pieces of data, it is possible to calculate and plot both inflow and tubing performance relations. For the typical case when the tubing shoe (inlet) reaches the perforation depth, wellbore flowing pressure and tubing intake pressure are considered at the same depth. When at a specific rate these two pressures are equal, the flow system is in equilibrium and flow is stable. The intersection of the IPR and TPR curves determines the rate of stable flow that can be expected from the particular well. At the wellbore (i.e. at perforation depth) the available pressure—well bore flowing pressure—determined by the IPR equals the required pressure—tubing intake pressure—determined by the TPR. The equilibrium rate and pressure constitute what is called the natural flow point. The equilibrium rate is called the natural flow rate. Figure 2.15(a) illustrates the condition of natural flow, with a Cartesian plot of IPR and TPR curves.

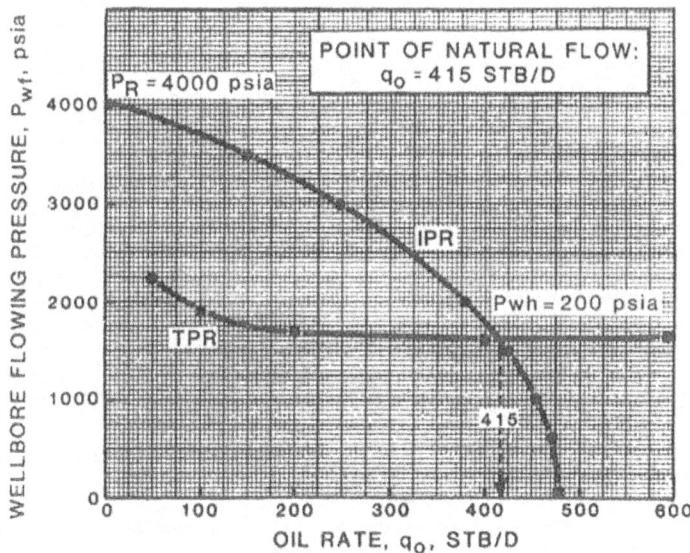

Fig. 2.15(a)

If inflow and tubing performance relations had simple algebric expressions, then the point of natural flow could be determined analytically or otherwise by trial and error. Generally, graphical determination of natural flow is necessary because IPR and TPR may not have simple algebric expressions. Also, graphical solution of natural flow is simple, and it illustrates the effect of different conditions of flow such as depletion, changing GLR, and wellhead pressure.

As indicated in Figure 2.15(a), there may be two points of intersection for multi-phase mixtures. One represents a stable flow condition and the other an unstable one. The stable point of natural flow is to the right. By analogy, it represents a ball locked in a concave circular container (having upward curvature). If the ball (i.e. rate) is pushed either to the left or right, then it will always return to its original point of equilibrium. The unstable point of natural flow is to the left. Once again by analogy, it represents a ball located on top of a convex surface (having downward curvature). If the ball is pushed to the right or left then it will continue its descent until a stable, upward-curving (or flat) surface is reached. Mathematically, stable point of natural flow exists when the two performance relations intersect with slopes (i.e. derivatives) of opposite sign. If the two curves have slopes of similar sign at the

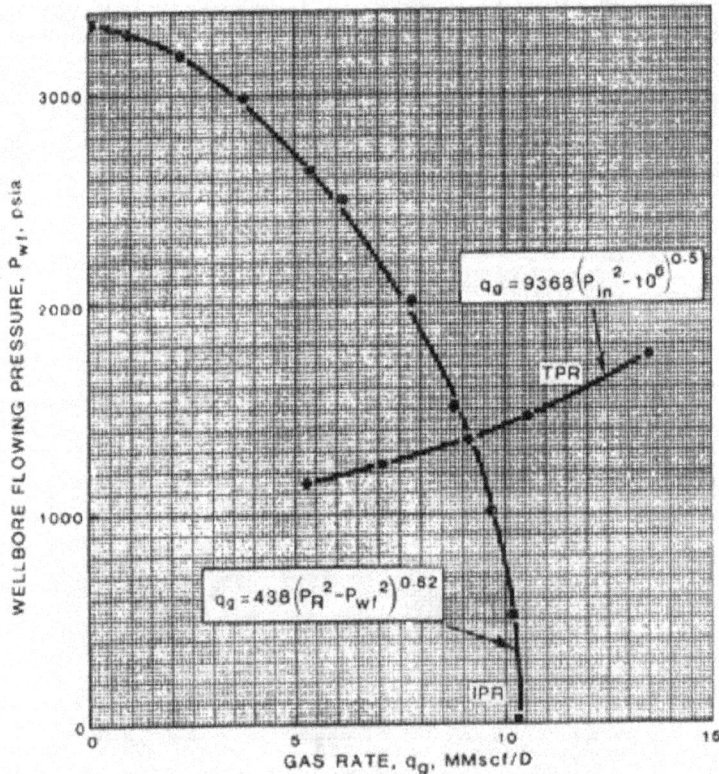

Fig. 2.15(b)

point of intersection, then only a small change in rate will cause the system to change its state of equilibrium, either killing the well or moving it toward the stable point of natural flow. Graphical determination of stable natural flow is illustrated by examples 2.9 and 2.10.

Example 2.9: Determining natural flow for an oil well.

The well No. 3 in example 2.8 has been tested at a rate of 202 STB/Day during 3-day period. Stabilized wellbore flowing pressure measured 3248 psia. Well No. 1 and 2 were tested earlier with a multi-rate sequence, which indicated the exponent in the IPR equation ranges from 0.77 to 0.81. A value of 0.8 is assumed for well No. 3.

$$q_o/q_{omax} = [1 - (p_{wf}/p_r)^2]^n$$
$$202/q_{omax} = [1 - (3248/4000)^2]^{0.8}$$

Hence, q_{omax} = 478 bbe

Thus, IPR equation is: $q_o = 478 [1 - (p_{wf}/4000)^2]^{0.8}$

Determine the rate of natural flow, assuming the tubing performance calculations in example 2.8 apply (i.e. p_{wh} = 200 psia, GLR = 600 SCF/STB, and 8000 ft of tubing having 3.5 in. nominal diameter).

Solution: Figure 2.15(b) shows a plot of the tubing and inflow performance curves on Cartesian paper. The intersection indicates a natural flow of about 415 STB/Day at a flowing bottomhole pressure near 1600 psia.

Example 2.10: Determining natural flow for a gas well.

Solve the following problem for gas well No. 3. Plot the gas IPR and the tubing performance curves on Cartesian coordinates. Determine the point of natural flow.

Solution: Figure 2.16(a) shows the IPR and tubing performance curves plotted on Cartesian coordinates, indicating a natural flow of about 9.2 MMSCF/Day at a flowing pressure of 1400 psia.

Natural flow rate and pressure usually change with reservoir depletion, depending on the variation in IPR and TPR resulting from changes in reservoir pressure and flow characteristics. Usually the change of natural flow is toward a lower rate, if all well parameters remain unchanged. To offset the natural decline in rate, it is possible to change equipment or operating criteria to maintain the desired rate of production. Lowering the wellhead pressure by choke manipulation or lowering of separator pressure is perhaps the simplest and most common of the adjustments made by the operator. Introducing artificial lift or treating wells by stimulation are both more

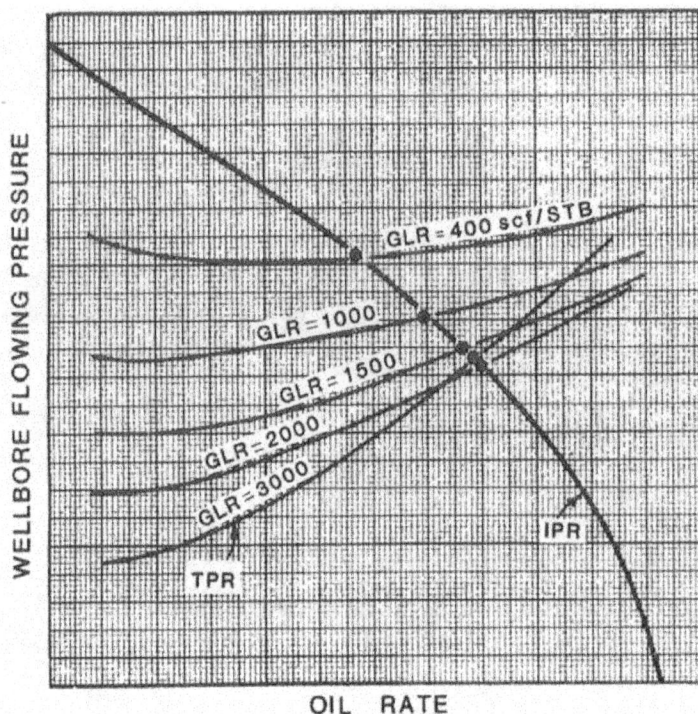

Fig. 2.16(a)

complicated and costly alternatives for maintaining a desired rate of production. The effect of changing well conditions is considered in detail as below:

Changing Wellhead Pressure: Decreasing wellhead pressure by increasing choke opening will usually shift the TPR curve downward to a lower intake pressure, consequently increasing the rate of natural flow (Figure 2.16(b)). If the wellhead pressure is reduced to atmospheric condition, then the well will produce at its maximum flow rate. Increasing the well head pressure by reducing the choke opening will shift the TPR curve upward, resulting in a decrease in rate. If the wellhead pressure is increased beyond certain point, then the well will stop producing because the required pressure exceeds the available pressure. Figure 2.16(b) shows the effect of changes in wellhead pressure on natural flow.

Changing Gas/Liquid Ratio: The effect of changing gas/liquid ratio is not as straightforward. It has different effects on the two components of pressure loss in tubing-friction and hydrostatic. Increasing GLR lightens the mixture density and therefore reduces the pressure loss due to hydrostatic forces. Larger quantities of gas will, however, usually result in larger pressure losses due to friction. The composite effect of gas on total pressure loss in tubing is illustrated in Figure 2.16(c). An increase in gas/liquid ratio tends to shift the TPR upward and to the right. The result is an

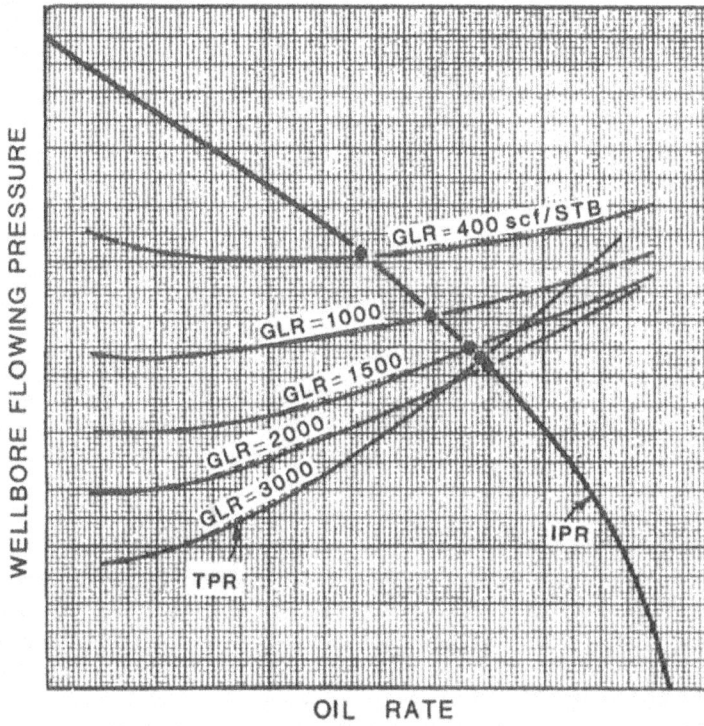

Fig. 2.16(b)

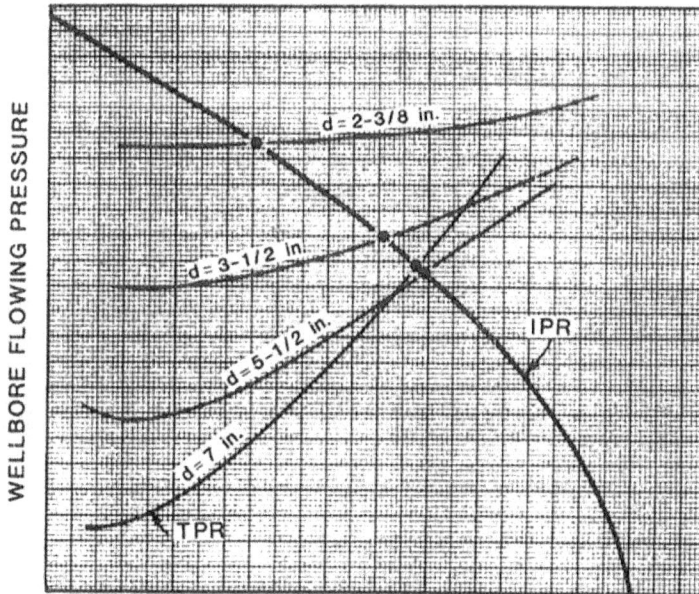

Fig. 2.16(c)

increase in natural flow rate. The trend continues up to a certain gas/liquid ratio, where the trend is then reversed. Injecting gas from the surface to the lower sections of the tubing is a conventional artificial lift method called gas lift. The rate of natural flow decreases with additional injection of gas (*i.e.* at increasing GLRs) once a critical gas/liquid ratio has been reached.

Changing Tubing Diameter: The effect of tubing diameter on natural flow is similar to the gas/liquid ratio. Increasing diameter increases the rate of natural flow until a critical diameter is reached. For higher diameters the rate will decrease. This is due to change of dominance in pressure loss from friction to gravity, and to holdup forces that occur with increasing pipe diameter. Figure 2.17(a) indicates the general effect of tubing diameter on natural flow. Natural flow is the primary criterion used to choose tubing size. Other criteria include price, availability, mechanical considerations, and future production characteristics.

Changing Inflow Performance: Deteriorating inflow performance is the natural result of reservoir depletion. First, average reservoir pressure decreases in the absence of artificial pressure maintenance or a strong natural water drive. Gas and water injection can be used to arrest the decline in IPR caused by depletion. Additional reductions in inflow performance may result from: (1) damage near the wellbore related to drilling and completion operations, (2) reduced drainage area due to infill drilling, (3) reduced permeability due to two-phase flow, compaction, or fines

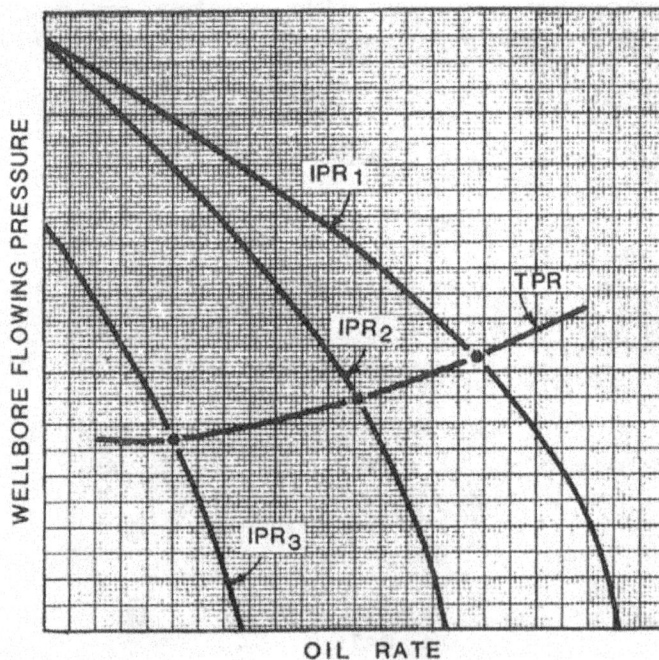

Fig. 2.17(a)

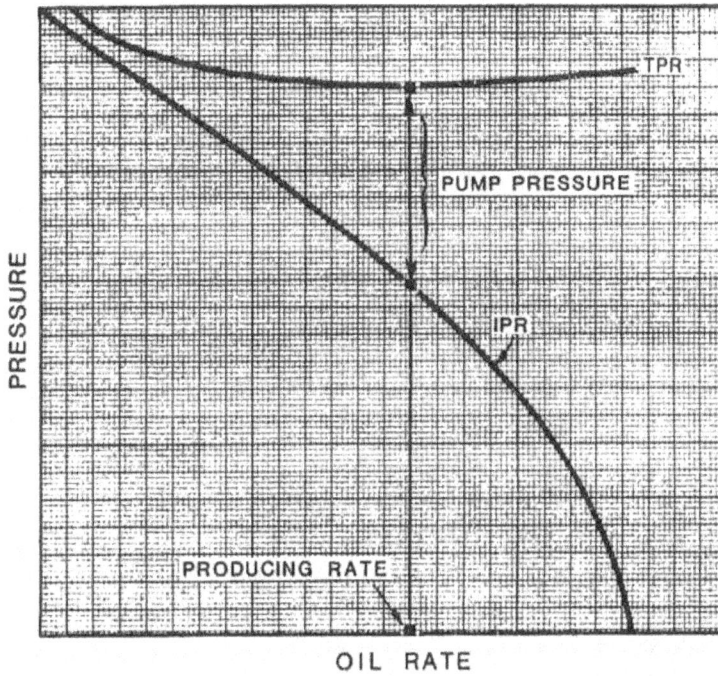

PUMP PRESSURE

IPR

TPR

PRODUCING RATE

OIL RATE

Fig. 2.17(b)

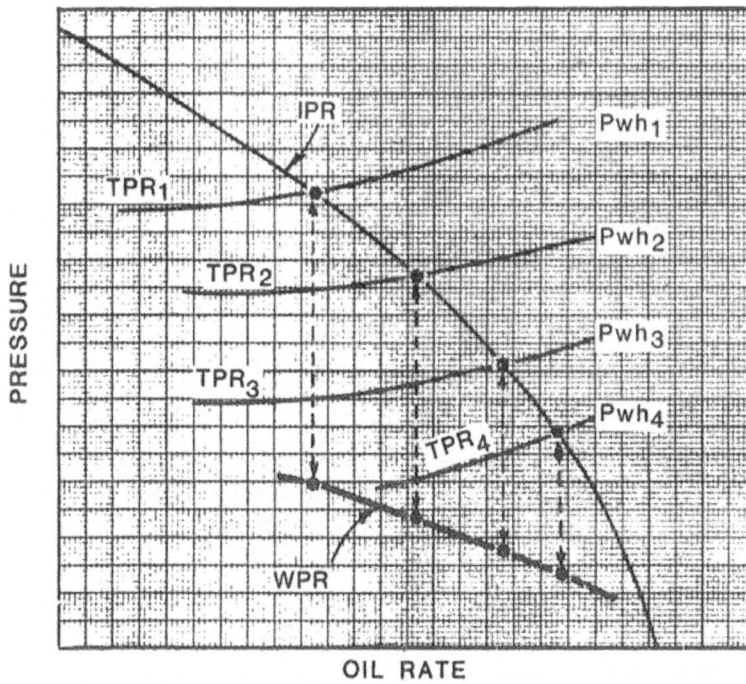

IPR

TPR₁

TPR₂

TPR₃

TPR₄

WPR

Pwh₁

Pwh₂

Pwh₃

Pwh₄

OIL RATE

Fig. 2.17(c)

migration, (4) increased viscosity due to gas liberation from reservoir oil, or (5) transient effects usually associated with low-permeability formations. Figure 2.17(b) shows the effect of changing IPR on natural flow.

Pumping Well: Installing a pump at the bottom of the tubing string creates an artificial lifting capacity and increases the available pressure to flow up the tubing. The effect of a downhole pump is shown in Figure 2.17(c). The pump adds a controlled amount of pressure to the IPR, thereby sustaining flow at higher than the natural rate. Wells producing with a pump may not have the capacity to produce in the absence of artificial lift, as indicated in the Figure 2.17(c). Other situations may arise where a pump is installed in a well that is producing by natural flow, but at a rate lower than desired. In either case, we consider the well to be producing by artificial lift.

Wellhead, Choke and Flowline Performance: Controlling production rate is often done by adjusting the choke size, which results in a change in wellhead pressure. Figure 2.17(b) shows the effect of wellhead pressure on production rate. For each wellhead pressure there is a different tubing performance curve, each intersecting the inflow performance at a different point. The general trend observed in Figure 2.17(b) is that flow rate increases as wellhead pressure decreases (resulting from an increasing choke size).

The choke can be manipulated to change flow rate. Determining the effect of choke size on production rate is an important design task. In doing so, however, one should recognize that there usually exists a maximum flowing wellhead pressure above which flow ceases. This is illustrated in Figure 2.17(c). At a critical wellhead pressure the TPR curve is tangent to the IPR, and at higher wellhead pressures the flow ceases altogether.

The points of intersection between TPR and IPR curves in Figure 2.17(b) can be plotted as rate versus wellhead pressure. The resulting curve is referred to as the *wellhead performance relation* (WPR). Figure 2.18(a) illustrates a WPR curve superimposed on the same graph as IPR and TPR curves. The distance between the TPR curves and WPR curve represents the pressure loss in the tubing.

The following procedure can be used to develop the WPR:

1. Calculate and plot IPR curve.
2. Tabulate the wellbore flowing pressures at rates for which gradient curves are available (e.g., 50, 100, 200, 400, and 600 STB/Day).
3. For the specific tubing configuration and the given rate and GLR, enter the gradient curve at the corresponding wellbore flowing pressure.
4. Move vertically upward along the length of the tubing, starting from the wellbore flowing pressure on the gradient curve.
5. Move horizontally to the left until the same GLR curve is found. Read the

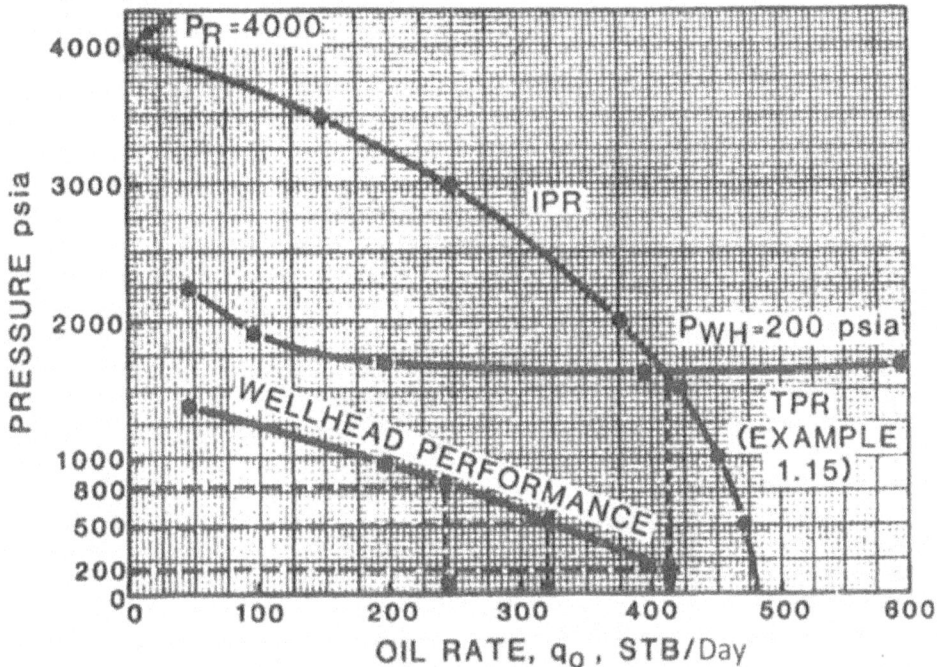

Fig. 2.18(a)

pressure at this point from the x-axis grid. This pressure represents the well-head pressure.

The procedure is shown in Figure 2.18(b). For more details of the steps listed previously, see example 2.11.

Example 2.11: Constructing a WPR Curve.

Alternative production plans for oil well No. 3 (examples 2.8 and 2.9) require the estimate of natural flow rate at three wellhead flowing pressures: 200, 500, and 800 psia. Perform the necessary calculations without determining tubing performance curves for each wellhead pressure.

Using IPR equation example 2.9, calculate and tabulate the wellbore flowing pressures for rates that correspond to the available gradient curves (50, 100, 200, 400, and 600 STB/Day). Rewrite the IPR equation in terms of p_{wf} as below:

$$p_{wf} = [p_r - (q_o/0.000828)^{1.25}]^{0.5}$$

The results are tabulated in Table 2.6.

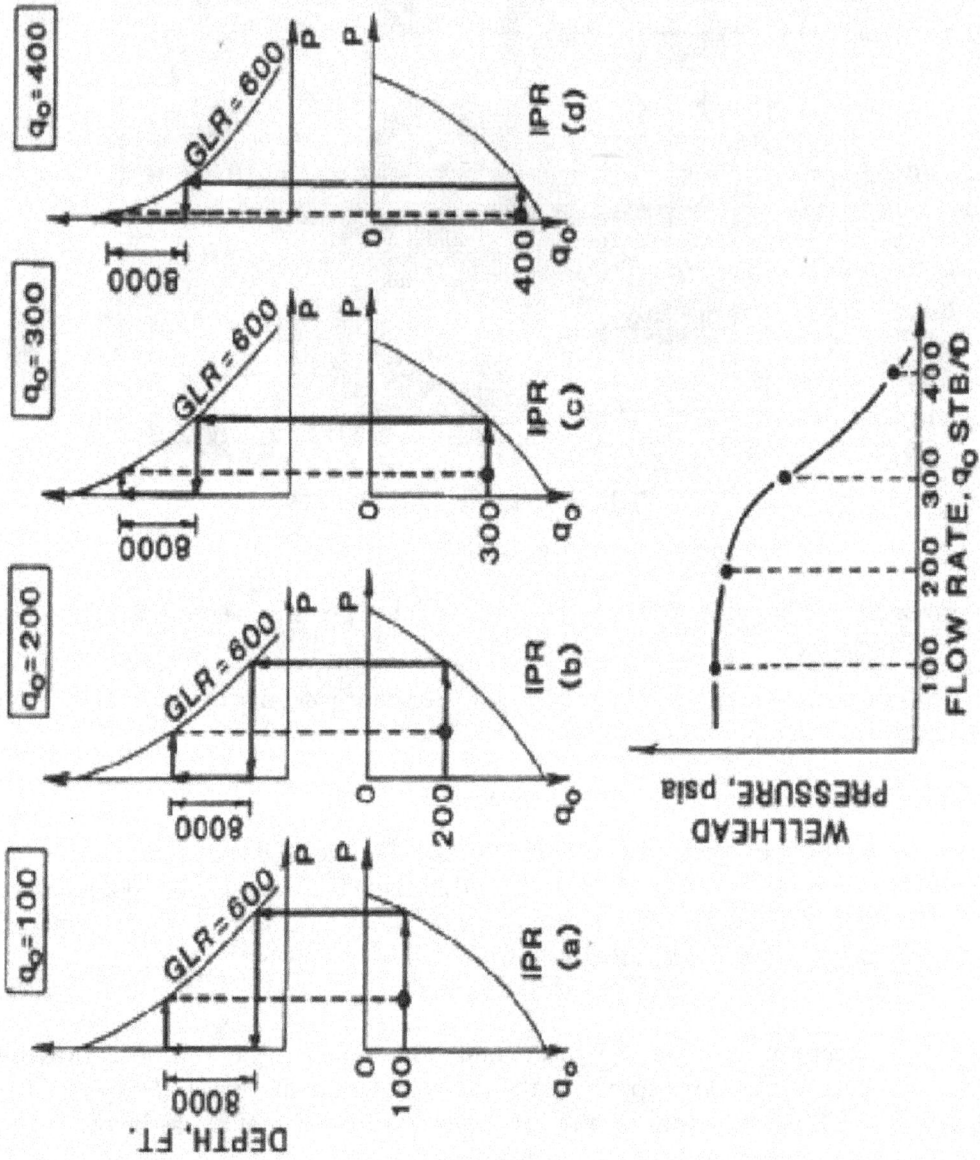

Fig. 2.18(b)

For a given rate, enter the gradient curve at the wellbore flowing pressure and note the intersection with the GLR of 600 SCF/STB. Moving vertically up the length of tubing (8000 ft) and then horizontally until the same GLR curve is found, read the flowing wellhead pressure from the x-axis. This procedure is repeated for each rate and

Table 2.6: Rate-Pressure Calculations

q_o (STB/Day)	p_{wf} (psia)	p_{wh} (psia)
50	3880	1350
100	3708	1150
200	3263	950
400	1807	250

results are tabulated in Table 2.6. A plot of wellhead flowing pressure versus rate. Entering wellhead pressures of 200, 500, and 800 psia, we read the corresponding flow rates of 415, 325, and 245 STB/Day.

The WPR is relation stating that for a given rate of flow there is a certain discharge pressure available at the wellhead. How this pressure is used or dissipated depends on the choke size and downstream choke conditions. Typically, the choke is located directly adjacent to the wellhead assembly. Offshore installations may require that the choke be located a considerable distance from the wellhead, for example, at the inlet to the manifold in subsea installation.

Choke: Several types of chokes are available but they generally fall into one of two categories: positive or fixed choke, and adjustable choke. Each type has its own advantages, and operator may choose one or the other based on field experience.

The *positive choke* is a replaceable, fixed dimension orifice threaded into an L-shaped housing. The end connections of the body may be flanged or threaded. The orifice is always made from a hard material, to resist erosion. It may be slightly rounded at the entrance. Total orifice length is usually 6 in., but it may be as short as 2 in. Bore sizes range from 1/8 in. to 2 in. in diameter, and they are usually expressed in 64th of an inch. For example, an 8/64 choke refers to a choke with an inner diameter bore of 1/8 in.

An *adjustable choke* allows for gradual changes in the size of the opening. The most common adjustable choke is a needle valve, which is calibrated to read in effective diameter openings (refer Figure 2.19). Another type of adjustable choke is the rotary positive choke, illustrated in Figure 2.20. This choke employs two circular rotating disks of hardened material, each with a pair of orifices. One disk is fixed in the valve body, while the other disk can rotate through 90 degree to expose all or part of the orifice flow area.

Choke selection involves mechanical, operational, and economical considerations. This discussion will be limited to selection of a choke to deliver specific flow rate. The main function of a choke is to dissipate large amounts of potential energy (e.g. pressure losses) over a very short distance. The design of a choke takes advantage of the flow regime resulting from a sudden disturbance in continuous flow through

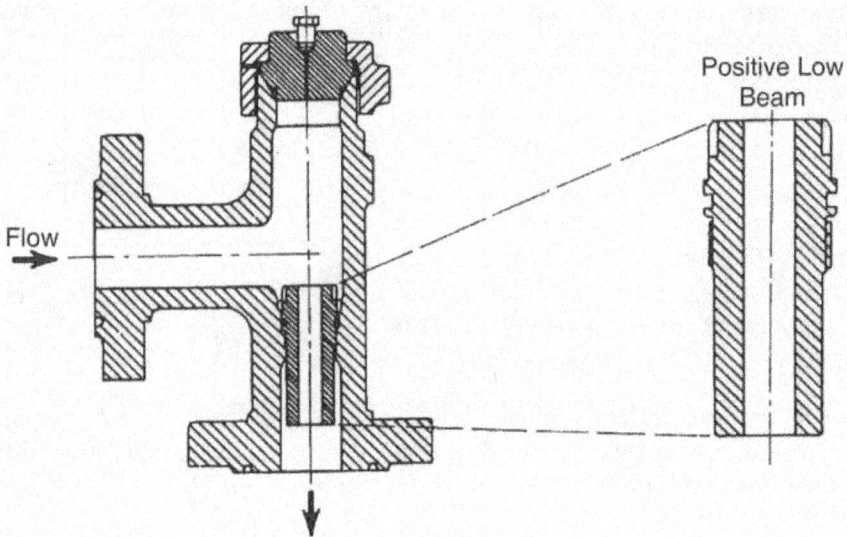

Fig. 2.19

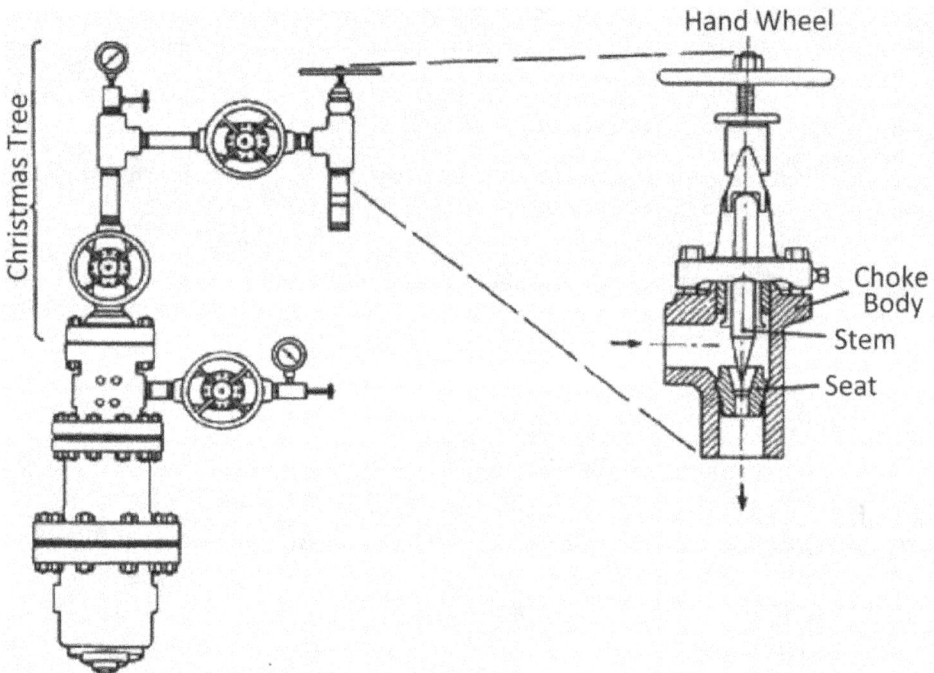

Fig. 2.20

a circular conduit. Figure 2.21 gives a schematic of the normal flow character of fluid passing through a fixed choke. It describes the combined effect of a sudden flow restriction, a small-bore flow tube, and an abrupt enlargement. As the fluid approaches the orifice, it leaves the pipe wall and contracts to form a high velocity jet.

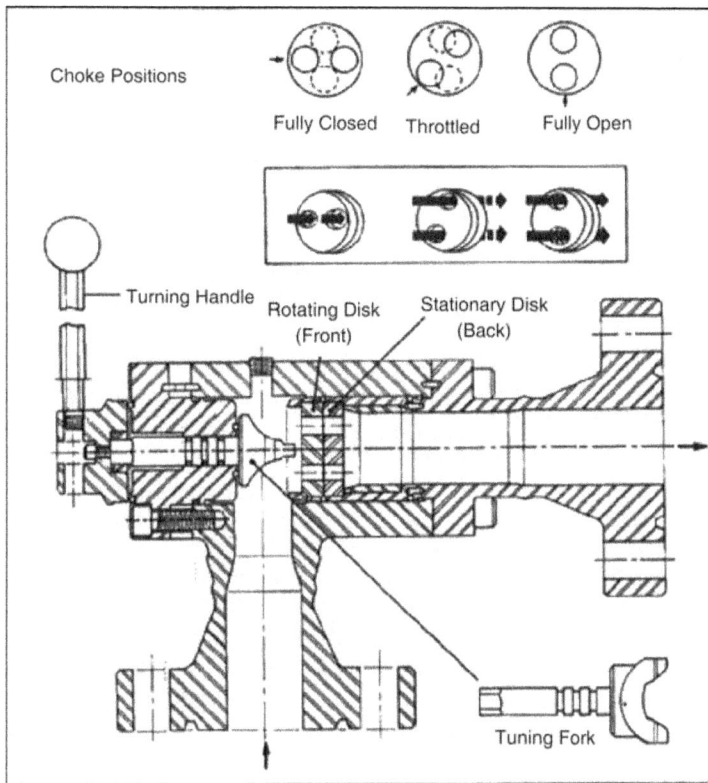

Fig. 2.21

The jet converges to a minimum called the throat or vena contracta, and then it expands towards the wall of choke bore. After leaving the choke, the stream of fluid expands and returns to a flow geometry similar to what it was before entering the choke. An area of turbulence just beyond the choke exit also contributes to pressure loss. Total irreversible losses are summarized in the following:

1. Friction throughout the choke and near choke areas.
2. Turbulence near the entrance and exit to the choke.
3. Slow eddy motions between the contracted jet and pipe walls.
4. Abrupt expansion at the exit to the choke.

For gas flow, the critical downstream pressure is reached when the velocity at the vena contracta (throat) reaches sonic velocity. The details of choke performance are not in the scope of this book.

The performance of a bean at the well head. To find bean size in a flowing well one designed in such a way that small variations in the downstream pressure should not affect the tubing head pressure and thus performance. The fluid flow through the bean at velocity greater than that of a sound, and the pressure requirement is able to meet the tubing head pressure is at least double the average flow line pressure. The theoretically, it will be possible to calculate the tubing head pressure,

$$P_{tf} = \frac{CR^{0.5}q}{S^2} \qquad\qquad\qquad ...(2.14)$$

When, P_{tf} = THP, KG/Sqcm

R = GLR, M³/M³

q = Gross liquid rate M³/Day

S = Bean Size, 1/64 inch

C = Constant

Example

Problem 2.1: A well is producing 100 M³/day gross liquid with a GLR of 60 M³/M³. If the bean size is half inch. Calculate THP.

Problem 2.2: A well is producing through 3/8 inch bean at a rate of 60 M³/day with a 10 Kg/ cm2 of THP. What will be the GLR.

Problem 2.3: What will be the size of the bean needed in the flow line to hold a THP of 10kg/cm2 when GLR 40 M3/M3 and liquid rate 20 M³/day.

Problem 2.4: Given that a well has declined from 100 stb/day to during a 1-month period, use the exponential decline model to perform the following tasks.

1. Predict the production rate after 11 more months.
2. Calculate the amount of oil produced during the first year.
3. Project the yearly production for the well for the next 5 years.

Solution: Production rate after 11 more months:

$$b_m = \frac{1}{(t_{1m} - t_{0m})}\ \ln\left(\frac{q_{0m}}{q_{1m}}\right) = \left(\frac{1}{1}\right)\ \ln\left(\frac{100}{96}\right) = 0.04082/month$$

Rate at end of 1 year,

$$q_{1m} = q_{0m}{}^{e-bmt} = 100\ e^{-0.04082(12)} = 61.27\ STB/day$$

if the effective decline rate b' issued,

$$b'_m = \frac{q_{0m} - q_{1m}}{q_{0m}} = \frac{100 - 96}{100} = 0.04/month$$

From,

$$1 - b'_g = (1 - b'_m)^{12} = (1 - 0.04)^{12}$$

$$b'_g = .3875/\ yr$$

Rate at the end of 1 yr,

$$q_1 = q_0(1 - b'_g) = 100(1 - 0.3875)$$

$$= 61.27\ STB/day.$$

CHAPTER 3

Sucker Rod Pump

3.1 Introduction

A sucker rod pumping (SRP) system consists of a surface pumping unit powered by an electric or gas prime mover, a rod string, and a positive displacement pump. Fluid is brought to the surface by the reciprocating pumping action of the surface unit attached to the rod string, which in turn, moves a travelling valve on the rod pump, loading it on the down-stroke and lifting fluid to the surface on the upstroke. The operation of SRP systems has a history of proven reliability while giving operators the flexibility to reuse various components in different well applications. Sucker rod pump, abbreviated as SRP is a very old oil exploitation technique in the oil industry for lifting of crude oil from the oil wells and in fact it is the most widely used mode of artificial lift system in the present day scenario. As per published data approximately 80 % to 90 % of artificial lift wells have been operating on SRP. SRP operated by beam pumping unit is more versatile and more common among other types of operating SRPs. The Sucker Rod Pump is believed to have been developed by the Chinese around 400 BC. At that time bamboo was used for the barrel and jade or ivory was carved into spheres to make the balls set on wooden seats. These pumps were used to pump water using a bamboo sucker rod string. Soon the Chinese were grappling with the same problems which we face in the oil industry today. The rod pump is the most commonly used artificial-lift system in land-based operations. The relatively simple down hole components and the ease of servicing surface power facilities render the rod pump a reliable artificial-lift system for a wide range of applications.

Although the sucker rod pumping system operation appears very simple, resembling a simple reciprocating tube well pump but in actual field practice, it has been found to be a very complex one owing to very deep installation of pump, lifting of a mixture of oil, gas and water which we technically term as multiphase fluid and other several factors, like rod/tubing- stretch/contraction, fluid viscosity, speed of pumping unit, length of stroke etc. The various factors which contribute to the complexity of the pumping must be thoroughly studied by the design engineer and therefore he needs to be very familiar with the distinguishing features and the complex function of sucker rod pumping system.

Therefore a superficial knowledge on the subject of sucker rod pump is not enough to understand operational complexities of pump and to make the running of the sucker rod pump efficiently.

It is therefore, a necessity for artificial lift engineer to have an in-depth knowledge of the total SRP system. Once an engineer knows fully the significance of the elementary principles of the pumping system, it will make him/her familiar with the complex functioning and distinguishing features of each part of the system and as such he/she will be in a position to operate the pumping system in a fool proof manner. As a straight forward and simple strategy let the whole pumping system be presented under three broad units (Figure 3.1), namely, surface unit, sub surface sucker rod pump, sucker rods.

It is important, in brief, to visualize the motion of each of the units before they are described in detail and how they tie them together into a unique pumping system. With the help of a prime mover, say an electric motor of comparatively low rpm (like 720 rpm) a rotating motion is generated. This rotating motion is then passed on to the surface unit by the V-belt transmission system. It effects a reduction in rpm. Thereafter, the onward rotating motion is further reduced to about 1:29 with the help of a double reduction gear box of the pumping unit. This very low rotary motion (say 6 rpm) is then converted with the help of different components of pumping unit to linear motion at the polished rod. This linear reciprocating motion is then transmitted to sub-surface sucker rod pump through the sucker rods, which is the linkage of the surface unit and sub-surface pump. In this way a sucker rod pump operates and lifts well fluids to the surface from the well.

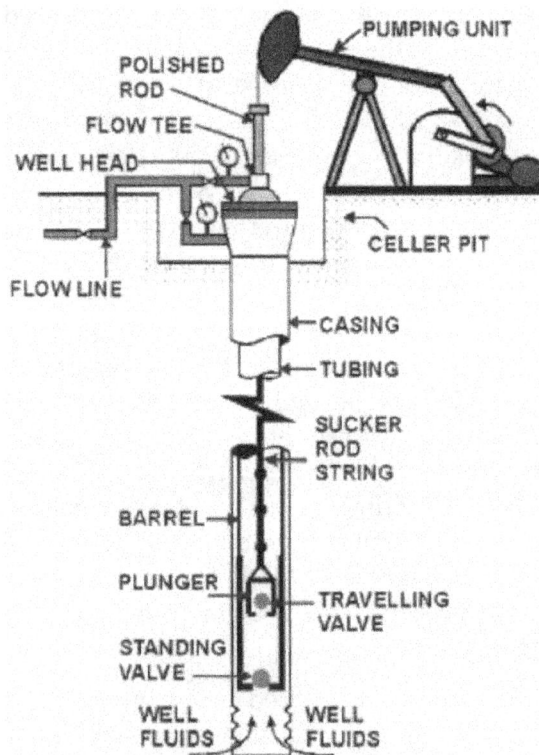

Fig. 3.1: Different Units of a Sucker Rod Pumping System

The advantages of Sucker Rod Pumping system are: high system efficiency, available optimisation controls, economical repairing and servicing, Positive displacement/ strong draw-down pressure, corrosion problems are reduced by upgraded materials etc.

3.2 Pumping Units

Pumping unit cum prime mover at the surface converts the rotary motion of the prime mover into the reciprocating/vertical motion with the help of several link arrangement. Majority of this pumping operation, world wide, utilize walking beam pumping unit and so it is named as beam pumping unit (Refer Figure 3.2a). The structural parts of a conventional beam pumping unit are as follows: walking beam, horse head, saddle bearing, equalizer bearing, equalizer, pitman arm, wrist pin or crank pin bearing, crank, counterweight (Figure 3.2b shows counter balancing effect), crankshaft, double reduction gear box, unit sheave, Sampson post, ladder, bridle (wire line hanger), carrier bar, electric motor, motor sheave or motor pulley, V-belt, belt cover, brake (its link and handle), reducer gear box, pumping unit base, motor base, grouting nuts and bolts.

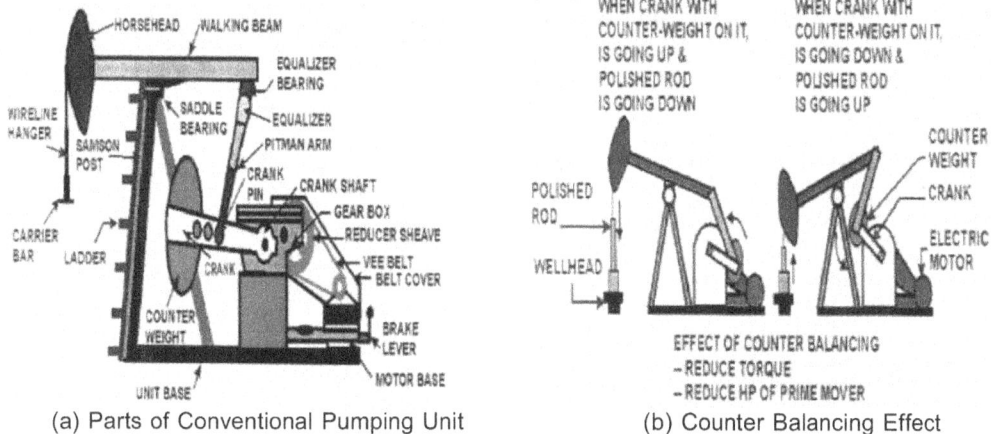

(a) Parts of Conventional Pumping Unit   (b) Counter Balancing Effect

Fig. 3.2a-b

When these are assembled, horse-head end of the walking beam hangs over the x-mass tree of the well, as such; the polished rod clamp on the polished rod is rested on to the carrier bar that is rigidly attached with the wireline hanger (bridle), which in turn is attached with the horse-head. The horse head which has a curvatureshaped surface and flexible hanger (i.e, bridle) together ensure that the polished rod is made to move in a vertical direction only. The other end of the surface unit has the prime mover, the pulley of which is connected to the gearbox reducer pulley with the help of V-belt.

Good operation of the pumping unit requires minimal friction losses in the structural bearings. These bearings are generally grease (graphite grease) lubricated as well as air tight sealed. Thus requiring less maintenance and ensuring their smooth and trouble-free operation. Two stage gear reducer is filled with proper lubricating oil up to the desired mark for lubricating the operating gears as well as all the roller bearings on the gear reducer shafts which get lubricated continuously with the same gear oil by splashing of moving gears.

The whole structure is based on a rigid steel base ensuring proper alignment of the components and this base is usually set on a concrete foundation. Two types of base systems are followed. One is the skid based where the skid is not grouted to the concrete foundation and the other is the base properly grouted by the grouting bolts with the concrete foundation. As the motor sheave rotates, the rotation is transmitted with the help of V-belt from the motor sheave to the gear reducer sheave. During this transmission, speed reduction takes place in relation to the ratio of the unit sheave to motor sheave diameter. Thereafter the speed is further reduced in the ratio of 1:29 approximately in the double reduction gear. Finally, the rotary motion of the gears is transmitted to the walking beam with the help of the connecting link called Pitman Arm. The walking beam then makes oscillating motion, moving the beam up and down on the pivot i.e. the saddle bearing located near the middle of the beam. This motion is finally transmitted to sub-surface pump through sucker rods.

3.2.1 *Different Pumping Unit Geometries*

The pumping units are broadly classified into two groups i.e. walking beam operates as a double arm lever (Class-l) and walking beam operates as a single arm lever (Class-III). The details are as under:

Walking Beam Operates as Double Arm Lever (Class-I): The conventional unit is perhaps the oldest and most commonly used beam pumping of this class. The schematic of this unit is given in Figure 3.3. The walking beam here acts as a double arm

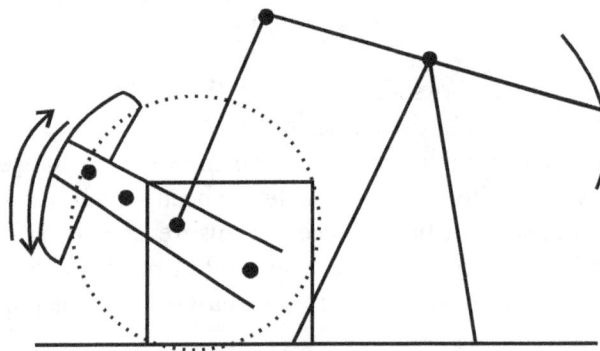

Fig. 3.3

lever on the two sides of pivot i.e. the Sampson post where pivot is near to the middle of the walking beam. The rear end of the walking beam is the driving end and the front end of the walking beam is the driven end. This is also called a "pull-up" leverage system. The walking beam is in the horizontal position.

The counterweights are positioned either at the rear end of walking beam or at crank arm depending on the load at the well to reduce the torque and horse power of the prime mover of the pumping unit. For less load, counterweights are placed on the beam and for the moderate to heavier load, counter weights types are placed on the cranks. As on date, all the sucker rod pumping units operating in various Indian oil fields are of conventional beam pumping unit with counterweight loading on cranks. In some of the earlier smaller capacity pumping units counterweights loaded walking beam were used.

Walking Beam Operates as a Single Arm Lever (Class-III): Air Balanced (Refer Figure 3.4): The air balanced unit was developed in the 1920s for operating the SRP in deep wells. This unit acts as a single arm lever (Class-III) system where the horse-head and Pitman arm are on the same side of the beam and the pivot at the extreme end of the beam. This is also called "push-up" leverage system. Counter- balancing is ensured by the pressure force of compressed air contained in a cylinder which acts on a piston connected to the bottom of the walking beam.

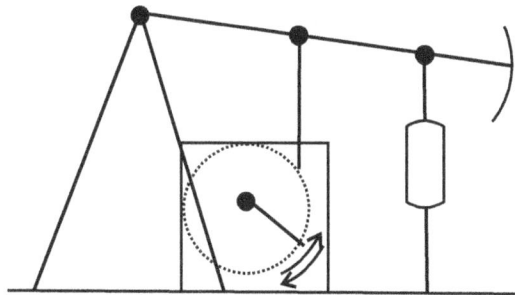

Fig. 3.4: Air-balanced Pumping Unit

Mark II Unit (Refer Figure 3.5): This unit was developed in late 1950s by J.P. Byrd. This is also a Class-III lever system where the pivot is at one extreme end of the walking beam. The main advantage of this unit is to decrease the torque and power requirements of the pumping units. It implies that the pumping unit of this type having less torque and power requirement can work for operating the pump at deeper depth in contrast to the heavier capacity conventional pumping unit required to operate at that depth. In Mark-II Unit the counterweights are placed on the counter balance arm that is on other side of the crank arm. This feature also ensures a more uniform net torque variation throughout the complete pumping cycle.

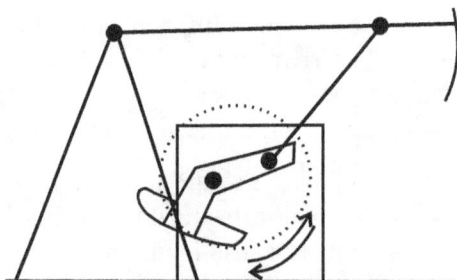

Fig. 3.5: Mark-II Pumping Unit

Torque Master (Refer Figure 3.6): Torque master is one of the latest developments of beam pumping of class-III type where the pivot is at the extreme end of the beam. Some of the distinguishing features in this pumping unit are that the gear reducer is located further away from the Sampson post and the rotary counterweights, which are placed on the crank arm, kept by an angle of about 8 to 15 degrees from the crank arm on the extended portion of the crank arm at one side.

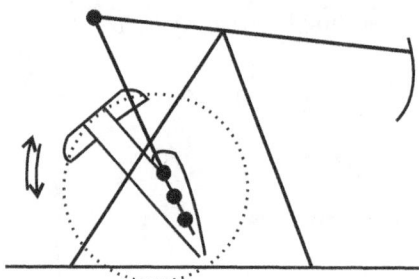

Fig. 3.6: Torque Master Pumping Unit

Long-Stroke Pumping Units: Weatherford's unique-designed Rotaflex® pumping units feature a stroke length of up to 25½-ft and an innovative 100% mechanical reversing mechanism design. These units have been phenomenally successful in providing greater efficiency and lower operating costs in deep, troublesome and high-volume wells. Rotaflex long-stroke pumping units are available in sizes 800 to 1100 with structural capacities of 25,600 to 50,000 lbs. Benefits such as 40 to 60% reduction in rod reversal for greater rod life and 20 to 50% reduction in electrical costs have proven successful with this unit.

3.2.2 *Designation of Pumping Units*

In order to identify a pumping unit, the following considerations are made:
- The geometry of the pumping unit.
- The maximum torque capacity of gear reducer.

Table 3.1: Pumping Unit Size Ratings

Pumping Unit Size	Reducer Rating, In-lb	Structure Capacity, lb	Max. Stroke Length, in.	Pumping Unit Size	Reducer Rating, in-lb	Structure Capacity, lb	Max. Stroke Length, in.
6.4-32-16	6,400	3,200	16	320-213-86	320,000	21,300	86
6.4-21-24	6,400	2,100	24	320-256-100	320,000	25,600	100
				320-306-100	320,000	30,600	100
10-32-24	10,000	3,200	24	320-213-120	320,000	21,300	120
10-40-20	10,000	4,000	20	320-256-120	320,000	25,600	120
				320-256-144	320,000	25,600	144
16-27-30	16,000	2,700	30				
16-53-30	16,000	5,300	30	456-256-120	456,000	25,600	120
				456-305-120	456,000	30,500	120
25-53-30	25,000	5,300	30	456-365-120	456,000	36,500	120
25-56-36	25,000	5,600	36	456-256-144	456,000	25,600	144
25-67-36	25,000	6,700	36	456-305-144	456,000	30,500	144
				456-305-168	456,000	30,500	168
40-89-36	40,000	8,900	36				
40-76-42	40,000	7,600	42	640-305-120	640,000	30,500	120
40-89-42	40,000	8,900	42	640-256-144	640,000	25,600	144
40-76-48	40,000	7,600	48	640-305-144	640,000	30,500	144
				640-365-144	640,000	36,500	144
57-76-42	57,000	7,600	42	640-305-168	640,000	30,500	168
57-89-42	57,000	8,900	42	640-305-192	640,000	30,500	192
57-95-48	57,000	9,500	48				
57-109-48	57,000	10,900	48	912-427-144	912,000	42,700	144
57-76-54	57,000	7,600	54	912-305-168	912,000	30,500	168
				912-365-168	912,000	36,500	168
80-109-48	80,000	10,900	48	912-305-192	912,000	30,500	192
80-133-48	80,000	13,300	48	912-427-192	912,000	42,700	192
80-119-54	80,000	11,900	54	912-470-240	912,000	47,000	240
80-133-54	80,000	13,300	54	912-427-216	912,000	42,700	216
80-119-64	80,000	11,900	64				
				1280-427-168	1280,000	42,700	168
114-133-54	114,000	13,300	54	1280-427-192	1280,000	42,700	192
114-143-64	114,000	14,300	64	1280-427-216	1280,000	42,700	216
114-173-64	114,000	17,300	64	1280-470-240	1280,000	47,000	240
114-143-74	114,000	14,300	74	1280-470-300	1280,000	47,000	300
114-119-86	114,000	11,900	86				
				1824-427-192	1824,000	42,700	192
160-173-64	160,000	17,300	64	1824-427-216	1824,000	42,700	216
160-143-74	160,000	14,300	74	1824-470-240	1824,000	47,000	240
160-173-74	160,000	17,300	74	1824-470-300	1824,000	47,000	300
160-200-74	160,000	20,000	74				
160-173-86	160,000	17,300	86	2560-470-240	2560,000	47,000	240
				2560-470-300	2560,000	47,000	300
228-173-74	228,000	17,300	74				
228-200-74	228,000	20,300	74	3648-470-240	3648,000	47,000	240
228-213-86	228,000	21,300	86	3648-470-300	3648,000	47,000	300
228-246-86	228,000	24,600	86				
228-173-100	228,000	17,300	100				
228-213-120	228,000	21,300	120				

- Gear reducer type—whether double reductions, triple reductions etc.
- The unit structural capacity.
- The longest polished rod stroke length available.

The API has standardized the above and accordingly designates pumping units. For example in a C-456-D-256-144 unit, 'C' means crank balanced conventional unit, 456 means gear box having a torque of 456,000 inch-lb maximum, 'D' means double reduction gear box, 256 means the unit having the PPRL (Peak Polished Rod Load) capacity of 25600 lbs and 144 means having maximum possible stroke length of 144 inch. In the above, the first letter can be either B, C, A, M or TM. 'B' is for beam balanced conventional unit, 'C' as described above, 'A' for air balanced unit, 'M' for Mark II unit and 'TM' for torque master unit.

Other than the above pumping units, many new innovations are being done in the geometry of pumping units for providing better results in the different applications.

Table 3.1 lists specifications for API standard units. Some manufacturers have additional intermediate sizes not shown in this table. Not all manufacturers make all sizes. The specification is a three-part code. A unit rating of 160-173-64 identifies a pumping unit having a gear reducer rating of 160,000 in.-lb of torque, a structural load bearing capacity as measured at the polished rod load of 17,300 lb force and a maximum stroke length of 64 in.

Each pumping unit size generally has two to four stroke lengths at which it can operate, the largest stroke length of the two to four being the one given in the unit rating. One should refer to the manufacturers' catalogues to find which smaller stroke lengths are available with each size of unit.

3.2.3 *Desirable Features of a Pumping Unit*

- A long and slow upstroke is most desirable.
- Downstroke should be faster than the upstroke.
- Low torque factors on the upstroke are desirable.

3.2.4 *Prime Mover*

Two types of prime movers are very common i.e. electric motor, and internal combustion engine.

Electric Motor: Most sucker rod pumping units are being run on electric motors. Low cost, ease of control, more compact and adaptability to automatic operations relative to other types of prime mover have endeared it most. The primary requirement is that these motors should have very low rpm say 720 rpm, 950 rpm etc. Four classes of electric motors within National Electrical Manufacturing Association (NEMA) are in common use for pumping unit prime movers.

- NEMA C normal slip (less than 5%): low initial cost, moderately high starting torque.
- NEMA D medium slip (5–8%): higher starting torque and slip than class C.
- NEMA D high slip (8–13%): widely used now, for having an attractive price and performance.
- Ultra-slip motors (slip upto 30%): relatively new; motor is more fully loaded during the entire pump cycle; designed specifically for beam pumping units. Motors with ultra high slip characteristics are more advantageous than those discussed as above. These ultra high slip motors are specially useful where cyclic loading exists.

Slip is usually defined as: $(N_s - N_f) \times 100/N_s$

Where, N_s is the synchronous speed, N_f is nominal full load speed.

N_f is less than the synchronous speed due to fully loaded condition. The main benefits of ultra high slip motor include reduction of pumping unit structural loads, peak torque and power consumption.

An induction motor that has a synchronous speed of 1200 rpm and loaded speed of 1050 rpm will have a slip of $= (1200 - 1050) \times 100/1200 = 12.5\%$.

Most oil field pumping unit electric prime movers are three-phase induction motors. Single phase motors are restricted to shallow low-volume pumping units.

Internal Combustion Engine: Internal combustion engines designed to run on natural gas or dissel/petrol and serve as pumping unit prime movers can be classified based on their speed, strokes per cycle and number of cylinders. Slow speed engines are those having a crankshaft rpm of 750 or less. High speed engines have speed upto 2000 rpm (these are multi-cylinder engines). Some engine designs employ two strokes per cycle and others use four strokes per cycle. The two stroke engines have been used widely. Two strokes engines are mostly slow speed engines in single or multiple cycliners having power rating from 15 to 325 HP.

Internal combustion engines are usually run on the well head gas from the casing head. A gas scrubber is installed at the well head to knock out the liquid such as oil and water from the gas before it enters the engine carburetor. Slow speed engines with operational speed between 200–800 rpm are petrol.

The four strokes (per cycle) engines is usually a multi-cylinder engine. These can run on natural gas, LPG or petrol.

The choice between electric or gas engine is based on several factors:
- Availability of gas or electricity at well site.
- Investment cost of a gas engine and an electric motor. Gas engines have higher initial cost.
- Service life of gas engine and electric motor. Gas engines have much higher service life.

- Energy costs of using electric motors. In comparison, gas, if available, can turn out to be more economical.

Selecting a Prime Mover Size: As per M/s Curtis and Showalter, the recommended prime-mover horsepower can be obtained by equation,

$$HP_{PM} = q \times D/PMF$$

Where, HP_{PM} = prime mover HP
          q = liquid production rate, b/d
          D = net lift of liquid, ft
        PMF = prime mover factor (for high speed internal combustion engines or normal slip NEMA C motors, PMF = 45,000; for high slip NEMA D motors or low-speed internal combustion engines, PMF = 56,000).

Example: Determine prime mover hp for lifting 100 b/d of oil and water production having a composite specific gravity of 1.0. The net lift is 9,800 ft. Assume a slow speed internal combustion engine or a high slip NEMA D motor.

$$HP_{PM} = q \times D/PMF = 100 \times 9,800/56,000 = 17.5 \text{ HP}$$

The actual engine or motor would be selected, with nearest larger name-plate rating at 20 HP.

3.3 Sub-Surface Pump

It is a sub-surface reciprocating pump, actuated by the up and down motion of sucker rods, which is a connecting link between surface unit and sub-surface pump. Its feature resembles a reciprocating tube-well water pump. It has five main components viz. barrel, plunger, standing valve, traveling valve, pump seat or nipple.

A conventional pump consists of a fixed "barrel" and a moving plunger, with "standing valve" fitted at the barrel end and with "traveling valve" fitted at the plunger end. The word "traveling" implies that the valve moves or travels along with the plunger. The standing valve is fixed with the stationary barrel hence the word 'standing valve'. Both the standing and traveling valves are unidirectional, means, both allow fluid to pass through them in the upward direction only. The fluid will not pass through them in the downward direction. The pump seat or nipple seals the annular area between the pump barrel and tubing and thus prevents the pumped out fluid falling back into the pump intake again. The pump seat also having other feature which make the pump to get locked in that depth.

3.3.1 *Pumping Cycle (Refer Figure 3.7)*

(a) *Plunger moving down and near bottom of stroke:* Traveling valve is open. Fluid is moving up through the traveling valve. Standing valve is closed due to weight of fluid column in the tubing.

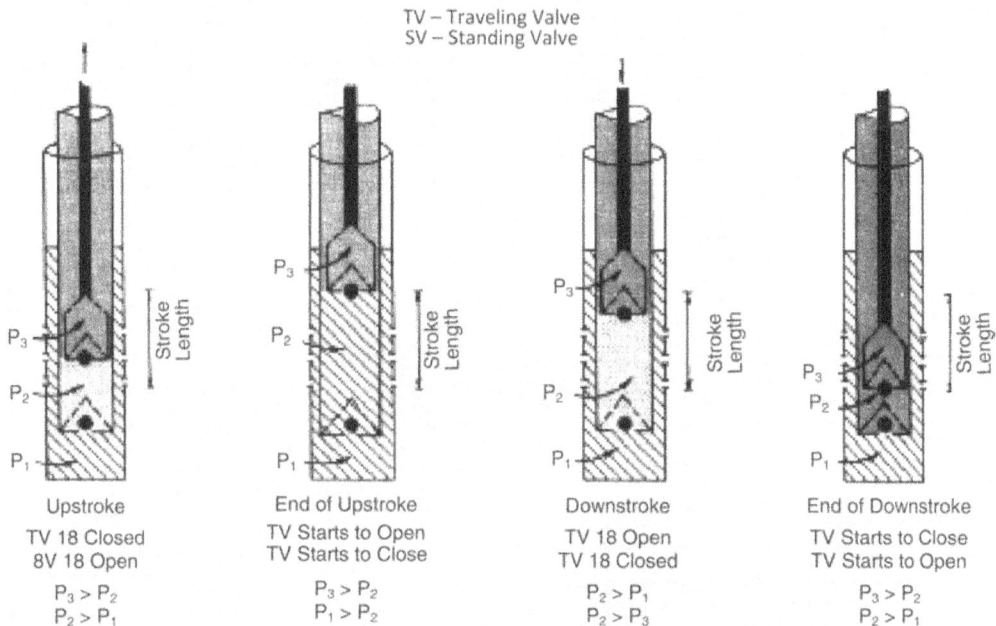

TV – Traveling Valve
SV – Standing Valve

Upstroke	End of Upstroke	Downstroke	End of Downstroke
TV 18 Closed	TV Starts to Open	TV 18 Open	TV Starts to Close
8V 18 Open	TV Starts to Close	TV 18 Closed	TV Starts to Open
$P_3 > P_2$	$P_3 > P_2$	$P_2 > P_1$	$P_3 > P_2$
$P_2 > P_1$	$P_1 > P_2$	$P_2 > P_3$	$P_2 > P_1$

Fig. 3.7

(b) *Plunger moving up and near the bottom of the stroke:* The traveling valve is closed due to load of fluid column on it. Standing valve begins to open to allow entry of fluid from the well bore.

(c) *Plunger moving up and near top of the stroke:* The traveling valve is closed due to fluid load above it. The standing valve is open and fluid entry from well bore into the barrel continues.

(d) *Plunger moving down and, near* top of stroke: The standing valve is closed by the increased fluid pressure resulting from the compression of the fluids in the barrel between the standing and the traveling valves due to downward movement of plunger. The traveling valve begins to open to allow the compressed fluid in the barrel to push through it against the tubing fluid load in the tubing. The plunger thereafter reaches the bottom of the stroke and another cycle is started.

3.3.2 *Types of Sub-Surface Sucker Rod Pumps*

The subsurface sucker rod pumps (Refer Figure 3.8) are mainly categorized in two principal groups viz. (a) insert (or rod) pump and, (b) tubing pump.

In the **insert or rod pump**, the barrel, plunger, traveling and standing valve are the integral part of entire sub surface assembly and is run as a unit on the sucker rod string.

In the **tubing pump** the working barrel is run as part of the tubing and is placed at the desired depth. The standing valve is then dropped into the well followed by running in plunger along with sucker rod strings and is placed inside barrel.

Insert (or rod) pump is the conventional choice and is more commonly used. As a general rule, tubing pumps are used where greater liquid volumes are required to be pumped out. Tubing pumps are especially useful for pumping from inclined wells. The pump itself is not as flexible as the tubing and therefore it is very difficult to push the pump through tubing with close tolerance in inclined/'S'-profile wells. In that case, tubing pump can be used since its barrel, as a part of tubing can pass easily in the inclined portion of the well having larger clearance between O.D of the pump barrel and I.D of well casing. The plunger having small length of say, four feet having sufficient clearance of plunger O.D and tubing I.D can easily be run through the inclined tubing to the desired depth. Many oil wells in Assam area which are inclined/'S'-profile, are being operated by tubing pumps.

3.3.3 *Basic Pump Classification as Per API*

* Stationary barrel top anchor (or top hold down) rod pumps.
* Stationary barrel bottom anchor (or bottom hold down) rod pump.
* Traveling barrel rod pump.
* Tubing pump.

Stationary Barrel Top Anchor Rod Pump (Refer Figure 3.8a): API has named this pump as RHA or RWA pump, where 'R means rod, 'H' means heavy wall barrel, 'W' means thin wall barrel and 'A' means top hold down pump. There is another pump, named by API as RSA which is a thin wall barrel and a soft-packed plunger type pump.

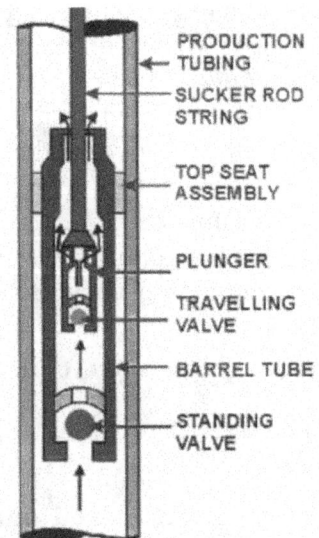

Advantages

* The top hold down is recommended in oil wells, which are producing sand because sand particles cannot settle in the pump-tubing annular space due to the top hold down mechanism.
* This pump performs better in gassy wells because of having comparatively larger opening of the standing valve, as such; larger opening of valve registers less friction.
* This configuration of pump and top seating nipple facilitates in the installation of a very simple and effective type of poor boy gas anchor where the barrel of the pump serves as pull tube of the gas anchor.

PRODUCTION TUBING
SUCKER ROD STRING
TOP SEAT ASSEMBLY
PLUNGER
TRAVELLING VALVE
BARREL TUBE
STANDING VALVE

Fig. 3.8a: Rod Pump (Type: RHA or RWA or RSA)

- The top hold down pump owing to its top hold-down arrangement at the top of the pump provides stability, while the pump is in operation.

Disadvantages

- Due to the top hold down system, the outside of barrel is at suction pressure while the inside of barrel experiences the pressure due to fluid load of total tubing length. The suction pressure can be as low as 20–30 kg/cm^2 or theoretically zero pressure, where as, depending upon the pump depth, say for 3000 meters, the pressure inside the barrel may be 300 kg/cm^2. This large differential pressure across the barrel wall may lead to the bursting of barrel, Therefore, the pump has a limitation of depth. Thin wall barrel can only be lowered in shallow wells or where very less differential pressure across the barrel is encountered. Manufacturer of pump specifies that say pump for 2 7/8″ tubing and 1.75″ plunger size with heavy barrel can be lowered upto 1500 meters. If pump is required to be lowered further deeper, then pumps with lower plunger sizes have to be used. But beyond a certain depth, the top hold-down pump will not be suited at all.
- Also the barrel is under high tensile load due to weight of liquid column. Therefore, the mechanical strength of barrel also limits the depth of installation of such pumps.

Stationary Barrel Bottom Anchor Pump (Refer Figure 3.8b): The API has named as RHB, RWB and RSB pumps where 'R means rod, 'H' means heavy wall barrel, 'W' means thin wall barrel, 'S' means a thin wall barrel with soft-packed plunger and 'B' means a bottom hold down pump.

Advantages

- The differential pressure across the barrel is much lesser in the case of Bottom Hold-Down pump as compared to the Top Hold-Down pump. During downward stroke of plunger pressure differential would be zero and during the upstroke, it would be equal to tubing head less by suction pressure. So this Pump can be used at greater depths.
- This pump as well is not subjected to the tensile stress as in the case of Top Hold-Down pump.

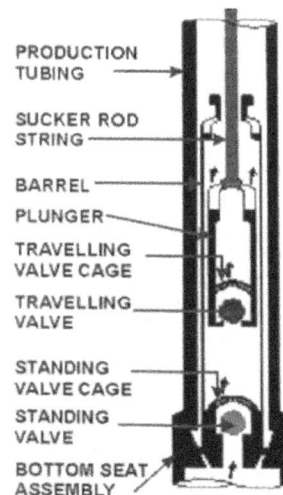

Fig. 3.8b: Rod Pump
(Type: RHB or RWB or RSB)

Disadvantages

During intermittent operation of pump, sand or other particles can settle in the barrel-tubing clearance space. Thus when it is required to pull out the pump, smooth pulling out of pump is obstructed.

Traveling Barrel Rod Pump (Refer Figure 3.8c): The API has named this type of pumps as RHT, RWT and RST. Where 'R' means rod or insert type pump, 'H' means heavy wall barrel, 'W' means thin wall barrel, 'S' means thin barrel with soft-packed plunger and 'T 'means traveling barrel pump. The traveling barrel rod pump has got the stationary plunger and moving barrel i.e. the plunger is held in place while the barrel is moved by the rod string. In this type of pump, there will be only one hold down or anchor which is at the bottom of the pump assembly. The plunger is attached to the bottom hold down arrangement by a short narrow pull tube through which well fluid enters the pump. The standing valve is located on top of the plunger.

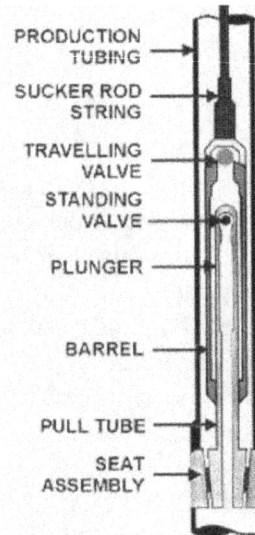

Fig. 3.8c: Rod Pump
(Type: RHT or RWT or RST)

Advantages

- Traveling barrel because of its larger diameter than the plunger creates a greater turbulence in the fluid motion around the hold-down, thus preventing sands/ solids from settling in the pump, during its operation.
- The pump has a rugged construction.

Disadvantages

- The size of the standing valve is less. Therefore, in case of the pumping of moderately high viscous fluid excessive pressure drop takes place at pump intake. That is why this pump is not recommended for lifting such type of fluids.
- Because of the restricted entry of fluid in the pump, more gas separation takes place, which may cause gas locking of the pump.
- In deep wells, high hydrostatic pressure acting on the standing valve may cause the pull tube to buckle. This limits the length of the barrel that can be used in deep wells.
- During the idle time (i.e., when pump is not operating) sand, fines, coal particles etc. settle around and underneath barrel, causing jamming of barrel and failure of pumping operation.

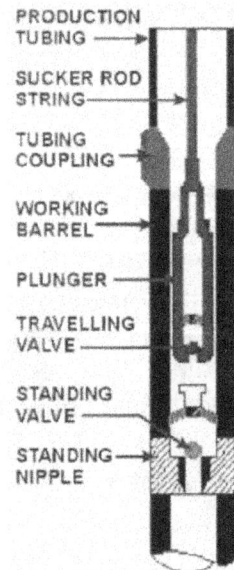

Fig. 3.8d: Rod Pump
(Type: TH or TP)

Tubing Pump (Refer Figure 3.8d): Tubing pumps are perhaps the oldest type of sucker rod pumps and have a simple but rugged type of construction. The API has named it as TH and TP where 'T' designates tubing pump, 'H' heavy wall barrel, 'P' heavy wall barrel with soft packed plunger.

Advantages

- Tubing pumps provide the largest pump sizes for a given tubing size. Thus this large size barrel allows more fluid to produce than with any other rod type of pump for the same size of tubing.
- The tubing pump is stronger in construction than any rod pump. The barrel is an integral part of tubing string and is slightly thicker than normal tubing string.
- The rod string of suitable size (i.e. compatible to barrel I.D.) is directly connected to plunger top without the necessity of an intermediate valve rod thus making the connection more sturdy and reliable. The large standing valve size results in low pressure losses in the pump and this facilitates pumping viscous and comparably higher GLR fluids.
- In inclined well, with moderate or small diameter casing, it is sometimes difficult to lower insert pump since the entire rigid pump body of barrel and its extensions can not match the profile of the bend tubing. It is in this respect, the barrel of tubing pump, having larger barrel—well casing annular area, can easily be lowered being it a part of tubing (i.e. along with tubing as tubing part) and subsequently plunger having much shorter length and less diameter with larger plunger O.D. than tubing I.D. clearance, can easily be pushed through tubing and placed in the barrel.

Disadvantages

- Work-Over-Operations require pulling out of tubing if pump barrel needs to be replaced.
- Large amount of rod and tubing stretch/contraction are expected because of large standing valve and therefore, setting depth of pump is limited. However, high strength sucker rods can be used whenever it is required to place tubing pump at greater depths.

3.3.4 *API Pump Classifications (Subsurface Pumps)*

The American Petroleum Institute has adopted a classification system for subsurface pumps. These classifications are mentioned in API Recommended Practice 11 AR. These are as follows:

RHA : Rod, stationary heavy wall barrel, Top Anchor pump.
RHB : Rod, stationary heavy wall barrel, Bottom Anchor pump.

RWA : Rod, stationary thin wall barrel, Top Anchor pump.

RWB : Rod, stationary thin wall barrel, Bottom Anchor pump.

RSA : Rod, stationary thin wall barrel, Top Anchor, soft packed plunger pump.

RSB : Rod, stationary thin wall barrel, Bottom Anchor, soft packed plunger pump.

RHT : Rod, Traveling heavy wall barrel, Bottom Anchor pump.

RWT : Rod, Traveling thin wall barrel, Bottom Anchor pump.

RST : Rod, Traveling thin wall barrel, Bottom Anchor, soft packed plunger pump.

TH   : Tubing, heavy wall barrel pump.

TP   : Tubing, heavy wall barrel, soft packed plunger pump.

Complete pump designations are as follows:

(1)	(2)	(3)	(4)	(5)	(6)	(7)	(8)	(9)
XX	XXX	X	X	X	X	X	X	X

1. Tubing size 15 means 1½″ Nom. Size Tubing i.e. 1.900″ OD (48.3 mm)

   20 means 2″   Nom. Size Tubing i.e. 2 3/8″ OD (60.3 mm)

   25 means 2½″   Nom. Size Tubing i.e. 2 7/8″ OD (73.0 mm)

   30 means 3½″   Nom. Size Tubing i.e. 3½″ OD   (88.9 mm)

   and so on

2. Pump bore (basic):   125–1¼″   (31.8 mm)

                          150–1½″   (38.1 mm)

                          175–1¾″   (44.5 mm)

                          178–1 25/32″ (45.2 mm)

                          200–2″   (50.8 mm)

                          225–2¼″   (57.2 mm)

                          250–2½″   (63.3 mm)

                          275–2¾″   (69.9 mm)

   and so on

3. Type of pump: R - Rod

                     T - Tubing

4. Type of barrel: H - Heavy wall pump

                     W - Thin wall pump

                     S - Thin wall pump with soft packed plunger

                     P - Heavy wall pump with soft-packed plunger

5. Location of seating assembly:

                    A - Top Anchored

                    B - Bottom Anchored

                    T - Bottom Anchored (for Travelling Barrel pump)

6. Type of seating assembly:

                    C - Cup type

                    N - Mechanical type

7. Barrel length: In feet, like 10′, 12′, 14′, 16′, 18′, etc.

8. Nominal plunger length: In feet, like 3′, 4′, 5′, etc.

9. Total length of extensions, which includes top and bottom extension of barrel: In feet. like 2′, 3′ etc.

Example: 25–175 RHAM 16-4-3

Here, 25 means 2½″ tubing; 175 means 1¾″ plunger opening or pump bore; RHAM means Rod, Heavy wall, Top anchored, Mechanical type hold down; 16 means 16 ft barrel length; 4 means 4 ft plunger length; 3 means a total of 3 ft extension on both sides of the barrel say 2 ft on one side and 1 ft on the other side.

Calculation of Barrel Length and Barrel Extension Length: Given plunger length as 4 ft and surface stroke length (maximum) as 144 inches. Therefore the effective total length will be (4 ft × 12) + 144 in. = 192 inches. So, barrel length = 192 inches = 16 ft.

Now in order to calculate barrel length, rod stretch, over-travel, necessary connections inside barrel like traveling valve's connection, top connection of plunger and with some allowable space known as dead space have to be taken into account. Therefore barrel length must be more than 16′. It is an usual practice to take barrel length of 16′ and then to add barrel extension on its two sides to keep the pump cost minimum. So, barrel length = 16 ft and say barrel extension = 3 ft (say 2 ft at the top and 1 ft at the bottom of the barrel). Therefore, the total effective length of the barrel = 16 ft + 3ft = 19 ft.

3.3.5 *Special Pumps*

Ring-Valve Pump (Refer Figure 3.9): The Ring-valve pump consists of a conventional top-hold down pump with an additional sliding check valve in the pump above the traveling valve. The ring valve slides up and down along the valve rod. It resembles a two stage pump.

During upstroke of the pump, the traveling valve pushes the liquid up as usually and the liquid passes through the ring valve in the tubing.

But during downstroke ring-valve gets closed at the outset due to the fluid load in the tubing with theoretically no tubing load effect on the traveling valve. This leads to drop in pressure in the space between traveling valve and ring valve and that ensures an early opening of traveling valve and smooth transfer of fluid in the barrel from below the traveling valve to above it.

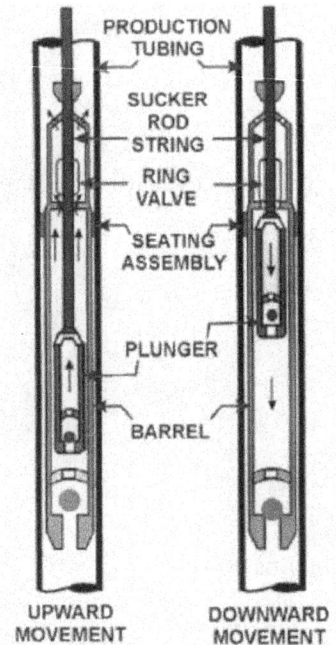

Fig. 3.9: Insert Pump - Ring Valve Type

The presence of the ring-valve in the insert pump has the following advantages:

- Conventional pumps generally operate with low efficiency for pumping fluids with substantial quantity of free gas. They also are susceptible to gas lock or fluid pounding. Both of these conditions can be attributed to the delay in the opening of the traveling valve. However, the ring valve pump enables the traveling valve to open early and thus eliminates or mitigates the occurrence of gas locking or fluid pound conditions. Thus, it helps to increase the efficiency of the pump.
- In case of sand laden fluids, the use of ring valve pump has a distinct advantage. For Top Hold-down pump, during the idle hours of the pump, sand settles on the top of the pump and some may get into the plunger barrel clearance specially when the plunger stays at the middle or at its lower most position. Ring-valve pump only allows the sand to settle on top of the ring-valve and thus arrests the sand to settle over the pump as well as in the plunger-barrel clearance. So, break-down of pumping action will not take place when the pump is restarted.
- In the case of high viscous fluid, ring valve pump has certainly an edge over the conventional pump because of its unique two stage pumping action.
- During the downstroke of the pump, the buckling of sucker rod string at its lower most ends can be avoided in the ring valve pump since this type of pump allows an early opening of the traveling valve. In other words, this type of valve action can keep the lower sections of sucker rod strings always in tension which thereby prevents to develop kinks in the rod and so helps increase the rod life.

Notwithstanding the above number of advantages, Ring valve pump can create fluid pound on the upstroke. During the upward movement of the sucker rod, the ring valve does not open at the very start of the upstroke. Since the space between the ring valve and travelling valve is only partly filled with liquid, so during upstroke, the liquid column above the traveling valve can pound on the ring valve. We can

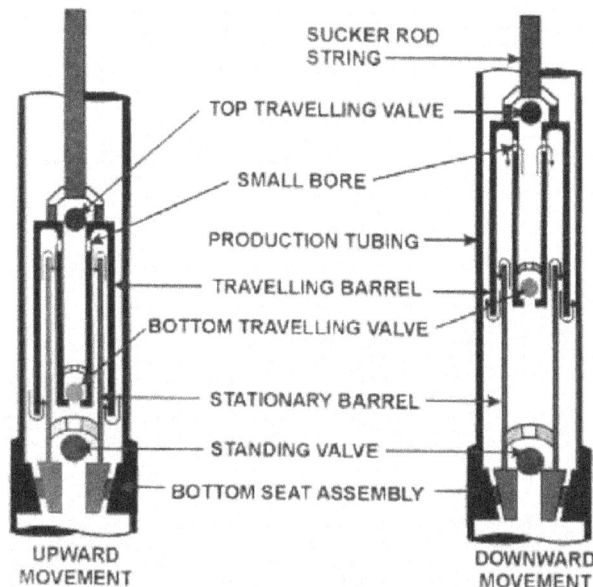

Fig. 3.10: Three Tube Sucker Road Pump

only logically say that this fluid pound occurring during upstroke will have a minor adverse effect on the operation of the pump.

Three-Tube Pump (Refer Figure 3.10): Three-tube pump has the combined features of a travelling and stationary barrel rod pump. The travelling barrel creates a fluid turbulence and thereby prevents the sand getting settled in the pump. The top travelling valve prevents sand entry in the barrel tubing clearance.

In the three-tube pump, the barrel tube with a standing valve is surrounded by two concentric travelling tubes that are joined together at the top of the pump with a travelling valve on top of them. The inner tube out of the two concentric traveling tubes acts as a plunger and has another travelling valve located at its bottom. The outer concentric tube is similar to a travelling barrel. These three tubes are loosely fitted to each other with about three times as much clearance as the largest fit between metal plungers and barrel of a conventional pump. Generally the clearance of the three tubes is of the order of 0.015 inch between each successive tube wall.

During upstroke both traveling valves are closed and standing valve is open. Because of the loose fit among the three tubes, well fluid can leak through the annular spaces from the high pressure area i.e. in the tubing above the pump towards the low pressure area that is below the bottom traveling valve. Again during the downstroke both traveling valves are open and standing valve is closed. Well fluids from high pressure area that is below the lower standing valve are displaced above the standing valves and a small fraction of the fluid escapes through the small bore at the top of the plunger tube as well as through the loose fit between the plunger and stationary barrel to enter into the tubing. The leakage rates are not appreciable because of comparatively large pressure drop developed in the clearances between the plunger and barrel fit.

We look at the conventional pump for pumping sand-laden fluid, we experience that abrasion wear is less if barrel plunger fit is increased but this, at the same time, entails a drastic drop in the quantity of fluid pumped. In this respect three tube pump results in negligible wear of the moving parts and at the same time creates sufficient turbulence inhibiting sand from settling into the pump.

Top and Bottom Anchor Rod Pump: In the top and bottom anchor rod pump is shown at Figure 3.11, the advantages of top hold down and bottom hold down pump are combined. Top hold down pump is a stable pump and it is more common. It prevents the settling of the sand in the pump. However, the barrel of this type of pump is subjected to high differential pressure and therefore this pump cannot work at deeper depths. Bottom hold down has the features which equalizes the pressures on two sides of the barrel and is therefore preferred for deeper installations but it is not as steady as the top lock pump and there is chance of sand accumulation in the barrel-tubing clearance. Therefore dual anchor rod pumps i.e. top and bottom anchor rod pumps are recommended when the pump needs to be installed at greater depths and specially when sand-laden fluid has to be pumped out.

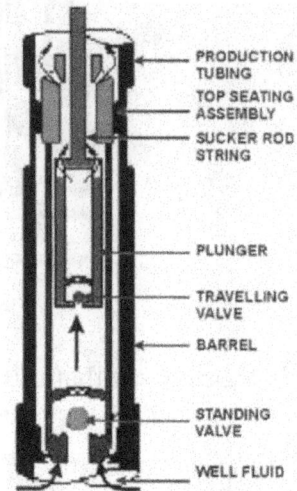

Fig. 3.11: Sucker Road Pump (Top & Bottom Anchor)

The bottom hold down is usually of the mechanical type and provides most of the necessary hold down force. The upper or top hold down is a cup type one which ensures a seal on the top of the pump and prevents vibration of the pump as well.

Pumps with Hollow Valve Rod (Refer Figure 3.12): The valve rod in stationary barrel rod pump connected to the plunger has a tendency to buckle during the down-stroke of pump due to compressive load operating there. The chances are more, specially in deep wells. Therefore the use of a hollow tube called 'pull tube' in place of a valve rod which either greatly reduces or totally eliminates the buckling problem.

The pump consists of a standing valve, fixed barrel, moving plunger, bottom traveling valve and bottom hold down

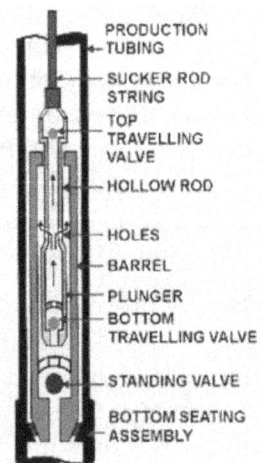

Fig. 3.12: Rod Pump—Hollow Two Stage Type

similar to a stationary barrel bottom hold down rod pump. The critical difference is that here the plunger is connected. With a hollow valve rod with a small port at its lower end and a top traveling valve at its uppermost end.

During upstroke, the bottom traveling valve in the plunger is closed due to the fluid load in the tubing. The standing valve opens and well fluid enters the barrel below the bottom traveling valve. As the upstroke proceeds, the plunger displaces out the fluid from the barrel chamber (i.e., fluid from the top barrel chamber). As the plunger starts the downward journey, the chamber space above the plunger starts increasing which results a drop in pressure in that area. This decrease in pressure causes an early closure of the top traveling valve and early opening of the bottom traveling valve. The standing valve is closed and the plunger displaces the fluid in the barrel below the bottom traveling valve into the barrel chamber above the plunger.

This type of pump operation resembles a two stage pump operation so it can be said to be a two-stage hollow valve rod pump. This pump provides good operation in wells with both gas and sand problems. Although the fluid path as provided inside the pump ensures a complete removal of sand particles during pumping, but settling of sand in between the tubing and barrel cannot be prevented. Therefore this two stage pumping action can be treated ideal for pumping fluids with high GORs and in sand cut wells. This type of pump can also find its application when the pump is run on a packer and the well fluid including the free gas is forced to enter into the pump.

3.4 Sucker Rod String

Sucker rod string, in fact, is the vital link between the sub surface pump and the pumping unit. These sucker rods are available as per API in three different lengths pump 25′, 30′ and 35′. These are connected to each other upto the depth of these are solid steel bars with forged upset ends with threads on it. API has standardized these solid steel sucker rods. The diameter of the rod body ranges from ½″ to 1 1/8″ with 1/8″ increments. Usually the rod body of diameters 5/8″, 3/4″, 7/8″ and 1″ are very common. At each ends of the sucker rod there is a short square section just before the sucker rod pin thread which facilitates the use of sucker rod tongs for connecting two sucker rods. These sucker rods are generally available with one coupling fitted at one end (So, one end of the sucker rod is called "pin end' and the other end is called "box end"). Sometimes due to limitations of tubing I.D., slim hole couplings are used.

For steel sucker rods, Table 3.2 gives data useful in calculations of sucker rod pumping problems. Table 3.3 gives useful data of tubing. Specifications for full size and slimhole couplings appear in Tables 3.4 and 3.5, respectively.

Table 3.2: Sucker Rod Data

Rod Size, in.	Metal Area, in.^{2}	Rod Weight in Air, lb/ft, W_r	Elastic Constant, in. per lb-ft, E_r
1/2	0.196	0.72	1.990×10^{-6}
5/8	0.307	1.13	1.270×10^{-6}
3/4	0.442	1.63	0.883×10^{-6}
7/8	0.601	2.22	0.649×10^{-6}
1	0.785	2.90	0.497×10^{-6}
1 1/8	0.994	3.67	0.393×10^{-6}

Table 3.3: Tubing Data Useful for SRP Design

Tubing Size, in.	Outside Diameter, in.	Inside Diameter, in.	Metal Area, in.^{2}	Elastic Constant, in. per lb-ft, E_r
1.90	1.900	1.610	0.800	0.500×10^{-6}
2 3/8	2.375	1.995	1.304	0.307×10^{-6}
2 7/8	2.875	2.441	1.812	0.221×10^{-6}
3 ½	3.500	2.992	2.590	0.154×10^{-6}
4	4.000	3.476	3.077	0.130×10^{-6}
4 1/2	4.500	3.958	3.601	0.111×10^{-6}

Table 3.4: Dimensions of Full Size Couplings and Sub-Couplings

Nominal Coupling Size*, in.	Outside Diameter, W, in.	Length min., N_L, in.	Length of Wrench Flat, W_L, in.	Distance between Wrench Flats, W_f, in.	Used with min. OD Tubing Size, in.
5/8	1 1/2	4	1 1/4	1 3/8	2 1/16
3/4	1 5/8	4	1 1/4	1 ½	2 3/8
7/8	1 13/16	4	1 1/4	1 5/8	2 7/8
1	2 3/16	4	1 1/2	1 7/8	3 1/2
1 1/8	2 3/8	4 1/2	1 5/8	2 1/8	3 1/2

*Also size of rod with which coupling is to be used.

Table 3.5: Dimensions of Slimhole Couplings and Sub-Couplings

Nominal Coupling Size*, in.	Outside Diameter, W, in.	Length min., N_L, in.	Used with min. OD Tubing Size, in.
1/2	1	2 3/4	1.660
5/8	1 1/4	4	1.990
3/4	1 1/2	4	2 1/16
7/8	1 5/8	4	2 3/8
1	2	4	2 7/8

*Also size of rod with which coupling is to be used.

3.4.1 *Sucker Rod Materials*

The material of steel sucker rods has more than 90% iron content. Other elements are added to increase strength and hardness and resist corrosion etc. Steel used for manufacturing of sucker rods can be categorized in two ways viz. carbon steel and alloy steel. Carbon steels contain carbon, manganese, silicon, phosphorous and sulphur, whereas alloy steel contains additional elements like nickel, chromium etc., as per the necessary requirements like more rod strength etc.

API has standardized different grades of sucker rods of which grade 'C' is the carbon steel sucker rods and is the least costly. Grade 'D' sucker rods is the chorine molybdenum alloy for higher range of strength; other than the above mentioned two grades, grade 'K' is a special nickel, molybdenum alloy used in moderate corrosive fluids.

A table indicating different tensile strengths of different rod grades viz. 'C', 'D', and 'K' has been given in Table 3.6.

Table 3.6

Rod Grade	Composition	Tensile Strength, psi	
		Min.	Max.
K	AISI 46	85,000	115,000
C	AISI 1536	90,000	115,000
D	Carbon or Alloy	115,000	140,000

Chemical and mechanical properties API sucker-rod Material according to API Spec. 1.1B.

The range of tensile strength available is accomplished by altering the chemical content of the steel and by the treatment process used in rod manufacturing such as tempering, normalizing, quenching, and case hardening.

In general, the maximum allowable stress on rods should not exceed 30,000 psi to 40,000 osi. In any case, the maximum allowable stress and range should be checked against the Goodman diagram maximum allowable stress. An exception is the "Electra" rods of the M/s Oilwell division of United States Steel that have ratings of 40,000 psi to 50,000 psi. The Goodman diagram analysis is not applicable to this section of rods.

Operating, sucker rods for a low fluid level at a much deeper depth of around

2500 m or more, the grade 'D' rod fails to perform because of the calculated stress in such cases exceeds its allowable stress. Due to this some Non-API high strength Sucker rods have come into the market [Some of these are Norris make '97' grade type and oilwell make 'EL' grade type].

Hollow sucker rod tubes can also be used to advantage in slim-hole completions. Since fluid lifting takes place inside the hollow sucker tubes, no tubing is needed in the well. So, the application of hollow sucker rods is restricted to lower fluid rates due to the obvious reason that large fluid volume involve greater pressure losses inside the tube. The hollow rods also require special well head consisting of hollow polished rod and flexible hose connected to flow line. Sometimes high sand production can be well tackled with this system. Injection of corrosion inhibitors can also be accomplished properly with the use of hollow rods.

Fibre glass sucker rods figure in the API specifications. These rods are low weight, corrosion resistant, non-metal and therefore have definite advantages over steel sucker rods specially when the pump is to be operated in very deep corrosive wells. A long steel rod string is normally heavy weight material and the weight increases with the increase in area of the plunger due to the liquid load on the plunger. The fibre glass rods became commercially available from 1977. The individual fibres have high tensile strength and depending on the resin/glass ratio, during the process of the manu-facture of the final rods, the fibre glass rods develop a strength of 110,000–180,000 psi which is about 25% stronger than the steel sucker rods. At the same time, rod weight of fibre glass sucker rod is only about $1/3^{rd}$ of the weight of steel sucker rods. When subjected to an axial force, fibre glass rods elongate 4 times more than that of steel. This excessive rod stretch prohibits the use of only fibre glass rod string in the well, rather they are used in combination with steel sucker rod, where top portion of the rod string is of fibre glass rods and bottom portion is of steel rods. This combination of fibre glass and steel rods weigh only about ½″ of that of an all steel sucker rods. Fibre glass rods are available in nominal sizes ranging from 5/8″ to 1¼″. While installing the fibre gloss rods in the well, it must be ensured that the environment in the well is within the safe operational temperature for fibre glass rods to operate.

Allowable Rod Stress and Stress Range
A string of sucker rods when in normal operation is subjected to alternating high and low stress because of the nature of pumping cycle. On the upstroke, the rods bear a load that includes their own weight, the weight of the fluid they are lifting, friction and effects of acceleration. On the down stroke, the rods carry some friction load and load of their own weight diminished by effects of de-acceleration. The ratio of up-stroke to downstroke load and hence ratio of upstroke to downstroke stress can be 2 to 1 or often 3 to 1 or more. This cycle of alternating high and low stress occurs at a frequency at least equal to the pumping speed. A unit pumping at 20 strokes per minute goes from high stress to low stress every 3 seconds or 10,500,000 cycles per year. This process repeated on the rods over months and years can easily lead to metal fatigue.

In designing API steel sucker rod strings, it is recommended that the modified Good-man diagram is suited as the basis for stress analysis. The diagram can be reduced to an equation as below:

$$S_A = (T/4 + 0.5625 \times S_{min}) \times SF \qquad \qquad ... (3.1)$$

Where,

S_A = maximum allowable stress in the rods for a given value of minimum stress, S_{min} and service factor.

T = minimum tensile stress for rods of a given API rating (refer Table 3.6). Example for *C*-rods, T = 90,000 psi.

S_{min} = minimum stress to be experienced by the rods in the pump cycle.

SF = Service factor. This is a factor that adjusts, usually downward, the estimated allowable maximum stress to account for corrosive conditions. The suggested service factors for API grade C and D rods in corrosive conditions are:

Service	API-C	API-D
Non-corrosive	1.00	1.00
Salt water	0.65	0.90
Hydrogen sulfide	0.50	0.70

The service factors should be considered as guideline and not as highly precise universal parameters. Experience from a given area should be used to determine proper service factors for given type of corrosive environments.

Example Calculation

A-77 Grade-D rod string operating in a salt water/crude oil environment has a minimum polished rod load measured with dynamometer of 16,000 lbf. Calculate maximum allowable stress and load for this rod string?

Solution

Step-1: The cross section area of 77 rod (7/8″ rod) from Table 3.2 is 0.601 in.². Hence, S_{min} = (Min. load/Rod area) = 16,000/0.601 = 26,622 psi.

Step-2: Service factor for API grade D rod for salt water service is 0.9.

Step-3: Use Goodman diagram equation to compute S_A.

$$S_A = (T/4 + 0.5625 \times S_{min}) \times SF = (115,000/4 + 0.5625 \times 26622) \times 0.9$$
$$= (28750 + 0.5625 \times 26622) \times 0.9 = 39,352 \text{ psi.}$$

Step 4: Compute maximum allowable load = 39352 × 0.601 = 23651 lbf.

3.4.2 *Benefits of Fibre Glass Sucker Rods*

The advantages of using fibre glass rods are many. The most important advantage is that the production rate can be increased manifold from a well with the help of a smaller pumping unit. For example if the stroke length is 80 inches at the polished rod, then at the pump, the stroke length will be even 3 × 80 i.e 240 inches. Therefore while installing sucker rod pump care is to be exercised so that sub-surface pump should have sufficient stroke length to match the requirement of fibre glass stretches.

Fibre glass sucker rods, even today, are more expensive than steel rods as well as these require very careful handling and that is why their application is very limited.

So, when the wells are very deep say around 3500 m and above, static fluid level is high/low and PI is very high fibre glass sucker rods in combination with steel sucker rods will logically be an appropriate choice of lift system.

3.4.3 *Sucker Rods Joints*

Sucker rod joints are probably the most important aspect in making a perfect integrated system of the rod string. It is of utmost importance for an engineer to understand the mechanism of these joints. Sucker rods are joined with the help of couplings. These couplings are also API specified. The threads are plain and therefore it is easy to put a coupling on sucker rod pin and to rotate it just by hand. The end of the sucker rod coupling (shoulder) which is in contact with the pin flat face (shoulder face) should then be tightened (shoulder to shoulder joint) with a proper make-up torque between the two parts to enable the two sucker rods to form one integrated body with a coupling in between them. The distribution of stresses in a sucker rod joint has been given in Figures 3.13a, 3.13b and 3.13c, where upper portion of the coupling is in compression and pin-inside in tension. Unless and until the stresses between them are relieved, it will not be possible to disconnect the Sucker Rods. This ensures that the accidental unscrewing of sucker rods will not occur during the operation of the pump.

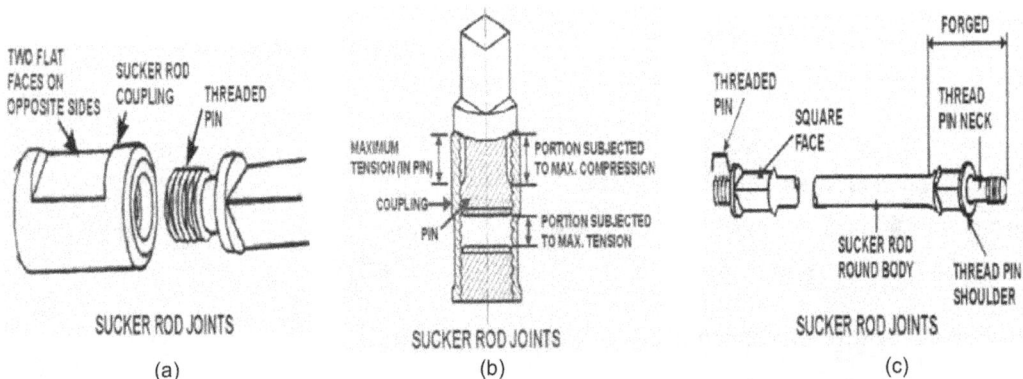

Fig. 3.13(a-c)

Rod loads during pumping cycle are cyclic. During upstroke, the rod loads are caused due to rod weight, fluid weight, load due to acceleration and friction and during downstroke rod load is due to rod weight only which is to some extent reduced due to negative acceleration. Therefore, the cyclic nature of rod weights demands a perfect coupling of sucker rod pin-joint.

Rod joints are usually made up with the using pneumatic or hydraulic power tongs. These power tongs exert a desired torque on the joint.

In many of the operating fields, a comparatively easy system to generate the required make up torque is being practiced.

At first the pin and coupling are made up to a hand tight position. Thereafter, two spring loaded tongs (jerk type tongs) are put over the square area of two sucker rods (one tong above the coupling and other is below it) in such a way that they make nearly 40 degrees between them in front of the person who is holding one tong with one hand and the other tong with his other hand. With the help of concurrent jerks from the two hands, approximately 3–4 times, a required makeup torque is created. This matter was discussed with Mr. R.H. Gault (Bob Gault) a renowned expert on sucker rod pumps. He was convinced and approved this system but at the same time he cautioned that this system be restricted to a shallower depth say within 1500 m. For greater depths, pneumatic or hydraulic power tongs are a must.

3.4.4 *Load to Pumping Unit*

The load exerted to the pumping system mainly depends on well depth, rod size, fluid properties etc. The maximum PRL and peak torque are major concerns for designing pumping system.

Maximum PRL: The PRL is the sum of weight of fluid being lifted, weight of plunger, weight of sucker rods string, dynamic load due to acceleration, friction force and the up-thrust from below on plunger. But is has been seen that no force attributable to fluid acceleration is required, so the acceleration term involves only acceleration of the rods. So the friction term and weight of the plunger is neglected. Let ignore the reflective force, which will tend to underestimate the maximum PRL. Consider up-thrust is zero and also assume the TV is closed at the instant at which the acceleration term reached its maximum value. With these above assumptions, the PRL maximum will be,

$$\text{PRL}_{\text{max}} = S_f + (62.4)D - \frac{A_p - A_r}{144} + \frac{\gamma_s\,D\,A_r}{144} + \frac{\gamma_s\,D\,A_r}{144}\left(\frac{SN^2M}{70,471.2}\right) \qquad \dots (3.2)$$

S_f = Specific gravity of fluid in tubing
D = Length of sucker rod string (ft.)
A_p = Gross plunger cross-sectional area (in²)
A_r = Sucker rod cross-sectional area (in²)

γ_s = Specific weight of steel (490 lb/ft^2)

$M = 1 + \dfrac{C}{h}$

C = Length of crank arm

h = Length of pitman arm

Equation 3.2 can be rewritten as below, after replacing,

$$\text{PRL}_{max} = S_f(62.4)\frac{DAP}{144} - S_f(62.4)\frac{DA_r}{144} + \frac{\gamma_s\, DA_y}{144} + \frac{\gamma_s\, DA_r}{144}\left(\frac{SM^2M}{70,471.2}\right) \quad \dots (3.3)$$

If the weight of the rod string in air is,

$$W_r = \frac{\gamma_s\, DA_r}{144} \qquad \dots (3.4)$$

which can be solved for A_r, which is,

$$A_r = \frac{144W_r}{\gamma_s D} \qquad \dots (3.5)$$

Substituting eqn. 3.5 in eqn. 3.3, then,

$$\text{PRL}_{max} = S_f(62.4)\frac{DA_p}{144} - S_f(62.4)\frac{W_r}{r_s} + W_r + W_r\left(\frac{SN^2M}{70,471.2}\right) \quad \dots (3.6)$$

This eqn. can be further reduced by considering the fluid density as an 50° APJ with $S_f = 0.78$. Then eqn. 3.6 becomes ($\gamma_s = 490$),

$$\text{PRL}_{max} = S_f(62.4)\frac{DA_p}{144} - 0.1\,W_r + W_r + W_r\left(\frac{SM^2M}{70,471.2}\right)$$

or,

$$\text{PRL}_{max} = W_f + 0.9\,W_r + W_r\left(\frac{SN^2M}{70,471.2}\right) \qquad \dots (3.7)$$

where, $W_f = S_f(62.4)\dfrac{DA_p}{144}$, i.e. fluid load

Then eqn. 3.7 can be as below,

$$\text{PRL}_{max} = W_f + (0.9 + F_1)\,W_r \qquad \dots (3.8)$$

For conventional units,

$$F_1 = \frac{SN^2(1+c/h)}{70,471.2} \qquad \qquad \text{... (3.9)}$$

For air-balance units $F_1 = \dfrac{SN^2(1-c/h)}{70,471.2}$ ... (3.10)

Minimum PRL: The minimum PRL occur while TV is open so that the fluid column weight is carried by the tubing and not by the rods. The minimum load is at or near the tap of the stroke. Not considering the weight of the plunger and friction term, the minimum PRL is,

$$\text{PRL}_{min} = -S_f(62.4)\frac{W_r}{\gamma_s} + W_r - W_r F_2$$

when 50° API oil, it works out to,

$$\text{PRL}_{min} = 0.9\, W_r - F_2\, W_r = (0.9 - F_2)\, W_r \qquad \text{... (3.11)}$$

where for conventional units,

$$F_2 = \frac{SN^2(1-c/h)}{70,471.2} \qquad \qquad \text{... (3.12)}$$

and for air balance units,

$$F_2 = \frac{SN^2(1+c/h)}{70,471.2} \qquad \qquad \text{... (3.13)}$$

Counter Weights: To reduce the power requirements for the prime mover, a counter balance load is used on the walking beam (small units) or the rotary crank. The ideal counter balance load C is the average PRL. Therefore,

$$C > 1/2\ (PRL_{max} + PRL_{min})$$

Using eqn. 3.8 and 3.11 in the above, then

$$C = \frac{1}{2}\, W_f + 0.9\, W_r + \frac{1}{2}\, (F_1 - F_2)W_r \qquad \text{... (3.14)}$$

or for conventional units,

$$C = \frac{1}{2}\, W_f + W_r\left(0.9 + \frac{SN^2}{70,471.2} - \frac{c}{h}\right) \qquad \text{... (3.15)}$$

and air balance units,

$$C = \frac{1}{2}\, W_f + W_r\left(0.9 - \frac{SN^2}{70,471.2} - \frac{c}{h}\right) \qquad \text{... (3.16)}$$

The counter weights can be selected from manufacturer catalog based on the calculated C value. Then relationship between the counter balance load C and total weight of the counter weights can be written as,

$$C = C_s + W_c \frac{\gamma}{c} \frac{d_1}{d_2} \qquad \qquad \ldots (3.17)$$

where, C_s = structural unbalance, lb
W_c = total weight of the counter weights, lb
r = distance between the mass center of counter weights crank shaft center, inch.

Peak Toque and Speed Limit: The peak torque exerted is usually calculated on the most severe possible assumption first, which is that peak load occurs when the effective crank length is also maximum (when the crank arm is horizontal).

Peak Torque T is, (Figure 3.2)

$$T = c[C - (0.9 - F_2)W_r] \frac{d_2}{d_1} \qquad \qquad \ldots (3.18)$$

Substituting this in eqn 3.14 to 3.17, then,

$$T = \frac{1}{2} S [C - (0.9 - F_2)W_r] \qquad \qquad \ldots (3.19)$$

or,

$$T = \frac{1}{2} S \left[\frac{1}{2} W_f + \frac{1}{2}(F_1 + F_2) W_r \right]$$

or,

$$T = \frac{1}{4} S \left[W_f + \frac{2SN^2 W_r}{70,471.2} \right] \text{(in - lb)} \qquad \qquad \ldots (3.20)$$

The pumping unit is not always perfectly balanced ($C_s \neq 0$), the peak torque is also affected by structure unbalance. Torque factors are used for correction,

$$T = \frac{\frac{1}{2} \{ PRL_{max}(TF_1) + PRL_{min}(TF_2) \}}{.93} \qquad \qquad \ldots (3.21)$$

where, TF_1 = Max upstroke torque fuels
TF_2 = Max downstroke torque fuels
0.93 = System efficiently.

For symmetrical conventional and air balance units,

$$TF = TF_1 = TF_2$$

The relationship between stroke length and cycles per minute. As given earlier, the maximum value of the downward acceleration is equal to,

$$a_{max/min} = \frac{SN^2 g\left(C \pm \dfrac{c}{h}\right)}{70{,}471.2} \qquad \dots (3.22)$$

Tapered Rod Strings: For deep wells, it is necessary to use a tapered string to reduce the PRL at the surface. The largest diameter rod is placed at the top of the string, then the next largest, then least largest. Usually there are in sequence up to four different rod sizes. The tapered rod strings are designated by 1/8 inch (dia.) increments.

Tapered strings are designed for static loads with a sufficient factor of safety to allow for random low level dynamic loads. The two criteria are used in the design of tapered sucker rod string design.

1. Stress at the top rod of each rod size is the same throughout the string.
2. Stress in the top rod of the smallest set of rods should be the highest for (~ 30,000 psi) and the stress progressively decreases in the top rods of the higher sets of rods.

Pump deliverability and power requirements:

Liquid flow rate delivered by the plunger pumps can be expressed as,

$$q = \frac{A_p}{144} N \frac{S_p E_v}{12 B_0} \frac{(24)(60)}{5.615} \text{ (bbc/day)} \qquad \dots (3.23)$$

or,

$$q = 0.1484 \frac{A_p \, NS_p \, Ev}{B_0} \text{ (stb/day)}$$

S_p = effective plunger stroke length (in.)
E_v = volumetric efficiency of the plunger
B_0 = Formation volume factor of the fluid.

3.5 Gas Anchors

Presence of free gas in the pump barrel (Figure 3.14) not only reduces the efficiency of the pump but presents some typical operational problems. On the down stroke, the traveling valve does not open at the start of the plunger making downward journey, rather, the opening of it is delayed considerably. The free gas collected over the liquid surface in the barrel, at first, starts dissolving in liquid with very little increase of pressure in the barrel. When all the free gas goes into solution, the fluid in the barrel turns into an incompressible fluid and as a result the pressure in the barrel starts building up with the downward movement of the plunger. As and when fluid pressure in the barrel exceeds the tubing load (fluid load), the traveling valve opens and the transfer of fluid from barrel to tubing above traveling valve takes place.

Secondly, on the upstroke, the standing valve does not open immediately at the start of the upstroke. Its opening is again delayed.

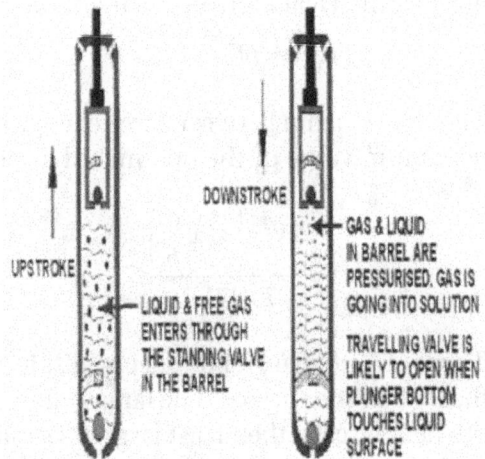

Fig. 3.14: Gas Interference in SRP Well

Due to dissolved gas in the liquid in the dead space of the barrel, the pressure of the fluid in the barrel falls gradually with the liberation of gas from it. When the pressure from below the standing valve exceeds the pressure in the barrel, the standing valve opens which allows fresh fluid to enter into it. This clearly demonstrates that the plunger's effective stroke length is savagely cut. As a result, a considerable reduction of pump displacement, takes place, resulting in lowering of pump efficiency. The decrease of pump efficiency is quite unpredictable and may be in the range of 30 %, 40 %, 50 %, etc. depending upon fluid intake and free gas generation. In extreme case, a complete 100 % reduction of efficiency occurs. It is then said that pump is "gas-locked". In this gas-locking case, only the expansion and contraction of a compressible fluid take place during the upstroke and downstroke of plunger with no fluid transfer from barrel to tubing.

The presence of gas in the barrel creates operational problem in the sense that, the barrel is then partly filled with liquid and gives rise to some peculiar problems like "gas pound" and "fluid pound". These can cause unscrewing of rods or create kink or bend in the rod resulting in pump failure.

Free gas occupation of barrel space in the pump depends upon its depth of installation. If the pump is placed at greater depth, the free gas generation will be less and, if it is placed at shallower depth, the free gas generation in the pump will be more.

Considering the above facts, it is essential to adopt some types of remedial measures to prevent gas generation in the barrel. The pump of longer stroke length, lesser SPM, more plunger diameter, lesser dead space, some other special pump like ring valve pump etc. are some of the measures to mitigate free gas interference in the pump. But the most effective method to prevent the gas interference in the pump is to install downhole gas separator called gas anchor before the intake of fluid in the pump. All the gas anchors operate on the principle of gravitational separation. Liquid and free gas, owing to their large difference in gravity, are separated in the casing tubing annulus. Gas goes up in the annulus and is let out through a non-return valve fitted on the well head annulus flow line. Also, the produced free gas from the annulus may be utilized to operate the gas engine as prime mover of the pumping unit.

Gas anchors are broadly classified into three main categories as: (a) Natural Gas anchor; (b) Packer type gas anchor; and; (c) Poor boy gas anchor.

3.5.1 *Natural Gas Anchor (Refer Figure 3.15)*

The simplest and most effective gas anchor is the natural gas anchor. The very meaning of the natural gas anchor is to facilitate gas separation from the fluid by gravity before it enters into the pump. In order to do the natural gas separation, the pump is set a few feet below the level of the lower most casing perforation, so that when liquid and free gas enter the well bore, gas goes up the annulus and finally finds its way out through a non-return-valve at the surface into the flow line and gas-free oil goes into the down-hole pump intake. The following are the necessary requirement conducive enough to create a Natural Gas Anchor.

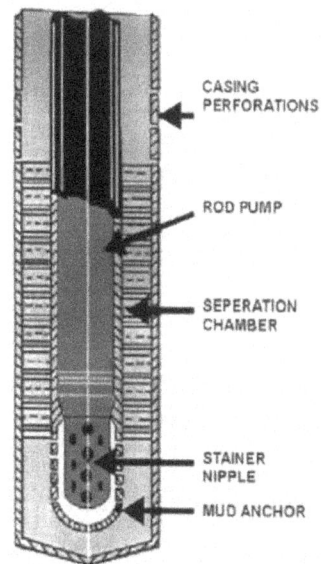

Fig. 3.15: Natural Gas Anchor

- The well produces more free gas than the acceptable limit of the pump.
- Surface unit, rods and pump types must permit to lower the pump at the required deeper depth.
- There should be adequate sump to place the pump below perforation.

3.5.2 *Packer Type Gas Anchor (Refer Figure 3.16a)*

Packer type gas anchor is considered next to natural gas anchor as per gas separator efficiency is concerned. When Natural gas anchor can not be provided, because of the reasons as explained above. Packer type gas anchor can be considered as a suitable alternative. The salient features of this Packer type gas anchor is that the packer is installed below the pump intake point and above the perforation. Formation is directed through a small by-pass pipe extending through the packer upto a certain comfortable distance above the pump intake point. As the fluid ejects out from the by-pass pipe into the much larger cross-sectional area of casing-tubing annulus, a good separation of free gas from liquid results. Free gas travels up the annulus and finds its way into the flow line. Liquid then falls back and enters the pump intake.

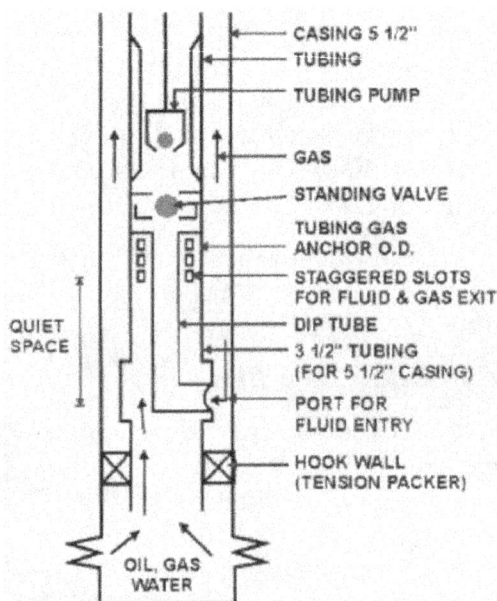

Fig. 3.16a: Packer Type Gas Anchor
(Packer without Bypass Tube)

Because of the very presence of packer, well completion is not simple. Sometimes conditions do not favour the packer completion. Because of these factors the packer type gas anchor is not very popular and is limited to certain categories of wells. Also, small by-pass pipe may restrict the fluid flow rate.

3.5.3 *Poor Boy Gas Anchor*

Next to the packer type gas anchor, poor boy gas anchor is considered, which is shown at Figure 3.16b. It basically consists of a mud anchor perforated at the top and a dip tube called the pull tube inside the mud anchor. The gas anchor is run immediately below the pump where the pump suction has direct access to the pull tube.

The formation fluid, at first, rises through the gas anchor-casing annulus. It then enters the mud anchor through its lower perforations with some free gas part travels up the annulus. As the mixture travels downwards to the intake of the pull tube during pump operation, the free gas separates out in the pull tube-mud anchor annulus and finds its outlet through the top perforations of the mud anchor, into the annulus and finally the gas is bled in the flowline like in other types

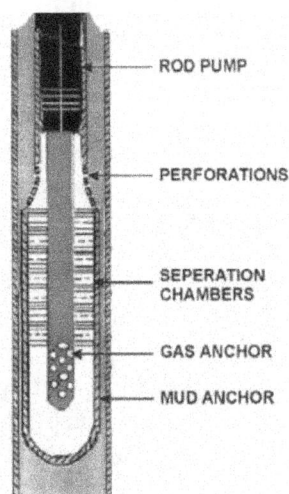

Fig. 3.16b: Poor Boy
Gas Anchor

of gas anchors. The gas-free oil goes into the suction of the pump through the pull tube.

While making an effective poor boy gas anchor, the primary care should always be taken for proper sizing of every component of the poor boy gas anchor. The technical requirement of each component of the gas anchor are enumerated below:

Mud Anchor: Mud anchor is made out of the available tubular goods for example 3½″ tubing or 4″ tubing or 4½″ tubing etc., primarily governed by the well casing diameter.

Pull Tube: Pull tubes are usually 1″ or more in diameter and is limited by the I.D. of the mud anchor.

Required annular area between Pull Tube and Mud Anchor: More the annular area, better the separation of gas from oil. Therefore care should be exercised so that larger annular area is made available for effective separation.

Length of the Quiet Space: More the length of the quiet space better is the separation.

Diameter of the Pull Tube: If the diameter of the pull tube is more, then the liquid (i.e. gas-free liquid) during the course of its movement towards pump inlet encounters less friction and this will minimize further free gas generation to minimum.

Length of the Pull Tube: More the length of pull tube, more will be friction inside the pull tube and so there would be greater chance of free gas breakthrough from the oil body, during the liquid travel through the pull tube into the pump barrel.

By considering entire scenario along with the escape velocity in mind, a suitable design of gas anchor can be worked out for each casing size, pump type and fluid parameters. Some published empirical relations are considered to calculate the proper size of poor boy gas anchor.

3.5.4 *Poor Boy Gas Anchor without Pull Tube*

Since the advantage of a poor boy gas anchor is negated to some extent due to presence of pull tube as discussed in the foregoing statement, the poor boy gas anchor has been made simple by omitting the pull tube and in place of that, pump barrel itself is utilized as the pull tube. By looking at the Figure 3.17 it is clear that this type of gas anchor is applicable only for top holddown

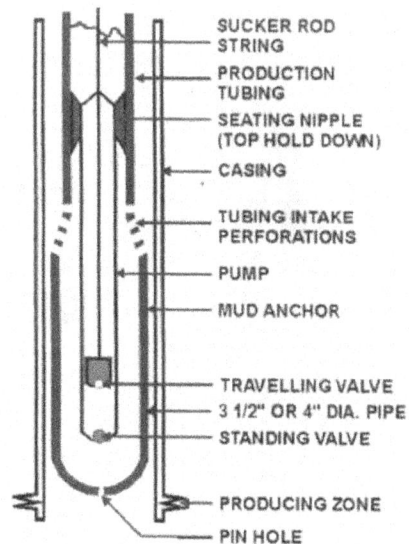

Fig. 3.17: Poor Boy Gas Anchor Without Pull Tube

insert pump. By this measure an efficiency of the gas separation can be expected to increase further by about 20–30%.

In the light of the above discussion, it is clear that natural gas anchor is the most preferred type, where it is applicable or feasible. Next to that, for reasons of simplicity, poor boy gas anchor is preferred and wherever the pumps are insert and top hold down type, the poor boy gas anchor without pull tube is, invariably the first choice over the general poor boy gas anchor. The Packer type gas anchor can be utilized when neither Natural gas anchor is feasible nor Poor boy gas anchor is effective.

3.6 Sinker Bar

A few number of sinker bars just above the pump provide the stability of the down-hole SRP operations due to their increased weight on the pump. These are the bottom-most part of the sucker rod string.

During down-stroke, the rod string as well as the plunger move downward but generally both do not move at the same speed. The lower part, that is the Plunger, is subjected to compression because of the impact of fluid pressure built-up in the barrel.

Also since barrel-plunger clearance is very small and as such, the movement of plunger takes place with close tolerance, a viscous drag force results due to fluid trapped in the clearance space. The total effect of these up-thrust is highly detrimental to the operation of the pump. This may lead to buckling, developing kinks or unscrewing of sucker rods. This problem becomes more severe for the kinds of combination of rod string, where the lowermost size of the tapered sucker rod is 5/8″.

Besides, some loss of liquid production can take place due to reduced plunger stroke, as a result of this phenomena explained above.

Therefore in order to overcome these undesirable effects, heavier sucker rods or sinker bars, which are heavy solid steel rods are run as part of rod string at its lower most end i.e., they are located just above the pump. These heavy sucker rods or sinker bars absorb the upthrust to a great extent as well as impose an extra load on the plunger to speed up the plunger during its downward journey. Thus it provide a stability of the operation of pump.

In erstwhile USSR, most of the sucker rod pump wells are completed with a few heavier sucker rods installed just above the pump. For example, a sucker rod pump well is completed with 1″, 7/8″ and 3/4″ and in that well about 60 meters or so 1″ rods are installed at the lowest end of the string just above the pump. In USA, many pumping wells are completed with a few 1½″ wireline sinker bars with matching threads. Each sinker bar is about 4–5 feet in length. In many of the Indian oil wells a few number of heavier rods (about 6–18 in number) are lowered just above the pump. Of late, a new type of sinker bar has been designed which is approximately 12 feet in length and most of its part is of 1½″ dia with the top neck of 1″ dia for making the room to hold the sinker bar with 1″ elevator with the top end having similar end connection of sucker rod ¾″ pin end and the bottom of ¾″ pin thread. In fiber glass rod completion, heavy rods are an absolute necessity.

3.7 Rod Guides, Scrapers, etc. Long Sucker Rod Coupling and Use of a Number of Pony Rods in the Sucker Rod String Combination Rod

When sucker rod pump is in operation, whole rod string moves up and down as well as tubing also moves a little up and down because of the expansion and contraction of tubing/sucker rods.

It invariably results in rod tubing friction. Since, metallurgy of the rod is inferior to that of tubing, so most of the wear and tear results on rods. The friction becomes severe when the wells are crooked and inclined. Due to the wear of the metal parts, strength of the rod decreases and eventually failure of rods result.

In order to overcome such rod tubing friction, several common field practices being used for a long time have been discussed below:

- *Use of a number of small length pony rods* at the vulnerable points where friction is more, help save the abrasion of sucker rod body Here only the sockets get worn out. Due to wearing out of only socket (higher O.D. than the body sucker rod), the frequency of failure of sucker rod string is reduced.
- *Extra long sucker rod coupling* is more effective than the normal coupling. So, extra long couplings with pony rods at the vulnerable points can further reduce the frequency of sucker rod failure.
- *The scrappers* are fitted on to the sucker rods at the specified interval. These scrapers not only protect the sucker rod body from getting worn out but also creates turbulence to prevent the settling of sand and mitigate the paraffin build- up in the tubing. These scrappers are either made of hard plastic material or metal.

Metal scrapers are available in two hemispherical shapes. These two hemispherical shapes are placed around the sucker rods and then press-fitted with the help of specially designed vice. These press fitted parts are then welded with each other (welding is not done with the sucker rod). Therefore scrapers fitted on to the sucker rods are fixed at that point and are not liable to slide. Plastic scrapers are placed around sucker rods with long heavy duty pipe wrench. They are generally of blade-shaped.

These scrapers however have not been found very effective. Generally they get dislodged and slide along with the up and down movement of rod and often they get jumbled at one place. Plastic scrapers have been found to even get detached from the rods and get accumulated around the sucker rod over the pump. These often creates typical pump operational problem.

Rod Guides: The sucker rods with a few built-in rod guides located at suitable interval of the length of the sucker rod string has been found most effective. These rod guides are factory-mounted metallic guides and moulded permanently to rods. The perceptible advantage of this guide is that because of its very nature of fitting (factory moulding) with the rods they do not slide over the length of the rods and therefore drastically reduces rod failure by minimizing the wear of sucker rod body. In many oil fields, moulded guides are most preferred. The moulded-rod guides system is considered the best for inclined wells (L or S type) and in the dog-legged portion of the well.

There are other rod guides like wheeled rod guides where several wheels are placed in special couplings. These wheels are placed at 45 degrees angle with each other and roll on inside surface of tubing during the up and down movement of the sucker rod and thus because of the rolling action the rod encounters a minimum friction. Although the design feature seems very attractive but it cannot work very effectively when the inclination is very severe dog-legged and well profile is in the shape of 'S'. The wheels get flattened at one side of sucker rod string where excessive friction takes place. These ultimately affect pump operation.

3.8 Well Head Equipment

The sucker rod wellhead is not like the normal wellheads which explain at Figure 3.18a and 3.18b. The material difference is the presence of polished rods projecting out of wellhead in case of sucker rod pumping system.

A normal wellhead consists of a master valve, flow Tee with wing valve and crown valve. In case of sucker rod pumping wells, it is ideal to have one stripper in place of master valve and stuffing box in place of crown valve. The flow Tee and wing valve are similar to that of normal wellhead. The stripper in SRP well can work as a master valve. When the stripper valve is closed, it closes the polished rod-tubing annular space and thus cuts off the fluid flow.

Every care is to be exercised to prevent any leakage from the stuffing box. Many improved designs of stuffing boxes are available in the market. 2-tier, 3-tier stuffing boxes are manufactured by many USA company. The stuffing box packing is made of rubber element with metallic supports and needs to be replaced as and when these are necessary. These rubber elements get worn out quickly, if the polished rod is not properly centred. Also, if the stuffing box is very tightly packed or is not having the proper lubrication by the well fluids, the rate of wear of rubber packing becomes very fast. In order to overcome this, a small lubricator box, made of aluminium inside of which is lined with flannel type cloth material, is placed just above the stuffing

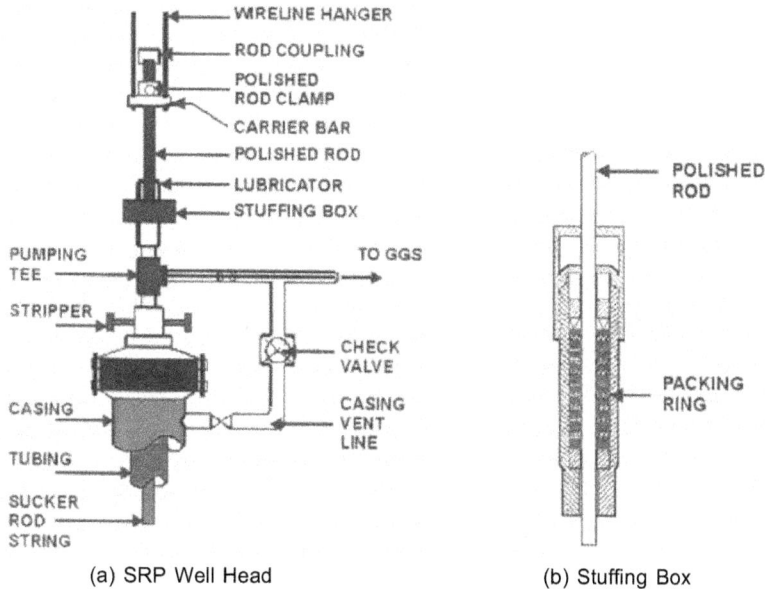

(a) SRP Well Head   (b) Stuffing Box

Fig. 3.18(a&b)

box. It stays on the stuffing box around polished rod during polished rod movement. This lubricator is filled with lubricating oil like engine oil. The engine oil stored in the lubricator drips slowly into the stuffing box. It has been experienced that approximately once or twice in a month the lubricator is required to be filled with engine oil.

Self-aligned type stuffing box is also available. It aligns easily with the off-centred polished rods. This type of stuffing box has the less damaging effect on the rubber elements. This along-with the polished rod lubricator will apparently be the ideal choice of a stuffing box.

It is equally important that stuffing box cap should never be over-tightened since over-tightening can squeeze the lubricating material out of the packing elements and subsequently packing elements get dry and in such a case early damage of packing material occurs. Therefore, it is advisable to adjust the tightness of packing elements periodically to have a longer trouble-free functioning of packing elements.

The stuffing box and its packing sizes should confirm to the size of the polished rod, i.e., the diameter of the polished rods.

During the operation of the sucker rod pumps, free gas in the well accumulates in the casing and finally finds its way out through the annulus to the surface either in the well-fluid flow line or in the gas engine prime mover at the wellhead for running the surface unit. Usually the flow line and casing vent line are connected through a check valve which allows the gas to vent into the flow line and prevents well-fluid to flow back in the well through the annulus.

3.9 Tubing Anchors

When the pump is in operation, the tubing string and rod string are successively subjected to a varying load (refer Figure 3.19). On the upstroke, the rod is loaded. The traveling valve is closed, and the fluid load is on the travelling valve. On the down stroke, the fluid load is transferred to the tubing. The standing valve is closed and traveling valve is opened and the liquid load in the tubing is on the standing valve. Therefore, in each pumping cycle, a freely suspended tubing string periodically stretches and contracts. This results into the buckling of tubing string, which sometimes becomes very severe. Due to this effect, not only the pump displacement is reduced because of reduction of stroke length, but also causes operational problems due to alternate stretching/contraction of tubing string and sucker rod. The friction between the rod and tubing can lead to rod or tubing failure. There are several ways to contain the buckling of the tubing caused due to movement of rods.

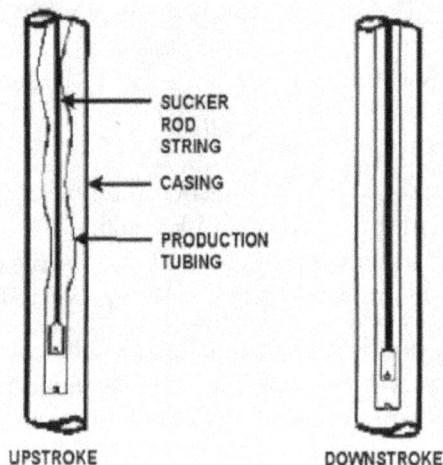

Fig. 3.19: Tubing Stretch Contraction for Freely Suspended Tubing during Pumping Operation

The most effective way is to anchor the tubing string by a tension type anchor. This tension type anchor can be set at any depth in the casing and the tubing is always kept in tension such that the rod movement will not create any up or down movement of tubing.

Two types of tubing anchors are known i.e. mechanical type tubing anchor catcher, and hydraulic type tubing anchor catcher.

Mechanical Type Tubing Anchor Catcher: The mechanical type tubing anchor catcher has the advantages of both the tension and compression anchors. This anchor is set by doing the left hand rotation of string and at the same time by moving the string up and down. By this process the cones pushes the slips out of the catcher body, which subsequently is pressed against the casing tightly. Thereafter, the required surface pull on the tubing string equivalent to the desired tension in the tubing strings is given and with the tension in the tubing string, the wellhead is fitted at its place.

In case, the unseating of anchor is required then the tubing string is given a right hand rotation and along with it the tubing is made to move up and down slightly. By doing so, cone retrieves back and the slips get back inside the catcher body. Thus, lower end of the tubing is made free before the tubing is pulled out of the well or to readjust the length of the tubing string.

Drawbacks: This anchor has certain drawbacks, which are as follows:
- Due to left hand rotation there are chances of opening up of tubing joints, since tubing joints are right hand rotation type.
- It is very difficult to release anchor-catcher even by doing the right hand rotation of tubing string when anchor is kept in well for a prolonged period.
- The special type of wellhead which should house slips to anchor and support the tubing at its stretched position, is required.

Hydraulic Type Tubing Anchor Catcher: Hydraulic tubing anchor is set in the well automatically by creating a difference of tubing and annulus pressure (approx. 100–200 psi, where tubing pressure is more than the annulus pressure) and when it is required to be released, the pressures in tubing and annulus have to be made equal. Since hydraulic tubing anchor-catcher operates due to pressure difference between tubing and annulus, it is always to be placed above the pump in the tubing string.

Drawbacks

Due to alternating stress in tubing, the packing element of the piston of hydraulic anchor catcher gets damaged and that point becomes a source of leakage of fluid in tubing string. Thus the pump fails to lift any liquid to the surface.

Because of these shortcomings of both these type of anchors i.e., mechanical and hydraulic tubing anchor catcher, installation of tubing anchor catcher is discontinued in SRP wells in India. Instead, in many SRP wells some additional tubing (if possible tubing of higher ppf) known as tail pipes are added below the pump catcher. This will always keep the tubing string as stretched as possible and prevents tubing from getting buckled. Also, it helps to unload accumulated bottom water from the well and there by increase draw-down.

3.10 Regular Checking, Monitoring and Troubleshooting

3.10.1 *Regular Checking*

Once in everyday for about 5–10 minutes the person from a group gathering station should visit each sucker rod pumping well and try to detect any type of disorder by a visual inspection. If there is none, then there is no need of doing any changes. But if there is any disorder then the pump should be stopped and necessary steps must be taken for rectification.

During the visual inspection, the following basic things should always be observed:

- There should not be any abnormal mechanical noise of continuous or periodic in nature.
- To take a look at a few important connecting bolts and nuts like crank-pin bearing nuts, saddle bearing nuts and bolts, base nuts and bolts etc. These should not be loose. Once the pump is installed and commissioned, it should be mandatory that in between 10–15 days of operation of the pump, all the connecting nuts and bolts are to be retightened by applying leverage. Special care should be taken so that crank pin bearing nuts remain adequately tight and locked. This part should be checked again after 15 days or so.
- Motor V-belts should be in proper tension to prevent slippage of V-belts on the pulleys. The crank should rotate in a direction as earmarked on the gear box by the manufacturer of surface unit.
- The person visiting a particular pumping well should have a record of position of polished rod clamp. If the polished rod clamp is displaced, then immediate measures must be taken for reinstallation of it at original position.
- The polished rod should always be more or less perfectly centred with respect to the well. If found otherwise, measures must be taken to correct it.

Sucker rod pumping units should be stopped for one hour once in every month and all of their bearings must be thoroughly greased and the level of oil in the gear box must be checked and topped up with oil, if required, to ensure that the oil level in the gear box is in the maxima-minima range. Within about six months of operation from the date of the commissioning of the pumping unit, there is no need to change the gear box oil. However, approximately in the seventh month, the gear box oil should be changed and after flushing thoroughly the gear box oil chamber, the gear box should be filled with new specified gear oil. Thereafter, for about five years there is no need to change the gear oil. Also, the lubricator fitted over the stuffing box should be checked and refilled by engine oil once or twice in every month.

3.10.2 *Monitoring and Troubleshooting—Acoustic Surveys*

A packer is not normally run in a rod pumped well. Thus well fluids which fill the casing-tubing annulus can be viewed as the reservoir to feed the fluid in to the pump. The height of the fluid column in the annulus above the mid-perforation is a measure of the well's actual static bottom-hole pressure when the pump is not in operation for a reasonably long period and also the height of the fluid column is a measure of the flowing bottom hole pressure when the pump has been steadily and continuously operating for a reasonably long period.

The acoustic survey instrument i.e. echo-sounder is the potent instrument for measuring the liquid level in the annulus. This instrument consists of two basic components i.e. (a) well head assembly; and (b) recording and processing assembly.

Wellhead Assembly: The wellhead assembly is connected to the casing annulus by means of a threaded nipple. It contains a mechanism that creates a sound wave and a microphone attached with it picks up the signal. The conventional well sounders utilize blank cartridges which are fired either manually or by remote control. Some other well sounder units employ gas gun which provide the required pressure impulse by suddenly discharging a small amount of high pressure gas. The microphone converts this to electrical signal which comes to the recording and processing unit.

Recording and Processing Assembly: In this unit, the electrical signals are filtered and amplified. The processed and amplified signals are recorded on a chart recorder as a function of time.

A number of peaks in the chart will be visible. From these peaks the tubing collars are easily identified and the length of the tubing is estimated from the number of peaks and average length of each tubing (refer Figure 3.20). The sound wave which hits the liquid level has characteristically a larger deflection in comparison to those of reflected waves from the tubing collars. Therefore, the liquid level peak can be easily distinguished from collar peaks. The repetition of collar peaks and the liquid level peak will be recorded in the chart with the diminishing intensity waves. In this figure, every tubing collar is identified by a small peak of the signal and the liquid level by a larger deflection i.e. larger peak. Therefore, the depth of the liquid level is determined by counting the number of tubing collar signals and multiplying the same by the average length of each tubing.

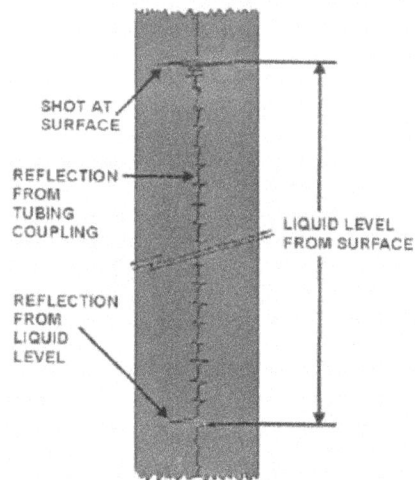

Fig. 3.20: A Typical Acoustic Survey Result

3.10.3 *Bottom-Hole Pressure Calculations with the Help of Echo-Meter*

Initially, annulus of the well is filled with the technical water (water for subduing the well). During pumping, initially the fluid in the annulus enters the pump and in this process, the annulus level comes down to a point when formation fluid starts coming into the well bore. With the continuous operation of the pump and steady inflow of formation fluid in the well bore, an almost identical gradient of fluid in the tubing and in the annulus sets up. With the fluid gradient in consideration (gradient is calculated on the basis of fluid rate measured at surface) and fluid level in the annulus, flowing bottom hole pressure can be calculated.

Calculation: Say gradient in the annulus is 0.7 kg/cm² per 10 meters. With the half of acoustic survey during operation of pump and in the stabilized flow conditions the depth of liquid level is found to be say approximately 800 meters. Also let the depth of perforation is 1200 meters.

Therefore, the depth of fluid in the annulus = 1200 – 800 = 400 meters.

Therefore, the pressure at the sand face in the well bore during flowing condition is 0.7 × 400/10 = 28 kg/cm².

For calculating the static bottom-hole pressure, the pump has to be stopped. During this period, the filling up annulus with the incoming formation fluid takes place. After of 24 hours or 45 hours or more depending how fast the static pressure is reached, the further build-up of annulus level becomes negligible.

Then the liquid level is found with the help of echo-meter. The calculation of pressure in the well bore at the sand face can be made in a similar manner, as has been explained above.

3.10.4 *Dynamometer*

Surface and Pump Card: Surface card is the recording of polished rod load over a complete pumping cycle at the polished rod on the surface. The dynamometer instrument is placed in between the polished rod clamp and carrier bar so that whole load on the sucker rod gets communicated in the dynamometer equipment.

Surface card is a combined reflection of surface unit, sub surface pump operation, rod and fluid dynamics besides various other unpredictable situations like frictional forces, vibrational aspects (especially when pumping speed is synchronous with natural frequency of rod string), sticking of plunger, dragging effect of buckled/twisted rod on the surface of tubing etc. These above have made surface card a very complex one and at times it becomes very difficult to interpret even by a very experienced analyst of dynamometer card. The interpretation of surface card has been demonstrated as follows starting from the very simple to very complex card.

(i) *Simplest/Ideal Surface Dynamometer Card:* Ideal conditions area as follows:
- Sucker rod pump is operative at a very slow speed and as such there are no acceleration forces in the sucker rods.
- There are no vibrational forces within the overall pumping system.
- There are no frictional forces in the surface unit part as well as in the down-hole (between plunger and barrel; rod and tubing).

- The standing valve (SV) and traveling valve (TV) opens/closes at the appropriate plunger position instantaneously (without any delay).
- There is no stretch/contraction of rod and tubing due to the cyclic transfer of fluid load.

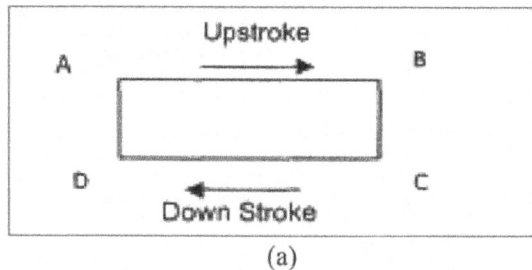

(a)

The shape of the dynamometer card will be a rectangle.

(ii) In addition to the ideal conditions as given in (i) first four points, let there be stretch/contraction in the tubing/sucker rod due to the cyclic transfer of fluid load.

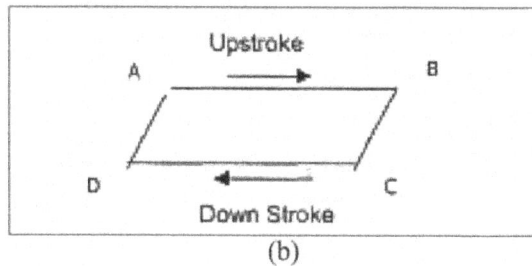

(b)

Then the shape of the surface dynamometer card will be a parallelogram, as indicated below:

The above two shapes are rarely encountered in the fields. Actual very complex and numerous shapes of actual card are available. One of the typical shapes are given as under.

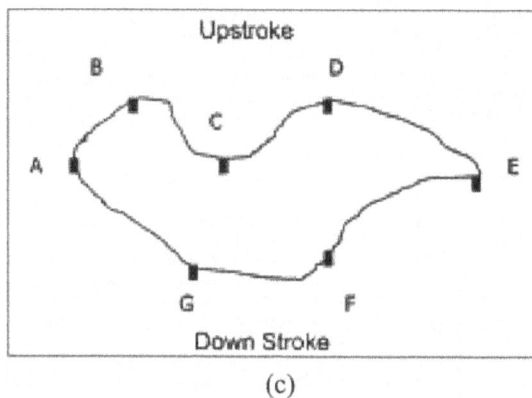

(c)

Point "A" indicates the end of the down stroke and the beginning of the up stroke for the polished rod.

- *Line A to B*—The traveling valve closes due to the fluid load on it, as the plunger started its upward journey and the polished rod begins to pick up the fluid load. This accounts for increase in polished rod load from A to B.
- *Line B to C*—The momentary decrease in polished rod load from B to C is the result of rod stretch that occur, when the rod takes over the fluid load completely.
- *Line C to D*—As the rod moves upward in approximately simple harmonic motion (because of the action of surface unit), the acceleration load is increased. Unit reaches a maximum at point 'D' which is theoretically near the middle of the up stroke.
- *Line D to E*—From point 'D' to point 'E' the acceleration load decreases (because of the action of the surface unit as mentioned above), as the rod velocity decreases to zero.

- *Point 'E'* represents the end of upstroke and beginning of downstroke.
- *Line 'E' to 'F'*—As the rod falls, the fluid pressure in the barrel increases which opens the traveling valve and closes the standing valve. At point 'F' the fluid load is transferred on to the standing valve, i.e., the fluid load is transferred on to the tubing. This is indicated by a marked decrease in polished rod load from 'E' to 'F'.
- *Line 'F' to 'G'*—This represents the negative acceleration load as a result of the action of the surface unit, which decreases the polished rod load further. 'G' is the point where minimum polished rod load occurs and this point is approximately near the middle of downstroke.
- *Line 'G' to 'A'*—This represents the decrease of negative acceleration load due to the action of the surface unit. This effects an increase in polished rod load.

Thus, in this manner a full cycle of dynagraph (A-B-C-D-E-F-G-A) is made on the dynamometer chart at the polished rod.

Special Analytical Points

- No account has yet been taken of the effect of rod vibration and frictional forces between the plunger and the barrel/between the rod and the tubing on the shape of the dynamometer card. They are generally present and at times, significantly contribute to total polished rod load.
- When the pumping speed is synchronous with the natural frequency of rod, the rod vibration becomes severe. This has been explained by Dr. Slonneger in his book on sucker rod dynagraph analysis.
- The other abnormalities like sticking of plunger, fluid pound, gas pound, gas locking, delay in opening and closing of standing and traveling valves etc., have not been taken into account in the above representative diagram.

Thus the real surface dynagraph is a complex one and is difficult to be *analyzed properly*. Surface dynamometer card may have many shapes but in contrast, Pump card (down-hole card) has only 45 to 47 number of standard shapes. So it is preferable to convert surface dynagraph to pump card and then the pump card is matched with the standard pump card shapes to determine the condition of the pump operation.

The standard shapes pump card are given in page number 90.

Dynamometer Card (Dynagraphs)

The dynamometer card (load displacement curve) is a continuous record of the resultant of all the forces acting on the polished rod at any instant during the operation of the pump. The dynamometer card is recorded with respect to polished rod position. The rod position (stroke length) is recorded on the abscissa (X-axis) and loads on the ordinate (Y-axis).

With the help of these dynamometer cards, the following information is generated.
- Peak and minimum loads at the polished rod.
- Torsional load on the speed reducer and prime mover.
- Work done by the polished rod in lifting the fluid load against friction.
- Proper counterbalance.
- Peak and minimum rod loads, rod stress and load range.
- Number of rod load fluctuations per crank cycle.

Types of Surface Dynamometer (Dynamometer Instruments): The most common types are—mechanical dynamometer and hydraulic dynamometer. In addition to these two dynamometers, various types of electronic dynamometers are also available.

These dynamometers are placed in the space between the carrier bar and the polished rod clamp. Either these are placed by creating sufficient gap between the polished rod clamp and carrier bar with the help of crane or by alternate switching off and switching on the sucker rod pump by concurrent application of pumping units brake. The other way is to make a permanent space for the dynamometer with the help of metallic props/blocks to accommodate dynamometer instrument there.

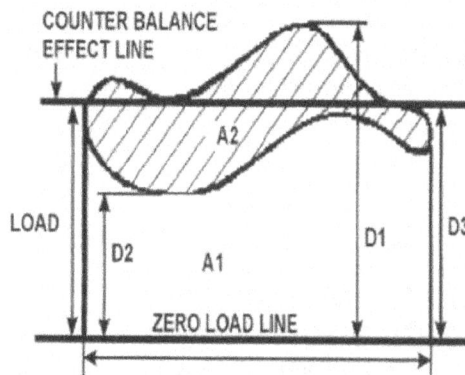

Fig. 3.21: Dynamometer Card showing Stroke Length, Load and Areas

Loads from Dynamometer Cards (Refer Figure 3.21)

Max. Load	$C \times D_1$	... (3.24)
Min. Load	$C \times D_2$	... (3.25)
Avg. Upstroke Load	$= \dfrac{C(A_1 + A_2)}{L}$	... (3.26)
Avg. Downstroke Load	$= \dfrac{C \times A_1}{L}$	... (3.27)
Approx. Correct C.B.	$= \dfrac{C(A_1 + A_2/2)}{L}$	... (3.28)

$$\text{Polished rod H.P.} \quad = C(A_2/L) \times \frac{S \times N}{33,000(12)} \qquad \dots (3.29)$$

Where, S = Stroke length, in.

N = Strokes per minute.

C = Calibration constant of the dynamometer, pounds per inch of card height.

D_1 = Max. deflection (along Y-axis), in.

D_2 = Min. deflection, in.

A_1 = Lower Area of card, Sq.in.

A_2 = Upper Area of card, Sq.in.

Torque from Dynamometer Cards: An accurate method for determining the instantaneous torque throughout the pumping cycle is done by the torque factor method. This uses the torque factors' at the corresponding positions of the polished rod supplied by the manufacturer.

The API - RP - 11-E stipulates that the manufacturer is required to supply the purchaser, on request, the stroke and torque factors for each 150 positions of the crank. The net torque on the speed reducer is the difference of torque due to well load and torque due to rotary counterweight.

$$T_{net} = TF\ (W{-}B) - M\ \text{Sin}\theta \qquad \dots (3.30)$$

Where,

T_{net} = Net Torque

TF = Torque factor

W = Well load at specific crank angle

B = Load due to structural imbalance of pumping unit (either plus or minus value)

M = Maximum moment of crank and countenweights (about the crank shaft supplied by manufacturers)

q = Crank-position, degree.

Factors Affecting Shape of Dynamometer Cards: Following factors can cause a change in the basic shape of the dynamometer cards:

1. Speed and pumping depth.
2. Fluid conditions.
3. Type of pump and its conditions including anomalies due to various factors at the time of taking dynagraph.
4. Friction factors.
5. Pumping unit geometry.

These factors can contribute singularly or collectively in presenting different shapes of dynagraph. A few peculiar shapes of dynagraph are given as under:

1. Plunger Undertravel or Overtravel.

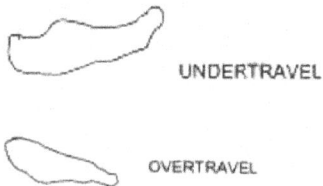

UNDERTRAVEL

OVERTRAVEL

2. Fluid pound.

3. Gas pound.

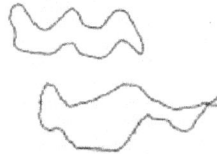

4. Gas lock.

5. Excessive friction of the sucker rod with the tubing wall.

6. Sticking plunger.

7. Excessive friction in the pumping system.

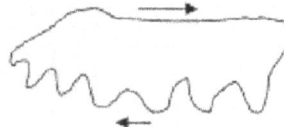

at two different times.

8. Vibrations.

3.11 Selection of SRP Installation

3.11.1 *Computation Procedures*

Selection of the proper equipment for new well follows an orderly sequence based on computation of pumping unit performance. Simpler microcomputer based programs are now available free of cost to perform calculations described here for manual use. One such SRP Calculator™ is available at www.unicous.com/oilgas/srpcalc.php internet site free of cost (refer Table 3.7). The manual calculation API - RP11L procedure is for design of conventional geometry (Class I) units pumping in non-devia ted wells at a depth of less than 12,000 ft. The method is based on use of API sucker rods. The following information must be available for the well:

(a) Expected oil and water production, in b/d and their specific gravities.
(b) Casing and tubing sizes.
(c) Anchoring depth (if any) for tubing.
(d) Pump setting depth.
(e) Fluid level during production (when pumping).

Table 3.7: SRP Calculator™ Sucker-Rod Pumping System Design Tool

Well Name	Example		
Pump Speed (spm)	6.5	Fluid Production Rate (bpd)	323
Pump Depth (ft)	8000	Oil Production Rate (bpd)	65
Fluid Depth (ft)	7000	Polished Rod Power (hp)	26.1
Tubing Pressure (psi)	75	Minimum Motor Size (hp)	60.0
Casing Pressure (psi)	75		
	Energy		
Pump Utilization (hr/day)	24	Electrical Input Power (kW)	26.6
Gearbox Efficiency (%)	95.0	Effective Lift Power (kW)	14.2
Motor Efficiency (%)	90.0	System Energy Efficiency %)	53.5
Electrical Energy Cost ($/kWh)	0.060	Oil Lift Energy Cost ($/bbl)	0.592
		Monthly Energy Cost ($)	1164
Water Cut (%)	80.0	Oil Specific Gravity	0.887
Water Specific Gravity	1.050	Fluid Specific Gravity	1.017
Oil API Specific Gravity	28.0	Fluid Column Load (lb)	9865
Class	Class 1	Max. Gearbox Torque (M in-lb)	958.91
Stroke Length (in)	169.80	Existing CB Effect (lb)	23557
Structural Unbalance (lb)	−1500	Required CB Effect (lb)	22344
Existing CB Moment (M in-lb)	2000.00	Required CB Moment (M in-lb)	1897.00
Type	Coupled	Buoyant Rod Load (lb)	15911
Total Rod Length (ft)	8000	Maximum Rod Load (lb)	30925
Upper Section Length (ft)	2624	Minimum Rod Load (lb)	10763
Lower Section Length (ft)	2712	Upper Section Stress (psi)	39131
Sinker Bar Length (ft)	0	Middle Section Stress (psi)	37311
Upper Section Diameter (in)	1.000	Lower Section Stress (psi)	35330
Lower Section Diameter (in)	0.750		
Sinker Bar Diameter (in)	0.000		
Drag Friction Coefficient	0.100		
Stuffing Box Friction (lb)	100		
Tubing Diameter (in)	2.875	Tubing Stretch (in)	1.07
Tubing Anchor Distance (ft)	7500		
Plunger Diameter (in)	2.000	Pump Stroke (in)	126.46
Pump Efficiency (%)	85.0		
Pump Fill Ratio (%)	100.0		
Pump Friction (lb)	200		

Fluid Production: The daily pumping rate of a sucker rod pump is the product of liquid displaced during one pumping cycle and the number of cycles per day. In equation form, it is expressed as,

$$q = 0.1484 \times N \times E_v \times A_p \times S_p \qquad \qquad \ldots (3.31)$$

where,

q = pumping rate (b/d)
N = pump speed, strokes/min (SPM)
E_v = volumetric efficiency

A_p = plunger area (in.²)
S_p = effective plunger stroke (in.)

The volumetric efficiency reflects small leakage between the plunger and barrel of the pump and is defined as,

$$E_v = \frac{\text{Actual Pumping Rate}}{\text{Pump Displacement}}$$

The primary reason for leakage is that it is difficult to maintain a perfect seal between two moving metal surfaces. Even so, the small leakage provides a necessary lubrication to the sealing surface between the plunger and the barrel. Volumetric efficiency is usually in the range of 0.7 to 0.8. Obviously, if leakage becomes severe due to bad sealing, the pump's efficiency decreases rapidly.

To simplify the calculations, the terms A_p, and constant 0.1484 are frequently lumped together to give a pump constant K. The equation of volume of fluid produced at surface is simplified as,

$$q = K \times N \times E_v \times S_p$$

Table 3.8 gives pump constants (K) for various sizes of plungers and can be used to compute pump displacement.

Table 3.8: Pump Factors or Constants

Plunger dia, in.	Area of Plunger, in.² (A_p)	Pump Constant (K)
5/8 (non-API)	0.307	0.046
3/4 (non-API)	0.442	0.066
15/16 (non-API)	0.690	0.102
1	0.785	0.117
1 1/16	0.886	0.132
1 1/8	0.994	0.148
1 1/4	1.227	0.182
1 1/2	1.767	0.262
1 3/4	2.405	0.357
1 25/32	2.488	0.370
2	3.142	0.466
2 ¼	3.976	0.590
2 ½	4.909	0.728
2 ¾	5.940	0.881
3 ¾	11.045	1.640
4 ¾	17.721	2.630

Example: A 1½″ plunger is being used to pump oil at 16 strokes per minute and an effective downhole stroke of 51 in. The pump volumetric efficiency is 70%. Calculate daily surface production of oil?

Solution: Value of K from Table 3.7 is 0.262.

$$q = K \times N \times E_v \times S_p = 0.262 \times 16 \times 0.7 \times 51 = 149.7 \text{ b/d (at surface)}$$

PD (pump displacement) $= 0.262 \times 16 \times 51 = 214$ b/d

The pump can displace the computed pump displacement only, if the formation can produce this much oil. Optimal performance is obtained when the pump production at the surface matches the ability of formation to produce. The estimate of the formation's ability to produce should be based on one or more well tests whenever possible. A productivity index or inflow performance relation should be used to determine the volume of fluid which the formation can produce under various conditions of flowing bottom hole pressure.

The designer may achieve the desired pump rate from several combinations of A_p, S_p and N. It is also possible to change the combination, if the pump is already installed. The adjustments are not complicated but do require some resources and efforts. Changing pump size requires pulling the pump out of the well with a service rig and running a new pump. Changing the pump stroke requires a change in the geometry of the surface beam pump unit. This is done by a maintenance crew assisted by a crane or other lifting equipment. Changing pump speed is the simplest task. It is done by changing sheaves in the V-belt drive between the prime mover and the reduction gearbox.

Casing and Tubing Sizes: Casing size is usually determined long before the decision is made to install a SRP unit. Frequently tubing is also already in place. If tubing is already in place, it is not cost effective to change out the tubing for a large size as a means of installing a larger pump. Where a pump is needed that is larger than possible with respect of existing tubing, a casing pump should be considered.

Tubing anchors have the advantage of keeping tubing from buckling and oscillating during pump cycle. The anchors increase the net effective bottom hole stroke over the unanchored tubing. Anchors can create problems in wells with severe scale, sand, or corrosion conditions as these conditions may damage the anchor and can obstruct pulling out of tubing.

Pump Setting Depth: Wherever possible, the pump should be set at a depth in the well where pump intake pressure is above bubble point of the oil as it is produced. This would suggest locating pump one to three tubing joints below perforations. Sometimes this is not possible because of total well depth, nor desirable because of sand production. Alternatively, the pump should be placed out of and above the turbulence zone near perforations (3–6 joints above perforations). In any case, the pump should have significant submergence, 50 to 100 ft of fluid, to assist in rapid fillage of the pump barrel. For low productivity wells, this amount of submergence may not be either possible or economically feasible but the consequence will be that wells with lower submergence may experience pump-off and some type of pump-off controller would be needed.

Expected Pump Fluid Level: This is one of the needed parameters about which there will be much uncertainty in a new installation. If a well test and an inflow performance relationship is available for the well, one can possibly convert flowing bottom hole pressure to flowing liquid height. Such a calculation would be fairly accurate for nearly dead oils or for wells producing mostly water. An alternative, conservative design approach is to assume that the flowing liquid level is at the pump depth. Once a unit is in operation, one of the commercially available acoustical well sounders can be used to measure the working fluid level. This value would be useful in any comparison of the well's current performance with the predicted by API-RP-11 L procedure.

Preliminary Design Data: Most of the pumping equipment in the world is manufactured and purchased according to API standards. API spec - RP - 11AX provides dimensions of subsurface pumps with bore sizes for use in 2 3/8", 2 7/8" and 3½" tubing. API spec. 11B provides dimensions of sucker rods, and API-RP. 11E covers dimensions, design and rating of sucker rod pumping units. In addition to above information, user of API-RP-11 L calculation procedure must select initial or tentative values of each of the four parameters i.e. plunger diameter, pumping speed, stroke length, sucker rod string design. The guidelines for these are given below:

Plunger Diameter: Plunger sizes (pump bores) covered by the API vary from 1¼" to 2¾". The maximum API plunger size for a given tubing size is listed in Table 3.9. The maximum stroke depends on the size of surface pumping unit. API specifies pumping units of increasing size with maximum stokes from 16 in. to 300 in. For a given unit it is possible to change stroke length by adjusting the position of the pin that connects the crank and the pitman.

Table 3.9: Maximum Pump Size and Type

Pump Type	Tubing Size (in.)			
	1.9	*2 3/8*	*2 7/8*	*3½*
Tubing one piece, thin-wall barrel (TW)	1½	1¾	2¼	2¾
Tubing one piece, heavy-wall barrel (TH)	1½	1¾	2¼	2¾
Tubing liner barrel (TL)	–	1¾	1¼	2¾
Rod one piece, thin-wall barrel (RW)	1¼	1½	2	2½
Rod one piece, heavy-wall barrel (RH)	1 1/16	1¼	1¾	2¼
Rod liner barrel (RL)	–	1¼	1¾	2¼

Courtesy: Kobe, Inc. 1961.

An additional guide to plunger size selection for a given fluid volume and pumping depth is given in Table 3.10 (pumping strokes upto 74 in. considered).

Table 3.10: Pump Plunger Sizes Recommended for Optimum Conditions

Net Lift of Fluid, ft	Fluid Production—Barrels Per Day (80% efficiency)									
	100	200	300	400	500	600	700	800	900	1000
2000	1½	1¾	2	2¼	2½	2¾	2¾	2¾	2¾	2¾
	1¼	1½	1¾	2	2¼	2½				
3000	1½	1¾	2	2¼	2½	2½	2¾	2¾	2¾	2¾
	1¼	1½	1¾	2	2¼	2¼	2½			
4000	1¼	1¾	2	2¼	2¼	2¼	2¼	2¼		
	—	1½	1¾	2						
5000	1¼	1¾	2	2	2¼	2¼				
		1½	1¾	1¾	2					
6000	1¼	1½	1¾	1¾						
		1¼	1½							
7000	1¼	1½								
	1 1/8	1¼								
8000	1¼									
	1 1/8									

Pumping Speed: For any given selection of stroke length there is maximum pumping speed. If this speed is exceeded, the rods are likely to "float" or go into compression on the downstroke. Rods are normally under tension on the downstroke as well as upstroke. If they are put under compression, they will buckle, causing wear on rods and tubing. The range of stress will be large, thus reducing Goodman diagram maximum allowable stress.

The alternating tension and compression will cause severe stress on rod threads and coupling and accelerate rod parting from fatigue and accelerated effects of corrosion. Dynamometer tests can detect floating rods. The maximum speed the maximum pumping speed for given stroke lengths is given in Figure 3.22 for all 3-types of widely used pumping units [4]. The speed N is varied by changing the surface driving speed. There is no limit on the slowest speed, though it is difficult to operate below 6 SPM with conventional motors. The maximum speed, on the other hand, is limited by two factors. The first is the minimum time needed to permit satisfactory liquid filling of the pump during upstroke.

Usually the filling does not present a practical limit unless the oil is viscous and the suction pressure is low. The second factor is the need to avoid resonance by reciprocating at a frequency different from the natural frequency of the rod. Most wells are pumped at speeds of 6 to 20 SPM.

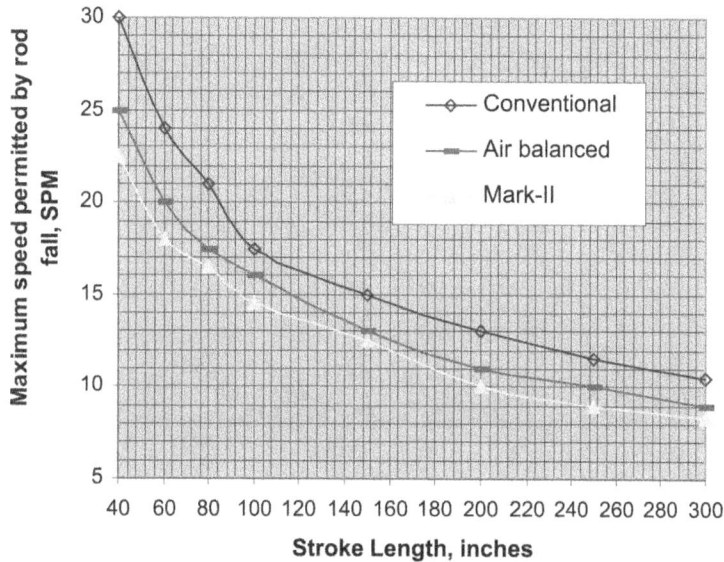

Fig. 3.22: Maximum Practical Pumping Speed

Stroke Length: Stroke length is a primary variable in determining pumping unit size. Because of the small number of stroke lengths (usually 2 to 4) available for any pumping unit, it is wise to select an available stroke length. The selected stroke length and speed combination should be such that the rods do not float (refer Figure 3.22). The effective stroke of a downhole pump S_p may be substantially different (less) from surface stroke. This results from elasticity of the tubing and rod string, both subjected to alternating pumping load. During the upstroke the rod string is loaded by displaced fluid and therefore tends to stretch. During downstroke (when the travelling valve opens) liquid load is relieved from the rod string, and the string recoils. Tubing is loaded in an inverse sequence. During downstroke it is loaded by column of fluid and stretches. Alternative stretch and recoil of rod and tubing affect the relative movement of the plunger in the barrel and thus effective stroke of the pump. It is difficult to predict effective plunger stroke by simple analytical methods because of complex dynamic loading of the rods during pumping cycle. Figure 3.23 illustrates effective plunger stroke compared with surface stroke for a particular pump assembly, as calculated by an analog simulator. The effective pump stroke in Figure 3.23 is considerably smaller than the surface stroke. In most applications, it is in the range of 40 % to 90 % of the surface stroke. It may, however, approach surface stroke if the system is properly designed and tuned.

Sucker Rod String Selection: For shallow wells (less than 2000 ft), most pumping installations will use sucker rods of the same diameter from top of the well to the pump. Since load on the rods is at its greatest at the top of the rod string on the upstroke, for single size rod string one needs only to check the stress at this point

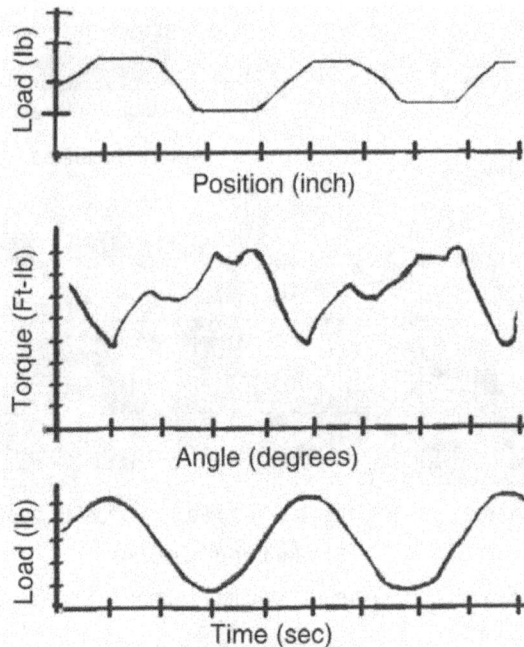

Fig. 3.23

using Goodman diagram to see if the rods are satisfactory. For deeper wells, one usually uses a larger diameter rod near the surface and smaller diameter rods further down the well. Such multiple size rods strings are called tapered rods. A coding system for designating the sizes has been adopted. A 76 rods string consists of 7/8″ rods near the top and 6/8″ rods near bottom. A 75 rod string has 7/8″ rods near the top, 6/8″ rods near the middle and 5/8″ rods near the plunger. This is called three way tapered rod strings. Four way tapers are also possible. This coding scheme does not specify what percent of each rod size is used in the string.

The API has given rod size and percent length of rods for available rod combinations and plunger diameter. For example, a 76 rod string using 1.5″ plunger would have 33.8% 7/8″ rods and 66.2% 6/8″ rods. An 85 rod string with 1.75″ plunger would have 29.6%, 30.4%, 29.5% and 10.5% of 1″, 7/8″, ¾″ and 5/8″ rods, respectively.

Example Problem 1

Given that a well has declined from 100 stb/day to 96 stb/day during a 1 month period, use the exponential decline model to perform the following tasks.

1. Predict the production rate after 11 more months
2. Calculate the amount of oil produced during the first year
3. Project the yearly production for the well for the next 5 years.

Solution

1. Production rate after 11 more months:

$$b_m = \frac{1}{(t_{1m} - t_{0m})} \ln\left(\frac{q_{0m}}{q_{1m}}\right)$$

$$= \left(\frac{1}{1}\right) \ln\left(\frac{100}{96}\right)$$

$$= 0.04082/\text{month}$$

Rate at end of 1 year

$$q_{1m} = q_{0m}e^{-bmt} = 100_e - 0.04082(12) = 61.27 \text{ stb/day}$$

If the effective decline rate b' issued,

$$b'_m = \frac{q_{0m} - q_{1m}}{q_{0m}} = \frac{100 - 96}{100}$$

$$= 0.04/\text{month}$$

From $1 - b'_y = (1 - b'_m)^{12} = (1 - 0.04)^{12}$

One gets $b'_y = 0.3875/\text{yr}$

Rate at end of 1 year

$$q_1 = q_0 (1 - b'_y) = 100(1 - 0.3875) = 61.27 \text{ stb/day}$$

2. The amount of oil produced during first year,

$$b_y = 0.04082(12) = 0.48986/\text{year}$$

$$N_{p,1} = \frac{q_0 - q_1}{b_y} = \left(\frac{100 - 61.27}{0.48986}\right)365 = 28{,}858 \text{ stb}$$

or

$$b_d = \left[\ln\left(\frac{100}{96}\right)\right]\left(\frac{1}{30.42}\right) = 0.001342 \frac{1}{\text{day}}$$

$$N_{p,1} = \frac{100}{0.001342} (1 - e^{-0.001342(365)}) = 28{,}858 \text{ stb}$$

3. Yearly production for the next 5 years:

$$N_{p,2} = \frac{61.27}{0.001342} (1 - e^{-0.001342(365)}) = 17{,}681 \text{ stb}$$

$$q_2 = q_i e^{-bt} = 100_e 0.04082(12)(2) = 37.54 \text{ stb/day}$$

$$N_{p,3} = \frac{37.54}{0.001342} = (1 - e^{-0.001342(365)}) = 10{,}834 \text{ stb}$$

$$q_3 = q_i e^{-bt} = 100_e{}^{-0.04082(12)(3)} = 23.00 \text{ stb/day}$$

$$N_{p,4} = \frac{23.00}{0.001342} (1 - {}_e{}^{-0.001342(365)}) = 6639 \text{ stb}$$

$$q_4 = q_i e^{-bt} = 100_e{}^{-0.04082(12)(4)} = 14.09 \text{ stb/day}$$

$$N_p5 = \frac{14.09}{0.001342} (1 - e^{-0.001342(365)}) = 4061 \text{ stb}$$

In Summary,

Year	Rate at End of Year (Stb/day)	Yearly Production (stb)
0	100.00	–
1	61.27	28,858
2	37.54	17,681
3	23.00	10,834
4	14.09	6,639
5	8.64	4,061
Total		68,073

Example Problem 2

The following geometric dimension are for the pumping unit C-320D-213-86

d_1 = 96.05 in
d_2 = 111 in
c = 37 in
c/h = 0.33

if this unit is used with a 2½ in plunger and 7/8 in rods to lift 25° API gravity crude (formation volume factor 1.2 rb/stb) at depth of 3,000 ft. answer the following questions:

(a) What is the maximum allowable pumping speed if L = 0.4 is used?
(b) What is the expected maximum polished rod load?
(c) What is the expected peak torque?
(d) What is the desired counterbalance weight to be placed at the maximum position on the crank?

Solution

The pumping unit C-320D-213-86 has peak torque of gearbox rating of 320,000 in. –1 b, a polished rod rating of 21,300 1b, and a maximum polished rod stroke of 86 in.

(a) Based on the configuration for conventional unit the polished rod stroke length can be estimated as,

$$S = 2c\frac{d_2}{d_1} = (2)(37)\frac{111}{96.06} = 85.52 \text{ in}$$

The maximum allowable pumping speed is,

$$N = \sqrt{\frac{70{,}471.2\,L}{S\left(1-\dfrac{c}{h}\right)}} = \sqrt{\frac{(70{,}0471.2)(0.4)}{(85.52)(1-0.33)}} = 22 \text{ SPM}$$

(b) The maximum PRL can be calculated with Eq. (12.17). The 25 API gravity has an $S_f = 0.9042$. The area of the 2½ in. plunger is $AP = 4.91 \text{ in}^2$. The area of the 7/8 in. rod is $A_r = 0.60$ in. Then

$$W_f = S_f\,(62.4)\frac{DA_p}{144} = 0.9042 \times 62.4 \frac{3000 \times 4.91}{144}$$

$$= 5770 \text{ lbs}$$

$$W_v = \frac{\gamma_s DA_r}{144} = \frac{490 \times 3000 \times 0.6}{144} = 6138 \text{ lbs}$$

$$F_1 = \frac{SN^2\left(1+\dfrac{c}{h}\right)}{70471.2} = \frac{85.52 \times 22^2(1+0.33)}{70471.2} = 0.7940$$

Then the expected maximum PRL is,

$$PRL_{max} = W_f - S_f(62.4)\frac{W_r}{\gamma_s} + W_r + W_rF_1$$
$$= 5.77 - 0.9042 \times 62.4 \times (6138/490) + 6138 + 6138 \times 0.794$$
$$= 16076 \text{lbs} < 21300 \text{ lbs}$$

(c) The peak torque calculated is,

$$T = 0.25\,S \times \left(W_f + \frac{2SW_rN^2}{70471.2}\right)$$

$$= 0.25 \times 85.52 \times \left(5770 + \frac{2 \times 85.52 \times 6138 \times 22^2}{70471.2}\right)$$

$$= 280056 \text{lbin} < 320000 \text{ lb in}$$

(d) Accurate calculation of counterbalance load requires the minimum PRL.

$$F_2 = \frac{SN^2\left(1-\dfrac{c}{h}\right)}{70471.2} = \frac{85.52\times22^2(1-0.33)}{70471.2} = 0.4$$

$$PRL_{min} = -S_f(62.4)\,\frac{W_r}{\gamma_s} + W_r - W_r F_2$$

$$= -0.9042 \times 62.4 \times (6138/490) + 6138 - 6138 \times 0.4 = 2976 \text{ lbs}$$

$$C = 0.5 \times (PRL_{max} + PRL_{min}) = 0.5 \times (16076 + 2976) = 9526 \text{ lbs}$$

A product catalog of LUFKIN industries indicates that the structure unbalance is 450 lb and 4 No. 5 ARO counterweights placed at the maximum position (*c* in this case) will produce an effective counterbalance load of 10160 lb i.e.,

$$Wc \times \frac{37\times96.05}{37\times111} + 450 = 10610$$

Which gives $Wc = 11221$ lb. to generate the ideal counterbalance load of $C = 9526$ lb, the counterweights should be placed on the crank at,

$$R = \left(\frac{9526\times111}{311221896.05}\right) \times 37 = 36.3 \text{ in.}$$

CHAPTER 4

Gas Lift

4.1 Introduction

In gas lift liquid gets lifted with the aid of gas. Before gas lift was introduced in oil industry as a very effective artificial mode of lift, a similar form was in vogue as early as in eighteenth century. Water was being lifted with the help of air. Air was conveyed through tubing and water received on the surface through tubing—wellbore annulus. The same system of lifting, i.e. with the air was in the beginning adopted to oil industry for lifting oil. It continued in this fashion to round about mid 1920's. The people started realizing the problems involved in the use of air as a lifting medium for oil, as mixing of air with hydrocarbon not only may form explosive mixture but also causes corrosion because of the presence of oxygen. So, from then onwards compressed natural gas or high pressure natural gas is being used in general to lift oil.

Early applications of gas lift adopted the simple "U"-tube, pin-hole principle in producing oil from shallow wells. Then, with the advent of gas lift valves, the principle of gas lift system was spread to deeper wells.

In a typical gas lift system, high pressure gas is injected through gas lift mandrels and valves into the production annulus string. The injected gas lowers the hydrostatic pressure in the production string to reestablish the required pressure differential between the reservoir and wellbore, thus causing the formation fluids to flow to the surface. Proper installation and compatibility of gas lift

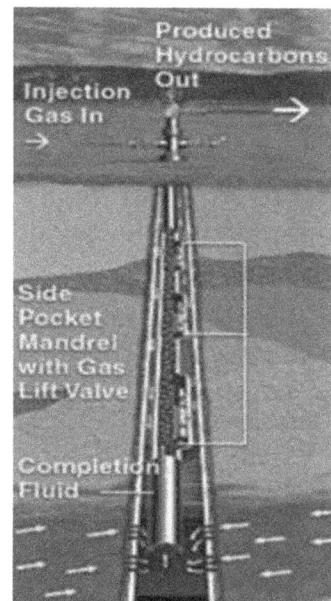

Fig. 4.1

equipment, both on the surface and in the wellbore, are essential to any gas lift system.

Gas lift system is now broadly classified into two categories as—continuous gas lift, and intermittent gas lift. The advantages of gas lift include—high degree of flexibility and design rates, wireline retrievable, excellent handling of sandy conditions, full bore tubing drift, minimum surface wellhead requirements, surface control of production rates, multi-well production from single compressor, minimum moving parts etc.

4.2 Continuous Gas Lift

The basic principle underlying the natural flow and continuous gas lift is same. The only difference between them is the source of gas. In the case of natural flow, gas comes into the well bore either along with oil or in the dissolved condition in the oil and gets separated in the wellbore, whereas, in the latter case, the gas is conveyed down the hole and is injected into the oil body. That is why continuous gas lift can be seen as an extension of the self flow period of oil well. A typical continuous gas lift operation is shown in Figure 4.2.

The basic principle of continuous flow gas lift is to inject the gas in the oil body at some predetermined depth at a controlled rate to aerate the oil column above it and as a result the density of oil column gets reduced to a point where a flowing bottom hole pressure for a desired rate of production is sufficient to lift the oil to the surface. Thus, oil is produced continuously from the well to the surface.

Gas injection is done at a slow rate continuously. Because of this reason, the port size of the gas lift valve is smaller in comparison with port sizes of the gas lift valves for intermittent gas lift. Generally, the port sizes for continuous gas lift are 3/16″, ¼″ and 5/16″.

It is also generally intended and the accepted practice that in the continuous gas lift, only one valve will be accomplishing the gas injection work and that this location of valve should be as deep as possible as per the available normal gas injection pressure. This valve is termed as 'operating valve'. The valves above it are used to unload the well to initiate the flow from the reservoir. Once the gas injection begins through the operating

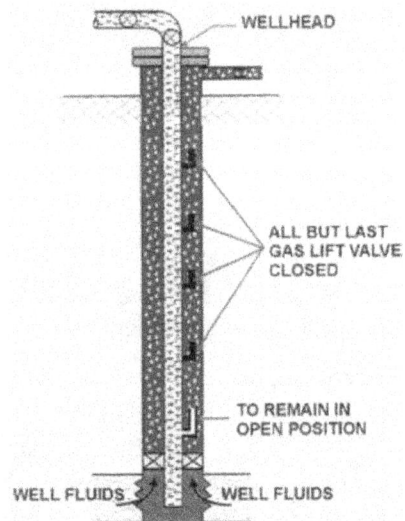

Fig. 4.2: A Typical Continuous Gas Lift Operation

valve and the upper valves, termed as "unloading valves" are closed. In case there is disruption in gas injection, the well will be loaded. So, when gas lift is resumed, the well is required to be unloaded with the help of unloading valves.

4.3 Intermittent Gas Lift

In intermittent gas lift sufficient volume of gas at the available injection pressure is injected as quickly as possible into the tubing annmulus under a liquid column and then the gas injection is stopped. The volume of gas expands and in the process it displaces the oil on to the surface. So, the assistance of flowing bottom hole pressure is not required when gas displaces oil. Static bottom hole pressure, flowing bottom hole pressure and productivity index of the well govern the fluid accumulation in the tubing. Typical intermittent gas lift operation is shown in Figure 4.3.

In this system, a pause or idle period is provided, when no gas injection takes place. In this period the well is allowed to build up the level of liquid which depends upon the reservoir pressure and PI of well. Then again, next gas injection cycle is initiated to produce oil. In this manner, as the name suggests, intermittent gas lift works on the principle of intermittent injection in a regular cycle. It is to be noted that in the cycle, injection time should be as short as possible, so that a large volume of gas can be injected quickly underneath the oil slug. As a result oil slug above the point of gas injection will acquire the terminal velocity (maximum velocity) within shortest possible time, which would minimize the liquid fall back in the tubing string. Less fluid fall back will not only increase production but also help reduce the paraffin accumulation problem in the tubing, if oil is paraffinic in nature.

For injecting large amount of gas, large ported gas lift valves are required. That is why gas lift valves having port sizes 1/2″, 7/1 6″, 3/8″ or 5/16″ are preferred. In intermittent gas lift application, two different gas injection flow rates are considered. One is the normal gas injection rate required for a well and other is the instantaneous gas injection rate, commonly called per minute demand rate of gas injection. The high rate of gas injection is calculated on the

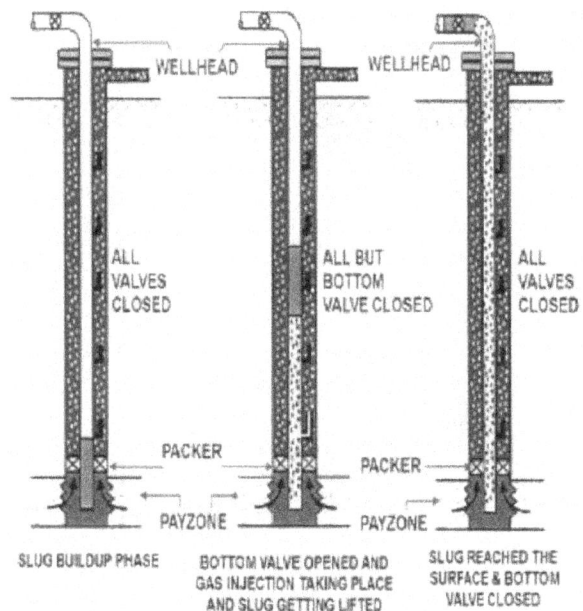

Fig. 4.3: A Typical Conventional Intermittent Single Point Gas Lift Operation

basis of short duration of gas injection. It helps to reduce the injection gas breakthrough and through this valve first and as such, it is also known as operating valve. The upper valves may or may not operate, when the liquid slug crosses the operating valve during its upward travel. If the upper valve opens as the slug crosses the valve, the additional gas further arrests fluid fall back and thus results in more oil production.

In the light of the above deliberations, it can be comprehended that continuous gas lift system should be employed when well has moderate to high reservoir pressure and PI. Continuous gas lift characteristically provides high volume of oil production.

Intermittent gas lift system should be deployed when the well has a poor PI and 70% of the hydrostatic bottom hole reservoir pressure. Intermittent gas lift produces comparatively much lower quantity of oil production than that of continuous gas lift.

4.4 Type of Installations

The gas lift can be categorized based on type of installation i.e. closed, semi-closed or open as shown in Figure 4.4.

4.4.1 *Closed Installation can be defined as*

- When there is a packer in the tubing—casing annulus, below the deepest gas lift valve and
- When there is a standing or non-return valve in the tubing at the tubing shoe.

4.4.2 *Semi-Closed Installation*

It can be defined as when there is only packer in the tubing annulus as described above.

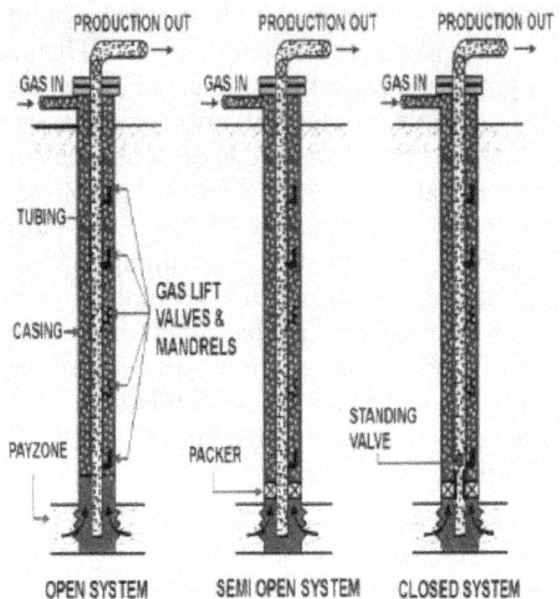

Fig. 4.4(a): Different Types of Gas Lift Installations

4.4.3 *Open Installation is defined as*

When there is neither any packer in the tubing-casing annulus nor any standing valve in the tubing shoe.

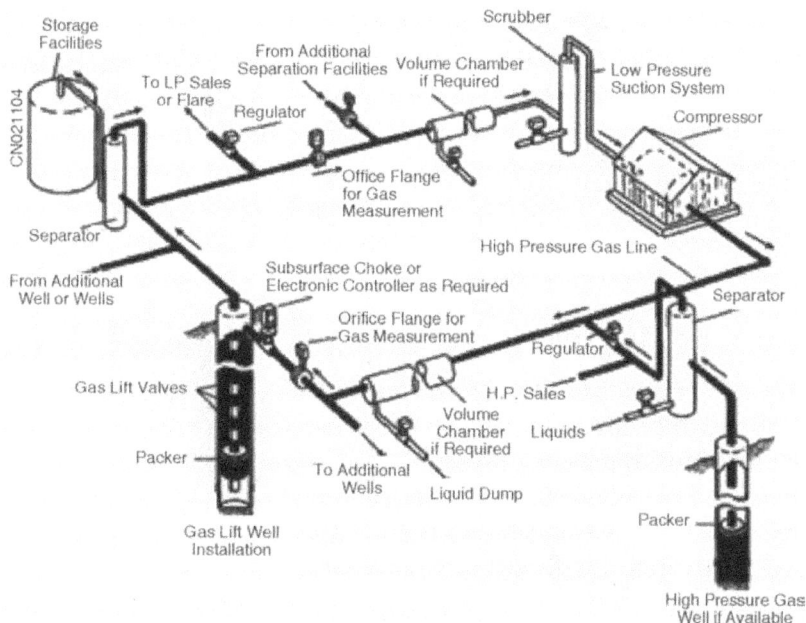

Fig. 4.4(b): A Typical Total Gas Lift Network
Courtesy: Parveen Industries

The installation of packer is recommended:

- To prevent U-tubing through the tubing, especially when bottom hole pressure is very low and the deepest gas lift valve is very near to the perforated part of the casing.
- To prevent rise of fluid level in the annulus, especially when there is an idle period of intermittent gas lift. So, the same liquid is to be U-tubed again through the gas lift valve before the normal gas injection is resumed.
- To prevent production casing coming in contact with the well fluid.
- In offshore wells, it is mandatory to have packer in the annulus. This is primarily due to safety aspect for offshore wells (API-RP-14C) to prevent accidental leakage of oil and gas in the sea through leaked casing.

The installation of standing valve is recommended when reservoir pressure is low and PI is in the range of moderately high to high. Generally the lowering of standing valve is decided afterwards during the production phase of the well. For this reason, in most of the places, the general practice is to lower A-nipple or D-nipple or equivalent along with the tubing string in the initial installation period. If required, the standing valve is either dropped or is lowered with the wireline on the A or D-nipple. It is also likely that with the production of fluid from the well, the sand slowly gets settled on the standing valve making the standing valve non-operative. So, to avoid this problem the deepest gas lift valve should always be placed just above or very near to the standing valve. The turbulence created due to gas injection at that place inhibits the build up of sand on the standing valve.

Generally, semi-closed type of installation is the standard practice for intermittent gas-lift wells, whereas open or semi-closed are for continuous gas lift wells.

4.5 Gas Lift Valve Mechanics

A gas lift valve is analogous to a down hole pressure regulator. The surface areas of the gas lift valve are exposed to tubing and casing pressures. So, in response to casing or tubing pressure the gas lift valve opens, which allows injection gas to enter the production string to lift fluid to the surface.

In the course of improvement of gas lift system several types of gas lift valves were developed.

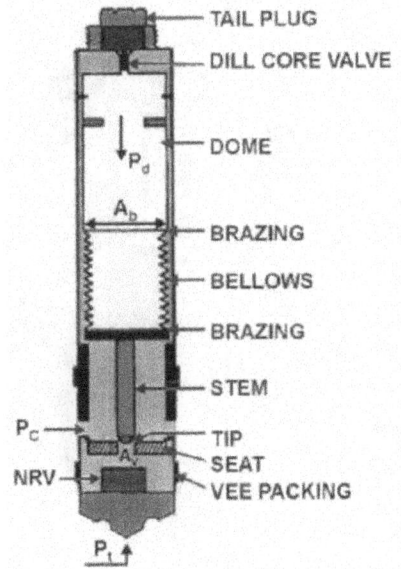

Fig. 4.4(c): Gas Lift Valve

Probably the differential type of valve was a very early development and this type of valve was very prevalent before the World War II. Advent of metallic bellow for making the gas lift valve has revolutionised the gas lift system. The bellow operated nitrogen pressure loaded gas lift valve is the most common type of gas lift valves being used by oil industries.

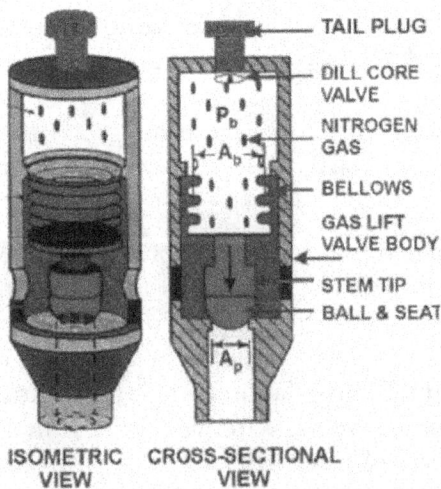

Fig. 4.5a: Gas Lift Valve in Closed Position with Nitrogen Pressure

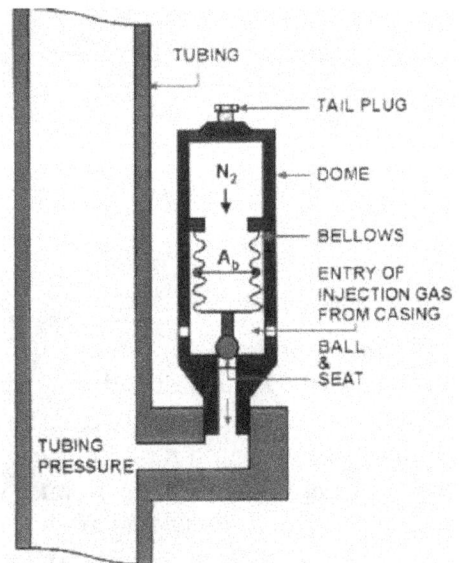

Fig. 4.5b: Casing Pressure Operated Gas Lift Valve with Conventional Mandrel

In Indian oil fields, whether it is in offshore or onshore, casing pressure operated, nitrogen loaded, unbalanced type gas lift valve is now being used.

A gas lift valve has five basic components (Refer Figs. 4.5a, 4.5b and 4.5c). They are— Body, Loading element, Responsive element, Transmission element, Metering element.

4.5.1 *Body*

The body is the outer cover of the gas lift valve and is generally of 1½″ O. D. or 1″ O.D. Some pencil type of gas lift valve is also there, which has an O.D. of 5/8″. The body of the gas lift valve is generally of S.S. –304 or 316. For the conventional type of gas lift valve, one end of it is threaded and that is screwed with the mandrel. For wire-line type, the "O" ring or VEE seal rings are

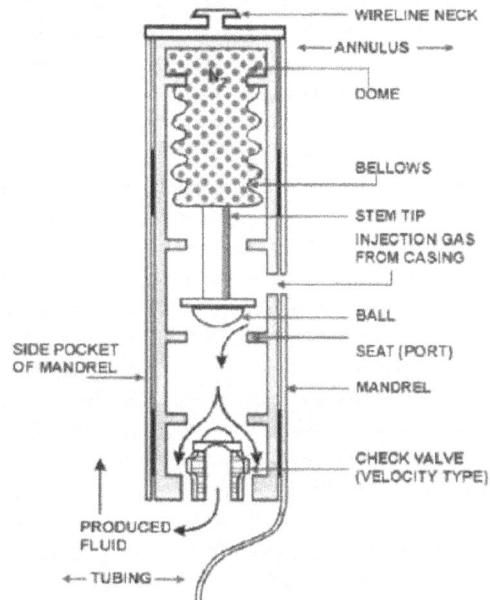

Fig. 4.5c: Casing Pressure Operated Wireline Retrievable Valve with Side Pocket Mandrel

provided on the body for isolating the required portion of the gas lift valve from the adjacent areas. The length of the gas lift valve i.e. its body varies usually from merely a feet to around three feet.

4.5.2 *Loading Element*

The loading element can be spring, gas (N_2 gas) or a combination of both. The spring or gas charge provides a required balancing force so that the valve can be operated at a desired pressure. It means that above this pressure the valve opens and below that it gets closed automatically.

Spring provides the required compression force, so when spring-loaded valve is required to open, the external pressure should be sufficient to overcome the compression force of the spring. In case of gas-charged valve i.e. nitrogen-loaded valve, the external pressure is required to overcome force due to nitrogen pressure to make the valve open for gas injection.

4.5.3 *Responsive Element*

Responsive element can be metal bellows or piston. Bellows type of gas lift valve is most prevalent. The bellow is made of very thin metal tube preferably of 3-ply monel metal. Its thickness is approximately 150th of an inch. This is hydraulically formed into a series of convolutions. This form makes the tube very flexible in the axial direction and can be compared with a similar rubber bellows.

The bellow is regarded as the heart of the gas lift valve. If bellows are properly strengthened, the gas lift valves become very strong. Different gas lift valve manufacturing companies follow their own method of bellows preparation. So, it can be said that if bellows is of high quality, it is reflected in the quality of gas lift valve.

4.5.4 *Transmission Element*

The transmission element is generally a metal rod, whose one end is fitted with the lowermost portion of the bellow and the other end is rigidly attached with the stem tip.

4.5.5 *Metering Element*

It refers to the opening or port of the gas lift valve, through which casing gas passes into the well fluid in the production tubing.

Force Balance Equations

Let A_b =   Effective area of bellows (in²)

A_v =   Area of valve port (in²)

P_{Id} =   Tubing pressure at valve depth (psig)

P_{bd} =   Bellows charge pressure at well temperature (psig)

P_b =   Bellows pressure at 60°F test bench

P_{sp} =   Spring pressure effect

P_{od} =   Operating casing pressure at valve depth

P_{obd} =   Valve opening pressure when there is no pressure exerted over the valve port area from the other side i.e. when tubing pressure is zero

P_{cd} =   Valve closing pressure in casing at valve depth

P_{TRO} =   Valve opening pressure at 60°F in the test rack

c_t =   Temperature correction factor

C_t = $\dfrac{\text{Gas lift valve dome pressure at 60°F}}{\text{Gas lift valve dome pressure at well temperature}}$

T_d =   Temperature at valve depth (°F)

P_{ik} =   Maximum injection pressure available at the surface (i.e. kick off pressure)

P_{wht} =   Flowing wellhead pressure

G_{st} =   Static fluid gradient

P_1 =   Normal gas injection pressure available at the surface

C_{gt} = Valve correction factor for specific gravity and temperature at the valve depth

$$= 0.0544\sqrt{((S.G)(T_{L1} + 460°F))}$$

Two types of situations can be envisaged in the well:

 (i) When valve is closed and ready to open
(ii) When valve is open and ready to close.

4.5.5.1 *Valve is closed and ready to open*

(a) For the valve (without spring)

Closing force $\quad\quad = P_{bd} \times A_b$

Opening force $\quad\quad = P_{od} \times (A_b) - P_{od} \times (A_v) + P_{Id} \times (A_v)$

When opening force = closing force

$$P_{od}(A_b - A_v) + P_{id}A_v = P_{bd}A_b$$

$$\Rightarrow P_{od} - (1 - A_v/A_b) + P_{id} A_v/A_b = P_{bd}$$

$$\Rightarrow \quad\quad P_{od} = \frac{P_{bd}}{(1 - A_v / A_b)} - P_{td}\left[\frac{A_v / A_b}{(1 - A_v / A_b)}\right]$$

$$\Rightarrow \quad\quad P_{od} = P_{otd} - P_{td} \text{ [T.E.F.]}$$

Where $\quad\quad P_{otd} = \dfrac{P_{bt}}{(1 - A_v / A_b)}$

And T.E.F. (Tubing effect factor) $= \dfrac{A_v / A_b}{(1 - A_v / A_b)}$

$$\Rightarrow \quad\quad P_{od} = P_{otd} - T.E.$$

Where $\quad\quad$ T.E. = Tubing effect = $P_{td} \times$ T.E.F.

Note: Every gas lift manufacturer is supposed to supply A_b and A_v for each type of valve.

Then, A_v/A_b, $(1 - A_v/A_b)$ and T.E.F. $= \dfrac{A_v / A_b}{1 - A_v / A_b}$ can either be calculated or the same would be provided by the manufacturer for each types of valve.

For spring loaded gas lift valve the manufacturer has to provide the spring pressure effect (P_{sp}) (in psi say) of the valve.

When the same equation is used to find P_{bd} the form of the equation is arranged as:

$$P_{bd} = P_{od}(1 - A_v/A_b) + P_{td}(A_v/A_b)$$

(b) For the valve (with spring)

Closing force $\quad\quad = P_{bd} \times A_b + P_{sp}(A_b - A_v)$

Opening force $\quad\quad = P_{od}(A_b - A_v) + P_{td}$

Since opening force $= $ closing force

$$P_{od}(A_b - A_v) + P_{td}A_v = P_{bd} \times A_b + P_{sp}(A_b - A_v)$$

$$\Rightarrow P_{od}(1 - A_v/A_b) + P_{td}(A_v/A_b) = P_{bd} + P_{sp}(1 - A_v/A_b)$$

$$\Rightarrow P_{od} + P_{td}\frac{A_v/A_b}{1 - A_v/A_b} = \frac{P_{bd}}{1 + A_v/A_b} + P_{sp}$$

$$\Rightarrow \quad\quad\quad P_{od} = \frac{P_{bd}}{1 - A_v/A_b} + P_{sp} - P_{td}\left[\frac{A_v/A_b}{1 - A_v/A_b}\right]$$

$$\Rightarrow \quad\quad\quad P_{od} = P_{otd} - P_{td} \times (\text{T.E.F.})$$

Here, $\quad\quad\quad P_{otd} = \dfrac{P_{bd}}{(1 - A_v A_b)} + P_{sp}$

Also in terms of P_{bd} the equation can be re-arranged as,

$$P_{bd} = (P_{od} P_{sp})(1 - A_v/A_b) + P_{bd}(A_v A_b).$$

4.5.5.2 When valve is open and ready to close

(a) For the valve (without spring)

Closing force $\quad\quad = P_{bd} \times A_b$

Opening force $\quad\quad = P_{cd} \times A_b$

Since, opening force $= $ closing force

$$P_{cd} \times A_b = P_{bd} \times A_b$$

or $\quad\quad\quad P_{cd} = P_{bd}$

(b) For the valve (with spring)

Closing force $\quad\quad = P_{bd} \times A_b + P_{sp}(A_b - A_v)$

Opening force $\quad\quad = P_{cd} \times A_b$

Since, opening force $= $ closing force

$$P_{cd} \times A_b = P_{bd} \times A_b + P_{sp}(A_b - A_v)$$

$$\Rightarrow \quad P_{cd} \times A_b / A_b = P_{bd} \times A_b / A_b + P_{sp}\left(\frac{A_b}{A_b} - \frac{A_v}{A_b}\right)$$

$$\Rightarrow \quad\quad\quad P_{cd} = P_{bd} + P_{sp}(1 - A_v/A_b)$$

The equation can be rearranged in terms of P_{bd} as:

$$P_{bd} = P_{cd} - P_{sp}(1 - A_v/A_b)$$

In the open bench calibration of valve, the valve is closed with the force of N_2 gas in the bellows with or without spring. Thereafter the pressure is applied open the gas lift valve. It is a very convenient way of calibrating the gas lift valve. It is case of where valve is closed and ready to open. So, with the little modification of the equation, along with converting some terms to surface condition, we get P_{OTB} in place of P_{od} and since there is no tubing pressure in the open test bench, so the term containing P_{td} is zero. Thus the expression without spring,

$$P_{OTB} = \frac{P_b}{(1 - A_v / A_b)}$$

For the valve with spring, it will be $P_{OTB} = \frac{P_b}{(1 - A_v / A_b)} + P_{sp}$

Here P_{sp} does not change with temperature.

4.6 Other Common Valve Types

Other than the casing pressure operated unbalanced nitrogen charged bellows type with or without spring, two more types of valves are common for oil field use are:

(a) *Fluid Operated Gas Lift Valves (or Tubing Pressure Operated Gas Lift Valve):* As the name implies the fluid operated Gas Lift valves operate predominantly with the pressure of tubing. So, its larger surface of opening and closing mechanism i.e., the bellows area is directly exposed to tubing and not the casing pressure. That is, the tubing pressure acts on the bellows and casing pressure on the downstream side of the seat. Due to this, the force balance equations as described for casing pressure operated valves are reversed. If it is required to reduce the influence of the casing pressure, it is required to reduce the port size to as minimum as possible. The schematic is shown in Figure 4.6.

(b) *Pilot Operated Gas Lift Valve:* This is a casing pressure operated gas lift valve, but with some fundamental differences in the construction of the valve as well as in its operating mechanism. The principle behind the construction of this type of

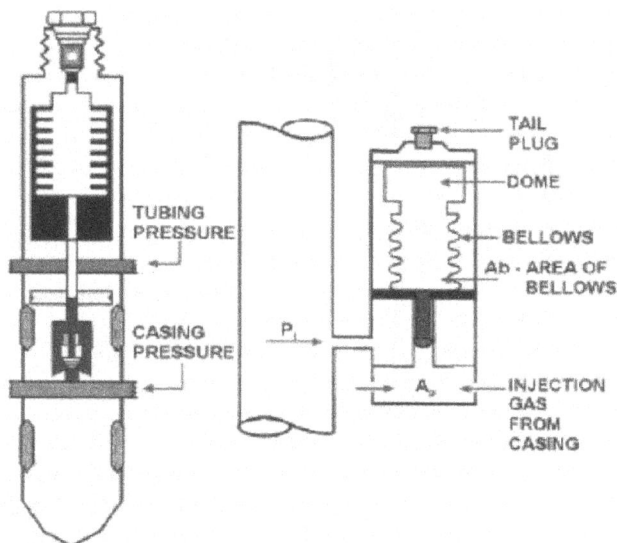

Fig. 4.6: Fluid Operated Valve
or
Tubing Pressure Operated Valve

valve is to separate the gas flow capacity from the pressure control system. The pilot valve power section has two distinct sections. One is pilot section and the other is power section. The pilot section is very similar to an unbalanced type of valve, with the exception that injection gas does not pass through the pilot port into the tubing.

The power section consists of a piston, stem, spring and the valve pod through which injection gas enters into the tubing. As the casing pressure reaches the opening pressure of the valve, at first, the pilot section port opens. The gas through the pilot port, then, exerts pressure over the piston in the main valve section. The piston is, then pushed

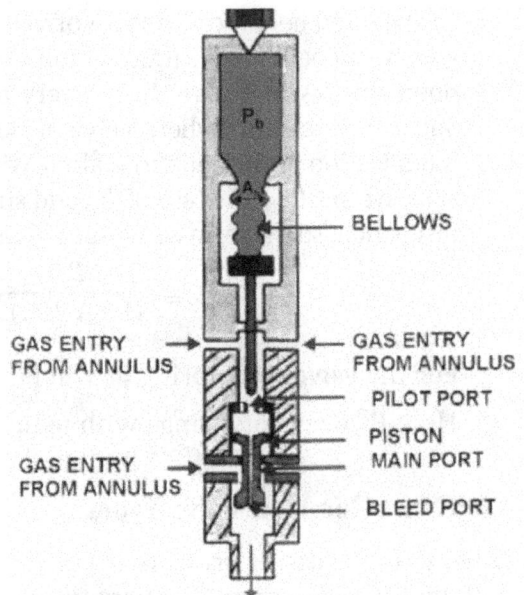

Fig. 4.7: Pilot Operated Gas Lift Valve

downward against the compressive force of spring. This causes the downward movement of the stem and the valve gets opened. Casing gas then, passes through the main section port to find entry in the tubing. When the casing pressure decreases below the closing pressure of the pilot section valve, the pilot section, like in the normal casing pressure operated valves, gets closed. Then, the trapped gas between the pilot port and piston is bled in the tubing through a specially constructed bleeder line in the main valve section. The schematic is shown in Figure 4.7.

The same force balance equation is applied to the pilot section only for the opening and closing of the valve, since it is the main functional area.

4.7 Merits and Demerits of Different Categories of Gas Lift Valves

4.7.1 *Casing Pressure Operated N$_2$ Charged Bellows with or without Spring for Continuous and Intermittent Gas Lift*

Advantages

- The valve is of a very simple design and is rugged.
- The calibration of the valve is done very easily
- Valves can be repaired easily
- Valve can be suited both to continuous and intermittent gas lift by differing the port sizes only. Small ported valve generally is suitable for continuous gas lift and bigger ported for intermittent gas lift.

- N_2-charged bellows with spring is not affected by the temperature. So, the opening pressure of the valve is not changed with varying temperature in the well.

Limitations

- Excessive valve spread (difference of pressure at which valve opens and closes) characteristic of the valve can result in an excessive injection gas volume to be used in one cycle in intermittent gas lilt.
- For dual installation of source of gas injection, gas lift valves in one well, with a common it is very difficult to control the gas injection.
- Only bellows type of valve is temperature sensitive. It affects the closing and opening pressure of valves.
- N_2-charged bellows type valve with spring has restricted gas passage, therefore, this valve is not suitable for intermittent gas lift.

4.7.2 *Tubing (Fluid) Pressure Operated Gas Lift Valve*

Advantages

It has got many advantages when used in intermittent gas lift design. The most important application is for dually completed gas lift wells, i.e., when two parallel tubings in a well both fitted with gas lift valves are used in a well for producing two zones through two different tubings with the help of gas lift.

Disadvantages

- It is not a good valve for use in the well with low flowing bottom hole pressure.
- In absence of any control from the surface, optimum/capacity oil production may not take place.
- This type of valve is not recommended for continuous flow. While trying to control the volume of gas injection, it has happened that the upper valve opens and desired point of gas injection is not maintained. This results in lower production.

4.7.3 *Pilot Operated Valve*

Advantages

- Control lift gas during cyclic operation is enhanced by this valve's ability to control spread.
- Maximum bellows travel stops prevents the bellow from sacking, providing greater reliability.
- Power stem shock absorber increases the service life of the valve.

Disadvantages

- It is complicated in design and in case, the bleed hole in the power section gets plugged, the valve, then, remains in open position.
- It is costlier than the conventional gas lift valve, because of its complex configuration.
- Pilot valve is not recommended for continuous gas lift since the discharge of gas in the tubing is very large and for a very short period.

4.8 Selection of Gas Lift Valve

From the foregoing discussions relating to merits and demerits of different types of gas lift valves the simple solutions for the selection of proper gas lift valves for continuous and intermittent gas lift are:

- Unbalanced, bellows operated N_2-charged with bigger port *viz.* 7/16″, 3/8″, 1/2″ and 5/16″ can be preferred for intermittent gas lift.
- Unbalanced, bellows operated, N_2-charged with smaller port opening *viz.* 1/8″, 3/16″ and 1/4″ can be preferred for continuous gas lift. However where because of the high volume production, greater port area is required, the gas lift valve of the required port area larger than 1/4″ should be utilized.
- Unbalanced N_2-charged, bellows operated with spring can be used for continuous lift wells, where production to be obtained is moderate and when well temperature is very high with high geothermal gradient. For high volume of production, this is again not suitable.
- For dually completed wells, tubing pressure operated valves are certainly better. With unloading valves as tubing pressure operated and the operating valve for both the string as casing pressure operated valves, are a better proposition. Though, many a times casing pressure operated valves are preferred for dually completed wells.
- Pilot valve as the operating valve for intermittent gas lift is sometimes a better choice. However, for application of PAIL (Program Assisted Intelligent Lift), pilot valve as the operating valve is recommended by M/s, Dapsco (manufacturer of PAIL system).

4.9 Reverse Flow Check Valve

A reverse flow check valve is either coupled with the gas lift valve or in-built with the gas lift valve. Its function is to prevent the backflow of fluids from the tubing to the casing. The back flow of fluids from the tubing to annulus is not desirable because:

- Back flow of fluid has to be stopped during setting of hydraulic packer with gas lift valves in the tubing string.
- It may damage the gas lift valve seats.

- It may result in deposition of sand etc. above the packer making the servicing of well with workover difficult.

Two types of reverse flow check valves are available i.e. velocity type and weak-spring loaded. The check valves ensure the tubing pressure to act below the seat, since this is one of the requirements for proper functioning of gas lift valve. The force balance equations involve the pressure from the tubing side below the valve seat, when the valve is closed.

In the velocity type check valve, the valve is normally open and gets closed, when there is a flow from tubing to annulus. So, when velocity type valve is lowered, it is to be lowered in upside down fashion (i.e. after connecting with the gas lift valve), so that it becomes a normally open check valve.

In the second category of reverse flow check valve, the only type is weak-spring loaded check valve. It remains normally closed. Because of weak spring action, even though the check valve is closed, it ensures the tubing pressure to act on the valve port-from below.

The opening of all the reverse flow check valve is kept slightly more than ½″, so that it should not restrict any amount of flow (maximum port size of the gas lift valve is ½″).

4.10 Gas Lift Mandrel

Gas lift mandrel is the part of tubing string. It houses the gas lift valve and check valve. The mandrel's length is very short—it ranges from 4′ to say 7′ to 8′ depending upon the length of the gas lift valve and check valve.

There are two general types of mandrels in use—one for conventional or for fixed valve and the other is for wireline retrievable valves. In the conventional mandrel gas lift valves and check valves are fitted on to the exterior side of the mandrel with the valve attachment lugs. The lower or inlet lug has a threaded connection to attach the check valve and gas lift valve (generally gas lift valve is coupled with the check valve and check valve is screwed to the lower lug of the mandrel). The lower lug is rigidly welded with the tubing part of the mandrel body in a perfectly seal-proof manner. It has a number of small holes to connect with inner side of the tubing. The upper lug, which is also called a guard lug or protective lug and this lug with a proper chamfering protects the gas lift valve from getting damaged during lowering and pulling out of tubing strings with gas lift valves. The lower lug also has a proper chamfer at its bottom side.

The mandrel for wireline retrievable valve is of a different type. The gas lift valve is housed inside mandrel (instead of being on to the outside). The outer shape of mandrel's tubing body looks oval shaped with its eccentric end. It has a pocket, welded inside the eccentric portion, which is intended to house the gas lift valve.

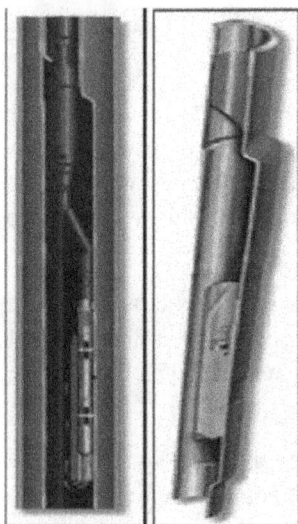

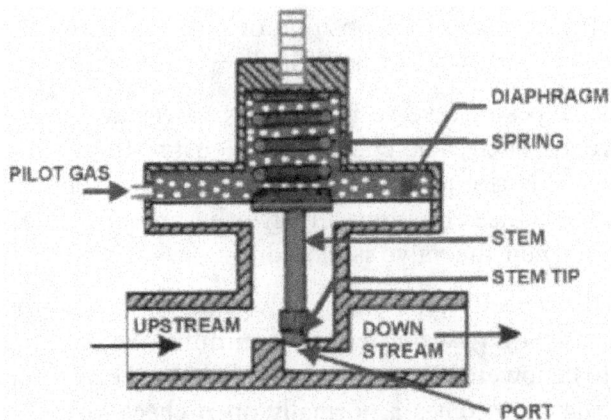

A PRESSURE REGULATOR TO BE USED IN GAS INJECTION LINE AT WELLHEAD (NORMALLY OPEN TYPE: CLOSES WHEN PILOT GAS PRESSURE APPLIED ON DIAPHRAGM)

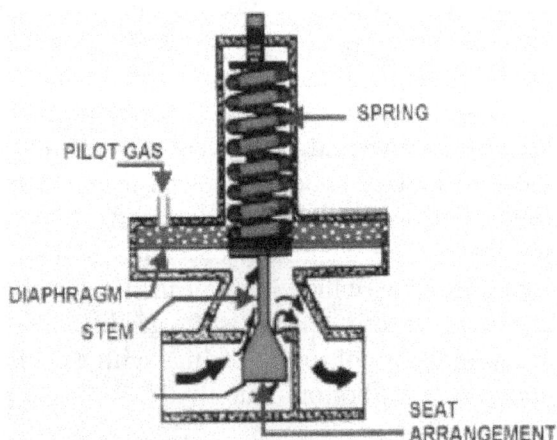

Fig. 4.8: Kick-over Tool & Gas Lift Mandrel
(Courtesy: Weatherford)

The pocket has drilled holes to connect the pocket bore with the tubing i.e., with the mandrel and separate drilled holes to connect with the inside of the tubing. The eccentric form of the mandrel is required to ease the wireline job for the selective setting and retrieval of the gas lift valve. The typical wireline retrievable gas lift mandrel along with kick-over tool is shown in Figure 4.8.

Generally, the conventional mandrel is having much less cost than the wireline mandrel. But at the same time if the servicing of gas lift valves is required or resetting of pressure valve is required, for conventional mandrel, entire tubing string is to be pulled out, whereas, only with the help of wireline job (using kick-over tool), redressal job of the gas lift valve is carried out by wireline tool. So, in this sense, wireline mandrels are more effective, since every effort is made to minimize the cost of workover job operations workover job operations. In this respect wireline retrievable mandrels are very attractive and the same is practiced all through the world. In onshore oil fields a majority of the mandrels are of conventional type.

A Pressure Regulator to be used in Gas Injection Line at Wellhead (Normally Open Type: Closes when Pilot Gas Pressure Applied on Diaphragm).

Fig. 4.9: A Pressure Regulator to be used in Gas Injection Line at Wellhead (Normally Closed Type: Opens when the Pilot Gas Pressure is applied on the Diaphragm)

Many times, it has been experienced that wireline job is extremely difficult in a well with high paraffin deposition in the tubing. Also, scale deposition inhibits the movement of wireline tools. So, with the high initial cost of the wireline mandrel coupled with the problems like paraffin, scale in the tubing (onshore oil wells are mostly equipped with conventional gas lift mandrels).

Proper identification with respect to its size is most important for a mandrel. It means how big a mandrel with gas lift valve in position (for conventional one) can go into the well for a given the wells minimum I.D. of lowered casing I.D. So, maximum diameter of mandrel is taken into account with tubing string coupled at its two ends with respect to casing drift diameter.

4.11 Surface Equipment

Surface equipment for the gas lift wells are equipped on the gas injection line which is leading to a gas lift well.

For continuous lift well, two nos. of equipment are required—one is the adjustable or fixed choke to regulate the volume of gas injection, the other is the pressure controller fitted to the upstream of the choke to regulate the upstream pressure.

For intermittent lift, it is a usual practice to install a time-cycle controller. This controller periodically opens and closes by itself with the pre-set time and so periodically injects gas into the casing tubing annulus. So, a time-cycle controller has two basic functions—one is idle period of the cycle or time between two injection cycles like 15 min, 20 min, 30 min, 40 min, 1 hr, so on, when it will be

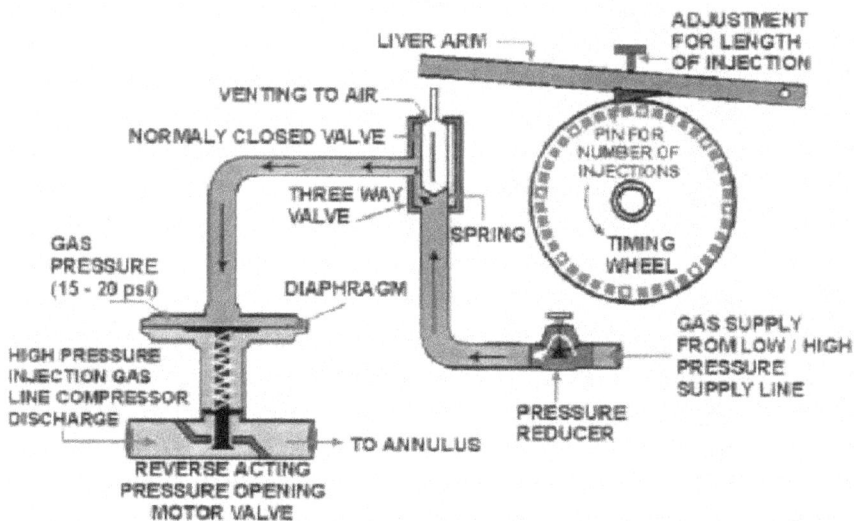

Fig. 4.10: Typical Time Cycle Controller for Intermittent Gas Lift Operation

closed and there will be no injection of gas in the well. The other is the injection duration, say 1 min, 1 min 30 sec, 2 min, 2 min 15 sec, 2 min 30 sec and so on. With the shortest possible injection time, the required volume of gas should flow into the casing. The time cycle controller operates with the pilot pressure of 25 psi-40 psi which is obtained either by tapping the injection line gas with pressure reducing equipment or from compressed low pressure air line. The typical time cycle controller is shown in Figure 4.10.

Many times, intermittent lift well is operated with the help of bean or orifice fitted in the injection line similar to continuous lift. The bean or orifice continuously allows the injection gas to flow into the casing. As and when the pressure in the annulus reaches the valve opening pressure, the valve opens and gas enters into the tubing. As the casing pressure goes down, the gas lift valve closes and again the casing pressure slowly builds up by the slow incoming gas through the orifice. This type of system works as a stop gap arrangement, whenever, time cycle controller is not available, as intermittent lift through the orifice is not considered efficient. It makes gas lift valve throttle and thus results in large fluid fall back.

4.12 Optimization and Troubleshooting of Gas Lift Operation

4.12.1 *Optimization*

Figure 4.11 is a typical layout of a well and the surface facilities in a field producing by continuous gas lift. This figure is again reproduced. It shows gas compression and gas distribution facilities at the surface and a typical injection arrangement at the bottom of the well. The basic objective of gas-lift design is to "equip our wells in such a manner as to compress a minimum amount of gas to produce a maximum amount of oil". Oil production by gas lift can be controlled by changing gas volumes, injection depth, wellhead pressure, and tubing size.

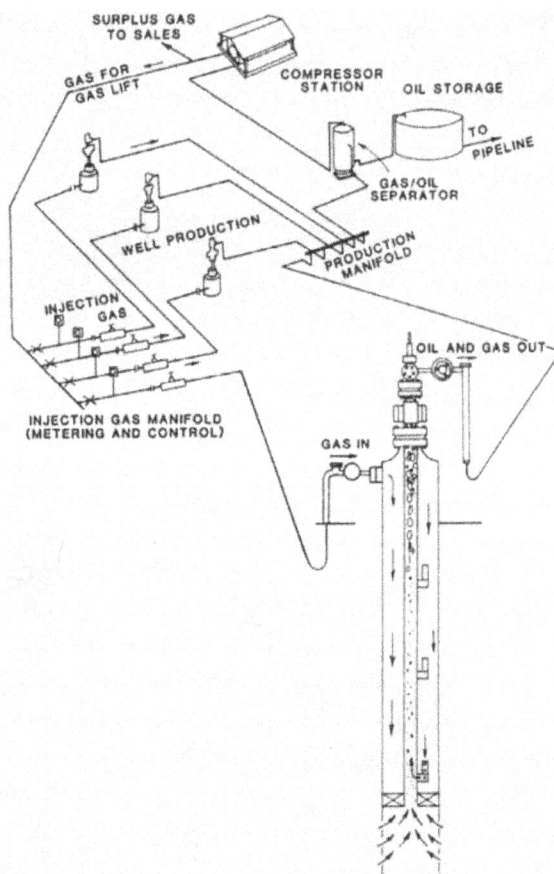

Fig. 4.11

The effect of gas injection on production rate is illustrated by pressure-depth diagram in Figure 4.12. The diagram referred to as a gas-lift diagram, relates IPR of the well, gas injection rate, surface injection pressure, and production rate. It is a snapshot of steady state condition established in the well after it has been unloaded and inflow has been stabilized.

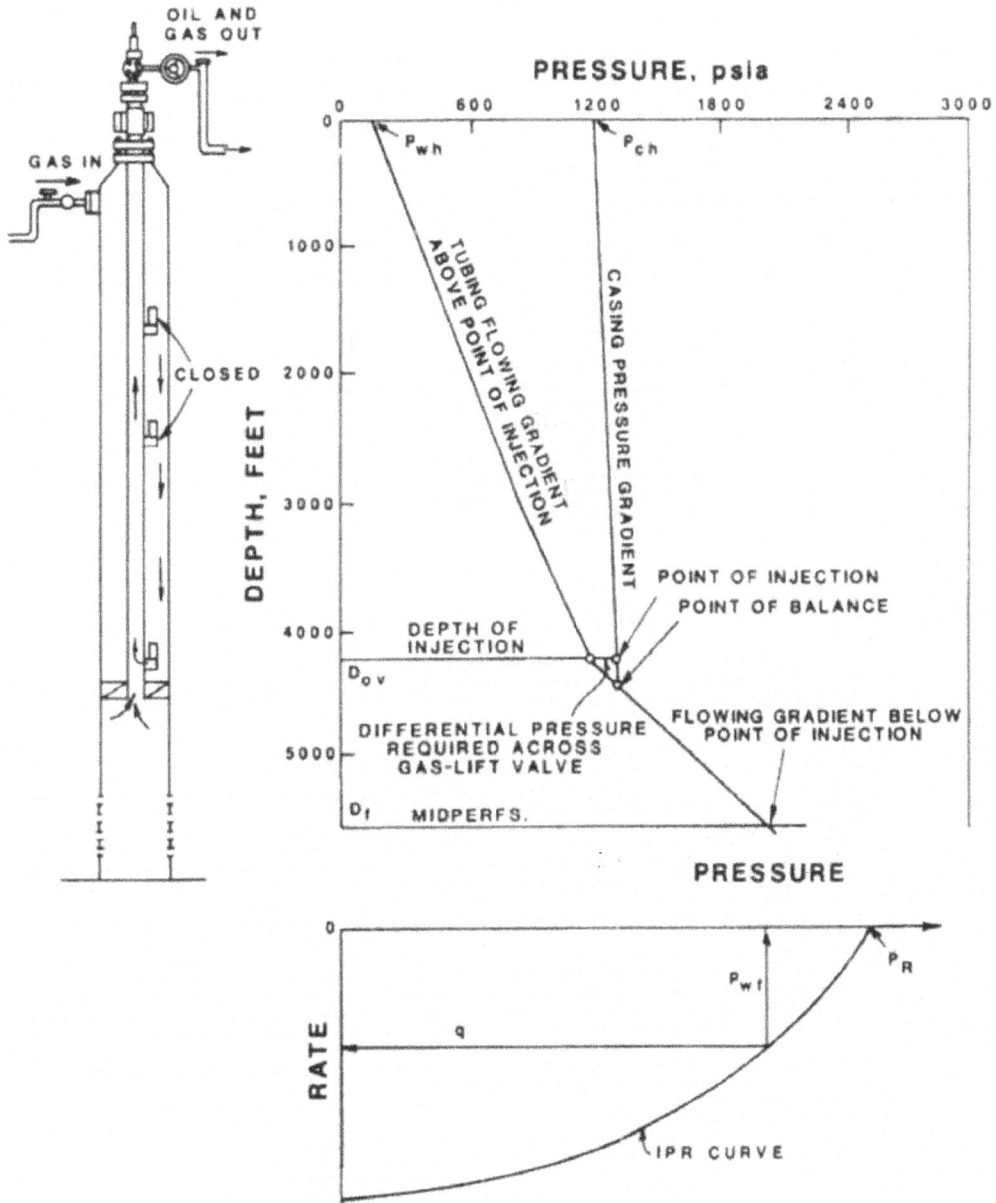

Fig. 4.12

The diagram in Figure 4.12 is relevant to conventional gas lift arrangements where gas is compressed into the casing at the surface and flows from annulus into the tubing through a single gas injection valve close to the bottom of the well. The conventional gas lift valve is merely an orifice to restrict and control the passage of gas from the casing to tubing. As the diagram indicates, wellbore flowing pressure is determined by pressure traverse in the above and below the injection point. Assuming linear pressure traverse below and above the injection point, the well bore flowing pressure can be expressed as:

$$P_{wf} = P_{wh} + G_{av} D_{ov} + G_{bv} (D_f - D_{ov}) \qquad\qquad ... (4.1)$$

Where

D_{ov} = depth of injection valve (ft)

D_f = depth of formation, mid-perforation (ft)

G_{av} = average pressure gradient above injection point, a function of gas rate injected (psi/ft).

G_{bv} = average pressure gradient of lowing formation fluid below injection point (psi/ft).

Two parameters in equation (a) the injection depth and the flowing pressure gradient above injection point may be varied independently by designer in a given well. The ability to control the bottom hole flowing pressure and production rate in a gas-lift well thus amounts to the ability to control the depth of injection and flowing pressure gradient.

Though a deep injection point implies, efficient gas lift, the injection point may be selected at any depth up to a maximum determined, primarily, by the maximum possible surface-injection pressure. For a given surface injection pressure, there is a depth where the casing pressure equals flowing tubing head pressure. This point is referred to as the pressure balance point. To account for pressure drop across the injection valve it is located a short distance above the balance point, so that the pressure drop across the valve plus the casing pressure are equal to the tubing pressure at that depth. The size of orifice in the injection valve is slected to give a 50 to 200 psi pressure drop.

The second independent variable in the diagram, the flowing gradient in the tubing, is controlled by the gas injection rate. Increasing injection rate increases the gas-liquid ratio in the tubing, and up to a certain limit, decreases the flowing gradient. Beyond this limit, the flowing pressure gradient is increased by larger gas-liquid ratios.

Expressing the interrelations between the variables governing gas-lift production as a gas-lift diagram is essential to all methods used to design and control gas-lift performance. Whether displayed graphically, or expressed as a set of equations, the following elements are the essence of gas-lift engineering:

- reservoir inflow performance (IPR)
- approximate flowing gradient in the production tubing below the injection point
- surface-gas-injection pressure
- gas gradient in the casing
- amount of injected gas
- approximate flowing gradient in the production tubing above the injection point.

A procedure to construct a gas-lift design is demonstrated in examples 4.1 and 4.2, following a presentation of certain considerations and simplifying assumptions used in the procedure.

Surface-injection pressure depends on the gas compressor rating. It is usually in the range of 700–1000 psia; together with injection rate, it determines the pressure traverse in the casing/tubing annulus. Unless the gas rate is very high, or the casing/tubing annulus is small, gas density governs the pressure gradient in the annulus.

The pressure traverse can he calculated using conventional pipe flow calculations where the annular cross section area is translated to an apparent (effective) radial pipe cross section. Annular restrictions due to external upset tubing connections or tubing couplings (collars) can be practically ignored in the calculations. For quick design calculations, the friction component is usually ignored and the gas pressure gradient in the casing is approximated by the weight of a static gas column. For further simplification the gradient is assumed constant, giving a linear pressure traverse. The surface and downhole pressures can be related by a simple equation suggested by Gilbert (1954):

$$P_{downhole} = P_{surface} \, (1 \, + \, H/40000) \qquad\qquad \text{... (4.2)}$$

Where,
   P = casing/tubing annulus pressure (psia)
   H = the distance from the surface to the gas-lift valve (ft).

If the injection point is unspecified, the casing traverse may be extended until it intersects the pressure gradient in the tubing below the injection point, thereby establishing the pressure balance point. The flowing gradient below the injection point can be calculated from multiphase flow correlations or estimated from gradient curves, if available, for flow of the given reservoir fluid at the particular rate assumed in the diagram. Similarly, the flowing gradient above the injection point can be calculated from correlations or estimated from a gradient curve using the gas-liquid ratio in a mixture of reservoir fluid plus the injection gas. Example 4.1 illustrates the first part of the procedure for constructing a gas-lift diagram where production rate by gas lift is related to injection depth.

Example 4.1: Estimating wellbore flowing pressure and continuous production rate with a given injection depth.

A gas-lift study is performed for the wells of a particular oil field, which produces from the sandstone formation. An appraisal well S-5 is representative of the wells in the field for gas-lift design purposes. The relevant well data are as below:

mid perforations	:	8000 ft
formation pressure (at 8000 ft)	:	2650 psi
wellhead pressure	:	200 psia
tubing	:	8000 ft × 3.5 in.
formation gas-liquid ratio	:	600 scf/bbl
- water fraction	:	0%
- well IPR	:	$q = 0.2 \times (2650 - P_{wf})$

Estimate the continuous production rate from a single well when injecting gas at 900 psia surface pressure and the injection valve is just above the perforations. Calculate the injection depth required to produce the well at a rate of 200 slb/day. The estimated pressure drop in the gas-injection valve, ΔP_v, is 100 psi. For design purposes, gas pressure in the annulus is represented by a straight line of equation (b) from the surface to 8000 ft.

Solution: The tasks in this example are concerned with pressure conditions in the annulus and in the tubing below the gas-injection point. When injecting at 8000 ft, the flowing bottom-hole pressure equals approximately the flowing pressure in the tubing at the injection point. It is, therefore, related to the annulus pressure by,

$$P_{wf} = P_{annulus} - \Delta P_{valve} \qquad \qquad ... (4.3)$$

which can be further expressed in terms of surface-injection pressure and pressure of gas column, calculating,

$$P_{wf} = 900 (1 + 8000/40000) - 100 = 980 \text{ psia}$$

The production rate is then calculated from the IPR as,

$$q = 0.2 \times (P_r - P_{wf}) = 0.2 \times (2650 - 980) = 334 \text{ STB/Day}$$

To produce the well at a rate of 200 STB/Day the required bottom-hole pressure, calculated from IPR, is,

$$P_{wf} = 2650 - 200/0.2 = 1650 \text{ psia.}$$

Expressing the tubing pressure at the injection point in terms of P_{wf} and tubing flowing gradient, and the annulus pressure in terms of annulus gas pressure gives,

$$900(1 + D_{ov}/40000) - 100 = 1650 - 0.39(8000 - D_{ov}), \text{ where } G_{bv} = 0.39 \text{ psi/ft}$$

Solving for D_{ov}, the injection depth is calculated as 6449 ft.

The tubing pressure at the injection point is then calculated as:

$$1650 - 0.39(8000 - 6449) = 1045 \text{ psia.}$$

It is important to realize that pressure conditions in the annulus and in the tubing below the injection point are not sufficient to confirm the feasibility of gas-lift production. It is also necessary to verify the conditions in the tubing above the injection point, as shown in example 4.2.

The production rate calculated in example 4.1 was based only on the flow conditions at and below the injection point. It is obvious, however, that to sustain the calculated production rate, it is necessary to maintain a particular pressure traverse above the injection point. This pressure traverse corresponds to a particular gradient curve that matches two points on the pressure traverse—the wellhead pressure and the tubing pressure at the injection depth. Figure 4.13 illustrates a quick manual procedure for matching pressure conditions in the tubing above the injection point to a gradient curve. The matched gradient curve determines the required gas-oil ratio in the tubing. Knowing the reservoir GOR and the required tubing GOR allows calculation of the gas needed for injection.

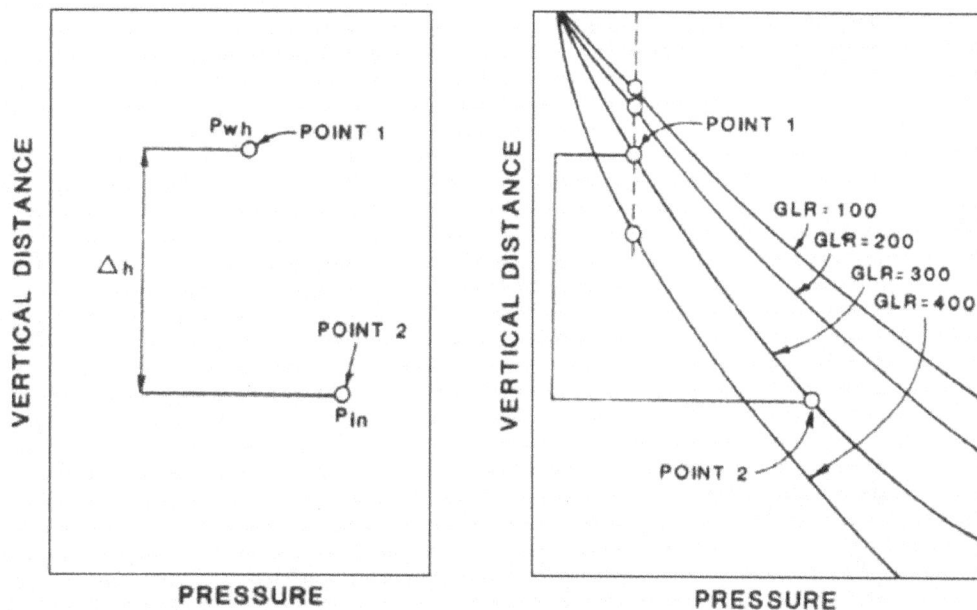

Fig. 4.13

Example 4.2 illustrates the matching procedure to identify a gradient curve and the calculation of the required gas-injection rate. In fact, matching the gradient curve in example 4.2 completes the procedure for constructing the gas-lift diagram, started in example 4.1.

Example 4.2: Gas-injection rate to maintain a particular production rate.

Complete the pressure diagram for the S-5 well (example 4.1) and determine the amount of injected gas needed to produce the well at a rate of 200 STB/Day. Estimate the per-well power requirement for compression with a compressor suction pressure of 65 psia. A quick estimation of compression power requirements can be obtained from the relation,

$$P = 2.23 \times 10^{-4} q_g [(p_2/p_1)^{0.2} - 1] \qquad \ldots (4.4)$$

where,

P = power (HP)
q_g = gas rate (scf/day)
p_1 = compressor inlet pressure (psia)
p_2 = compressor output pressure (psia).

Solution: The production conditions calculated in example 4.1 determine a particular pressure drop in the tubing between the gas injection point and the wellhead. The pressure drop implies a particular gradient curve with a particular gas-liquid ratio. The gradient curve and the corresponding gas-liquid ratio can be identified by matching a gradient curve to the two known pressure points in the tubing:

* wellhead $h = \Delta.\ p = 200$ psia
* injection point $h = 6449$ ft, $p = 1045$ psia

The gradient curve of GLR = 1000 scf/STB matches the two points. The injection gas-liquid ratio is then calculated as the difference between the obtained flowing GLR and the formation GLR i.e. 1000 – 600 = 400 scf/STB. The gas injection rate q_g is then,

$$q_g = (1000 - 600)\ q_o = 400\ (200) = 80 \times 10^3 \text{ scf/day}.$$

The required compressor power per well is estimated as,

$$P = 2.23 \times 10^{-4} \times (80 \times 10^3) \times [(900/65)^{0.2} - 1] \qquad \ldots (4.5)$$
$$= 12.34 \text{ HP}$$

The state of equilibrium described by the gas-lift diagram is not a stable one. If for any reason, the pressure in the tubing across the valve drops momentarily, more gas will be injected into the tubing, causing the tubing pressure to drop even further. It may he stabilised, however, by controlling the surface injection rate. Constant surface injection rate implies a drop of casing pressure if injection rate across the gas injection valve increases. This, in turn, tends to reduce pressure drop across the valve and thus decreases injection rate. Another control capability is included in certain types of injection valve that automatically reduce injection rate in response to tubing pressure drop.

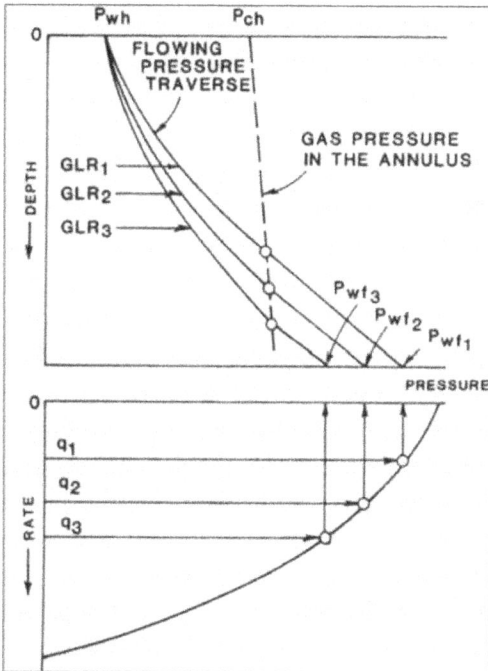

Fig. 4.14

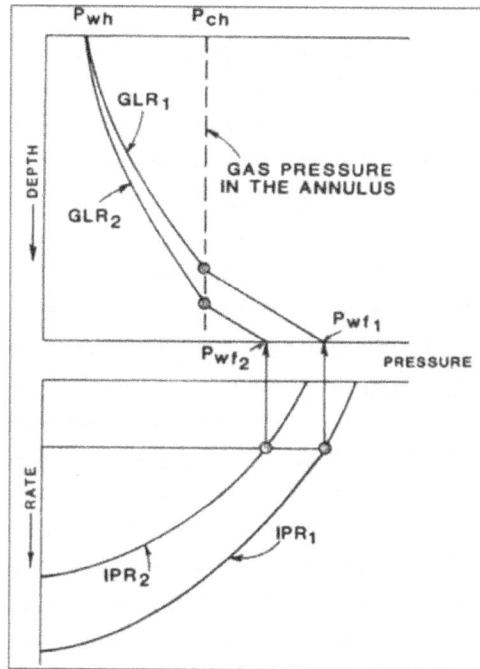

Fig. 4.15

The gas-lift diagram can be used to study the functional relationship between variables in design and operation of a gas-lift system. For example, in Figure 4.14 it is used to investigate the possibility of increasing production by increasing the gas injection rate. As demonstrated in the figure, wellbore flowing pressure drops and production rate increases as more gas is injected. It illustrates also that a higher gas-injection rate allows injection at a deeper point without increasing injection pressure or changing the pressure traverse in the annulus.

Figure 4.15 investigates the change in injection depth and gas-injection rate needed to maintain a constant production rate as the reservoir depletes and the IPR deteriorates. The figure explains the advantage of controlling both the injection depth and the amount of injected gas. Controlling these two factors by proper selection and spacing of continuous gas-lift valves, production rate can be adjusted as needed during the life of a well.

Basic assumption applied in Figures 4.14 and 4.15 is that a higher gas-liquid ratio in the tubing results in a smaller pressure gradient. This assumption is correct only up to a limiting GLR. From the set of gradient curves in Appendix A. it can seen that the pressure gradient decreases with increasing GLR, to a certain limit. Increasing GLR above this limit causes an increase in gradient. The minimum gradient curve signifies minimum flowing bottom hole pressure and maximum production rate.

Increasing gas-injection rate, therefore, may increase the production rate to a maximum level beyond which increasing gas rate results in decreasing production rate. Poettmann and Carpenter (1952) and Bertuzzi *et al.* (1953) discussed this behavior and its application to the design of efficient gas-lift systems. Bertuzzi *et al.* translated this behavior to a curve relating injection rate to production rate. The curve, referred to as the gas-lift performance curve, has become the standard expression of gas-lift deliverability.

Gas Lift Performance Curve

A step-by-step procedure to construct the gas-lift performance curve is illustrated in Figure 4.16. Figure 4.16(a) illustrates the general trends observed in gradient curves as GLR varies. This trend implies that at a given rate and constant wellhead pressure, the tubing intake pressure varies with GLR. A plot of tubing intake pressure, given in Figure 4.16(b), indicates that for each flow rate in a given tubing size there is a particular GLR that yields minimum tubing intake pressure. We refer to this GLR *as favourable GLR.* A plot of favourable GLR versus the corresponding rates in a given tubing size is given in Figure 4.16(c). Favorable GLR decreases as oil rate increases. The favorable GLR is seldom equal to reservoir GLR, and it may be achieved only by adding gas to the tubing. The amount of gas required to achieve a favourable GLR is indicated also in Figure 4.16(c).

Figure 4.16(d) is a plot of the locus of all minimum tubing intake pressures versus their corresponding flow rates. As mentioned before a minimum tubing intake pressure at a given flow rate is obtained with the favorable GLR. The locus line divides the plane of the graph into two regions. The region below the line is where intake pressure is less than the minimum pressure required to sustain flow, and thus flow in the tubing cannot exist. No matter how much gas is injected, the tubing intake pressure cannot sink below the minimum indicated by this line. The region above the line is where all possible flow situations in the tubing occur, with or without gas injection. In fact, the intersection between the IPR of the well and the locus of the minimum intake pressures (Figure 4.16(d)) gives the point of maximum liquid rate possible from the particular well. Figure 4.16(e) illustrates the significance of this intersection point in terms of a tubing performance curve. It shows that tubing performance curves for any GLR higher or lower than the preferable GLR will intersect the IPR at a lower liquid rate.

For the particular well plotted Figure 4.16(d), formation GLR is lower than the favorable ratio, and therefore injection of gas rate increases the production. On the other hand, in wells where formation GLR is higher than the favorable GLR there is no gain in production by gas lift.

Oil rate versus favorable GLR is shown in Figure 4.16(f). Calculating the injection GLR as the difference between the favorable and formation GLRs and further

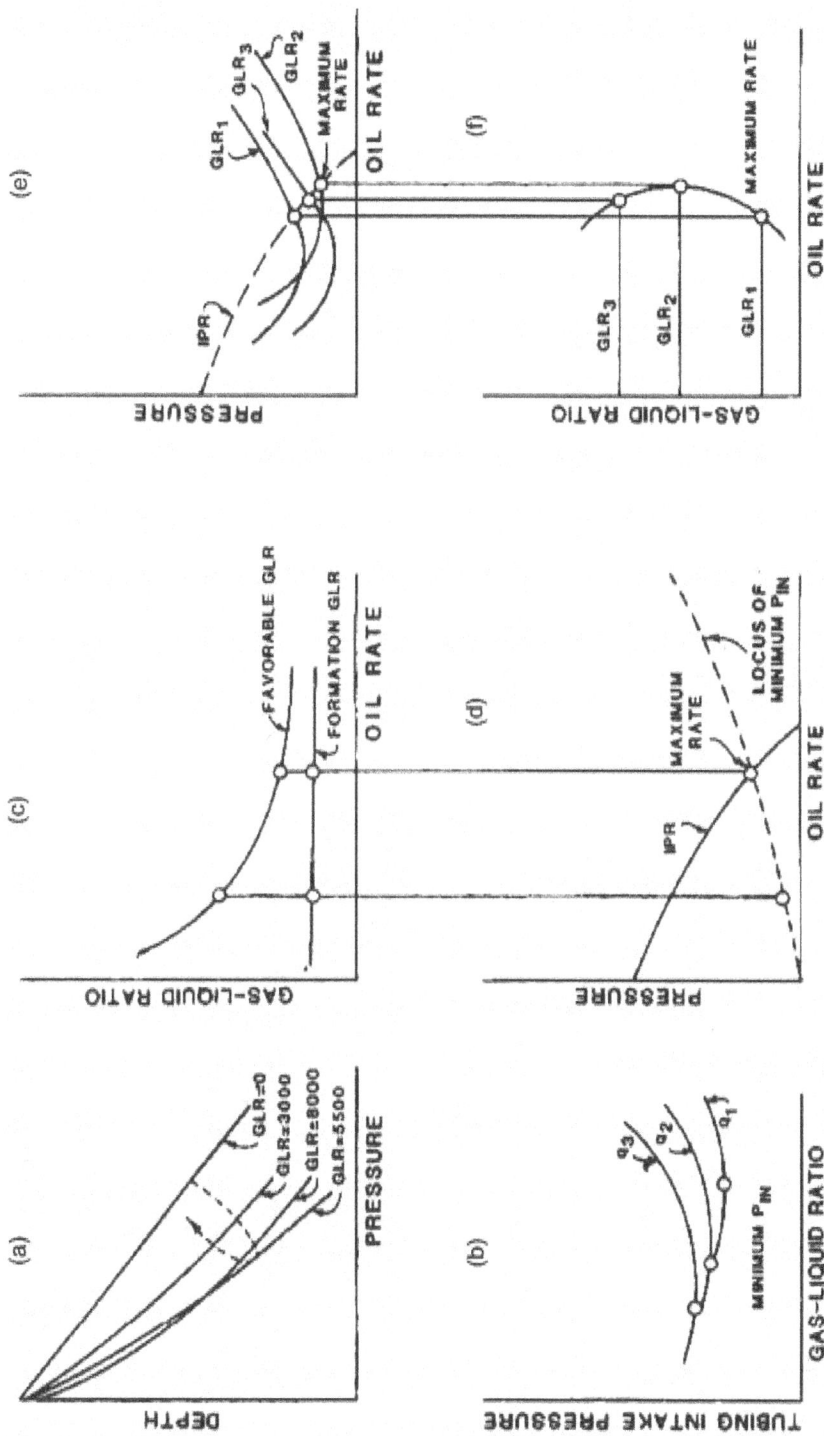

Fig. 4.16(a–f)

computing the corresponding injection rate yields the plot of the gas-lift performance curve can be seen in Figure 4.18. Example 4.3 illustrates the procedure to construct the gas-lift performance curve and to determine maximum liquid rate possible by gas lift.

The procedure in example 4.3 is valid for reservoir conditions existing at a particular stage of depletion. Similar calculations can be made for any stage of depletion later in the life of the well. Figure 4.17 adds the time dimension and illustrates the decline of the maximum rate at each stage of depletion. As shown in Figure 4.17(a), the favorable GLR for a given liquid rate is independent of reservoir behavior. Therefore, in spite of depletion, the locus of favorable GLRs in Figure 4.17(b) does not change. On the other hand, IPRs shown in Figure 4.17(b) deteriorate with depletion and therefore intersect the locus at progressively decreasing rates. The result is a decreasing maximum liquid rate, which is plotted versus cumulative production in Figure 4.17(c).

The injection rate required to maintain maximum liquid rate as the reservoir depletes is the difference (Figure 4.18) between the favorable GLR and formation GLR. Summarizing the time behavior of a gas-lift system, Figure 4.17(d) plots the changes of reservoir GLR, favorable GLR and gas injection GLR needed to produce the maximum liquid rate, versus cumulative production at each stage in the well's life. Figure 4.17(d) indicates that for solution gas-drive reservoirs, the needed gas-injection

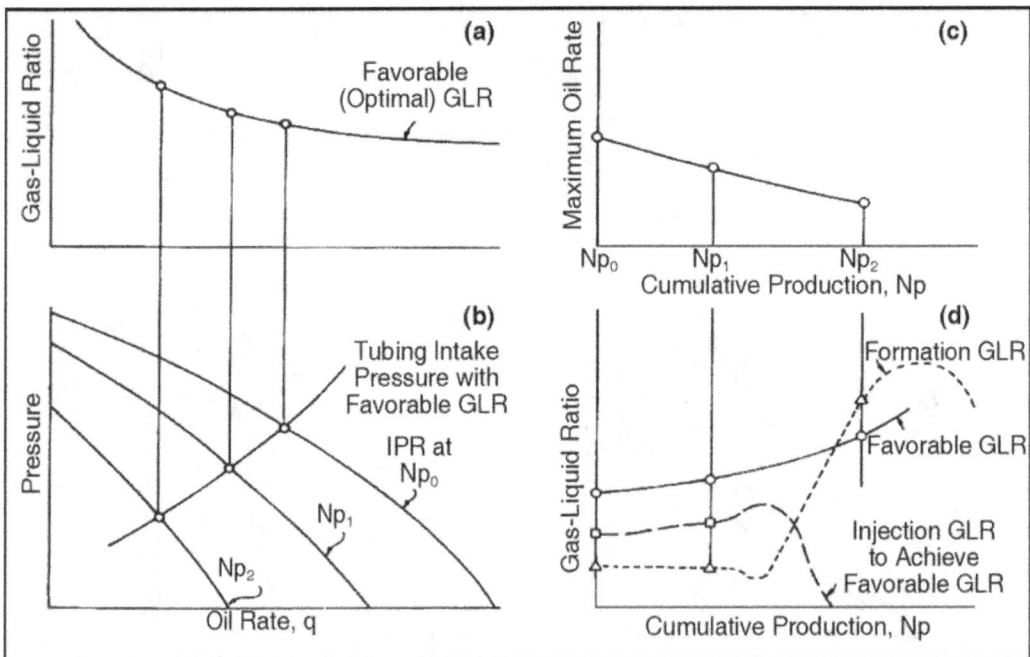

Fig. 4.17

GLR increases at early stages but drops rapidly as reservoir GLR increases. Example 4.4 illustrates the procedure to determine depletion effect on gas-lift performance.

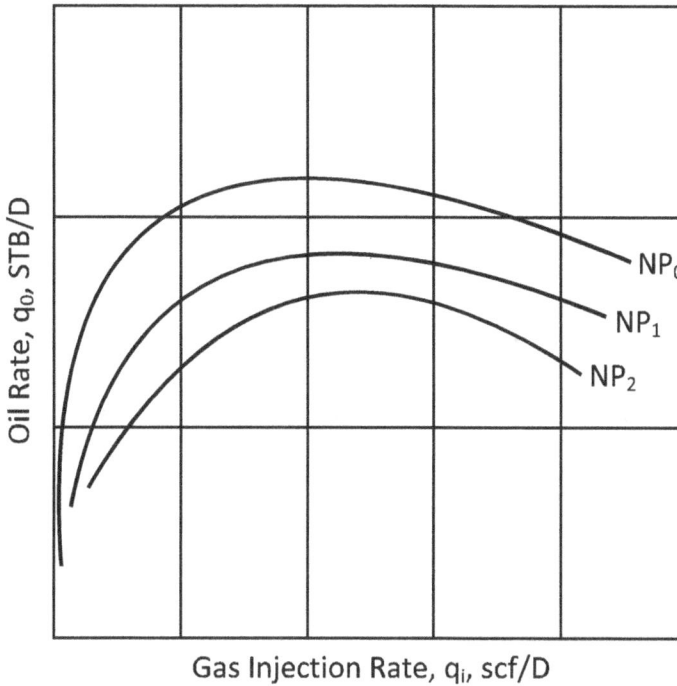

Fig. 4.18

Example 4.3: Calculating maximum possible production rate by gas lift.

Gas lift is one of the artificial lift options considered for a field. Data for a typical well in the field are as below:

Reservoir pressure : 2000 psia
Wellhead pressure : 65 psia
Reservoir solution gas-oil-ratio : 400 scf/stb
Tubing : 2 7/8" × 5000 ft
IPR : $q = 0.2 \times (2000 - P_{wf})$

Two cases of gas-lift production are investigated:

Case 1: An unlimited amount of injection gas is available to produce the well at or near its maximum production rate.

Case 2: A limited amount of injection gas. 180 Mscf/Day, is allocated to each well.

Investigate the performance of a well producing under each case.

Solution:

Case 1: The gas-liquid ratios that yield minimum pressure loss in $2\frac{7}{8}''$ tubing (favourable GLR) are listed in Table below:

Table E4.4(a): Tubing Intake Pressure with Favourable GLR

q (STB/Day)	Favorable GLR (scf/stb)	p_{in} (psia)
50	8800	180
100	6300	210
200	4300	250
400	3250	300
600	2400	380

These favorable GLRs are plotted versus rate in Figure E4.3. With 65 psia wellhead pressure the minimum tubing intake pressure at 5000 ft is determined from the favorable gradient curve and listed also in table above. Plotting p_{in} from table versus q in Figure E4.3 gives the locus of minimum intake pressures.

The straight-line IPR of the well is also plotted in Figure E4.3. The intersection of IPR with the locus of $(p_{in})_{min}$ gives maximum flow rate from this well by means of gas lift. Reading from the graph, maximum rate is q_{max} = 260 STB/day. The favorable GLR corresponding to this maximum rate is indicated in Figure E4.3. GLR = 3750 scf/stb. The required injection rate is calculated as:

$$(q_g)_{inj} = (GLR - R_s)\, q = (3750 - 400)\, 260 = 871 \times 10^{-3} \text{ scf/D} = 871 \text{ Mscf/day}$$

Case 2: The GLRs obtained by injecting 180 Mscf/day in addition to 400 scf/stb of reservoir oil are calculated and listed in Table E4.4(b) below:

Table E4.4(b): Tubing Intake Pressure with Limited Gas-Injection-Rate

Q (STB/Day)	Injection GLR (scf/stb)	Total GLR (scf/stb)	p_{in} (psia)
50	3600	4000	280
100	1800	2200	250
200	900	1300	400
400	550	950	550
600	300	700	600

Given 65 psia wellhead pressure, the tubing intake pressure with a flowing GLR is obtained from the gradient curves and listed versus rate in Table E4.4(b). The p_{in} values are plotted versus q and form the intake pressure curve in Figure E4.3. The

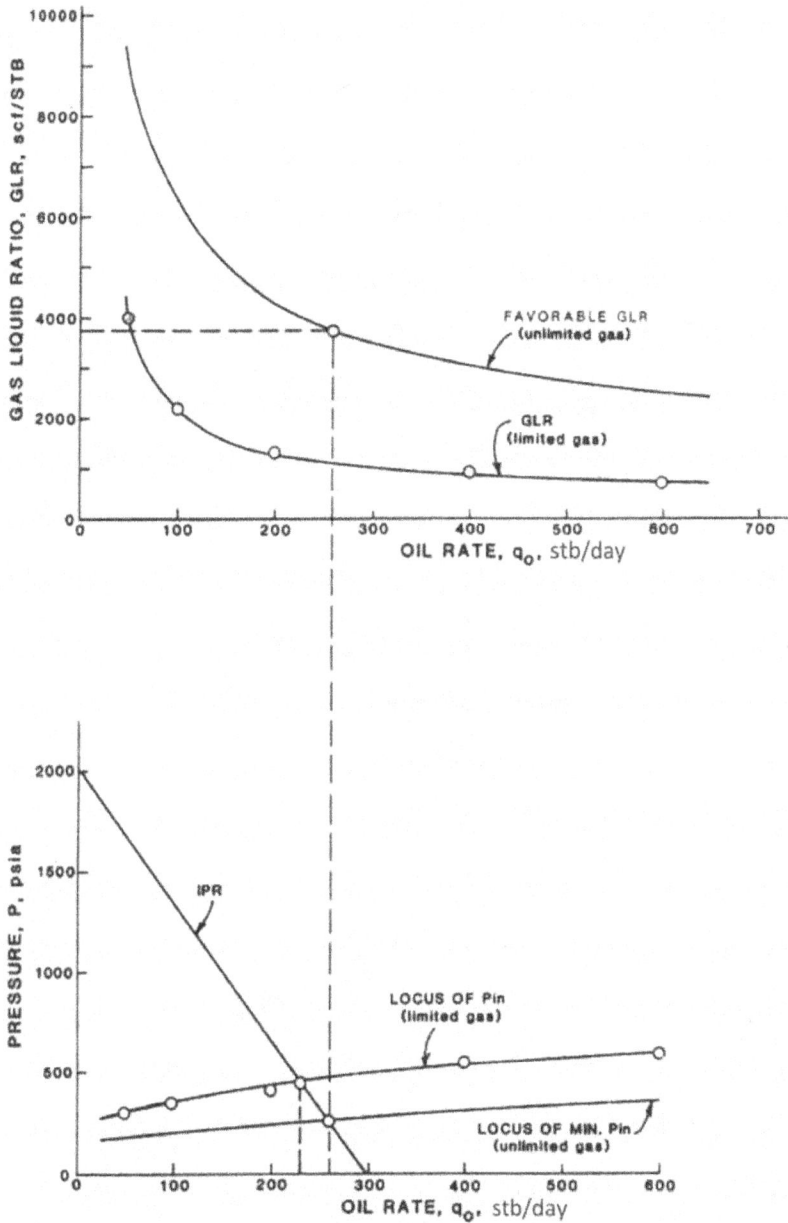

Fig. E4.3

intersection of this curve with IPR defines the flowing conditions for 180 Mscf/day gas injection. The intersection is at a rate of 230 stb/day. An interesting observation is that the calculated gas injection rate in case 1 is 3.5 times higher than in case 2, whereas the production rate is only 10% higher.

Example 4.4: Changes in gas lift and gas-injection requirements with reservoir depletion.

A gas-lift production system is perhaps the only practical possibility to increase production of a typical offshore field. The first two tasks of a study to evaluate the performance of a gas lift system are:

1. Determine the maximum possible flow rates with gas lift and the favorable GLR to achieve these rates at each depletion stage.
2. Calculate the amount of gas needed for injection to achieve the maximum production rate.

The study assumes that gas is injected at the bottom of the tubing (8000 ft). All other well and reservoir data are as below:

Tubing   : $2\frac{7}{8}''$
Mid perforation depth   : 8000 ft
Casing   : $7''$
Reservoir press p_r : 4000 psia
q_{max} : 480 stb/day
IPR is given by   : $q/q_{max} = [1 - (p_{wf}/p_r)]^{0.9}$

The IPR of well No. 4 is plotted in Figure E4.4(a). For a given rate, the minimum possible intake pressure for the tubing is obtained from the favorable GLR line in the gradient curve. The favorable GLRs are listed in Table E4.4(c) and plotted versus rate in Figure E4.4(a). For a given wellhead pressure, $p_{wh} = 200$ psia, the minimum

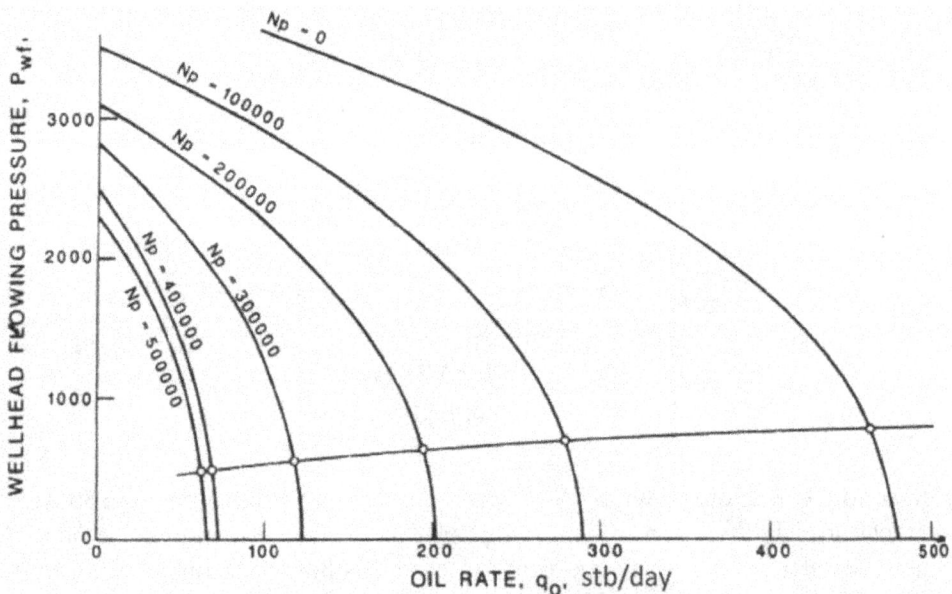

Fig. E4.4(a): Maximum Possible Gas-lift Production for the Ina No. 4 Oil Well

---

intake pressure is read from the favorable gradient curves and listed in Table E4.4(c). The locus of minimum intake pressures is also plotted in Figure E4.4(a). The intersection points with IPR curves indicate maximum rates with gas lift. These maximum rates are given, versus cumulative production, in Table E4.4(d). The corresponding favorable GLRs are obtained from Figure E4.4b and listed in Table E4.4(d). The required injection rate to achieve maximum production rate is calculated as,

$$(q_g)_{inj} = (GLR_{favorable} - GLR_{reservoir})\, q_o$$

The calculated results are listed in Table E4.4(d).

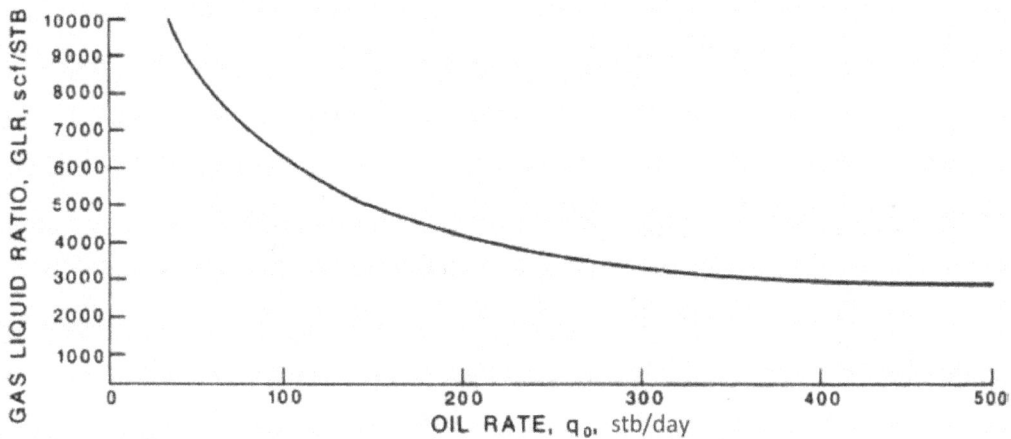

Fig. E4.4(b): Flow Rate *versus* Gas-Liquid Ratio for the Tubing of Ina No. 4

Table E4.4(c): Minimum Intake Pressure for Well No. 4

q (STB/Day)	Favorable GLR (scf/stb)	p_{in} (psia)
50	8800	450
100	6300	530
200	4300	640
400	3250	760

Changes in reservoir conditions with depletion imply that the gas lift characteristic curve also changes. In fact, for each depletion stage there is a different curve given together with a family of performance curves, as illustrated in Figure 4.18. For any given stage of depletion, the oil rate curve increases rapidly with increasing GLR at the low GLR range, and then tends to level off before reaching the maximum oil rate. The particular shape of a gas-lift performance curve indicates that, beyond a certain point, substantial increase in gas injection is required to achieve relatively small

Table E4.4(d): Maximum Gas Lift Rates and Corresponding Gas-Injection-Rate

N_p (bbl)	q_{max} (STB/Day)	Favorable (scf/stb)	GLR q_{inj} (MMscfd)
0	460	2900	1.058
100,000	279	3750	0.766
200,000	194	4300	0.524
300,000	118	5750	0.384
400,000	70	7500	0.175
500,000	63	7850	0.337

increases in oil rate. In other words, very little production is gained by increasing gas injection beyond the maximum rate point.

The high cost of the gas compression and separation equipment needed to separate large gas quantities suggests that the maximum oil rate is not necessarily the most economical one (Simmons 1972; Redden *et al.* 1974). Translating produced oil and injected gas into monetary quantities using produced oil (and gas) revenues and the cost of gas injection, the operator may express the gas-lift performance curve in terms of revenue versus expenses, and obtain the curve illustrated in Figure 4.19. The curve is continuous and smooth. It assumes gradual increase of injection cost. Such an assumption is only valid when a compression system has the capacity to increase injection rate with only incremental increases in cost. If increasing injection rate is achieved by upgrading the capacity of an existing compression system, the upgrading cost is reflected as a discontinuity and sometimes even a change of trend in the curve.

Various criteria have been proposed for using cost-revenue analysis to select the "optimal" operating conditions of a gas-lift system. They generally refer to one of two cases:

- An individual well of field with an unlimited supply of injection gas.
- A group of wells with limited gas supply to be distributed among individual wells.

For the case of unlimited gas supply Simmons (1972) and Redden *et al.* (1974) suggest that for a given depletion stage the most profitable rate occurs at a unit slope on the curve in Figure 4.19. This point represents the situation when the incremental revenue gain from increasing oil rate is equal to the incremental increase in gas-injection expenses. This point is referred to as *maximum daily operating cash income* (maximum OCI). Up to this point, increasing gas injection yields a gain in profit. At the maximum OCI, incremental gain in profit is zero. Beyond this point the incremental gain becomes negative and the total profit starts to diminish. Figure 4.19 illustrates the maximum OCI at a particular stage of depletion. Similar procedures can calculate the maximum OCI for a series of future depletion stages. Maintaining maximum OCI during the life of a field requires a continuous increase in injection

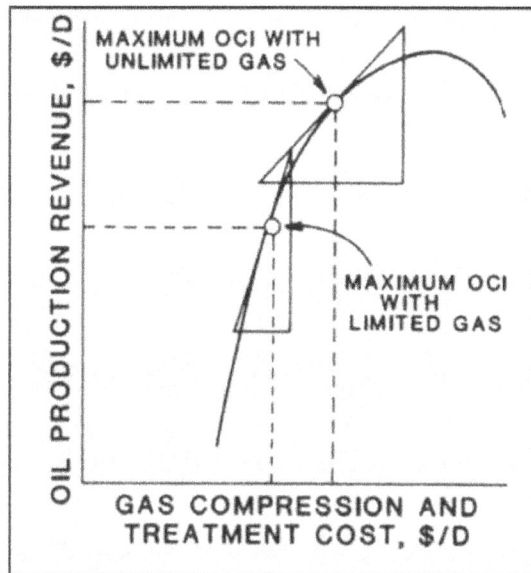

Fig. 4.19

rate and progressively less total profit. This is not necessarily the most economical production schedule. Only a detailed economic analysis, including cost, revenue, and initial investment, can calculate a gas injection schedule that maximizes the operator's economic returns.

Often a gas-lift study is carried out when a compression system of given capacity is already installed. In such cases the designer needs to prepare a program for distributing the available gas among individual wells. In some gas-lift systems the available compressed gas is less than the rate needed to produce all the wells at their most economical rates. A new criterion should therefore be established to determine the most profitable gas allocation.

The simplest approach for allocating available gas in such cases is to maximize the total daily oil rate achievable from the field with the available gas injection rate. Kleyweg *et al.* (1983) present a program for optimizing the gas-lift system in the Claymore field, summarizing their allocation approach by *"Once performance curves have been obtained for each well on the offshore platform gas-lift optimization can be achieved. Two methods are possible:*

- Analytical method, where each curve is represented by a polynomial, and then applying a linear-programming technique to calculate true optimised distribution of the available lift gas.
- Step-by-step method, where all wells are supplied with enough gas to kick them off and from there on the performance curves are scanned to find the curve with the maximum slope. The gas-lift rate to the corresponding well is then increased by one step and so on until all the available gas is distributed."

Figure 4.20 illustrates the application of the maximum rate approach for the simple case of a two-well system. Injection gas is plotted against produced oil rate for each individual well. A procedure formulated by Clegg (1982) suggests that maximum total production from the two wells is the sum of the rates q_{o1} and q_{o2} where the two performance curves have equal slope $m_1 = m_2$ and the total available injection gas equals $q_{i1} + q_{i2}$. For fields with only a few wells the procedure in Figure 4.20 can be performed manually by trial and error. Large fields with many wells may need computerised procedures.

All the discussions so far assume constant wellhead pressure. This assumption is usually valid when the wells are near the separator and producing against a constant separator pressure. In cases when the wells are away from the separator, wellhead pressure varies due to pressure loss along the flowline. The procedure to establish the relationship of oil rate to gas-injection rate in such cases is illustrated schematically in Figure 4.21.

The gas-lift diagram in Figure 4.21(a) illustrates GLR changes corresponding to changes in wellhead pressure p_{wh} when oil rate and injection point are held constant. The relationship p_{wh} versus GLR can be similarly developed with other rates; the result is illustrated in Figures 4.21(b). The flowline performance curve is

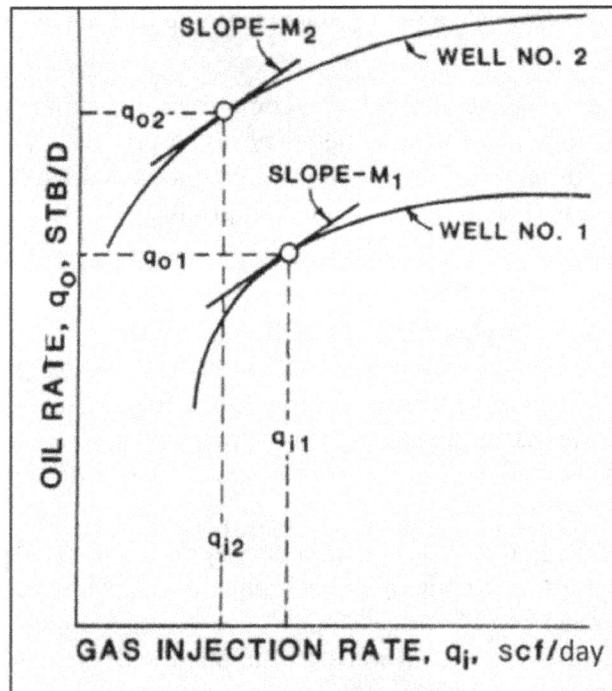

Fig. 4.20

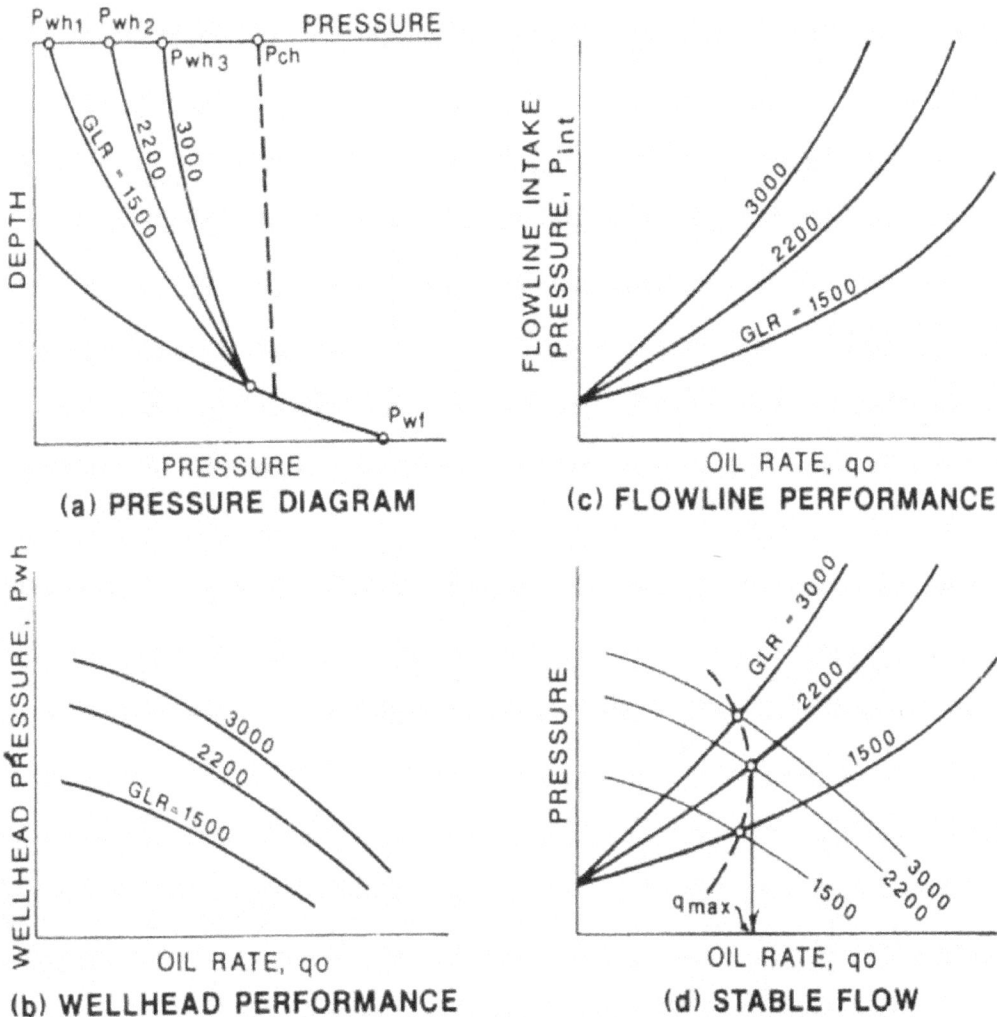

Fig. 4.21

plotted in Figure 4.21(c) as flowline intake pressure p_{in} versus flow rate. Recognising that flowline intake pressure and wellhead pressure are essentially the same, Figure 4.21(b) and (c) may be combined as in Figure 4.21(d), where the intersection points of the curves with similar GLRs indicate stabilized flow conditions for the well-flowline system.

The procedure in Figure 4.21 resembles a procedure for naturally flowing wells. The only difference is the controllable gas-liquid ratio in gas-lift wells. In fact, the entire treatment of continuous gas lift is an extension of techniques and procedures developed earlier for naturally flowing wells. Continuous gas injection is a major, but not the only aspect of gas-lift technology. A great deal of effort is required to design efficient start up and unloading of wells. Techniques for unloading gas-lift wells are also covered in this book.

Thus, the analysis, has concentrated on calculating the maximum possible oil rate by a continuous gas-lift method and the corresponding gas injection quantities. This information is important because it allows the determination of production limits, and thus the viability of gas lift as a potential artificial lift method in a given field. In the following, an extension is discussed of the analysis needed to relate the main parameters controlling the operation of a gas-lift well to its performance. This extension will allow a detailed design of gas-lift wells. It will also prepare the background for applying computerized procedures for gas-lift design and analysis. The base case for the following discussion is where gas is injected through the annulus and the well produces through the tubing, as illustrated in Figure 4.22. The discussion, however, is valid also for the arrangement where the gas is injected through the tubing and the production occurs through the annulus.

The flow equilibrium concept for naturally flowing wells is also valid, with few modifications, for continuous gas-lift wells. Figure 4.22 summarizes the idea and the application of this concept for gas lift.

Figure 4.22a indicates that fluids enter the production tubing from two separate flow paths. One path is flowing reservoir oil and gas from the perforations upwards, and another path discharges compressed gas from the annulus into the tubing through a gas injection valve. Thus, a node in the tubing in front of the entry point is, in fact, a joining point of three flow paths: the reservoir, the discharged gas, and the tubing. The corresponding pressure versus rate relationships for the three paths are the IPR expressing reservoir inflow, the TPR expressing flow through the tubing correspondingly, and the Discharge Performance Relationship (DPR) expressing the discharge through the gas-lift valve.

The IPR and TPR have been extensively discussed in chapter 1. The DPR will be discussed later in this section. At this point it is sufficient to note that the DPR is the relationship between the flow rate of the injection gas discharged into the tubing, $q_{ginj,}$ and the gas stream pressure, $p_{vf,}$ downstream of the discharge point.

Two basic fluid mechanics principles can be stated for the considered node: the continuity of flow and the conservation of mass. The continuity principle implies that the flowing pressure in the node is equal in all the joining streams. The conservation of mass principle, when expressed in terms of flow volume, implies that the algebraic sum of the oil flow rates and the algebraic sum of the gas flow rates entering and leaving the node are zero. Mathematically, these conditions give three equations which are sufficient to solve for the oil production rate, the gas injection rate, and the gas injection pressure. These equations are referred to as flow *equilibrium equations*.

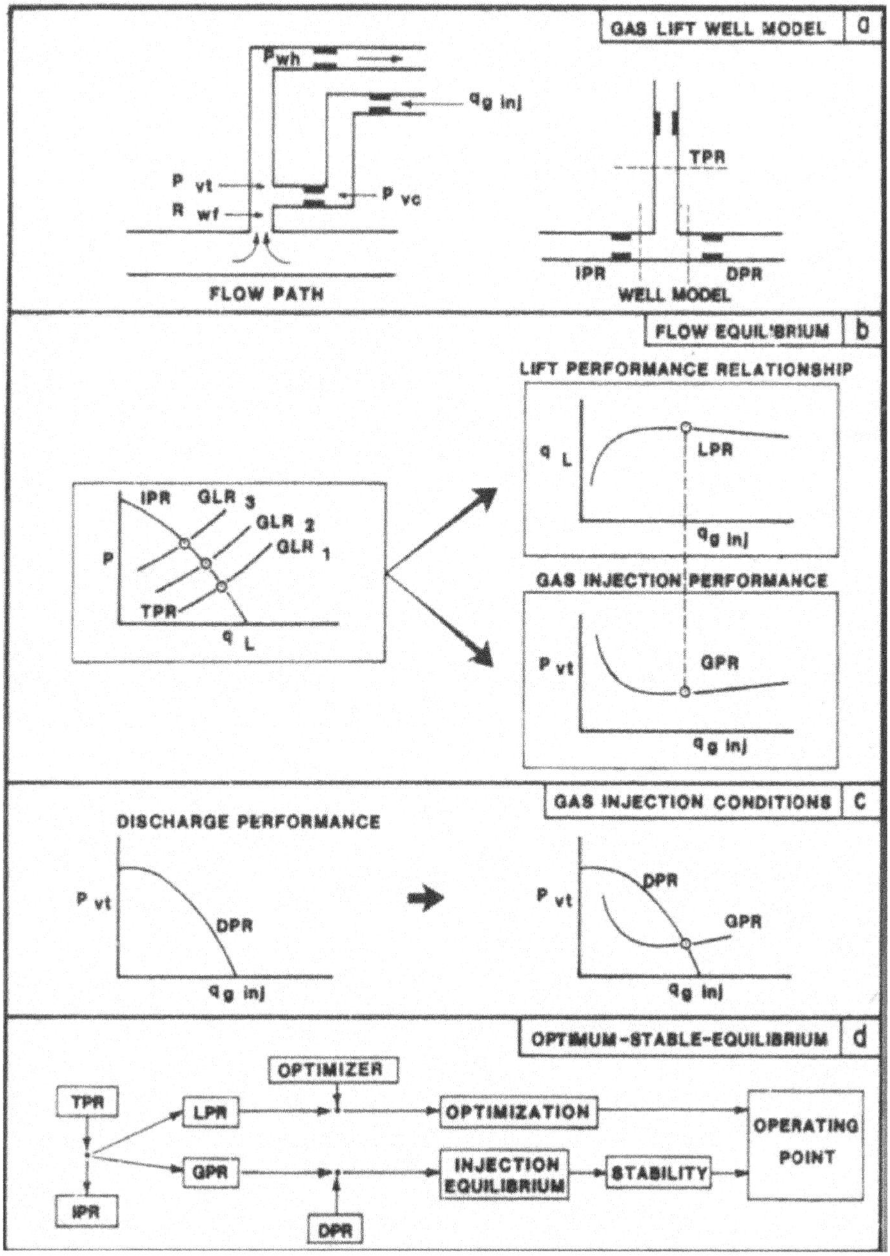

Fig. 4.22

Substituting the IPR, TPR, and DPR into the three equilibrium equations and solving simultaneously for the production rate, injection rate, and injection pressure of interest is a rather straightforward mathematical exercise (three equations with three unknowns). Graphically, the three equations are solved in two steps. First, two equations arc solved simultaneously assuming one unknown of the three is a parameter. The result is a curve representing the locus of all the solution points satisfying the first two equations. Then, the obtained locus is solved simultaneously with the third equation giving the particular solution that satisfies all three equations.

Figure 4.22b illustrates the first step of the solution where the IPR and the TPR are solved graphically assuming a variety of gas injection rates (variety of GLRs in the tubing). The obtained result is in the form of either one of two loci. The first one is called the Lift Performance Relationship (LPR), and the second one is the Gas-lift Performance Relationship (GPR). The LPR curve expresses the relationship between the liquid production rate, q_L and the gas injection rate, q_{ginj}. The curve was introduced earlier in this section in connection with gas-lift optimization analysis. The GPR, on the other hand, expresses the relationship between the gas injection rate and the pressure in the tubing in front of the gas injection valve p_{vf}. The GPR is introduced here for the first time (in this text), and it will be used for sizing a downhole gas injection orifice and to determine the well's operation conditions.

Note the typical shapes of the LPR and the GPR. For the LPR, with increasing gas injection rate, the liquid rate increases rapidly in the beginning and then flattens gradually until it reaches a peak, beyond which the liquid rate starts decreasing gradually. Correspondingly, the pressure of the GPR decreases rapidly and then flattens gradually until a minimum is reached. Beyond it, the pressure starts increasing gradually.

The trend of the production rate versus gas injection rate exhibited by the LPR made the curve a key instrument in gas-lift optimization analysis since it was originally observed by Bertuzzi *et al.* (1953). Unfortunately, the other role of the LPR and GPR to provide an input to the second stage of the solution of equilibrium equations was never fully recognized and thoroughly explored.

To complete the process of solving the equilibrium equations, the GPR obtained in the first stage is solved together with the DPR as illustrated in Figure 4.22c. This determines the particular gas injection rate, and the particular pressure in the tubing in front of the gas injection orifice that satisfy all the equilibrium conditions. The corresponding liquid production rate can be obtained from the IPR of LPR.

The significance of completing the solution of the three equilibrium equations is indicated in Figure 4.22d. The explanation of this figure requires further discussion.

By definition, the DPR is the relationship between the gas flow rate and the downstream pressure in an orifice that operates with a constant upstream pressure, p_{vc}. The characteristics of the DPR are shown in Figure 4.23. Figure 4.23a illustrates the effect of annulus pressure, p_{vc}, on the DPR of a given orifice. Figure 4.23b illustrates a situation where three different orifices operating with different upstream pressures inject the same rate with the same downstream pressure. Figure 4.23c illustrates three different orifices operating with a common annulus pressure. Finally, Figure 4.23d illustrates the DPR of a widely used valve with a built-in throttling mechanism that reduces the orifice as flow rate increases. This is in contrast to the conventional orifice valve which is essentially a fixed size orifice.

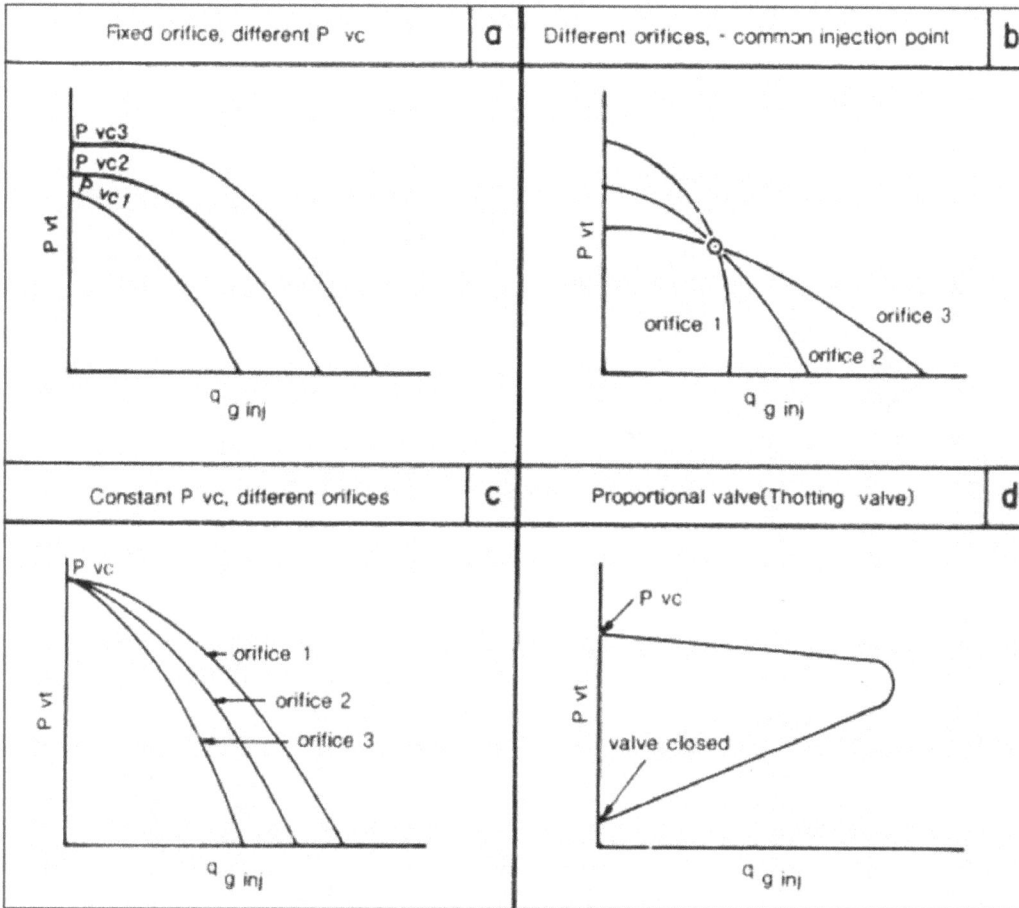

Fig. 4.23

Considering the shape of the GPR and the DPR, their simultaneous solution may yield two points. Applying the same reasoning as in the multiple solution case in flowing wells, one of the solutions here represents an unstable state.

As in natural flow, the stability can be determined by checking the relative trend of the upstream and downstream pressure-rate relationships. More specifically for the gas-lift case, it can be determined by checking the relative trend of the DPR (available pressure) and the GPR (required pressure).

Figure 4.24 illustrates a case where a particular injection rate can be achieved by two different orifices. The small orifice operating with a high injection pressure yields stable production, while the larger orifice operating with considerably lower annulus pressure yields unstable flow equilibrium.

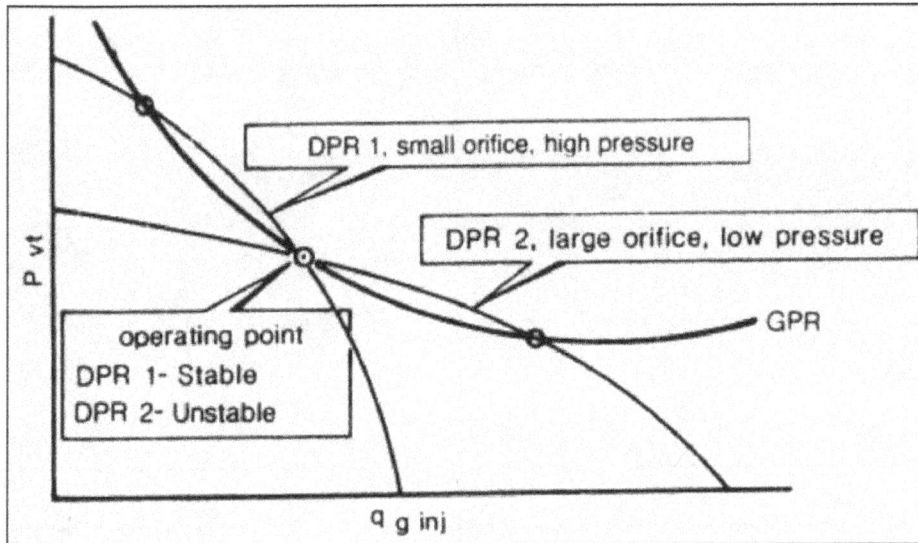

Fig. 4.24

From Figure 4.24, it is obvious that a combination of the size of the downhole orifice, as well as the orifice's upstream pressure, governs the stability of the designed state of operation. As indicated in the figure, a high injection pressure with a small size orifice yields a stable production point that lies to the left hand side of the GPR.

In many gas-lift wells, operating with a large orifice size or large unrestricted gas supply, the stable operating point lies to the right hand side of the peak of LPR curve (Figure 4.22b). Such an operation is very inefficient, as the same amount of oil can be produced with much less gas if the operating point could be stabilized to the left-hand side of the peak.

Truly, operators of wells producing with continuous gas lift are, in certain cases, facing difficulties in regulating and maintaining the production rate or the gas injection rate at a desired level (Kleyweg *et al.* 1983). In many cases, stable liquid production rates are achieved only by injecting gas at considerably higher rates than originally planned. Otherwise, the production and injection can be stabilized only by replacing the downhole orifice with a smaller one which, in turn, requires higher injection pressure.

Operators are very reluctant to replace a downhole orifice in an attempt to stabilize the production of a gas-lift well. Rather, they are tempted to adjust the surface gas injection choke first. Throttling the injection gas at the inlet to the annulus, however, can lead to another undesirable phenomenon, namely the *annulus heading.* It is a severe pressure and rate oscillation that may occur if the size of the surface orifice is too small in relation to the size of the downhole one.

The reason for the heading phenomenon is the time lag between the discharge from and refill of gas to the annulus. In a steady state flow, the rate of surface injection into the annulus equals the rate of gas discharged from the annulus into the tubing, and the pressure is steady. When a sudden decrease in the tubing pressure promotes an excessive discharge of gas at the downhole orifice, the pressure at the bottom of the annulus will drop. Due to the compressibility of the gas and the large volume of the annulus, there will be a delay in affecting the pressure at the top of the annulus. Only after the pressure at the top of the annulus starts to drop, the excessive pressure difference across the surface choke will increase the gas injection rate.

If the surface orifice is small, the gas supply to the annulus cannot compensate fast enough for the high discharge rate. Thus, the pressure at the bottom of the annulus will continue to drop. Drop of annulus pressure will start to reduce the discharge rate into the tubing. Once again, the surface choke will delay to respond and continue supplying excessive gas. This in turn will result in a build-up of annulus pressure. Under certain conditions, the described heading cycle will be repeated resulting in periodic alternation of production and injection rates. In extreme situations, the well may even die and flow alternately, similar to the behavior of wells producing with *intermittent gas lift.*

Though this section addresses only steady state aspects of continuous gas-lift production, the heading phenomena was brought up here to emphasize the importance of a proper sizing of the downhole orifice. It strongly supports the conclusion that a proper orifice sizing should account for the stability, as well as for an optimal relationship between production and injection rates. A design process combining these two design considerations is illustrated in Figure 4.22d.

Another useful tool for designing continuous gas-lift wells is the *equilibrium curve* illustrated in Figure 4.25. It has been used for many years by Shell Oil Company but received wider acceptance only after it was published in the API Gas-Lift Book (API, 1984). The curve displays the relationship between the downhole injection depth and the corresponding pressure downstream of the downhole orifice. It is, in fact, a locus of points obtained by constructing a set of gas-lift diagrams, each for a different production rate. As explained at the beginning of the section, a gas-lift diagram determines the injection depth and the corresponding downhole injection pressure for a considered production rate (Figure 4.25a). In the considered set, all the diagrams and thus all the points of the locus (Figure 4.25b) are for the same well (the same IPR) and are constructed with the same wellhead pressure and for the same gas injection rate.

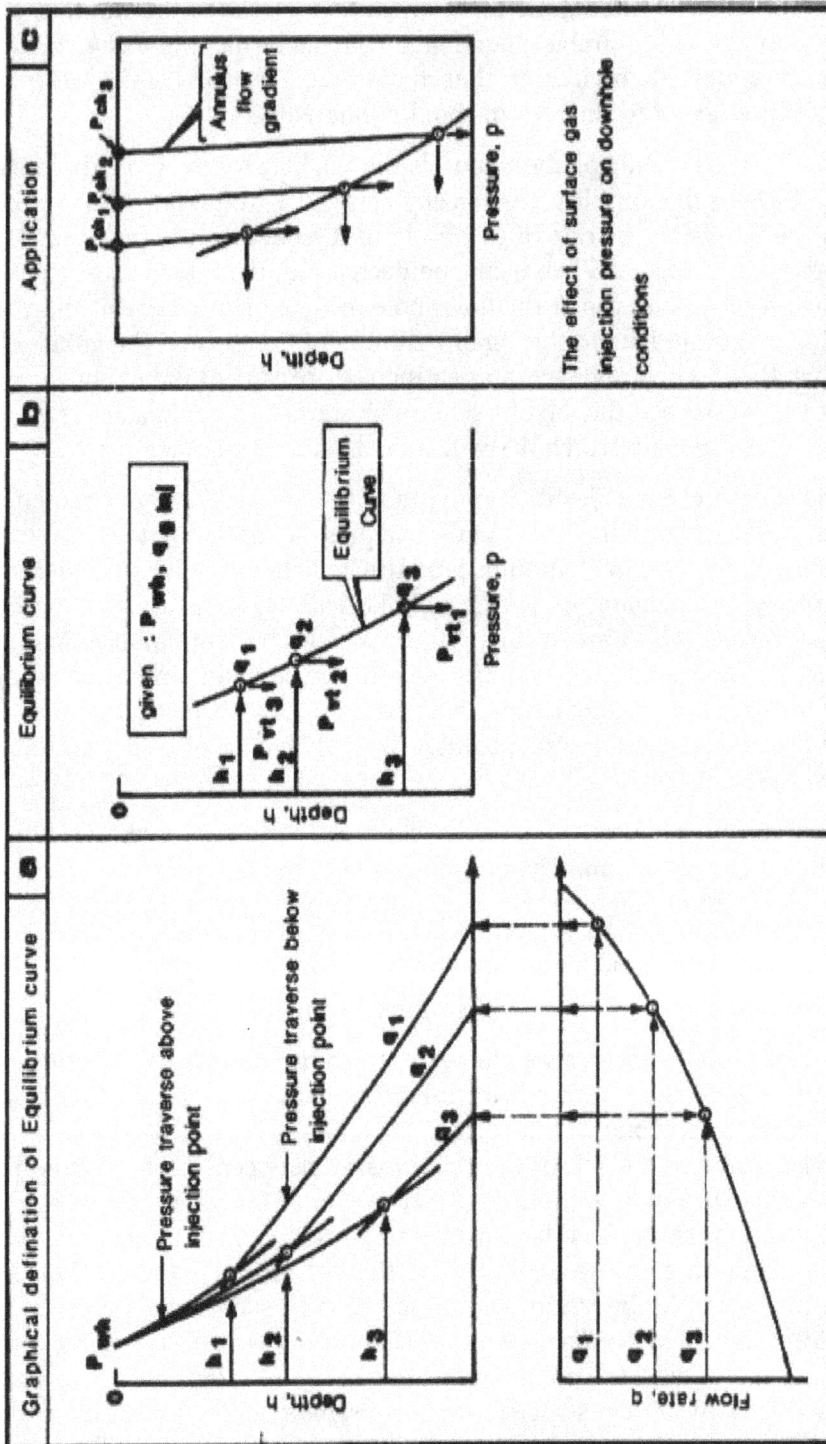

Fig. 4.25

In addition to the downhole injection pressure, a gas-lift diagram relates the injection depth to a particular production rate and a particular surface injection pressure (Figure 4.25c). Thus, each point on the locus of the injection depth versus injection pressure points corresponds uniquely to a production rate and a surface injection pressure.

The main use of the curve is in designing the start-up and the unloading of gaslift wells, a transient process beyond the scope of this book. Regarding steady state operations, the curve is very useful as a basis for selecting a continuous injection depth in gas-lift well design.

Though, by definition, the curve is a locus of solutions obtained by preparing a set of gas-lift diagrams, there is a shorter procedure to develop it that docs not require the actual construction of a set of gas-lift diagrams. This procedure is illustrated in Figure 4.26. It is based on the fact that the sought locus of injection points is related to the locus of solution points obtained by solving a set of IPR curves at various nodes along the tubing, with a set of TPR curves at the corresponding nodes. The special feature about this procedure is that the considered nodes represent the gas injection points.

Therefore, the GLR of the TPR curves in Figure 4.26 is the sum the formation GLR and the injection GLR. Each intersection point thus represents the equilibrium conditions at a corresponding gas injection point. As depth is a parameter in the plot of the locus of the equilibrium conditions at the various nodes, it is very easy to transfer the rate versus pressure locus into a depth versus pressure locus, namely equilibrium curve.

Using a sheet of tracing paper together with a set of gradient curves can speed up the construction of the equilibrium curve by the described method. A tracing paper marked with the set of horizontal lines representing the various depths under consideration allows multiple reading of pressures along the gradient curve. A step by step procedure is illustrated in Figures 4.26b and 4.26c. It consists of the following steps:

TPR

1. Draw on tracing paper a set of parallel horizontal lines spaced according to the depth scale of the available gradient curves.
2. Calculate the GLR above the gas injection point and identify the corresponding gradient curve.
3. Mark the point of the wellhead pressure on the relevant gradient curve. Start, for example with the gradient curve of the lowest rate (50 stb/day).
4. Overlay the tracing paper on the gradient curve such that the zero depth line of the tracing paper intersects the gradient curve at the point marked in step 2.
5. Read from the gradient curves the pressures at the intersection points with the depth curves of the tracing paper. These are the intake pressures at the considered depth. List the intake pressures in the upper table in Figure 4.26b.

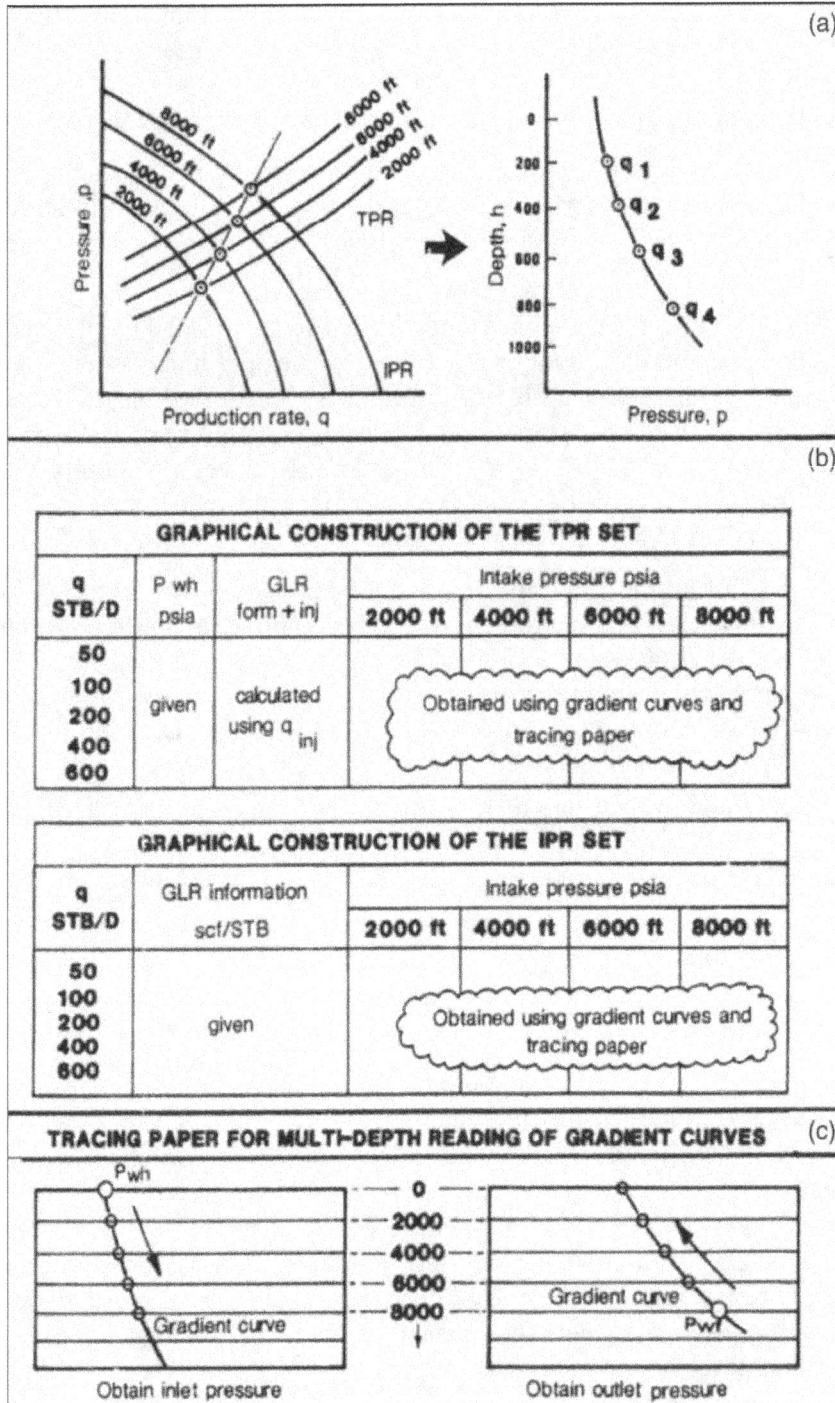

Fig. 4.26

6. Repeat the same procedure with the gradient curves of other rates (100, 200, 400, 600 stb/day) and complete the upper Table in Figure 4.26b.
7. Plot the data listed in the upper table in Figure 4.26b as a set of TPR curves for various depths.

IPR

1. Draw on tracing paper a set of parallel horizontal lines spaced according to the depth scale of the available gradient curves.
2. Plot the IPR of the well at mid-perforation depth.
3. Determine from the TPR the p_{wf} corresponding to the rate of the lowest rate gradient curve (50 stb/day).
4. Determine the gradient curve with the formation GLR and mark on it the p_{wf} pressure point.
5. Overlay the tracing paper on the gradient curve such that the mid-perforation depth line of the tracing paper intersects the gradient curve at the point marked in step 5.
6. Read the pressures at the points of intersection with the depth lines of the tracing paper and list the data in the lower table in Figure 4.26b.
7. Repeat the same procedure with the other rate gradient curves and complete the lower table in Figure 4.26b.
8. Plot the data listed in the table as a set of IPR curves.

4.12.2 *Troubleshooting*

Gas lift problems are usually associated with three areas: Inlet, outlet, and downhole. Examples of Inlet problems may be the input choke sized too large or small, fluctuating line pressure, plugged choke, etc. Outlet problems could be towards high back pressure due to a flowline choking or closed or partially closed wing or master valve, or plugged flow line. Downhole problems, of course, could include a cut-out valve, restrictions in the tubing string, or sand covered perforations. Further examples are included in following text. Often, the problem can be found on the surface. If nothing is found on the surface, a check can be made to determine whether the downhole problems are well bore problems or equipment problems. Thus the well should be trouble shoot before calling rig for workover.

4.12.2.1 *Inlet Problems*

Choke Sized Too Large

Check for casing pressure at or above design operating pressure. This can cause reopening upper pressure valves and/or excessive gas usuage. Approximate gas usages for various flow rates are included in the section on "tuning in" the well.

Choke Sized Too Small

Check for reduced fluid production as a result of insufficient gas injection. This condition can sometimes prevent the well from fully unloading. The designed gas liquid ratio can often give an indication of the choke size to use as a starting point.

Low Casing Pressure

This condition can occur due to the choke being sized too small or the choke being plugged or frozen up. Choke freezing can often be eliminated by continuous injection of methanol in the gas lift gas. A check of gas volume being injected will separate this case from low casing pressure due to a hole in the tubing or cut out valve. Verification of gauge readings to be made to confirm whether the problem is genuine.

High Casing Pressure

This condition can occur due to the choke being sized too large. Check for excessive gas usuage due to reopening upper pressure valves. If high casing pressure is accompanied by low injection gas volume, it may be possible that operating valve is partially choked or high tubing head pressure is reducing the difference between the tubing and casing (remove flow line choking or restriction). High casing pressure accompanying low injection gas volumes may also be caused by higher than anticipated temperatures raising the set pressures of pressure operated valves.

Verify Gauges

Inaccurate gauges can cause false indications of high or low casing pressures. Always check the wellhead casing and tubing pressures with a calibrated gauge.

Low Gas Volume

Check to ensure that the gas lift line valve is fully open and that the casing choke is not too small, frozen or plugged. Check to see it the available operating pressure is in the range required to open the valves. Be sure that gas volume is being delivered to the well, as nearby wells may be robbing the system – especially intermittent wells. Some times a higher than anticipated producing rate and the resulting higher temperature will cause the valve set pressure to increase and thereby restrict the gas input.

Excessive Gas Volume

This condition can be caused by the casing choke sized too large or excessive casing pressure. Check to see if the casing pressure is above the design pressure causing upper valves to be opened. A tubing leak or cut-off valve can also cause this symptom but they will generally also cause a low casing pressure.

Intermitter Problems

Intermitter cycle time should be set to get the maximum fluid volume with the minimum number of cycles. Injection duration should then be adjusted to minimize "tail gas". Avoid choking an intermitter unless absolute necessary. Check to make sure that the intermitter has not stopped—whether it be a manual wind or battery operated model. Less than 1 bbl per cycle is an indication of cycling the well too rapidly.

4.12.2.2 *Outlet Problems*

Valve Restrictions

Check to ensure that all valves at the tree and header are fully open or that an under-sized valve is not in the line (i.e. 1^2 valve in 2^2 flow line) Other restrictions may result from a smashed or crimped flow line. For example, check valve places where the pipe line crosses a road.

High Back Pressure

Wellhead pressure is transmitted to the bottom of the hole, reducing the differential into the well bore and thereby reducing production. Check to insure that no choke is in the flow-line. Even with no choke bean in a bean housing, it is usually restricted to less than full ID. Remove the bean housing, if possible. Excessive 90° turns can cause high back pressure and should be removed where feasible. High back pressure can also result from paraffin or scale buildup in the flow line. Hot oiling the line will generally remove paraffin. However, scale may or may not be able to be removed depending on the type. Where high back pressure is due to long flowlines, it may be possible to reduce the pressure by "looping" the flowline with an inactive line. The same would apply to cases where the flowline I.D. is smaller than the tubing I.D. Sometimes a partially open check valve in the flowline can cause excessive back pressure. Common flowlines can cause excessive back pressure and should be avoided, if possible. Check all possibilities and remove as many restrictions from the system as possible.

Separator Operating Pressure

The separator pressure should be maintained as low as possible for gas lift wells. Often a well may be flowing to a high or intermediate pressure system when it dies and is placed on gas lift. Insure that the well is switched to the lowest pressure system available. Sometimes an undersized orifice plate in the meter at the separator will cause high back pressure.

4.12.2.3 *Downhole Problems*

Hole in Tubing

Indicators of a hole in the tubing include abnormally low casing pressure and excess gas usage. A hole in the tubing can be confirmed by the following procedure: Equalize the tubing pressure and casing pressure by closing the wing valve with the gas lift gas on. After the pressures are equalized, shut off the gas input valve and rapidly bleed-off the casing pressure. If the tubing pressure bleeds as the casing pressure drops, then a hole is indicated. The tubing pressure will hold if no hole is present since both the check valves and gas lift valves will be in the closed position as the casing pressure bleeds to zero. A packer leak may also cause symptoms similar to a hole in the tubing.

Operating Pressure Valve by Surface Closing Pressure Method

A pressure operated valve will pass gas until the casing pressure drops to the closing pressure of the valve. As a result, the operating valve can often be estimated by shutting off the input gas and observing the pressure at which the casing holds. This pressure is the surface closing pressure of the operating valve. Closing pressure analysis assumes the tubing pressure to be zero, and single point injection. These assumptions *limit* the accuracy of this method since the tubing pressure at each valve Is never zero, and multipoint injection may be occurring. This method can be useful when used *in combination* with other data to bracket the operating valve.

Well Blowing Dry Gas

For pressure valves, check to insure that the casing pressure is not in excess of the design operating pressure, thereby causing operation from the upper valves. Insure that no hole exists in the tubing by the previously mentioned method. If the upper valves are not being held open by excess casing pressure and no hole exists, then operation is probably from the bottom valve. Additional verification can be obtained by checking the surface closing pressure as indicated above. In the case where the well is equipped with fluid valves and a pressure valve on the bottom, blowing dry gas is a positive indication of operation from the bottom valve *after* the possibility of a hole in the tubing has been eliminated. Operation from the bottom valve generally indicates a lack of feed-in. Often, it is advisable to tag bottom with wireline tools to see if the perforations have been covered by sand. When the well is equipped with a standing valve, check to insure the standing valve is not stuck in the closed position.

Well will Not Take any Input Gas

Eliminate the possibility of a frozen input choke or a closed input gas valve by measuring the pressures upstream and downstream of the choke. Also check for closed valves on the outlet side. If fluid valves were run without a pressure valve on bottom, this condition is probably an indication that all the fluid has been lifted from the tubing and not enough remains to open the valves. Check for feed-in problems. If pressure valves were run, check to see if the well started producing above the design fluid rate as the higher rate may have caused the temperature to increase sufficiently to lock-out the valves. If temperature is the problem, the well will probably produce periodically then stop. If this is not the problem, check to make sure that the valve set pressures are not too high for the available casing pressure.

Well Flowing in Heads

This condition can occur due to several causes. With pressure valves, one cause is port sizes too large as would be the case if a well initially designed for intermittent lift were placed on constant flow due to higher than anticipated fluid volumes. In this case large tubing effects are involved and the well will lift until the fluid gradient is reduced below a value that will keep the valve open. This case can also occur due to temperature interference. For example, if the well started producing at a higher than anticipated fluid rate, the temperature could increase causing the valve set pressures to increase and thereby lock them out. When the temperature cools sufficiently the valves will open again, thus creating a condition where the well would flow by heads. With pressure valves having a high tubing effect on fluid operated valves, heading can occur as a result of limited feed-in. The valves will not open until the proper fluid load has been obtained, thus creating a condition where the well will intermit itself whenever adequate feed-in is achieved. Since over or under injection can often cause a well to head, try "tuning" the well in.

Installation Stymied and will not Unload

This condition generally occurs when the fluid column is heavier than the available lift pressure. Applying injection gas pressure to the top of the fluid column (usually with a jumper line) will often drive some of the fluid column back into the formation thereby reducing the height of the fluid column being lifted and allowing unloading with the available lift pressure. The check valves prevent this fluid from being displaced back into the casing. For fluid operated valves, "rocking" the well in this fashion will often open an upper valve and permit the unloading operation to continue. Sometimes a well can be "swabbed" to allow unloading to a deeper valve. Insure that the wellhead back pressure is not excessive, or that the fluid used to kill the well for workover was not excessively heavy for the design.

Valve Hung Open

This case can be identified when the casing pressure will bleed below the surface closing pressure of any valve in the hole but tests to determine, if a hole exists show that no hole is present. Try shutting the wing valve and allowing the casing pressure to build up as high as possible, then open the wing valve rapidly. This action will create high differential pressures across the valve seat, removing any trash that may be holding it open. Repeat the process several times, if required. In some cases valves can be held open by salt deposition, and pumping several barrels of fresh water into the casing will solve the problem. If the above actions do not help, a cut out or flat valve may be the cause.

Valve Spacing Too Wide

Try "rocking" the well as indicated when the well will not unload; this will some-times allow working down to lower valves. If a high pressure gas well is nearby, using the pressure from this well may affect unloading. If the problem is severe, re-spacing, installing a pack-off gas lift valve, or shooting an orifice into the tubing to achieve a new point of operation may be the only solution.

4.12.2.4 *"Tuning-in" the Well*

Continuous Flow

Unloading a well generally requires more gas volume than producing the well. As a result, the input gas volume can be reduced once the point of operation has been reached. Since excess gas usage can be costly in terms of compression cost, it is desirable in continuous flow installations to achieve the maximum fluid production with the minimum amount of input gas. This can be accomplished by starting the well on a relatively small input choke, such as an 8/64″, and then increase the input choke size by 1/64″ increments until the maximum fluid rate is achieved. Allow the well to stabilize for 24 hours after each change before making another adjustment. If, for some reason, a flowline choke is being used, increase the size of that choke until maximum fluid is produced before increasing the gas input choke. If the total gas liquid ratio exceeds the values indicated below, it is possible that too much gas is being used.

Intermittent Flow

In intermittent lift, the cycle frequency is normally controlled by an intermitter. The intermitter opens periodically to lift an accumulated fluid slug to the surface by displacing the tubing with gas. The same amount of gas is required to displace a small slug of fluid to the surface as is required to displace a large slug of fluid. As a result, optimum performance is obtained when the well produces the greatest amount of fluid with the least number of cycles. To accomplish this the initial injection gas volume must be slightly more than required and number of injection cycles more than required. A rule of thumb is to set the cycles based on two minutes per 1000 ft of lift with the duration of gas injection based on ½ minute per 1000 ft of lift. Reduce the number of cycles per day until the most fluid is obtained with least number of cycles, then decrease the injection time until the optimum amount of fluid production is maintained with the least injection time. If one barrel or less is produced per cycle, the cycle time should be increased. Be sure that the intermitter stays open long enough to get the gas lift valve fully open. This will be indicated by a sharp drop in casing pressure. Where a two pen recorder is used, this will give a "saw tooth shape to the casing pressure line.

4.12.2.5 *Troubleshooting Tools Diagnostic*

Calculations

One method of checking gas lift performance is by calculating the operating valve. This can be accomplished by calculating surface closing pressures or by comparing the valve opening pressures with the opening forces that exist at each valve downhole due to the operating tubing and casing pressures, temperatures, etc. Although this method may not be accurate as a flowing pressure survey due to inaccuracies in the data used, it can still be valuable tool in high grading the well selection for more expensive types of diagnostic methods.

Flowing Pressure Survey

In this type of survey, a pressure bomb is run in the well under flowing conditions. A no-blow tool is run with the tools and prevents the tools from being "blown up the hole". The no-blow tool is equipped with "dogs" or slips which are activated by sudden movement up the hole. The bomb is stopped at each valve for a period of time, and records pressure. From this information, the exact point of operation can be determined as well as the actual flowing bottom hole pressure. This type of survey is the most accurate way to determine a gas lift well's performance, provided that an accurate well test is run in conjunction with the survey. A detailed procedure follows.

Well Sounding Devices

The fluid level in the annulus of a gas lift well will sometimes give an indication of the depth of lift. This method involves firing a cartridge at the surface and utilizes the principle of sound waves to determine the depth of fluid level in the annulus. Acoustic devices are fairly inexpensive when compared to flowing pressure surveys. It should be noted that for the wells with packers, it is possible for the well to have lifted down to a deeper valve while unloading, then returned to operation at a valve up the hole. The resulting fluid level in the annulus will be below the actual point of operation.

Tagging Fluid Level

Tagging the fluid level in a well with wireline tools can sometimes give an estimation of the operating valve subject to several limitations. Fluid feed-in will often raise the fluid level before the wireline tools can get down the hole. In addition, fluid fallback will always occur after the gas lift gas has been shut off. Both of these factors will cause the observed fluid level to be above the operating valve. Care should be taken to insure that the input gas valve was closed prior to closing the wing valve or the gas pressure will drive the fluid back down the hole and below the point of operation. This is certainly a questionable method.

Two Pen Recorder Charts

In order to calculate the operating valve, it is necessary to have accurate tubing and casing pressure data. Two pen recorder charts give a continuous recording of these pressures, and can be quite useful, if accompanied by an accurate well test. The two pen recorder charts can be used to optimize surface controls, locate surface problems, as well as identify downhole problems.

4.12.2.6 *Procedure for Running Flowing Bottom Hole Pressure Test where Well is Equipped with Gas Lift Valves*

Intermittent Gas Lift Wells

1. Install crown valve on well, if necessary and flow the well to the test separator for 24 hours so a stabilized production rate is known. (Test facilities should duplicate as nearly as possible normal production facilities).
2. Put well on test before running bottom-hole pressure. Test is to be for a minimum of 6 hours. Test information, 2-pen recorder charts and separator chart should be sent in with pressure traverse.
3. Pressure bomb *must* be equipped with one, or preferably two, "No-Blow" tools. Use a small diameter bomb.
4. Install lubricator and pressure recording bomb. Let well cycle one time with the bomb just below the lubricator record the wellhead pressure and ensure that the "No Blow" tools are working. Run bomb, making stops 15 feet below each gas lift valve. Be *sure to record* a *maximum and minimum pressure at each gas lift valve. Do not shut well in* while rigging up or recording flowing pressures in tubing.
5. Leave bomb on bottom for at least two complete intermitting cycles.
6. High and low tubing and casing pressure should be checked with a dead weight tester or recently calibrated 2-pen recorder.

4.12.2.7 *Continuous Flow Gas Lift Wells*

1. Install crown valve on well, if necessary and flow the well to the test separator for 24 hours so a stabilized production rate is known (Test facilities should duplicate as nearly as possible normal production facilities).
2. Put well on test before running bottom-hole pressure. Test is to be for a minimum of 6 hours. Gas and fluid test, 2-pen recorder chart and separator chart should be sent in with pressure traverse.
3. Pressure bomb *must* be equipped with one, or preferably two, "No-Blow" tools. Use a small diameter bomb.
4. Install lubricator and pressure recording bomb. Make first stop in lubricator to record wellhead pressure. Run bomb, making stops 15 feet below each gas lift valve for 3 minutes (*Do not shut well in* while rigging up or recording flowing pressures in tubing).
5. Leave bomb on bottom for at least 30 minutes, preferably at the same depth that the last static bottom-hole pressure was taken.
6. Casing pressure should be taken with a dead weight tester or recently calibrated 2-pen recorder.

Gas Lift Troubleshooting Check List

I. Inlet

Problem

1. Choke sized too large, Popping upper valves, Excessive gas usage
2. Choke sized too small, Cannot unload, Insufficient gas quantity
3. Choke plugged, Choke frozen up
4. Faintly ressure gauges—causing insufficient or excessive casing pressure
5. Intermitter, stopped, Intermitted cycle or injection time incorrect
6. Intermitter on constant flow well
7. Intermitter malfunction, other
8. Gas lift supply gas shut off
9. Line pressure down, why?_____
10. Fluctuating line pressure, why?_____
11. Other problems/remarks: _____
    Corrective Action: _____

II. Outlet

Problem

1. Master valve or wing valve closed

2. High back pressure due to:
   Flow line choke • Flowline choke body • Excessive 90 degree turns
   Long flowline • Flowline plugged or partially plugged
   Excessive canal crossings • Flowline ID smaller than tubing string

3. Valve shut at header • Restricted ID valve

4. Check valve at header leaking causing back pressure

5. Separator operating press too high;

6. Sepr orifice plate sized too small

7. Other Problems/Remarks: _____
   Corrective Action: _____

III. Downhole

Problem

1. No feed-in: Fluid standing at or below bottom valve

2. Perforations covered;

3. Fluids too light to load valves

4. Restrictions in tubing string;

5. Spacing too wide to allow unloading

6. On bottom valve-not valved deep enough;

7. Cut out valve or tubing leak

8. Flat valve • Valve plugged

9. Valve pressures set too low   • Too high

10. Salt deposits or trash in valves;

11. Leaking pack off gas lift valve

12. Excessive back pressure popping valves up the hole

13. Working as deep as possible but:

    Back pressure preventing higher rate

    Low casing pressure preventing higher rate

14. Dual gas lift:

    • One side robbing gas; • Temperature affecting other string

15. Other Problems/Remarks: _____

    Corrective Action:_____

Where to install 2-Pen Recorder

Connect casing pen line.

1. At well; not at compressor or gas distribution header
2. Downstream of input choke so that the true surface casing pressure is recorded.

Connect tubing pen line.

1. At well; not at the battery, separator, or production header
2. Upstream of choke body or other restrictions. (Even with no choke bean, less than full opening is found in most chokes).

Interpretation of 2-Pen Recorder Charts on Gas Lift Wells: The two most significant forces acting on any gas lift valves are the tubing pressure and the casing pressure. The downhole valves can be calculated and compared to the operating characteristics of the type gas lift valves in service. From this information, it is possible to estimate the point of operation. Observing the surface pressures can also give valuable information on the efficiency of the system. The charts illustrating the types of information gained by the use of 2-pen recorders are included as below.

Continuous Flow

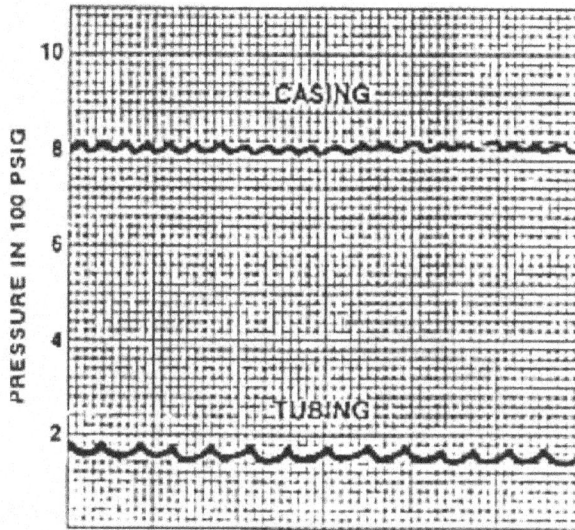

Trouble: Fluctuating gas lift line pressure. This can be caused by intermittent wells in the same system as the continuous flow wells.

Cure: This problem can be resolved by putting the continuous flow string in a separate gas supply system apart from the intermittent wells, increasing the system gas pressure, or lowering the set pressure of the gas lift valves in the continuous flow well, or increasing the storage capacity of the supply system to "dampen" out pressure fluctuations.

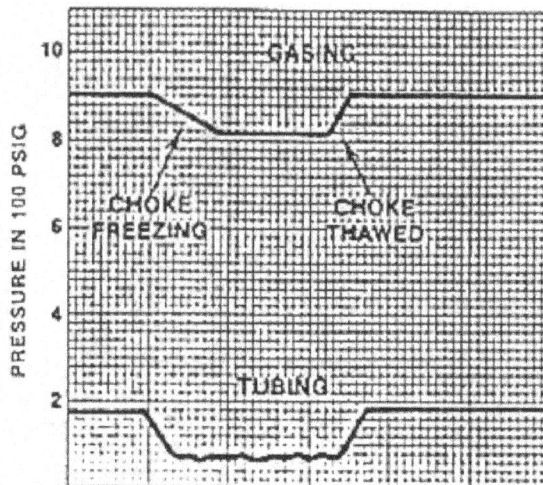

Trouble: Injection gas choke freezing.

Cure: Sometimes installing a slightly large input choke will reduce freezing. Dehydrating the lift gas, injection methanol upstream of the choke, or the use of heat exchangers may prove necessary in severe cases.

Continuous Flow

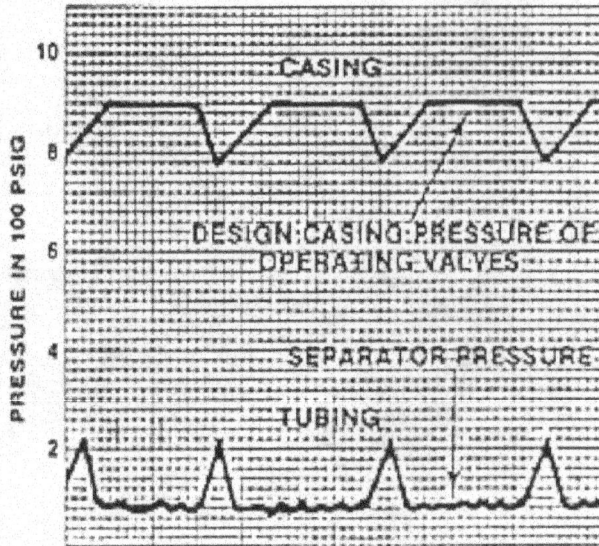

Trouble: Valve opening periodically on tubing pressure effect.

Cure: Correct well bore problems which are reselecting feed in, or redesign gas lift string for lower producing rate.

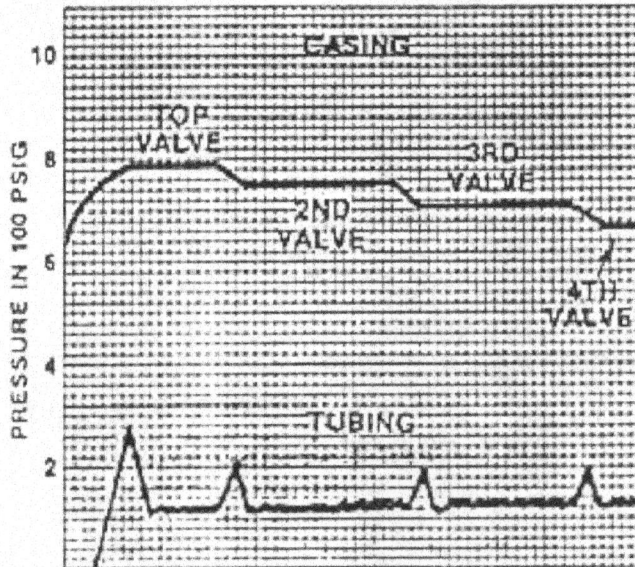

Trouble: None—well unloading.

Cure: Allow well to unload and get stabilized will test. Make adjustment based on test.

Continuous Flow

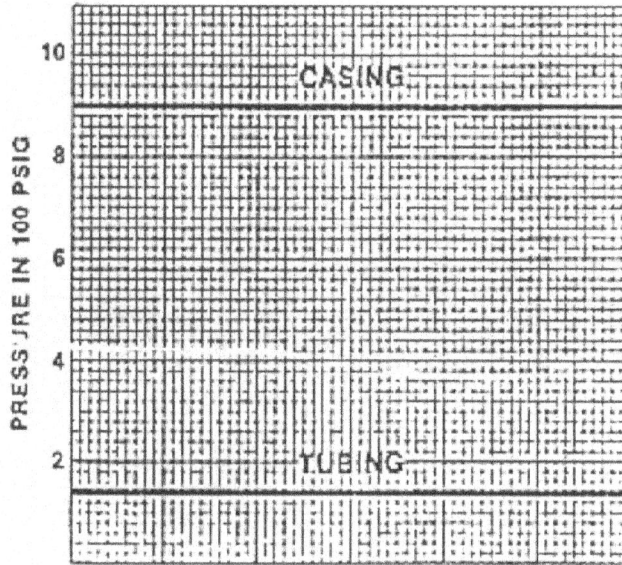

Trouble: None. Note the uniform tubing and casing pressures, and the relatively low back pressure. Horizontal flow curves are available which will indicate, if back pressure is above normal.

Cure: Leave well alone as long as production and gas liquid ratios are optimum.

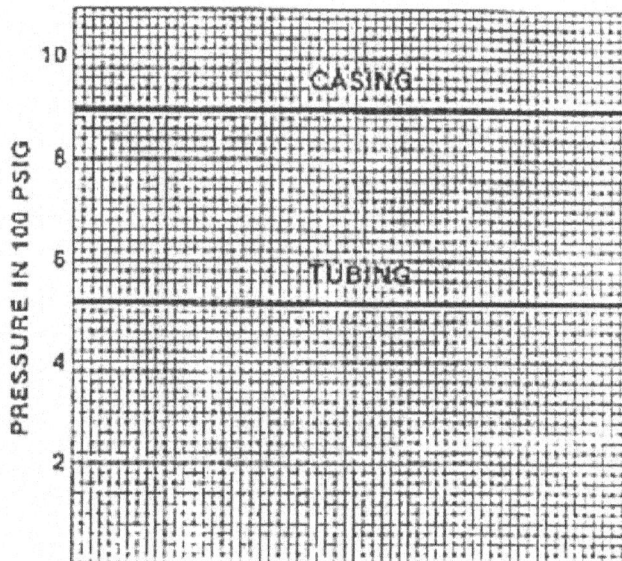

Trouble: Excessive back pressure.

Cure: Remove choke from flowline, excessive 90° turns, paraffin, scale or other restrictions flow. "Looping" or replacing existing line with a larger size line may be indicated in severe cases.

Continuous Flow

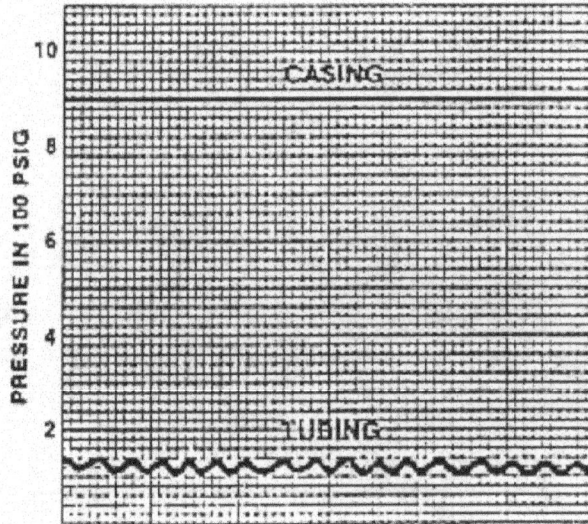

Trouble: Valve throttling.

Cure: The wavy tubing pressure line indicates valve thrilling. This condition is caused by the casing pressure being too near the valve closing pressure. A slightly larger gas input choke would eliminate the problem. If a larger input choke causes excessive gas usage, it is probably an indication of over sized ports in the gas lift valve.

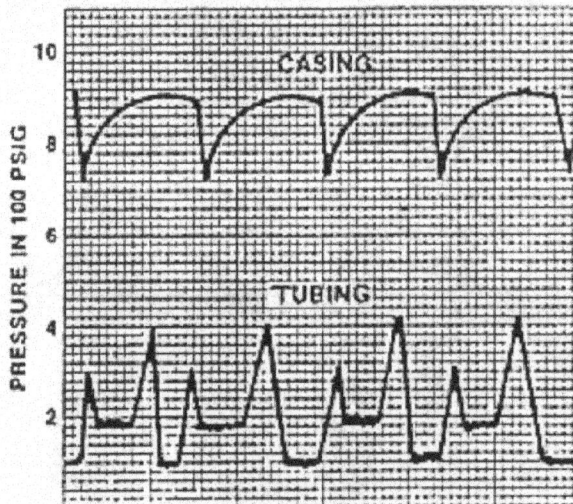

Trouble: Holes and/or parted tubing. Well produces continuously until hole or parted tubing is uncovered, causing the casing pressure to be dropped rapidly. Production is stopped until the casing pressure builds up.

Cure: Pull well and replace faulty tubing. It may be possible to locate the hole and isolate it by installing a pack off.

Intermittent Flow

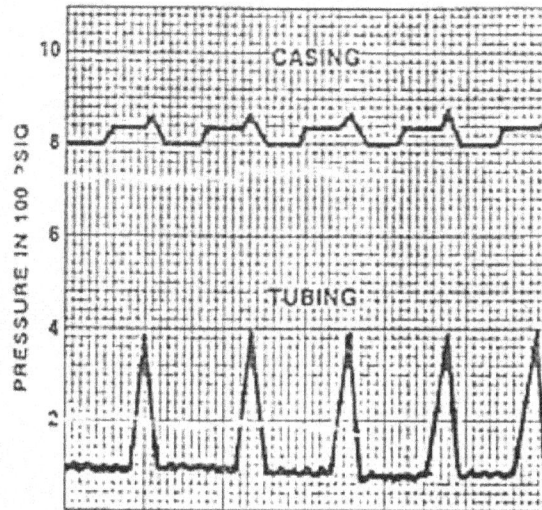

Trouble: Improper intermitter settling—the injection gas shuts off before the valve opening pressure is reached. As a result two intermitted cycles are required to open the valve. Tubing kicks show good fluid recovery.

Cure: Adjust intermitted cycle & duration of injection until maximum fluid with minimum cycles is achieved.

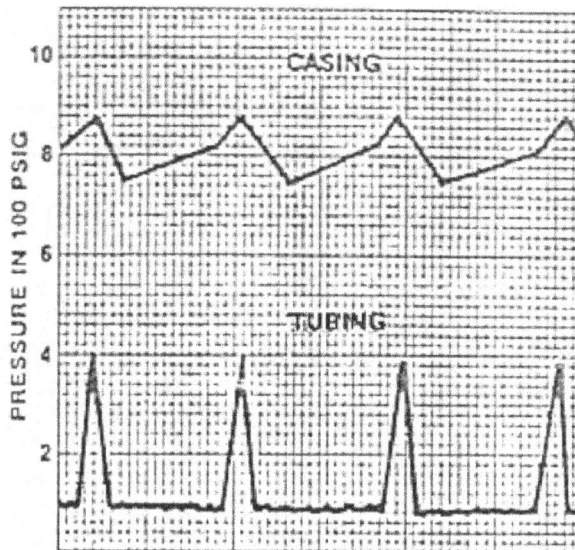

Trouble: Leaking intermitter—indicated by casing pressure build-up between cycles. Tubing kicks show good fluid recovery.

Cure: Replace seat in intermitter.

Intermittent Flow

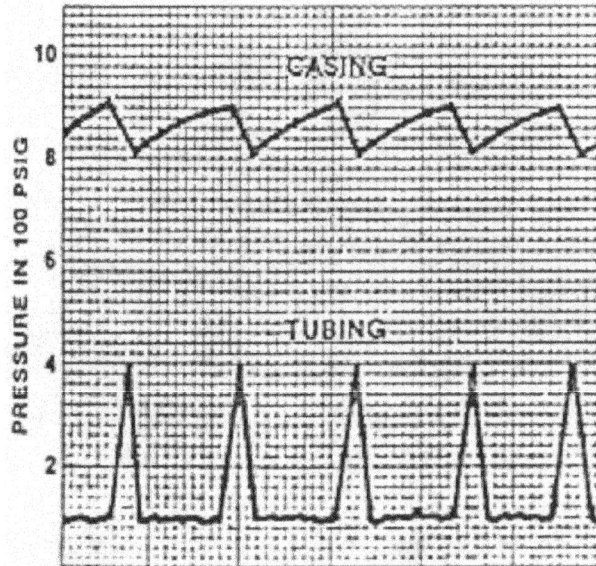

Trouble: None—Well intermitting with casing choke.

Cure: Leave well alone, if production and gas usage are optimum.

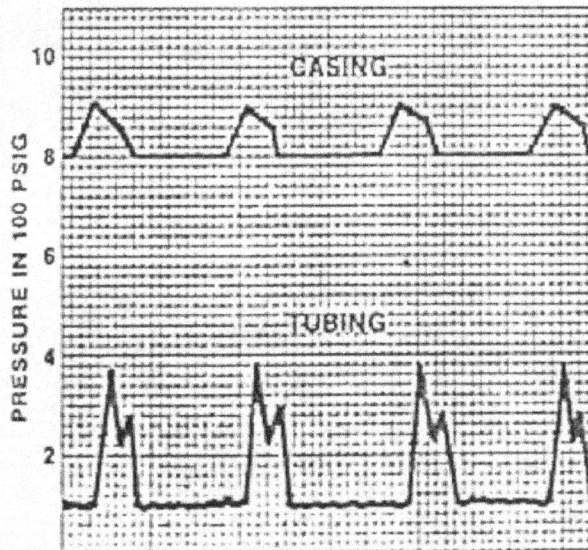

Trouble: Intermitter cycle not fast enough—well loading up dual tubing kicks and casing pressure drops indicate two valves at work.

Cure: Use faster injection cycle.

Intermittent Flow

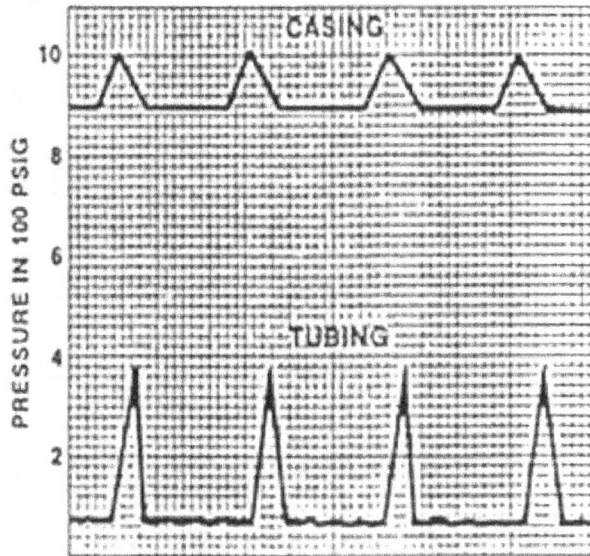

Trouble: None—Good operation. Rapid build-up and drawdown of casing pressure with a constant pressure between cycles indicates good valve operation. Thin sharp kicks on tubing pressure pen indicate good slug recovery.

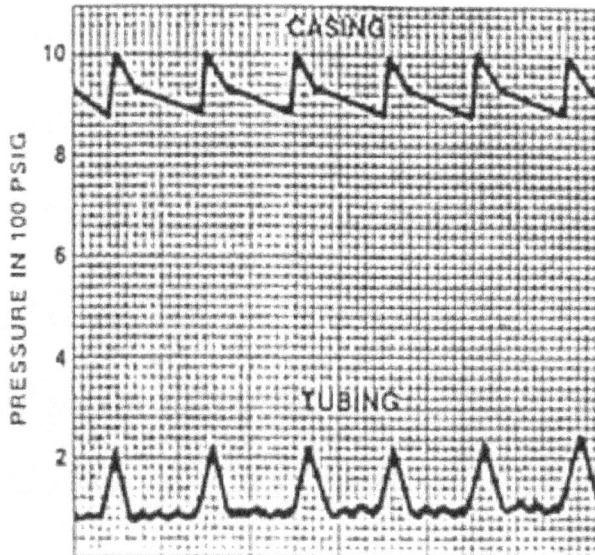

Trouble: Leaking valve indicated by casing pressure drawdown between cycles.

Cure: Attempt to clear trash (which may be preventing valve closure) from valve seat by means described in downhole problems section of this book. If that fails, it will be necessary to pull the valves, if the problem causes significant loss of production or excess gas usage.

Intermittent Flow

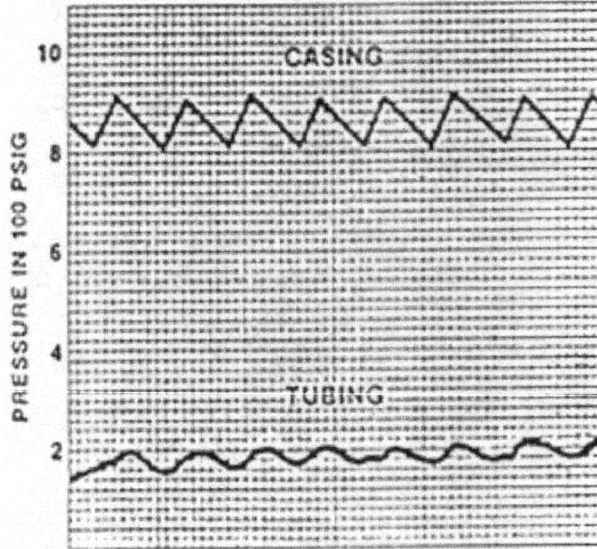

Trouble: Leak in tubing string—indicated by relatively flat tubing pressure line and excessive gas usage. Lack of tubing kicks indicates no valve action at all.

Cure: Pull and replace defective tubing.

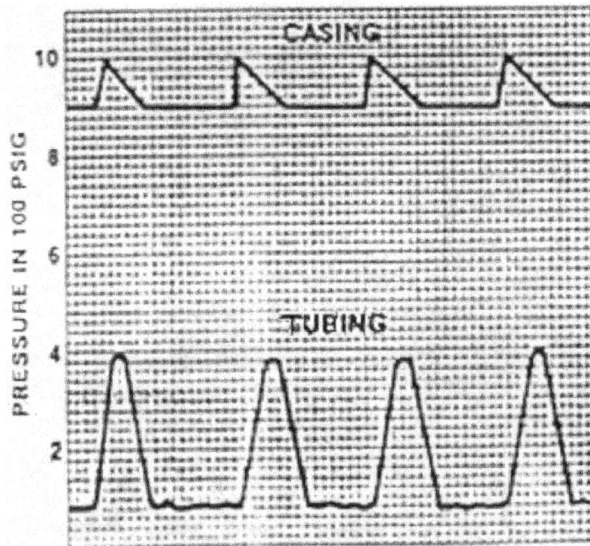

Trouble: Valve throttling close—indicates by slow casing pressure drawdown. Broad tubing pressure kicks generally are indicative of excessive gas usage and reduced fluid recovery.

Cure: This condition is generally caused by tunning valves with low come volumes of heavy springs. Try to select valves such as McMurry "SACS" which attend rapid opening and closing.

4.13 Pack-off Gas Lift Installation

The pack-off gas lift technique consists of installing a gas lift valve inside the tubing string between an upper and lower pack-off assembly (refer Figure 4.27). The two pack-offs are placed above and below a pre-perforated section of the tubing. Injection gas from the annulus enters through the tubing perforation and into the tubing string via the gas lift valve. The well fluids are produced up through the centre of the pack-off assembly. The pack-offs have sealing elements to prevent casing pressure from entering the tubing until the GLV opens. Any number of these pack-off assemblies can be placed in a well similar to a regular gas lift Installation.

The pack off gas lift system differs from the conventional gas lift in the following respects:

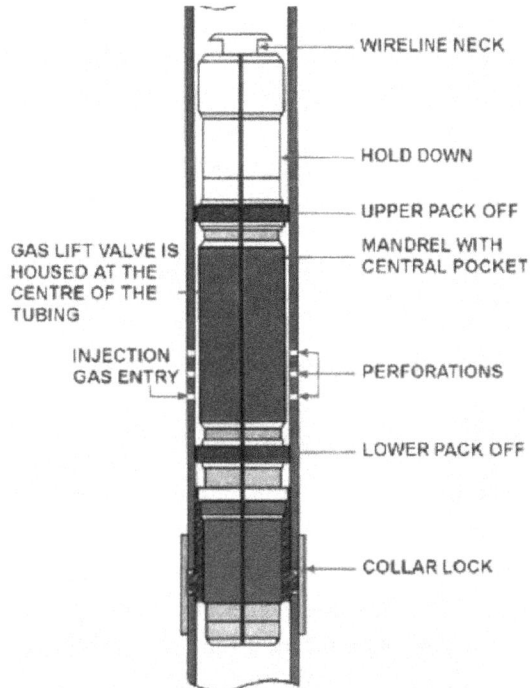

Fig. 4.27: Pack-off Gas Lift Assembly

(a) In the conventional gas lift system, side pocket gas lift mandrels are lowered with the tubing at different predetermined depths. Therefore, these mandrels are a part of the tubing. Here, the mandrels not only house the GLVS (gas lift valve), but also has a port facilitating communication of injection gas from annulus to the GLV system. The packoff gas lift mandrel, on the other hand, is not a part of the tubing string. It is placed inside the tubing with the necessary seal and stopping arrangements at its top and bottom. Although the mandrel houses the GLV, the injection gas is conveyed from the annulus to the GLV system by making a hole in the tubing at the appropriate location.

(b) In case of a wireline mandrel, GLVS are housed in one side pocket, built at one end of the mandrel. The GLV is generally of 1″ or 1½″ OD. In case of packoff installation, the GLV is placed, in general, at the center of the mandrel (hence this system is often called a "Concentric packoff gas lift system"). The GLV size ranges from less than 1″ OD to 2.5″ OD. However, in some models, the GLV is placed at the side and in such cases the mandrels are of macaroni size with less than 1″ OD GLV.

(c) The conventional gas lift system is initially installed in the wells with the help of the workover rig, whereas the pack-off gas lift system does not require manoeuvring the original tubing. For installation of pack-off system only wireline job is needed.

As per the current literature, the packoff gas lift installation system is being extensively used in offshore Gulf Coast area of USA and in other parts of the world, presumably due to the unavailability of workover rigs.

4.13.1 *A Typical Pack-off Gas Lift System*

A typical pack-off gas lift assembly consists of—upper tubing stop, upper pack-off assembly, gas lift mandrel with GLV/Check valve, Lower pack-off assembly, lower collar stop. The total pack-off gas lift system consists of a number of such assemblies (one for each GLV/Check valve) which are required to be placed in a well at the predetermined depths similar to the regular Gas Lift installation.

4.13.2 *Installation Procedure*

Installation of pack-off gas lift system is done starting from the bottom most GLV assembly to the top most one. The installation procedure for each pack-off assembly is as follows:

1. The bottom collar stop is run in and set with wireline just below the desired depth of gas injection.
2. A tubing perforator is then lowered which ultimately rests on the lower collar stop. With the aid of this perforator, a hole of dimensions around 3/16″ × 3/4″ or a circular hole of around 5/16″ is perforated in the tubing approximately 15″ above the bottom tubing stop.
3. The total assembly consisting of lower pack-off, mandrel with GLV, upper pack-off and upper tubing stop is made at the surface and the whole assembly is then run in and set on the lower collar stop. This ensures that the hole in the tubing has direct access to the GLV in the mandrel.

The successive assemblies are run in the well at the predetermined depths (above the previous set) in a similar fashion.

4.14 Gas Lift Design

Now-a-days all the gas lift designs are being done with computer. However, to understand the design of a gas lift system every engineer associated with gas lift application must understand thoroughly the intricacies and variations of gas lift design. He or she must do it either graphically or analytically by himself or herself without the assistance of computer. Then, subsequently he could accomplish the task of gas lift design for wells with the aid of computer. The design of continuous gas lift and intermittent gas lift installation is explained with the help of examples.

Data Required for Continuous Gas Lift Design: To design a continuous gas lift installation, data required are—Depth of perforation interval, Tubing and casing size, Inclination profile of the well, API gravity of oil, Formation gas-oil-ratio, Specific gravity of injection and formation gas, Specific gravity of water, Desired liquid production rate, Flowing wellhead pressure (FTHP), Injection gas pressure at well, Volume of injection gas available, Static bottom hole pressure (SBHP), Productivity index (PI) or Inflow Performance Relationship (IPR), Bottom hole temperature, Type of reservoir etc.

Type of Design Problems

In gas lift design, there are two distinct types of design problems.

1. In the first case, gas lift is to be designed and gas lift valves run with the tubing in an existing well.
2. Second case is encountered mainly in offshore operations where wireline mandrels are spaced in the tubing string on the basis of an appropriate design for later installation of gas lift valves. The dummy gas lift valves are replaced with gas lift valves when need arises to install gas lift system as per merits.

4.15 Quality Control of Gas Lift Valves

The concept of using unloading valves along with the operating valve and the various design methods are explained in this section. The quality of gas lift valves used in a gas lift well plays a very important factor in the efficient operation of the well. The failure of valves in the well due to various reasons could detrimentally affect the oil production from the well. This leads to the situation where an effective quality control is essential to evaluate the performance of different makes of gas lift valves. The two API standards applicable for gas lift valve testing are API 11 V 1 for static testing and API 11 V2 for dynamic testing.

4.15.1 *Static Tests*

The static tests can be normally considered as mandatory for accepting a lot of gas lift valves for field use. The tests which can be categorized under static are as follows:

1. Leakage through seat and stem
2. Ageing Test
3. Shelf Test
4. Probe Test.

(i) *Leakage through Seat and Stem:* In this test, the leakage through seat and stem of a gas lift valve is measured when the valve is in its closing condition. Each gas lift valve will have a closing pressure which depends on the opening pressure and port size of the valve. This test can be performed in a closed test hood facility as well as open test bench facility (Figure 4.28). In the open test bench, the valve is initially allowed to open and the opening pressure is noted down. The closing pressure is

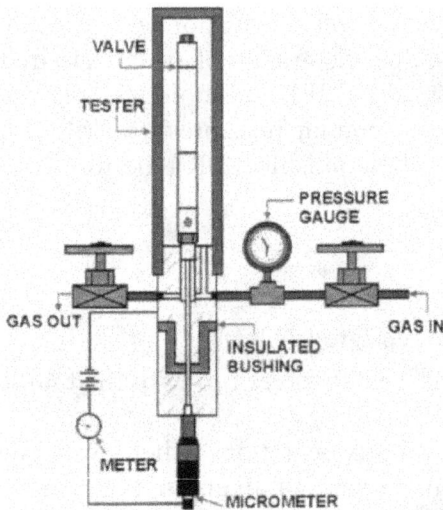

Fig. 4.28a: Typical Gas Lift Valve
Probe Test Fixture

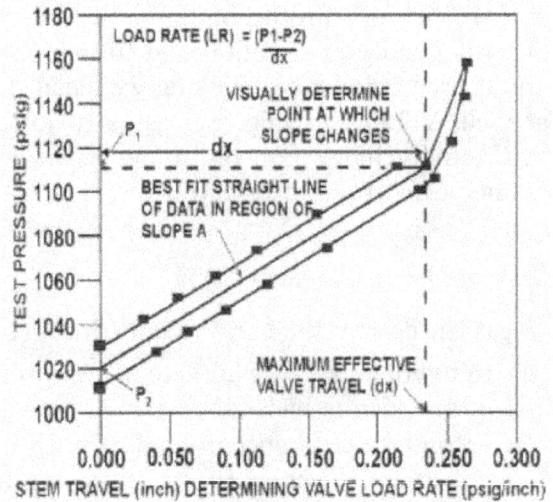

Fig. 4.28b: Stem Travel (inch) Determining Valve
Load Rate (psig/inch)

calculated based on the force balance equation i.e. $Pd = P_{TRO}(1- R)$, where $R = A_p/A_b$ The upstream pressure is down to that that of the valve closing pressure.

As per API standards, the leakage rate measured downstream of the valve should not be more than 35 scf/day for accepting a valve. Recently in India all operations is following much stringent criteria, where in the leakage in the formof gas bubbles down stream of the valve is measured by keeping a beaker of water below the valve. If the gas bubbles formed are more than 1 bubble in 5 seconds, the valve is rejected for leakage. The leakage rate for acceptance in this method is even less than API leakage criteria.

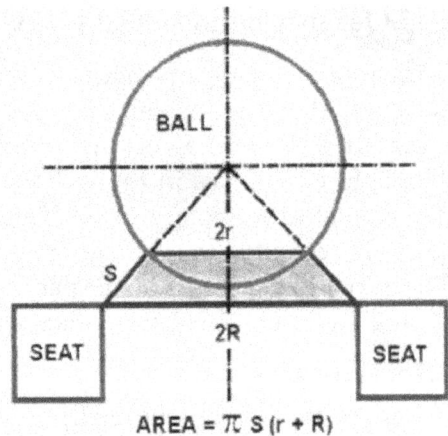

Fig. 4.28c: Lateral Surface Area of the
Frustum of a Right Circular Cone

Ageing Test

This test is mainly done to access the quality of bellows of the gas lift valve at the pressure it is to be used. The valve is initially charged to a predetermined dome pressure (normally the operating injection pressure of the field where the valves are to be used); and the corresponding opening pressure is noted. The valve are put in an Ageing chamber filled with water and then subjected to 5000 psi hydrostatic pressure for a minimum 15 minutes. The chamber is then depressurized and again pressurized to 5000 psi within a minute. This is to be repeated 3 times and then the valve opening pressure is again checked. Care should be taken to maintain the temperature of the valve during initial checking and final checking, by using a temperature controlled water bath.

The difference in opening pressures before and after this test should not be more than 5 psi for acceptance.

This test will also check the proper design of the top plug of the valve as well, since an improper design would allow water to enter the valve dome which will allow the dome pressure to increase.

Shelf Test

In this test, the opening pressure of the valves are noted down and the valves are kept on shelf for a minimum of 5 days. The opening pressure are again checked after 5 days and the difference should be within 1 % for acceptance (Pressures should be checked at same temperatures).

This test ensures that the different joints in the valve are proper and check for minor leakage which is not instantaneously detectable. The above mentioned three tests should be done on 100 % of the valves.

Probe Test

This test is to be done on randomly selected valves out of a whole lot offered for inspection. The API standard recommends this test on at least one valve of each valve configuration.

The probe test set up is as shown in Figure 4.28. In this test, a probe micrometer is used to measure the stem travel of a valve with incremental upstream pressure increase from the dome pressure, which is the closing pressure of a valve. The stem travel is measured till the maximum travel of the particular valve is reached, where the stem travel remains constant for further increase in pressure. The pressures are then reduced, preferably by the same increment till dome pressure is reached, where the valve stem should travel back to it's original close position. The upward and downward stem travel with respect to pressure is plotted and the slope of the plot gives the bellow load rate, measured in psi/inch (Figure 4.29).

This test is of utmost importance in assessing the quality of bellows by the way of load rate measurement and uniformity of bellow movement. The total travel should be higher than the equivalent stem travel for particular port sizes which ensure full area open for flow during normal operation. The stem travel with respect to different pressure ranges can be used to analyze the behaviour of a valve, i.e. whether it will operate as on orifice or it will throttle close during different casing pressure conditions.

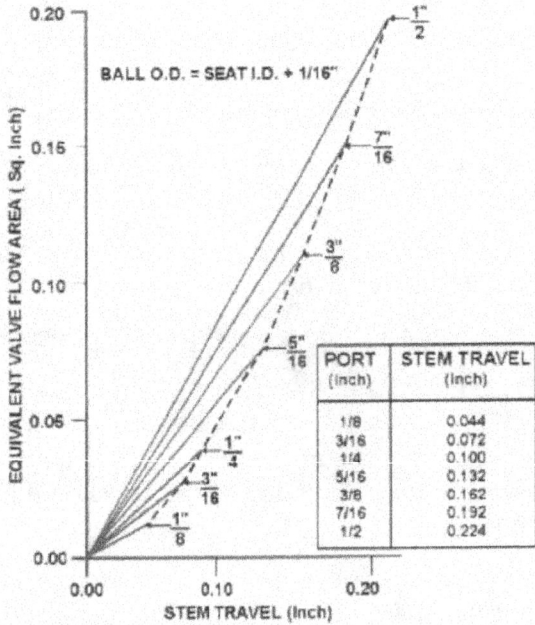

Fig. 4.29a: Equivalent Area Available *vs.* Stem Travel for Flow through Frustum of a Cone

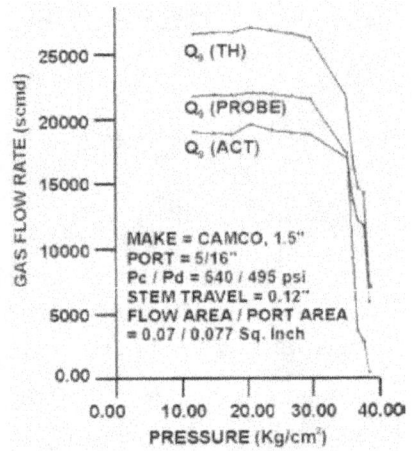

PORT (inch)	STEM TRAVEL (inch)
1/8	0.044
3/16	0.072
1/4	0.100
5/16	0.132
3/8	0.162
7/16	0.192
1/2	0.224

Fig. 4.29b: Dynamic Performance Curve of a Gas Lift Valve

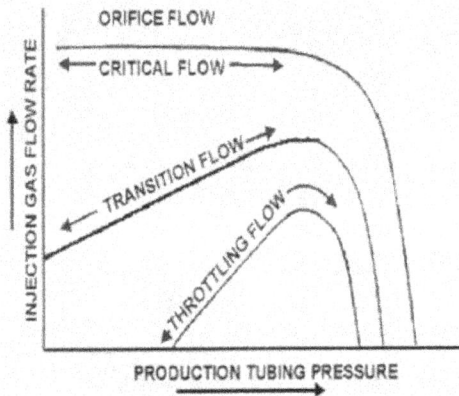

Fig. 4.30: Flow Regimes through a Gas Lift Valve

4.15.2 *Dynamic Test*

The dynamic test set up is an elaborate set up with upstream and downstream control valves, high pressure lines and different pressure, flow and temperature transmitters for online measurement of different parameters. The set up can effectively simulate different upstream and downstream pressure conditions which will vary the gas through put for a gas lift valve.

The test is done to generate flow performance curves for a GLV with different casing and tubing pressures, to predict the flow performance of the gas lift valve; i.e. whether they are actually behaving in orifice regime or throttling regime under particular conditions.

4.16 Continuous Gas Lift Design

Example

The following is one example of graphical design of a continuous gas lift system. This type of design is available in the literature of different reputed manufacturer of gas lift valve.

Flowing tubing head pressure	= 160 psi
Load fluid (or kill fluid) gradient	= 0.45 psi/ft.
Static wellhead temperature	= 74°F
Geothermal gradient	= 0.019°F/ft
Gas oil ratio (GOR)	= 300 scf/bbl
Water cut	= 50%
Gas liquid ratio (GLR)	= 150 SCF/bbl
Static Bottom hole pressure	= 2020 psi at 6000 ft.
Flowing Bottom hole pressure	= 1780 psi at 6000 ft.
Productivity Index	= 4.6 b/d/psi

Step 1

Well depth	= 6000 ft
Packer depth	= 59000 – 100 = 5800 ft.
Gas lift injection pressure	= 800 psi at zero depth.
Flowing tubing head pressure	= 160 psi
Static bottom hole pressure	= 2020 psi
Flowing bottom hole pressure	= 1780 psi
Static well head temperature	= 74°F at zero depth are marked properly on a graph sheet.

Step 2

With geothermal gradient = 19°F/100 ft. i.e. 0.019°F/ft., the bottom hole temperature at 6000 ft. = 74°F + (0.019 × 6000)°F
= 74°F + 114°F = 188°F

74°F at surface 188°F at 6000 ft. are joined to obtain temperature gradient line on the graph sheet.

Step 3

T_{avg} $= (74°F + 188°F) \times \frac{1}{2} = 131°F = (131 + 460)°R = 591°R$

Z_{avg} $= 0.84$ (assumed)

Using the formula for injection pressure at depth (P_d)

$$P_d = P_{sur} \times Ex\,P\left[\frac{\text{Gas gravity} \times \text{Depth}}{53.3 \times T_{avg} \times Z_{avg}}\right]$$

$$= 1100 \times Ex\,P\left[\frac{0.65 \times 6000}{53.3 \times 591 \times 0.84}\right]$$

$$= 1100 \times 1.158 = 1274 \text{ psi at } 6000 \text{ ft.}$$

1100 psi at the surface and 1274 psi at 6000 ft are joined to obtain injection pressure gradient line 1, on the graph sheet.

Step 4

Hagedorm and Brown curve is selected for 1500 bbl/day producing rate through 2 7/8″ tubing with producing fluid all water at average flowing temp. 140°F.

With the help of above vertical gradient curve, the minimum gradient line on the graph sheet is drawn with FTHP = 160 psi (or P_{wh}).

160 psi is located at 950 ft. on the vertical gradient curve and therefore at 6000 ft + 950 ft. 6950 ft. the pressure = 1355 psi is noted. This pressure is marked on the graph paper at 6000 ft. By joining P_{wh} and 1355 psi point at 6000 ft. the minimum gradient line is obtained on the graph sheet.

Data given:

Production rate	= 1500 BPD
Perforation interval	= 5900 – 6100 ft.
Depth	= 6000 ft.
Casing	= 5½″ : (17 – 20 ppf)
Tubing	= $2\frac{7}{8}''$; N-80, 8RD, EUE
Gas lift valve	= J-40 (Camco Co. 1″ O.D.) Port sizes: 3/16″. ¼″. 5/16″
Gas lift mandrel	= $2\frac{7}{8}''$
Injection gas quantity	= Unlimited
Specific gravity of gas	= 0.65
Gas injection pressure	= 1100 psi
Separator pressure	= 120 psi

Step 5

The zero GLR line is drawn on the graph sheet in the similar way. On the vertical gradient curve, 160 psi = P_{wh} is located at 325 ft. At 5325 ft. the pressure is read as 2520 psi. Therefore 2520 psi is marked at 5000 ft. on the graph sheet and 280 GLR line is drawn.

Step 6

The depth of the top valve (L_1) is located from,

$$L_1 = \frac{P_{inj} \text{ at surface } - P_{wh}}{\text{Kill fluid gradient}} = \frac{(1100 - 160) \text{ psi}}{0.45 \text{ psi/ft}} = 2088 \text{ ft} = 636 \text{ m}$$

L_1 is marked on the graph sheet; T_1 (Temp. at L_1) = 114°F, as obtained from temp. gradient line.

Step 7

From L_1 depth line, a second injection pressure gradient line (line 2), which is less by 50 psi than injection line 1, is drawn parallel to injection line 1.

Step 8

Approximate gas through - put (Q) through valve 1 and selection of valve port size:

From flowing gradient chart, the required GLR at L_1
= 500 scf/bbl is obtained. P_{min} at L_1 = 560 psi; P_{inj1} at L_1 = 1160 psi

$$\therefore Q \text{ calculated} = \frac{1500 \text{ b/d} \times 500 \text{ scf/bbl}}{1000} = 750 \text{ Mcfd}$$

$$Q \text{ corrected} = \frac{Q_{cal}}{0.0544\sqrt{S.G. \times (T + 460)}} = \frac{750}{0.0544\sqrt{0.65 \times (114 + 460)}}$$

$$= \frac{750}{1.05} = 714 \text{ Mcfd}$$

$$C \text{ (Port coefficient)} = \frac{Q_{corr}}{P_{up} \times K} = \frac{714}{1174 \times 047} = 1.29, \text{ where } K = 0.47$$

The valve of K is calculated from gas through-put chart as provided by P. EADS of M/s Camco:

Against $\dfrac{P_{down}}{P_{up}} = \dfrac{575}{1174} = 0.48$, K valve is obtained. Against the valve of C, the Port size of 3/16″ is calculated (For L_1) valve.

This valve can also be obtained from Thronhill-Craver equation.

Step 9

From P_{min}, a line parallel to zero GLR line is drawn, which cuts the injection line 2 at the depth of 2nd valve location L_2.

  L_2 = 3350 ft; T_2 = 137°F

Step 10

P_{wh} and P_{inj2} at L_2 are joined. It cuts L_1 depth line at P_{max} = 790 psi, at the depth L_1.

Step 11

The additional tubing effect (A.T.E.) for the valve 1 is calculated by the following formula,

  (A.T.E.)L_1 = (P_{max} at L_1 – P_{min} at L_1) T.E.F.

  [T.E.F. = Tubing effect factor, T.E.F. = 0.104 for 3/16″ size port gas lift valve]

          = (790 – 560) × 0.104 = 24 psi

Step 12

Operating pressure at the second valve, that is the valve opening pressure (P_{op}) at L_2 = P_{inj2} at L_2 – (A.T.E.) L_1 = 1150 – 24 = 1126 psi

Step 13

It is similar to step 8
Required GLR = 800 scf/bbl

  $$\frac{P_{down}}{P_{up}} = \frac{P_{min} \text{ at } L_2}{P_{inj2} \text{ at } L_2} = \frac{834}{1164} = 0.71$$

Therefore K = 0.445

  Q_{cal} = 1500 b/d × 800 scf/bbl × 1/1000 = 1200 Mcfd

  $$Q_{corr} = \frac{1200}{0.0544\sqrt{0.65(137+460)}} = \frac{1200}{1.07} = 1120 \text{ Mcfd}$$

∴ $$C = \frac{1120}{1164 \times .445} = 2.16$$

Therefore, ¼″ size for second gas lift valve is selected.

Step 14

Parallel to injection line 2, a line is drawn from L_2 depth, which is less than line 2 pressure by 24 psi. This line is injection line 3.

From P_{min} at L_2, a line is drawn parallel to zero GLR line, which cuts injection line 3 at the third valve location (L_3).

L_3 = 4060 ft.; P_{min} at L_3 = 960 psi; P_{inj3} at L_3 = 1145 psi T_3 = 151°F.

From P_{inj3} at L_3 and P_{wh}, P_{max} at L_2 = 970 psi is found as in earlier step.

Step 15

(A.T.E.) L_2 = $[P_{max}$ at $L_2 - P_{min}$ at $L_2]$ T.E.F.

$\qquad$ = $(970 - 820) \times 0.196$ = 30 psi

P_{op} at L_3 $\quad$ = P_{inj3} at L_3 − (A.T.E.) L_2 = 1145 − 30 = 1115 psi

Step 16

$$\frac{P_{down}}{P_{up}} = \frac{P_{min} \text{ at } L_3}{P_{inj3} \text{ at } L_3} = \frac{975}{1160} = 0.84$$

Therefore K = 0.4

$\qquad Q_{cal}$ = 1500 bbl/day × 800 scf/bbl × 1/1000 = 1200 Mcfd

$$Q_{corr} = \frac{1200}{0.0544\sqrt{0.65\,(151+460)}} = \frac{1200}{1.08} = 1111 \text{ Mcfd}$$

$\therefore \qquad C = \dfrac{1111}{1160 \times 0.4} = 2.39$

Gas lift valve port size ¼″ is selected (for L_3 valve).

Step 17

The injection line 4 is drawn from L_3 depth and parallel to injection line 3. It is less than injection line 3 pressure by 30 psi.

From P_{min} at L_3, the line parallel to zero GLR, line is drawn to obtain L_4 = 4450 ft., as in the previous step.

$\qquad P_{min}$ at L_4 = 1040 psi; T_4 = 158°F; P_{inj4} at L_4 = 1130 psi

From P_{inj4} at L_4 to P_{wh}, a line is drawn to find P_{max} at $L_3 = 1050$ psi.

Step 18

(A.T.E.) $L_3 = [P_{max}$ at $L_3 - P_{min}$ at $L_3]$ T.E.F.

$= (1050 - 960) \times 0.196 = 18$ psi

P_{op} at $L_4 = P_{inj4}$ at $L_4 - $ (A.T.E.) $L_3 = 1130 - 18 = 1112$ psi

Step 19

$$\frac{P_{down}}{P_{up}} = \frac{P_{min} \text{ at } L_4}{P_{inj4} \text{ at } L_4} = \frac{1054}{1144} = 0.92$$

Therefore $K = 0.27$

$Q_{cal} = 1500$ bbl/day $\times 800$ scf/bbl $\times 1/1000 = 1200$ Mcfd

$$Q_{corr} = \frac{1200}{0.0544\sqrt{0.65(158+460)}} = \frac{1200}{1.09} = 1100 \text{ Mcfd}$$

$$\therefore \quad C = \frac{1100}{1144 \times 0.27} = 3.56$$

Gas lift valve port size 5/16″ is selected (for L_4 valve).

Step 20

Parallel to injection line 4, injection line 5 is drawn from L_4 depth, which is less than injection line 4 pressure by 18 psi.

From P_{min} at L_4, the line is drawn parallel to zero GLR line, which cuts P_{inj5} at the fifth gas lift valve location L_5.

$L_5 = 4600$ ft.; P_{min} at $L_5 = 1060$ psi; $T_5 = 162°F$; P_{inj5} at $L_5 = 1100$ psi.

P_{max} at $L_4 = 1070$ psi is found as in the previous step.

Step 21

(A.T.E.) $L_4 = [P_{max}$ at $L_4 - P_{min}$ at $L_4]$ T.E.F.

$= (1070 - 1040) \times 0.342 = 11$ psi

P_{op} at $L_5 = P_{inj5}$ at $L_5 - $ (A.T.E.) $L_4 = 1118 - 11 = 1107$ psi

Step 22

$$\frac{P_{down}}{P_{up}} = \frac{P_{min} \text{ at } L_5}{P_{inj5} \text{ at } L_5} = \frac{1074}{1114} = 0.96$$

Therefore $K = 0.195$

$$Q_{cal} = 1500 \text{ bbl/day} \times 800 \text{ scf/bbl} \times 1/1000 = 1200 \text{ Mcfd}$$

$$Q_{corr} = \frac{1200}{0.0544\sqrt{0.65 \,(162 + 460)}} = \frac{1200}{1.09} = 1100 \text{ Mcfd}$$

$$\therefore \quad C = \frac{1100}{1114 \times 0.195} = 5$$

Gas lift valve (L_5) port size 5/16″ is selected.

Step 23

Since the difference of depth of 5th and 4th valve $= 4600 - 4450 = 150$ ft., the difference of depth of 6th and 5th valve will be less than 150 ft.

It is generally accepted that the difference of depth of successive valve should not be less than 200 ft.

A final table is made with all the relevant data as obtained from previous steps.

Step 24

The bellows charge pressure at well temperature (P_{bt}) is calculated by using equation, as below:

$$P_{bt} = (P_{op} \text{ at } L) \,(1 - A_v/A_b) + P_{min} \text{ at } L) \,(A_v/A_b)$$

Where, $(1 - A_v/A_b)$ and (A_v/A_b) values are supplied by gas lift valve manufacturer.

For, J-40; 3/16″ valve Port : $(1 - A_v/A_b) = 0.906$; $(A_v/A_b) = 0.094$
For, J-40; ¼″ valve Port : $(1 - A_v/A_b) = 0.836$; $(A_v/A_b) = 0.164$
For, J-40; 5/16″ valve Port : $(1 - A_v/A_b) = 0.745$; $(A_v/A_b) = 0.255$

Therefore,

P_{bt} for $L_1 = 1110 \times 0.906 + 560 \times 0.094 = 1005.66 + 52.64$
    $= \mathbf{1058.30}$ psi at 114°F.

P_{bt} for $L_2 = 1126 \times 0.836 + 820 \times 0.164 = 941.34 + 134.48$
    $= \mathbf{1075.82}$ psi at 137°F

P_{bt} for $L_3 = 1115 \times 0.836 + 960 \times 0.164 = 932.14 + 157.44$
$\qquad = \mathbf{1089.58}$ psi at $151\,°F$

P_{bt} for $L_4 = 1112 \times 0.745 + 1040 \times 0.255 = 828.44 + 265.2$
$\qquad = \mathbf{1093.64}$ psi at $158\,°F$

P_{bt} for $L_5 = 1107 \times 0.745 + 1070 \times 0.255 = 824.72 + 272.85$
$\qquad = \mathbf{1097.57}$ psi at $162\,°F$

Step 25

The bellows charge pressure (P_b) at test bench ($60\,°F$) is calculated from $P_b = P_{bt} \times C_t$.

Where, $C_t = \dfrac{P_b}{P_{bt}}$ and C_t values are provided by the manufactuer

Therefore,

P_b for $L_1 = 1058.30 \times 0.896 = 948.24$ psi
P_b for $L_2 = 1075.82 \times 0.858 = 923.05$ psi
P_b for $L_3 = 1089.58 \times 0.836 = 910.89$ psi
P_b for $L_4 = 1093.64 \times 0.826 = 903.35$ psi
P_b for $L_5 = 1097.57 \times 0.820 = 900.01$ psi

Step 26

The test rack opening pressure of the valve at $60\,°F$ (P_{TRO} or P_{vo}) is calculated

From $P_{TRO} = \dfrac{P_b}{(1 - A_v / A_b)}$

Therefore,

P_{TRO} for $L_1 = \dfrac{948.24}{0.906} = 1046.62$ psi; Adjusted P_{TRO} (at $60\,°F$) $\rightarrow 1050$ psi (at $60\,°F$)

P_{TRO} for $L_2 = \dfrac{923.05}{0.836} = 1104.13$ psi; Adjusted P_{TRO} (at $60\,°F$) $\rightarrow 1045$ psi (at $60\,°F$)

P_{TRO} for $L_3 = \dfrac{910.89}{0.836} = 1089.58$ psi; Adjusted P_{TRO} (at $60\,°F$) $\rightarrow 1040$ psi (at $60\,°F$)

P_{TRO} for $L_4 = \dfrac{903.35}{0.745} = 1212.55$ psi; Adjusted P_{TRO} (at $60\,°F$) $\rightarrow 1035$ psi (at $60\,°F$)

P_{TRO} for $L_5 = \dfrac{900.01}{0.745} = 1208.06$ psi; Adjusted P_{TRO} (at $60\,°F$) $\rightarrow 1025$ psi (at $60\,°F$)

Step 27

A 2nd table is made for the convenience of calibrating gas lift valves and numbering them before despatching the valves to the field for installation.

Critical Analysis of the Design

1. There is a scope of deeper installation of gas lift valves, upto a maximum limit of packer depth, provided the injection pressure is more than 1100 psi.
2. Also, lowering of P_{Wh}, valves can be lowered slightly more deep, if minimum gradient line is considered for gas lift design.

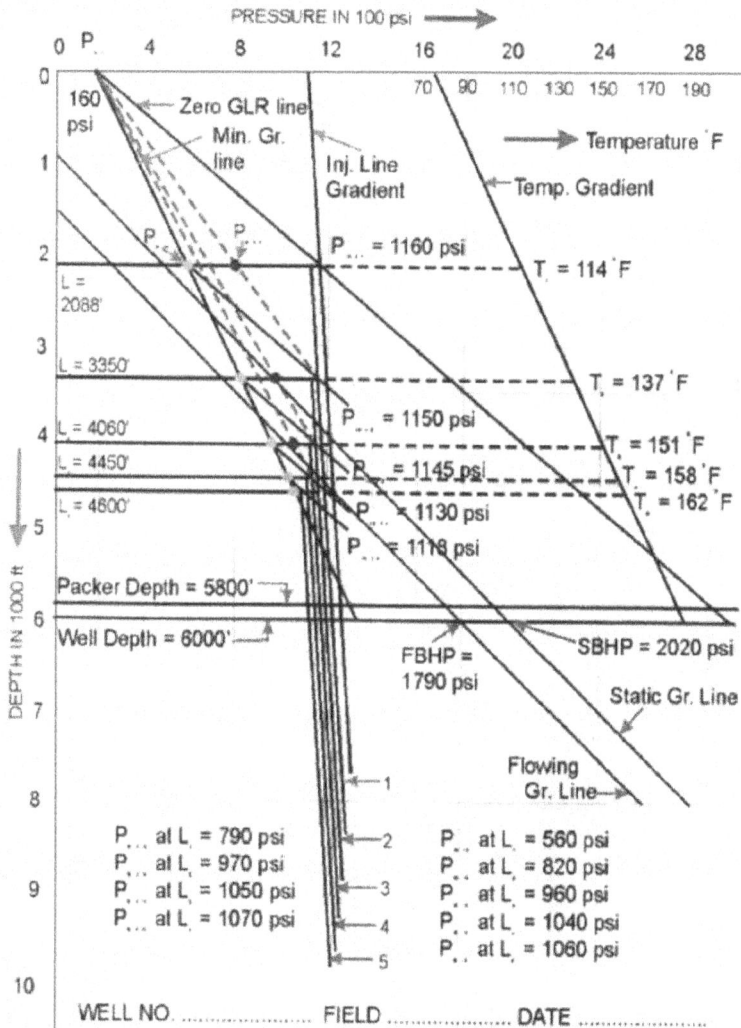

Fig. 4.31: Continuous Gas Lift Design

3. While operating the gas lift system against the designed $P_{Wh} = 160$ psi, if tubing pressure, say, can be reduced to $P_{Wh} = 100$ psi, there can be three likely possibilities.

   (i) Quantity of gas injection will be reduced.

   (ii) Quantity of fluid production will increase.

   (iii) Both the above can be achieved.

4. This gas lift valve design is for one rate of fluid production i.e. 1500 bbl/day. To make the gas lift design more flexible in handling two to three rates of production, for example, 1500 b/d, 800 b/d, 200 b/d, this can also be done in the similar way. In this case, it is required to proceed the design first with 1500 b/d, then 800 b/d and finally 200 b/d. A software developed in India by one of the authors based on this named "GLIDE" can provide this flexible design.

5. Flowing gradient line, as drawn from FBHP, indicates that L_3 will be the operating valve, with L_1 and L_2 closed and L_4 and L_5 opened. However, if the actual PI is less than the estimated PI, L_4 or L_5 will be the operating valve.

6. While the well will be on steady production with continuous gas lift, the injection gas quantity required minus wells gas.

7. While doing calibration of gas lift valves following must be ensured.

   (i) The gas lift valves must pass successfully the ageing test, shelf test, leakage test and through-put test.

   (ii) For connecting the valves on the conventional mandrels, the connecting joints of the gas lift valve and mandrel must be perfectly leak-proof.

   (iii) Gas lift valve number and depth of installation must be properly written on each mandrels before dispatching them to fields for installation.

Table 1

Well No.: X_1 Field: India   Type of Valve: ....   Date: 20-1-2004

Valve No.	Port Size in Inch	Depth of Valve in ft.	Temp. at Valve Depth °F	P_{min} at L psi	Temp. Correction Factor (C_t)	Inject. Press. at L psi	Inject. Press. Less by 50 psi at L psi	A.T.E. psi	Summation of A.T.E. psi	Pop. at L psi
L_1	3/16	2088	114	560	0.896	1160	1110	24	–	1110
L_2	¼	3350	137	820	0.858	1200	1150	30	24	1126
L_3	¼	4060	151	960	0.836	1219	1169	18	54	1115
L_4	5/16	4450	158	1040	0.826	1234	1184	11	72	1112
L_5	5/16	4600	162	1070	0.820	1240	1190	–	83	1107

Table 2: Continuous Gas Lift Design

Well No. ....   Field ....   Type of Valve ....   Date ....

Valve No.	Valve Type	Port Size in Inch	Depth (ft.)	Dpeth (mts)	P_{TRO} or P_{vo} psi
L_1	J-40 (1" O.D.)	3/16	2088	635	1050
L_2	J-40 (1" O.D.)	¼	3350	1019	1045
L_3	J-40 (1" O.D.)	¼	4060	1234	1040
L_4	J-40 (1" O.D.)	5/16	4450	1353	1035
L_5	J-40 (1" O.D.)	5/16	4600	1400	1025

4.17 Design of Intermittent Gas Lift

The design of Intermittent Gas-lift can be done either in the multipoint gas injection fashion or in the single point gas injection system. In the multipoint injection system, as the liquid slug moves up due to the gas injection from bottom valve, the upper gas lift valves will open allowing some gas entry into the tubing that helps in lift efficiency. In single point injection, only the bottom valve will operate during each cycle of injection gas, Generally for moderate volume of production, multipoint injection is preferred and for a very low producing well, single point injection system is adopted, Multipoint, however, has other advantages like it reduce paraffin accumulation in the tubing etc. The design of multipoint and single point differs materially with the assumed surface closing pressure (P_{vc}) of gas lift valve. If the values of P_{vc} for the successive lower valves are taken same or with a little difference, multipoint design system is obtained. If the values of P_{vc} for successive valves are comparatively large, then single point injection design will result.

The following problem illustrates the design calculation for an ideal multipoint gas injection system.

Data Given

Tubing size	: $2\frac{3}{8}''$ O.D. (1.995" I.D.): EUE 8RD: N-80
Casing size	: $5\frac{1}{2}''$ (20–23 PPF: N-80)
Perforation depth	: 6480 ft. – 6520 ft.
Mid-Perforation depth	: 6500 ft.
Operating injection pressure	: 800 psig
Well head flowing pressure (P_{wh})	: 100 psig
Temp. of gas injection at surface	: 86°F
Temp. bottom hole at 6500 ft.	: 160°F

Static fluid gradient	:	0.45 psi/ft.
Specific gravity of gas	:	0.65
Reservoir pressure (PR)	:	1700 psig
Bubble point pressure =		
Reservoir pressure	:	1700 psig
Water cut	:	30%
API gravity of oil	:	35°
Average P.I.	:	0.2 bbl/psi
Desired liquid production	:	160 b/d
Type of gas lift valve	:	J–20 (Camco schlmbegr, make)
Gas lift valve port size	:	7/8″

The stepwise design procedure for intermittent gas lift is as follows:

Step 1

Depth of mid-perforation and depth of packer are marked on the graph paper. Reservoir pressure (P_R) = 1700 psi is marked on the 6500 ft. depth line.

Step 2

From, $\quad \text{\AA{}PI} = \dfrac{141.5}{\text{Sp. Gr}} - 131.5$

Sp. Gr. of oil $= \dfrac{141.5}{\text{\AA{}PI} + 131.5} = \dfrac{141.5}{35 + 131.5} = 0.85$

Considering 30% water cut and with Sp.Gr. of water = 1, the composite specific gravity of the produced fluid $= \dfrac{0.85 \times 0.7 + 1 \times 0.3}{0.7 + 0.3} = \dfrac{0.595 + 0.3}{1} = 0.895$

That is,

Composite well fluid gradient = 0.895 kg/cm²/10 m $= \dfrac{0.895 \times 14.223}{10 \times 3.28}$ psi/ft.

$\qquad\qquad\qquad\qquad\qquad\qquad\qquad\qquad\qquad\qquad = \mathbf{0.388}$ psi/ft.

Static fluid gradient is drawn from 1700 psi upwards with 0.388 psi/ft.

Step 3

Flowing wellhead pressure (P_{wh}) = 100 psig is marked on the zero depth line.

The spacing factor (S.F.) in psi/foot is required to calculate the valve depths. During intermittent lift operation, liquid fall back, fluid transfer along with fluid production in the idle period, which determine the valves of S.F. Generally all gas lift valve manufactures use the valve of S.F. = 0.04 psi/ft. for locating the depth of 2nd valve and gradually increase the valves of S.F. for determining the successive lower valves. The maximum valve of S.F. depends on the estimated rate of fluid production and production tubing size. Sometimes, for a very low P.I. well, the depth of first two or three valves from 2nd valve onward are calculated by using the minimum valve of S.F., that is S.F. = 0.04 psi/ft.

A common practice is to use S.F. = 0.04 pis/ft. to obtain the depth of 2nd valve and to use increasing valves of S.F. for successively lower valves. For determining the maximum valve of S.F., it is considered safe to use a little higher valve of S.F. than will obtained from the published chart (S.F. for various fluid productions and tubing sizes).

A line is drawn form P_{wh} with S.F. = 0.04 psi/ft., which cuts the static gradient line at 'A'.

Step 4

For 160 bbl/day production and with average P.I. = 0.2 b/d/psi, and using equation, P.I. = $\dfrac{Q}{P_r - P_f}$,

where P_f = flowing bottom hole pressure.

$$P_r - P_f = \frac{Q}{P.I.} \text{ or, } P_f = P_r - \frac{Q}{P.I.} = 1700 - \frac{160}{0.2}$$

Or, P_f = 1700 – 800 = 900 psig

P_f = 900 psig is marked on 6500 ft. line and flowing gradient line is drawn parallel to the static gradient line.

Step 5

For determining the maximum S.F. at 6500 ft., it is approximately estimated that it is possible to give a maximum drawdown of 1500 psi, so by using the equation,

Q_{max} = P.I. × drawdown = 0.2 × 1500 = 300 b/d.

Using the chart S.F. (psi/ft.) vs. Q(b/d) for 2 3/8″ tubing size.

The max. S.F. = 0.086 ≅ **0.09**, which is the next whole number.

With 0.09 psi/ft gradient a line is drawn from P_{wh} which cuts 6500 ft. depth line at "*B*". It also cut P_f gradient line at "*C*".

Step 6

Point "*A*" and "*C*" are joined by a line which is called "Intermediate spacing factor line".

Step 7

Operating injection pressure = 800 psig is marked on the zero depth line. The gas pressure gradient is calculated by using,

$$P \text{ at depth } = P \text{ at surface Exp.} \left[\frac{(\text{Gas Gravity}) \times (\text{Length of gas column})}{(53.3) \times \text{Tavg} \times 2} \right]$$

$$T_{avg} = \frac{86 + 160}{2} = 123°F = (123 + 460)°R$$

Gas gravity = 0.65

Length of gas column = 65000 ft; P at surface = 800 psig = 814.7 psia

$$Z = 0.85 \text{ (assumed)}$$

$$\therefore P \text{ at depth} = 814.7 \text{ Exp.} \left[\frac{0.65 \times 6500}{53.3 \times (123 + 460) \times 0.85} \right] = 956 \text{ psia} = 941 \text{ psig at 6500 ft.}$$

The gas gradient line is drawn from 800 psig point.

Step 8

Temp. at zero depth = 86°F and 6500 ft. 160°F are marked and both are joined by a line to provide temperature gradient line (i.e. geothermal gradient line).

Step 9

Kill fluid gradient line = 0.45 psi/ft is drawn from P_{wh} which can be called zero GLR line.

Step 10

The depth (4) of the top valve (L_1 valve), is calculated by using

$$L_1 = \frac{\text{Operating Inj. Press.} - P_{wh}}{\text{Kill Fluid Gradient}} = \frac{(800 - 100) \text{ psi}}{(0.45) \text{ psi/ft}} = \mathbf{1555 \text{ ft.}}$$

A horizontal line is drawn at L_1 level.

Step 11

Another inj. Gas gradient line *MN*, which is De inj. Gas gradient minus 100 psi, is drawn parallel to *DE*. This '*MN*' line is for determining the depths of 2nd valve onward.

L_1 valve depth cuts P_{wh}–A line at q. From q_1 a line is drawn parallel to kill fluid gradient, which cuts MN at W. So. W point is the depth of 2nd valve. L_2 depth line cuts P_{wh}–C line at r. From r, a line is drawn parallel to kill fluid gradient, which cuts MN line at x. Point x is the dept of 3rd valve. In this manner parallal line are drawn from s, t, u and v and points y, z, g and h are obtained points y, z, g and h are the depth of 4th, 5th, 6th and 7th valve respectively.

It is noted that the location of 7th valve is located just below the packer and so, this valve is relocated just above the packer depth.

Step 12

All the depth of valves, temperature and injection pressure of valve depths are marked on the graph. A table is prepared accordingly.

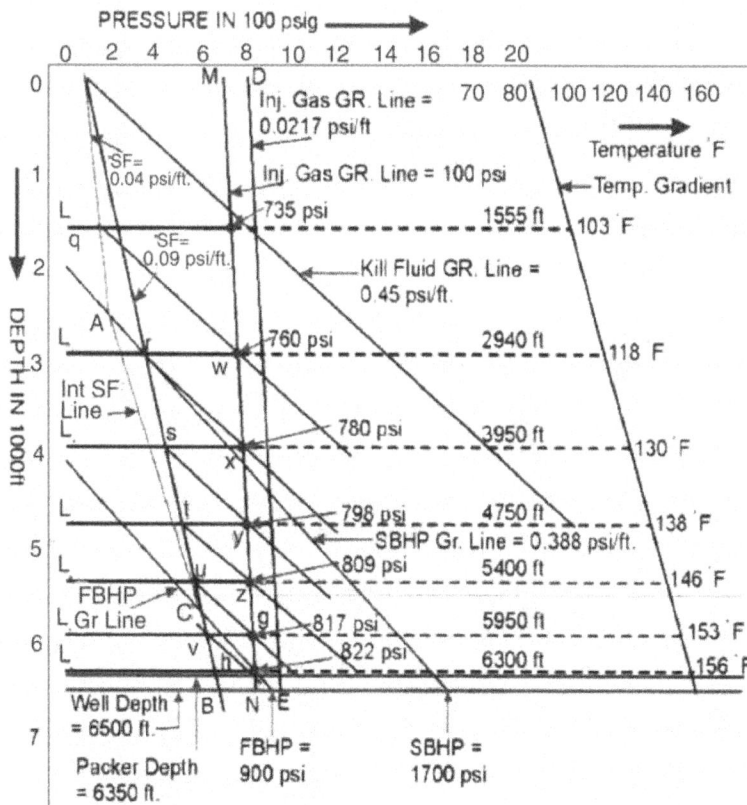

Fig. 4.32: Conventional Multi-Point Intermittent Gas Injection Design

Critical Observations/Analysis

1. Closing pressure (P_{vc}) at surface is same for all valves to effect the multipoint intermittent gas lift design.
2. First valve is located with the help of common pressure and gradient formula. All the depths of valves from 2nd valve onward is based on P_{vc}.
3. For single point intermittent gas lift, for eat valve depth, P_{vc} at surface is to be reduced by say 10 psig or 15 psig etc.
4. For locating the depths of gas lift valves, 0.04 psi/ft (S.F.) has been considered for the calculation of distance between the first valve and second valve. But for the remaining valves, the calculations have been done with uniform increase of gradient of S.F., where max. S.F. at 6500 ft = 0.09 psi/ft. This is very conservative design. Otherwise design can also be done Considering Intermediate S.F. line. (that is considering gradient P_{wh}- A – C – B, where A – C is the intermediate S.F. line.
5. Since the location of 6th valve (L_6) is below point 'C', (FBHP Gr. Line meets at C), the 6th valve, theoretically can be the operative valve to give desired rate of production (160 b/d). However, there appears no harm if another valve (7th valve) is located below the operating valve.
6. 1½" O.D. gas lift valves with bigger port size (7/16") have been considered for intermittent gas lift design. This valve has a lower load rate and allow to pass through it can adequate quantity of gas in a very short time for efficient intermittent gas lift operation.
7. All gas lift valves can well be calibrated with the calculated valves of P_{TRO} or P_{VO}. Here it is slightly modified to take into consideration in any aberration of considering various data for gas lift design.

Conventional Multipoint Intermittent Gas Lift Design Table

Well No. ....   Field ....   Date of Design ....

Type of gas lift valve : J-20 (Camco Make)

Gas lift valve port size: 7/16"

Valve No.	Depth of Valve from Surface (ft.)	Temp. (°F) at Valve Depth	Ct	P_{vc} Valve Depth = P_{bl} (psig)	P_{ve} C_1 = P_b (psig)		P_{TRO} or P_{vo} Considered for Calibration (psig)
L_1	1555	103	0.915	735	672	841	820
L_2	2940	118	0.889	760	675	844	818
L_3	3950	130	0.869	780	677	847	816
L_4	4750	138	0.856	798	683	854	814
L_5	5400	146	0.844	809	682	853	800
L_6	5950	153	0.833	817	680	851	795
L_7	6300	156	0.829	822	681	852	791

C_t = Temp. correction factor = $\dfrac{\text{Dome pressure at 60°F } (P_b)}{\text{Dome pressure at valve depth } (P_{bt})}$

$(1-A_w/A_b) = 0.799$

4.18 Advances in Gas Lift

Some of the latest development in Gas Lift equipment are described below.

4.18.1 *Surface Controllers*

Weatherford provides a complete line of CEO electronic controllers for gas lift applications. These unique controllers are designed to operate wells under various conditions. These units contain real time software with menu-driven formats to ensure smooth operation on time, off time and sales time. This series also includes battery voltage sensors and mandatory shut in features. Utilizing a rechargeable battery with a solar panel and sophisticated lightening protection and electronics, the CEO controller can be mounted on a wellhead in the most extreme environmental conditions. Should an operator so desire, a CEO controller can be operated with a Murphy Switch.

Additional features include the ability to operate wells through differential pressure plus the capacity to handle three transducers and four ports to operate motor valves and chemical or gas injection systems as needed. Programming for pressure-activated high and low shut in or openings is available with the CEO controller series. The CEO controller can be self-adjusting to provide constant monitoring, and to correct on-and-off time to provide maximum efficiency. Several models are available in the CEO series. Each model has a standard set of features and various options to address specific applications. All models can be upgraded for monitoring and automation via satellite, radio or mobile phone, and can be programmed in any language for global use.

4.18.2 *Motor Valves/Chokes*

From simple-to-install calibrated adjustable chokes and time-cycled control devices to automated control systems capable of optimizing a single well to a complete automation/information management system, Weatherford can provide the right answer to surface control requirements and system optimization.

4.18.3 *Motor Valves*

These diaphragm-actuated valves are used in cycle-type intermittent gas lift control, plunger lift, dump valve applications, differential control and other types of pneumatic wellhead controls for fluid or gas passage. Several material and trim sizes are available for either standard or H_2S service.

4.18.4 *Adjustable Chokes*

Our in-line variable choke is a calibrated adjustable choke used for regulating injection. This lightweight choke easily adjusts to changing well conditions. The process of changing out the choke *trim* and replacing internal parts is extremely simplified with the design of this valve.

4.18.5 *Side Pocket Mandrel*

Weatherford's *SBRO-DVX* side-pocket gas-lift mandrel, features unique dual-pocket communication channels that, when equipped with external backflow check assemblies, prevents well fluids from entering the casing annulus through gas lift valves or empty pockets. The mandrel can encompass multiple barriers to protect the annulus from corrosive fluids and pressure during operation and during valve change out. The mandrel will accept existing standard gas-lift valves and latches for maximum flexibility. It is available in 2-3/8- to 7-in. sizes, in a variety of materials for high-pressure and corrosive environments.

Applications include any well completion incorporating standard side-pocket gas-lift mandrels, including deepwater, high-pressure and/or highly corrosive well environments. Dual external valves prevent tubing fluids from entering the casing annulus. The dual-valve design also eliminates need for a positive check valve within the gas-lift valve, reducing required pressure drop at gas-lift valve depth. Thus, use of higher injection pressure lowers operating depth and improves system efficiency. Industry standard pocket configuration accepts 1.0- and 1.5-in.-OD gas-lift valves and latches from other manufacturers. No additional gas-lift equipment has to be purchased for upgrading to the new mandrels.

4.18.6 *Continuous Gas Pump*

This is a first new production system which uses a downhole injection, gas-powered booster pump assembly to eliminate flow gradient back pressure on continuous-flow wells. The new system allows gas lift to compete with rod pump and ESP to produce a well from initial flow to final tertiary recovery. The new device can produce gaseous fluids, high-viscosity liquids, and those with entrained solids better than either rod pump or ESP. It can be installed on coil tubing, jointed pipe or a combination of both, with wireline retrievable capability (*5,000-psi high pressure valve*). The second innovative development from BST Lift Systems is a 0 to 5,000-psi+ operating pressure gas-lift valve with an improved bellows design which prevents overpressure and over-travel in existing 1-1/2- or 1 in.-OD standard configured valves. This unique bellows design provides for long-life operation utilizing standard high-pressure mandrels and proven wireline tools.

Fig. 4.33

4.18.7 *High-Pressure Gas Lift System*

The new proprietary **XLift** high-pressure gas lift system from Schlumberger, Houston, extends the capability of present gas lift systems by increasing the range of operating pressures to 5,000 psi from 2,000 psi. Improving on field-proven Camco gas lift technology, this new system enables completion of high-pressure gas lift wells and operation with higher injection pressures and deeper injection points to increase well performance. With higher operating pressure, wells can be completed with fewer mandrels and valves, making the new valve the system of choice for deepwater and subsea gas lift.

To accommodate higher operating pressure, the 1-3/4 -in. valve incorporates an innovative edge-welded bellows system. Manufactured using corrosion-resistant materials, this new bellows reduces the internal gas charge while increasing operating injection pressure. During offshore operations, this allows the operator to inject high-pressure gas below the mudline and significantly improve depth of injection required to maximize drawdown and increase production.

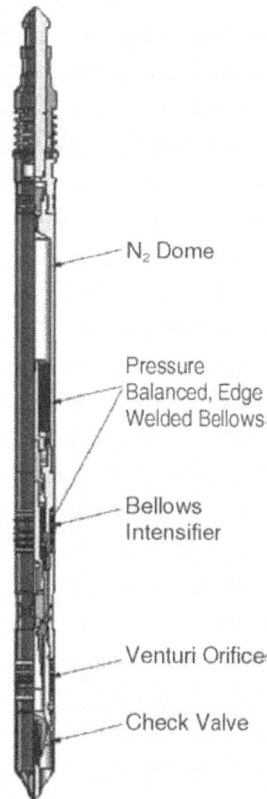

Fig. 4.34: High-Pressure Gas-lift Valve System for Deeper Wells, Both onshore and Offshore

The new valve is subsurface-controlled, with no physical linkage to surface. It features a venturi flow configuration for more efficient and stable gas throughput, and a positive check valve that eliminates potential leak paths to the casing/tubing annulus. A larger 1-3/4-in. OD improves the valves' geometry, resulting in improved performance. The venturi flow system with dual-bellows control is incorporated into the valve, mounted in the polished bore receptacle of the side pocket mandrel to form a single lift station.

Designed using proven retrievable gas lift and flow control valve (FCV) equipment technology, this innovative high-pressure system consists of a gas lift valve, valve latch, side pocket mandrel, and associated kickover tool with running and pulling tools.

4.18.8 *Fiber-Optic Gas Lift Monitor*

Schlumberger, Houston, Texas, has developed a SPOOLABLE/Retrievable downhole, fiber-optic Distributive Temperature System (DTS) that utilizes the same technology—and is an alternative to—its already available, permanent DTS. This system allows the operator the flexibility to spool-in the fiber optic DTS through tubing, log the producing temperature profile of the well for any given time and then remove the system.

The spoolable system is connected to the same surface instrumentation used in Schlumberger's permanent DTS installations, and utilizes a surface optic-electronics package that contains a laser and highly sensitive optical detectors. As with the permanent system, the application is used in monitoring gas-lift operations by making use of the temperature changes caused when gas expands through the gas-lift valve downhole. The DTS allows the operator to see, at the surface in real time, which gas lift valve is operating, or if there is a tubing leak, packer leak or some other form of undesirable downhole communication. The temperature profiling system offers monitoring of the well along the entire producing string during critical operations such as start-up, well unloading, gas-lift system optimization, or any time during the production phase life.

4.18.9 *Optimizing Gas-Lift Operations (GLO)*

eProduction Solutions (eP), Houston, Texas, has developed a unique solution for optimizing gas-lift operations by combining intelligence at both the well site and the desktop. At the well site, the GLO controller provides complete 24-hr local optimization. The controller performs well stability profiling, AGA 3 gas flow calculations, and constant injection control.

Built-in sequential start-up and shutdown functions are standard to assure proper casing unloading and well kick-off. A data logger provides historic information to the on-site operator, and near-continuous information for analysis in the desktop software. The sampling frequency is variable and can be set easily with the user-friendly keyboard interface. Parameters can be set at the well site through a multi-language local interface, laptop MMI, or remotely through the host software. The controller provides eight analog and 16 digital I/O ports for extensive expansion.

The desktop intelligence portion of the solution provides real-time awareness and understanding of individual well performance. Permanent records of well history and real-time information allow the user to control, analyze and design gas-lift wells.

The software can prioritize gas allocation to high-priority wells based on total gas available. The management-by-exception methodology provides the operator a list of trouble wells rather than requiring a lengthy search. Alarms draw attention to critical well situations preventing optimal production. To analyze the wells, the operator can choose from various pressure and PVT models. Trending, reports and charts provide current and predictive analysis.

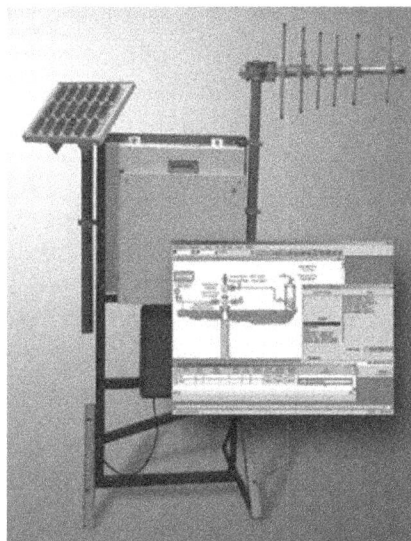

Fig. 4.35: Principal Components of Website Equipment and Visual Product of the Gas Lift Optimizer (GLO) Controller

4.18.10 *Multiphase Pump Lowers Surface Back Pressure*

Weatherford Artificial Lift Systems, Houston, has used its new RamPump to increase production by more than 700 bpd on Nexen Petroleum's Eugene Island Block 257 D platform, and produce a gas surplus from wells that previously had to borrow gas for artificial lift. The hydraulically operated, plunger-style, multiphase pump, and has been operating continuously for more than a year, in parallel with a gas-lift compressor, to lower the back pressure of four wells producing to the platform.

Fig. 4.36: Surface Pump Capable of Handling Gas-cut Liquid Lowers Back Pressure on Gas-Lifted Wells

Being able to lower the wellhead pressure by multiphase boosting at surface can enhance the application of artificial lift systems. Prior to pump installation, the well produced into a platform separation system operating at 230 psi—the pressure needed to move the liquids 1.5 mile downstream to Platform C for final separation. Evaluation indicated increased production could be obtained by lowering the pressure to 50 psi.

The pump was installed in March 2001. Pressure was lowered to 45–55 psi, and the pump began moving the liquids along with any entrained gas to Platform C. Free gas from the platform separator was diverted to the compressor where it was recycled for gas lifting the wells. Well productivity has been sustained, the pump has been mechanically reliable, and the operator is evaluating other properties that may capitalize on this method.

Additional installations are planned, with units of varying capacity—from 10,000 to 150,000 bpd, at discharge pressures as high as 1,250 psi. Special models can increase this discharge capability to well over 3,000 psi. The ability of the pump to manage the full wellstream across a broad pressure range has expanded its role to include well kick off and stimulation applications.

4.18.11 *Perforated-zone Gas Lift System*

The *PerfLift* perforated-zone gas lift system from Schlumberger is a cost-effective artificial lift system for low-rate, gas-lifted oil, liquid-loaded gas, and coal-bed methane wells. The system uses Camco gas lift technology in a simple completion architecture that enables gas lift across long, perforated intervals below a production packer.

The system uses a series of Camco gas lift products both above and below a ported or dual-bore production packer. Conventional or side-pocket gas lift mandrels are installed in the upper tubing string. Internal gas lift mandrels and valves are sized and installed on the lower tubing string across the perforated zone. The lower tubing string is capped with a bull plug.

The system can be used with a 4-1/2-in., 5-1/2-in or 7-in. ported or parallel flow packers. The tubing below the packer ranges in size from 1-1/4 in. to 3-1/2 in., depending on the packer and casing size.

Fig. 4.37: A New System Enables Gas Lift Across Long, Perforated Intervals below a Production Packer

During system operation, gas is injected down the upper casing-tubing annulus and into the lower tubing string, through the packer, lifting the fluid column across the perforated zone. Liquid then travels to the surface through the upper production string.

4.19 Plunger Lift

In plunger lift, a freely moving plunger falls through fluids in the tubing and is lifted back to surface with its slug of mostly liquids by use of formation, or injected, gas admitted from the tubing-casing annulus. Four recently introduced advances in plunger-lift (PL) technology include: a PL controller with added capabilities; a PL analysis system that locates the plunger at all times; an enhanced plunger-chamber lift performance; and an improved casing plunger.

Fig. 4.38: New Plunger-Lift Controller Model Extends Capability of Previous Systems to Collect Additional Well and Production Facility Data

4.19.1 *Plunger Lift Controller*

The latest in a line of production optimization solutions is the *CEO FOUR* controller eProduction Solutions. This new type of controller extends the control methods of the CEO THREE + controller. Optimization functionality has been added that allows operators to more accurately detect well conditions to produce wells at maximum efficiency levels.

As with the earlier system, the new controller provides time, pressure, rate and differential-based control modes. The new system adds: gas-flow calculations through AGA standards, a smart interface to industry standard tank level gauges, a generic Modbus scanner to poll local EFM RTU equipment and enhanced data-logging capabilities. These new features allow the operator to monitor/collect additional data from the well site and adjacent production facilities.

The addition of flowrate and tank-level gauge support allows the operator to optimize well production through tuning of the system for maximum efficiency. For example, the controller can poll a flow computer and store gas flow volume and rate information using the multi-channel data logger. Centralizing the information from all smart devices significantly reduces complexity and cost of well-site configurations and enables use of a single radio to retrieve all necessary well-site information.

The CEO FOUR rounds out plunger lift offering in conjunction with eProduction's production enhancement solution, to offer a range of options from a very simplistic controller, all the way through advanced control options.

4.19.2 *Plunger Lift Analysis*

Echometer has developed a new, innovative *Plunger Lift Analysis System* that takes the guesswork out of PL analysis, troubleshooting and optimization. One of the primary requirements for the operator, to efficiently and economically operate PL installations is: "Know at all times where the plunger is!"

Otherwise, the operator, even when using electronic controllers, has to guess at the proper shut-in/afterflow time and plunger velocities. The major problem of not knowing plunger location during the operation cycle has been overcome with this new technology. The system increases efficient PL operation by allowing the operator to see the plunger throughout its cycle and maximize oil/gas production from PL-lifted wells.

The portable system is a very sensitive acoustic/pressure monitoring system operated through a user-friendly graphical Total Well Management software application. The portable system, comprising a laptop computer and analog/digital converter, is programed to monitor the well during the plunger cycles, recording data at high speed from the microphone of a conventional acoustic fluid level gas gun and two pressure sensors installed on the tubing and casing head.

Shown in Figure below, is the acoustic signal, plus tubing/casing pressures acquired during the shut-in time period, [B] to [C], of a cycle. With this system, the tubing collar reflections are used to determine plunger velocities; it is now possible for the operator to virtually "see" the plunger at all times during its cycle and determine

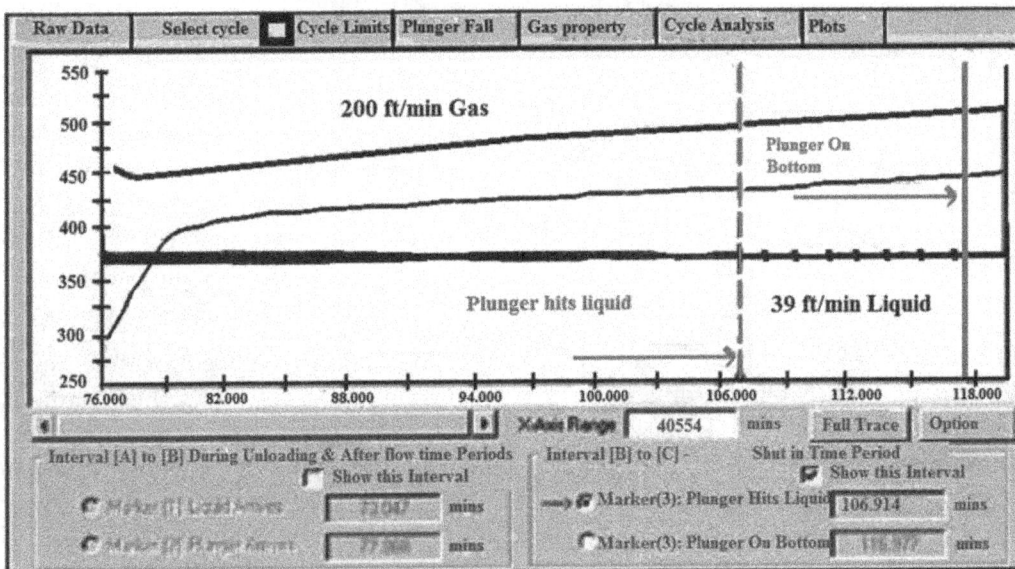

Fig. 4.39: Acquired Data from Pad Type Plunger during Cycle Shut-in Period

precise plunger fall velocity e.g., 200 ft/min. in gas and 39 ft/min. in liquid. Marker [3] identifies where the plunger hit liquid in tubing and how much liquid volume accumulated at the bottom of the tubing. Marker [4] identifies when the plunger reaches the bumper spring and what minimum shut-in time is required to maximize well production.

Having a detailed analysis of the well operation is a tremendous advantage that makes PL production optimization achievable, with a minimum of effort and time devoted to trial and error procedures. This advantage is equally important for timer-based as well as intelligent controllers; since, in both cases, it is necessary to make initial estimates of appropriate shut-in, plunger arrival and plunger fall times. In addition, the system's capability of calculating volumes of liquid and gas produced during the cycle and determining the well's potential gives invaluable information regarding the potential for improvement.

4.19.3 *Plunger-Enhanced Chamber Lift*

Wells with long perforated intervals and low-pressure reservoirs can often benefit from some form of chamber lift to improve production. By creating a chamber, the fluids that accumulate during the production cycle are dispersed over a larger cross-sectional area, creating less backpressure on the formation. Gas is injected into the annulus, displacing fluid into the production string; and intermittent cycles to deliver gas to the surface are repeated. Operation inefficiency over time creates excess fluid fallback, and well performance declines.

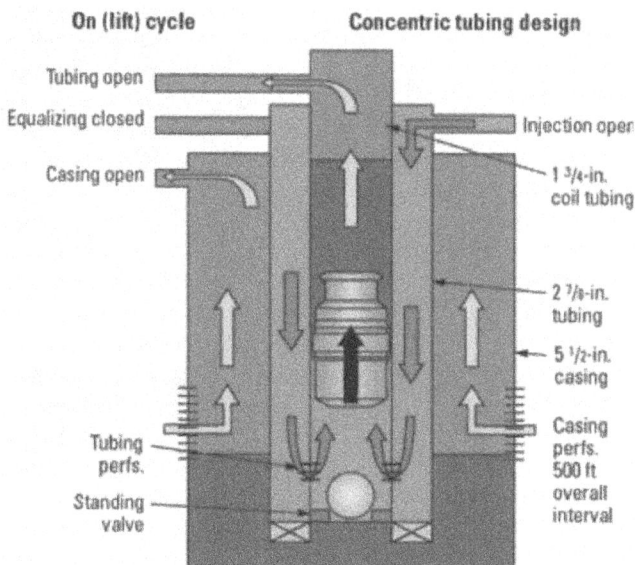

Fig. 4.40: Plunger-enhanced Chamber Lift, Step 4

With Plunger Enhanced Chamber Lift (PECL), a plunger is added to the chamber lift to improve efficiency and reduce liquid fallback. Surface and sub-surface mechanical designs vary, and one form uses coiled tubing and standard tubing to create a concentric tubing design. By sealing off the two strings at the bottom of the well, a secondary annulus or chamber is formed, and this conduit allows for the transfer of injection gas to efficiently remove liquids from the wellbore.

A typical operating process using this design would include the following steps:

- Precharge cycle
- Purge on cycle
- Purge off cycle
- On cycle, shown in Figure on page 194
- Afterflow cycle, and
- Closed cycle.

Key benefits of this type of operation allows the producing system to: 1) achieve continuous flow; 2) produce long perforated intervals with low BHP; 3) reduce friction through annular flow; 4) reduce formation and compression surge; and 5) allow total gas system management.

4.19.4 *Improved Casing Plunger*

Every producer needs a *"PAL"* Innovations, recently patented, in casing plungers for 4-1/2-in. and 5-1/2-in. casing have been developed and introduced by Oklahoma City based company. These radical departures from old technology offer production increases and reduced operating expenses for lower-pressure oil and gas wells requiring artificial lift. "PAL" cups utilize an entirely new concept of shape and function. Previous available models relied on cup diameters in excess of casing ID to secure the sealing contact with the casing wall.

The innovative new cup design is smaller than the casing ID and eliminates unnecessary wear of the cups on descent. The cups are first mechanically actuated to seal against the casing wall at the well bottom. Then, through the simple, but clever, design of the system, the cups are sealed pneumatically for ascent and the unloading of accumulated wellbore fluids that inhibit, or prevent, production.

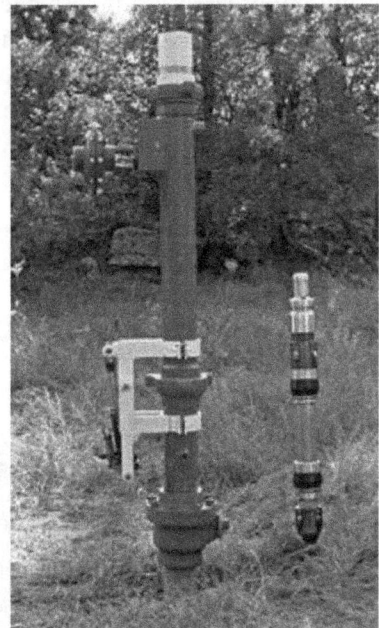

Fig. 4.41: Improved Casing Plunger with Cup Design that Falls Freely, then Seals Pneumatically during Ascent and Fluid Lifting

A surface lubricator designed to provide simple and convenient retrieval of the system's plunger can be configured for free fall, to catch and hold, for time release, or for release on monitored and pre-set wellhead pressures.

After five years of development, testing and actual production results, conclusive evidence verifies significant success in many areas. Production increases of five fold have occurred. Better utilization of surface equipment—through replacement of tubing plungers, rod pumps and pump jacks—has been achieved. Often, salvage value of the existing equipment exceeds the cost of installation.

In some cases, replacement of pumping equipment maintained the same production, but with substantial reductions in lease operating expenses. In other cases, wellhead operating pressures have been lowered, extending productive well life. Well depths have varied from 2,000 ft to over 8,200 ft. Multiple-weight casing strings and old squeeze cemented casing leak repair sections are easily accommodated. In addition, the presence of paraffin seems to be an advantage. Installations in all combinations of wellbore fluids, and BHPs from 60 psi to over 350 psi have been successful in operation.

4.19.5 *Continuous-Flow Plunger*

The recently introduced *RapidFlo* plunger, from Weatherford, Houston, is a high-speed, continuous-flow plunger designed for maximum flow-through while still maintaining strength in material. This plunger, when applied properly, will fall against flow to the bottom of the well, reset, then return to surface using only velocity of the flowing gas to form a seal to deliver fluid without any shut-in time.

4.19.6 *Optimum-Seal Plunger*

Compared to other plungers which either allow large amounts of gas to flow past the outer sealing areas or have slow fall times, the patented *Opti-Seal-Quick Trip* Plunger from Opti-Flow, a VSA based company, allows for the optimum seal on its trip to the surface and quick fall time to bottom. Combination of these two features will provide more efficient operations, improved travel times and increased production.

The sealing plunger comes with a unique large bypass for quick fall times. Patented interlocking pad fingers

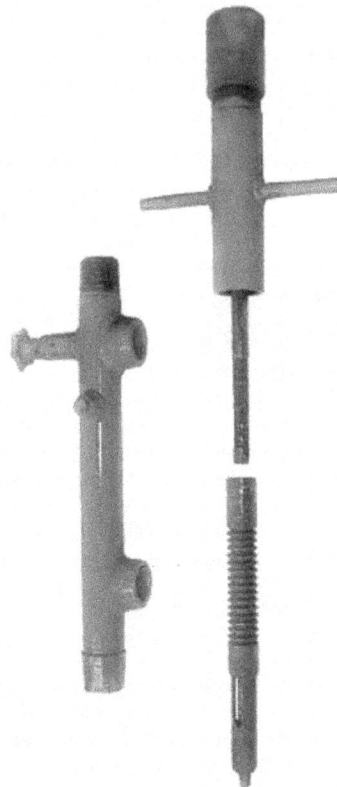

Fig. 4.42: High-speed, Continuous-flow Plunger System

provide the optimized seal, as gas flow trapped underneath the pads keeps them against the tubing wall, creating the seal between gas below the plunger and liquid above.

The patented bypass system allows the plunger to fall to bottom at a faster rate, incorporating a dart that goes on seat when the plunger falls to the bottom stop. Differential pressure from below causes the dart to stay on seat on its trip to surface; no springs, balls or O-rings are required. When the well is shut in, the dart comes off seat, allowing flow through the bypass for quick fall time. Made of corrosion resistant SS–304 materials for CO_2 and H_2S service, no elastomers are present. Pads are heat treated and hardened for extended life.

4.19.7 *Inexpensive Plunger-Lift Controller*

A new, inexpensive plunger-lift controller system is described below. The controller's simplistic design provides accurate control of plunger lift wells. It includes an intuitive user interface for system configuration that makes it easy to set parameters for both single- and two-valve modes. Users can easily configure high-and-low pressure overrides from a digital pressure transducer input. The unit also includes a single, high-pressure shut-down feature for use with an analog pressure transducer. The unit is designed for stand-alone operation and does not accommodate remote communication. While the unit does not include the SCADA interface that other eP controllers have, it does include an event logger that stores plunger cycle time and plunger arrivals that can be accessed through its HMI data port for analysis.

Fig. 4.43: Plunger Lift Controller

The unit is designed for use with 1-W or 2-W solar panels and operates on a 6-V battery system. A smart charger/regulator system that is built into the unit extends battery autonomy and life.

4.19.8 *Pressure-Relieving Standing Valve*

A new Pressure Relieving Standing system, with an integrated bumper spring assembly, for relieving potentially unwanted liquid from the tubing string. The idea of plunger lift is to remove as much liquid as possible from the wellbore, with the plunger arriving at surface every cycle. If an operator uses a standing valve with standard ball and seat, operations can be hindered if the fluid slug is too large.

Build-up pressure in the annulus may not permit enough lift gas to bring the plunger and slug to surface, thus causing a loaded condition. If this condition exists, having installed the pressure relieving valve will allow the operator to equalize the tubing and casing pressures at the wellhead and effectively drain liquid from the tubing. This is accomplished when the added pressure on the tubing column compresses the relief spring, moving the seat across slots that open the tubing to the annulus. A balance spring in the standing valve will cause a small fluid slug to be preserved inside the tubing string to ensure against a "dry run." During normal operations, fluid loads are supported by pressure working upward against the seating assembly.

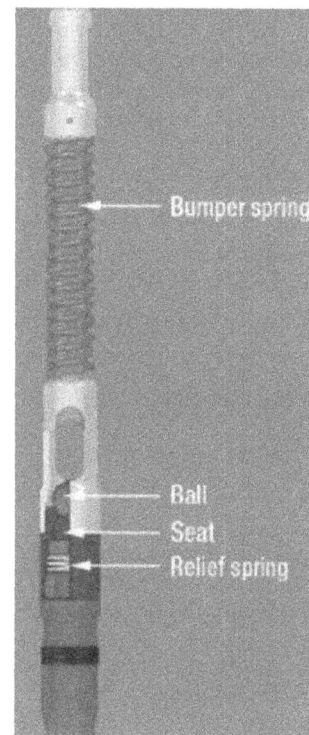

Fig. 4.44: Pressure Relieving Standing Valve with Downhole Spring

4.19.9 *Multi Control Plunger-Lift*

A new type of complete multiproduct plunger-lift pack, which provides its own gas supply and pressure to assure removal of liquids from any well with tubing installed. A safe, clean source of lift gas means that many wells in decline can be produced at higher production levels, with what has always been a low-cost artificial lift method.

The lack of gas to assist the plunger is no longer an issue of concern to making the plunger operate successfully. Generating nitrogen (N2) at a low cost per Mcf and compressing this back into the injection line supplies pressure to lift liquids up the tubing with the plunger. Total cost of a per-well package is below that of mechanical artificial-lift forms and provides electronic controls that monitor the operation and recycle the N2 for the next cycle.

The N_2 can be produced at as little as 16 cents per MCF—this cost is even lower when the gas is recycled. The N_2 generator will then be used to make-up gas for any losses to the tank or for the tank cover. The process begins with installation of the N_2 processing unit about the size of an office desk. The well is equipped with the plunger-lift equipment and the controller with the monitoring system. The N_2 is run through a compressor and injected into either the annulus or a side-string of small-diameter tubing.

The controls open the motor valve to the inject line and allow the pressure to accumulate to a set point in the well. They sense when lift pressure is present and open the motor valve to the separator. The separator will provide the sensing device to allow the N_2 to divert back into the closed loop system for reinsertion. Gas produced by the well is sent down the consumer line.

The N_2 is recycled to again provide the lift, and the gas is sold through the consumer line and meter. Multi provides the complete package from the N_2 generator to the meter to monitor and record, as well as the plunger lift with electronic monitoring. It also provides installations to various producers with differing types of operations to achieve a high level of what limitations and situations can occur. Current cost reductions occur in lift equipment and nitrogen production.

4.19.10 *Two-Piece Plunger*

A new two-piece plunger has been introduced in plunger design in an effort to solve some of the problems that are inherent in current plunger-lift technology. Conventional plungers require a "shut-in" period to allow the plunger to drop, and to build enough well pressure in the annulus to drive plunger and fluid to surface.

This two-piece plunger, is designed to trip to bottom while the well is producing gas at considerable rate. In some wells, the plunger falls to the bottom while the well is producing at 1,000 Mcfd or more. Both pieces of the plunger have considerable bypass area, allowing the well to produce around the bottom piece (the ball) and through the top piece (the piston). They join at the bottom and are held together by the flow from the zones below as it pushes the plunger (now one unit) and its fluid load to surface.

The surfacing plunger strikes a shifting rod and a gas-powered catch cylinder, which separates the two pieces and holds the piston for a short time. The ball falls back to bottom. When released, the piston arrives at the bottom of the well and joins with the ball, beginning the process again. The plunger can trip to bottom and back at speeds of 1,000 ft/min., or faster, while the well is flowing gas. The high round trip speed allows the plunger to lift smaller amounts of fluid with each trip, so it can lift more fluid per day with less average BHP than conventional systems.

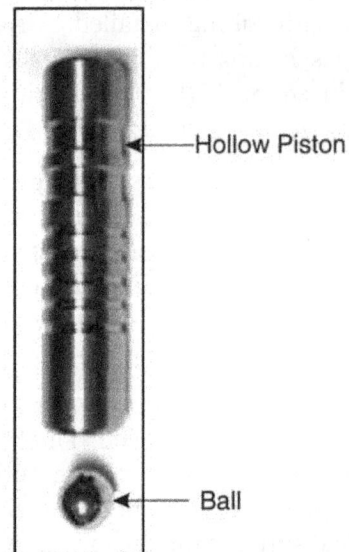

Fig. 4.45: Two-piece Plunger Separates at Surface. Ball and Hollow Piston Fall Separately at High Rate, Even while well is Flowing Gas

Another advantage is that the system performs well without using the casing/tubing annular volume for pressure storage. The new plunger relies more on volume than trapped pressure to trip the plunger to surface. It works in 1-7/8-in. slim hole or wells with a packer and no communication with the annulus. On-site compressors usually adapt easily to the new plunger because the "shut-in" time of only seconds has almost no effect on compressor suction pressure.

4.19.11 *Plunger–Lift Controller*

This is a revolutionary design for plunger-lift system control and optimization that incorporates the latest in ultra-low-power electronics technology and sophisticated RTU functionality, to provide a complete plunger-lift controller system. Four control states allow configurable optimization using both time- and pressure-based optimization. Incorporated to provide user resources are:

A flexible, configurable multi-channel data-logger; gas-flow calculations based on AGA Reports 3 and 8; a configurable Modbus polling utility; and an automatic control language (ACL) user programing tool that provides for customized system operation.

The controller is capable of polling local "intelligent" peripherals, using Modbus communication protocol, to retrieve and store data. For example, the controller can poll a local electronic flow measurement RTU system and store gas-flow volume/rate information using the multi-channel datalogger. By centralizing the information from all local intelligent devices, upfront cost for optimization is decreased. Up to four local devices can be polled together.

Modular auxiliary I/O is available to expand the base controller to provide interface for casing/tubing pressure sensors and other peripherals. Logging and trending of this information can be achieved through use of the multi-channel datalogger. All information is available remotely through optimization software available from eProduction.

4.19.12 *Casing Plunger System*

Multi Products Co. in Millersburg, Ohio, has developed an artificial lift system that operates in the well with no tubing to save the added cost of that tubing and additional lift equipment. Other casing-type plungers have been marketed. One drawback to these systems includes the dynamics of a large device flowing upward in a wellbore with high-pressure gas sealed below it.

Regulation of the upward-flow velocity with such a device has always been a concern. The reduced gas volume rate required to displace the tool safely to surface limits the gas sales. A solution to this problem was developed by Multi Products with the use of a metallic sensing device and an electronic controller, designed as the JetStar Fail Safe Controls system. The sensor cannot be deceived by noise or stray signals.

This prevents the large-ported valve from opening prematurely and damaging surface equipment. Upward plunger velocity is controlled by reduced orifice size in the outlet line from the lubricator to the flowline. Arrival of the casing plunger into the lubricator signals the controller that the lower, large-ported valve is now safe to open. This change of size in the flow orifice to the flowline can account for several-hundred-thousand cubic feet of increased gas sales. The resultant production not only creates a profitable scenario in gas volume, it also gives the well a system that can produce to "plug-and-abandon" status.

Fig. 4.46: Surface Control for JetStar Casing Plunger Artificial-Lift System that Controls Plunger Speed at Surface Lubricator Location

The plunger is released after a set time period based on well deliverability. The plunger must contact the bottomhole stop set in the casing before the internal bypass in the plunger closes. After the bypass closes, the plunger returns to the surface, with velocity controlled by the smaller orifice in the lubricator outlet.

Example: A well is producing from 500 meter depth in making 20 M^3/day in an intermittent gas lift, 40 cycle per day through 2-7/8 inch tubing. The gas injection time per cycle is 20 minutes. It is estimated that the static bottom hole pressure of the producing zone is 100 kg/cm² and PI of the well is 0.3 M^3/day/kgsqcm. Could the production rate from the well be improved by changing the number of cycle per day.

4.20 Plunger Lift Operation

This is applicable for high gas oil ratio wells. Plunger automatically keeps tubing clean of paraffin's and scale. They are suitable for low flow rate wells i.e. less than 150 BOPD.

Plunger lift uses a free position that travels up and down in the tubing of the wells. It minimise the liquid fall back and uses the energy of the well more efficiently.

The purpose of the plunger lift is to remove liquid from well bore so that the well can be produced at the lowest bottom hole pressure, whether in an oil well, gas lift well or gas well, the mechanics of plunger lift system are the same.

The plunger works like a piston, interfaces between liquids and gas in the well bore and prevents liquid fall back. By providing a 'seal' between the liquid and gas, a well own energy are used efficiently to lift liquids out of the well bore.

A plunger operation consists of shut-in and flow period. The flow period is further divided in to unloading period and flow after plunger arrival at surface. Lengths and duration of the periods depends on bottom hole pressure of the well. The well must be shut-in for long duration to build up reservoir pressure that will provide energy to lift both plunger and the liquid slug to the surface against back pressure and friction in the tubing when the plunger and pressure have been reached, the flow period starts and unloading begins. In the initial stages of the flow period, the plunger and liquid slug being to flow to the surface.

Gas above the plunger quickly flow from the tubing into the flow line above the plunger and liquid slug follow up the hole. The plunger reaches at the surface lubricator which is shown in figure and unloading of liquid starts. Initially, it flows at the high rate, later this rate goes down slowly. When rates drop below the critical value and liquid being to accumulate in the tubing. The well is shut in and the plunger falls at the bottom hole on bumper spring.

Upon shut in, the casing pressure builds up more rapidly. How fast, it depends on the inflow of lift gas. It will eventually track casing pressure (less the liquid slug). Casing pressure will continue to increase quickly from line pressure, as the flowing gas friction increases. It will eventually track casing pressure (less the slug). Casing pressure will continue to increase to maximum pressure until the well is opened again.

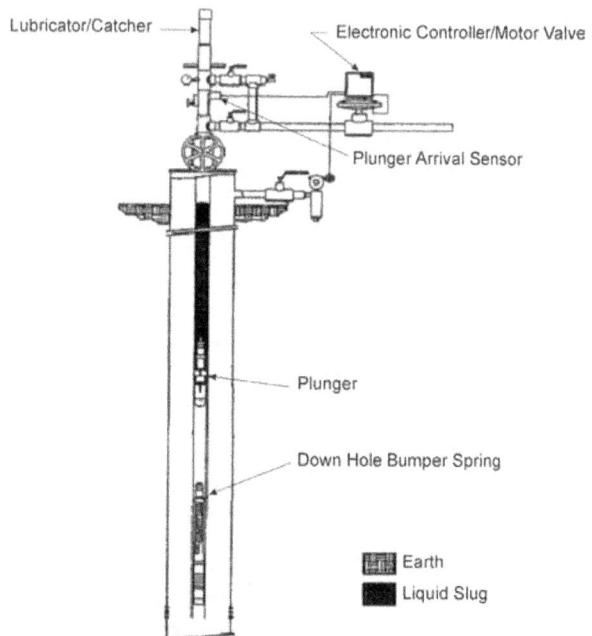

Fig. 4.47: Typical Plunger Lift System

As with most wells, maximum plunger lift production occurs when the well produces against the lowest possible bottom-hole pressure. Practical experience and plunger lift models demonstrate that lifting large liquid slugs requires higher average bottom-hole pressure. Longer shut-in periods also increase the average bottom hole pressure so that purpose of plunger lift should be to shut the well in the minimum amount of time and produce only enough liquids that can be lifted at this minimum buildup pressure.

The well must be shut-in in this length of time regardless of what other operating condition exist. Plungers typically fall between 200 and 1,000 ft/min in dry gas and 20 and 25 ft/min in liquid.

A well with a high GLR may be capable of long flow period without requiring more than minimum shut-in times. In this case, the plunger could operate as far as 1 or 2 cycles/day.

4.20.1 *GLR and Buildup Pressure Requirements*

There are two minimum requirements for plunger lift operations: minimum GLR and buildup pressure. For the plunger lift to operate there must be gas available to provide the lifting force, in sufficient quantity per barrel of liquid for a given well depth.

Rules of Thumb: The minimum GLR requirement is considered to be about 400 scf/bbl/1,000 ft of well depth, that is

$$GLR_{min} = 400\frac{D}{1000'}$$

where,

GLR_{min} = minimum required GLR for plunger lift, scf/bbl

D = depth to plunger ft.

An improved rule for minimum pressure is that a well can lift a slug of liquid equal to about 50–60% of the difference between shut-in casing pressure and maximum line pressure. This rule gives

where,

P_c = required casing pressure, psia

PL_{max} = maximum line pressure, psia

P_{sh} = slug hydrostatic pressure, psia

F_{sl} = slug factor, 0.5–0.6

Plunger lift models

The required minimum casing pressure is expressed as

$$PC_{min} = [Pp + 14.7 + pl + (plh + plf) \times V \text{slug}] \times \left(1 + \frac{D}{K}\right)$$

where,

$P_c \min$ = required minimum casing pressure psia

P_p = W_p/A_t, psia

W_p = plunger weight, lb_f

A_t = tubing inner cross-sectional area, in.^{2}

P_{lh} = hydrostat liquid gradient, psi/bbl slug

P_{tf} = flowing liquid gradient, psi/bbl/slug

P_t = tubing head pressure, psi

V_{slug} = slug volume, bbl

D = depth to plunger, ft

K = characteristic length for gas flow in tubing, ft.

Various authors suggested an approximation where K and $P_{lh} + P_{lf}$ are constant for a given tubing size and a plunger velocity of 1,000 ft/min.

Tubing Size (in)	K(ft)	$P_{lh} + P_{lf}$ (psi/bbl)
2$^{3/8}$	33,500	165
2$^{7/8}$	45,000	102
3$^{1/2}$	57,600	63

To successfully operate the plunger, casing pressure must build to PC_{max} given by

$$PC_{max} = PC_{min}\left(\frac{A_a + A_t}{A_a}\right)$$

The average casing pressure can then be expressed as,

$$PC_{avg} = PC_{min}\left(1 + \frac{A_t}{2A_a}\right)$$

Where, A_a is annuals cross-sectional area in square inch.

The gas required per cycle is formulated as,

$$V_g = 3\frac{37.14\, F_{gs}\, PC_{avg}\, V_t}{Z(T_{avg} + 460)}$$

where,

V_g = required gas per cycle, Mscf

F_{gs} = 1 + 0.02 (D/1,000). Modified Foss and Gaul Slippage factor

V_t = $A_t (D - V_{slug} L)$, gas volume in tubing, Mcf

L = tubing inner capacity factor in average tubing condition

T_{avg} = average temperature in tubing, °F.

The maximum number of cycles can be expressed as,

$$NC_{max} = \frac{1440}{\dfrac{D}{W} + \dfrac{D - V_{slug}}{V_{fg}} + \dfrac{V_{slug L}}{V_g}}$$

where,

NC_{max} = the maximum number of cycles per day

V_{fg} = plunger falling velocity in gas, ft/min

V_{ft} = plunger falling velocity in liquid, ft/min

V_r = plunger rising velocity, ft/min.

The maximum liquid production rate can be expressed as,

where,

NC_{max} = the maximum number of cycles per day

V_{fg} = plunger falling velocity in gas, ft/min

V_{ft} = plunger falling velocity in liquid, ft/min

V_r = plunger rising velocity, ft/min.

The maximum liquid production rate can be expressed as,

$$qL_{max} = NC_{max} V_{slug}$$

The required GLR can be expressed as,

$$GLR_{min} = \frac{V_g}{V_{slug}}$$

4.21 Chamber Lift

When BHP of a well goes down then well is placed on intermittent gas lift also where liquid falls to a low level, two unfavorable phenomena occur that can severely decrease the efficiency of fluid lifting efficiency. First, the slug volume that accumulates in the tubing string during one cycle considerably decreases and the well's production rate declines accordingly. Second, the injection Gas liquid Ratio inevitably increases since the tubing string below the liquid slug must always be filled with injection gas, no matter what the actual starting slug length is. The chamber lift is the last stage of artificial lift technique when both PI and BHP are low.

In a chamber lift installation, well fluids accumulate in a downhole accumulation chamber of a greater capacity than the tubing string. If the same starting slug length is allowed to accumulate, then the cycle volume is much greater in the chamber installation than in a conventional intermittent installation.

The operation of the intermittent cycle in a chamber lift installation of the two-packer type system the segment of operation. After the previous liquid slug has been lifted and the operating valve has closed, fluid inflow to the well starts with opening of the standing valve. Well fluids simultaneously rise in the casing-tubing annulus as well as in the tubing because of the perforated nipple situated close to the bottom packer. The bleed valve continuously vents formation gas into the tubing, and thanks to its operation, the chamber fills up fully with liquid. At this moment, gas injection at the surface occurs, forcing the operating gas lift valve to open and inject gas to the top of the chamber. The bleed valve immediately closes, and the liquid in the casing-tubing annulus is gradually U-tubed (displaced) into the tubing string. Gas injection below the liquid slug and into the tubing occurs only after all the liquid has left the casing-tubing annulus. Note that the point of gas injection is at the perforated nipple and not at the operating gas lift valve. The liquid slug starts its upward journey and attains a considerable velocity even before the actual gas injection takes place. This is a crucial feature that distinguishes chamber lifting from simple accumulation chambers sometimes used in intermittent installations.

4.21.1 *Principle*

Due to its advantageous features, chamber lift is recommended for gas lifting wells with very low formation pressures, low PI. It can be used in the last stage of a well's or an oil field's productive life and is usually the last kind of gas lifting before the well is finally abandoned. Additionally, chamber lift is ideally suited to produce deep, high productivity wells with low BHP.

The section of the type of chamber lifts installation from the many available versions is governed by many factors such as well completion type, casing size and condition, production rate, etc. Two-packer chambers are more expensive and are justified if the well's production rate is sufficient to cover the additional costs. This is shown in Figure 4.48. It is generally considered when the dynamic liquid level is above the top of the perforations. The possibility of using the maximum annular capacity of the casing and the utilization of a tubing section as a dip tube allow relatively high production rate to be attained. An added advantage is that all equipments (operating and bleed valves, standing valve) can be of the wireline-retrievable type.

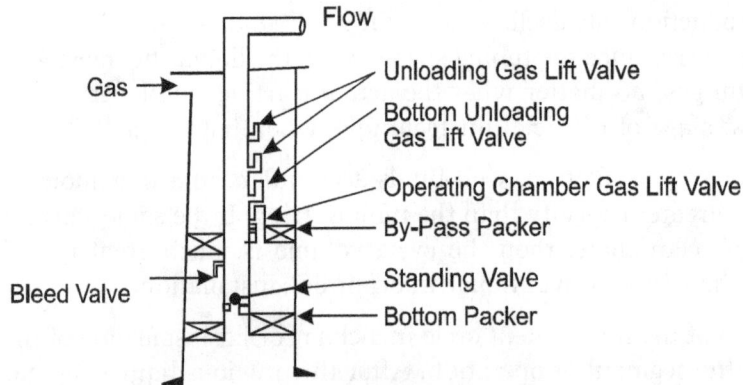

Fig. 4.48: Two Packer Chamber Lift

Chamber lift installation are used for accumulating liquid volume at the bottom hole of intermittent gas lift wells. A chamber lift is ideal for a low BHP and low PI well. Figure shows a standard two packer chamber lift. This type of chamber is installed Figure to ensure a large storage volume of liquids with minimum amount of back pressure on the formation so that liquid production rate is not hampered.

Insert chambers are required for open hole completions and cases when the dynamic liquid level is below the top of the perforations. This will result for shallow low capacity or a stripper well (small mechanical packer). This inexpensive solution uses a dip tube with diameter smaller than that of the tubing as well as standard gas lift mandrel and valve. In deeper wells with more production potential, more expensive bypass packer can be used. Advantages include the use of the tubing as dip tube and the possibility of using wireline retrievable operating and standing valves for easier servicing.

4.21.2 *Advantages*

Chamber lift installations offer many advantages over conventional intermittent gas lift installations. The more important ones are detailed as follows.

- Chamber lift produces the lowest possible FBHP for gas lift and can thus markedly increases the production rate of most intermittent lift wells.
- Wells with low formation pressures can be produced until ultimate depletion.
- Liquid fallback is considerably reduced when compared to conventional installations because gas injection takes place only after all of the accumulated liquid is U-tubed to the tubing and the liquid slug is in full motion. Gas breakout is thus greatly reduced, leading to a higher liquid recovery.
- Injection gas requirements are reduced due to the greater starting slug lengths possible.

- Since the point of gas injection in an insert chamber installation is at the bottom of the tubing, lift gas can be injected near the total depth in wells with a long perforated interval.

4.21.3 *Limitations*

The basic limitations of chamber lift are the same as those for intermittent lift in general, i.e. lower available flow rate, wasting of formation gas, etc. Additional disadvantages are follows:

- Well dimensions like a small casing size or a long perforated interval can severely limit the applicability of chamber installations.
- In wells with high sand production, wireline operations and the pulling of the chamber can be difficult.

4.21.4 *Selection of Equipment*

In order to ensure an efficient operation of chamber lift installations, the proper selection of the required equipment is necessary. The following presents a short description of the most important considerations.

- Preferably, upper packers should be of the bypass type, allowing the tubing to be used as a dip tube. Production packers, on the other hand, are less expensive but the use of small diameter dip tubes.
- The operating or chamber valve can be any IPO valve suitable for intermittent lift operations. Selection of the valve's spread is very important because tubing pressure at valve depth is extremely low so valve spread is near its maximum value. Therefore, pilot-operated valves are preferred since their spread can be selected independently of the main port size.
- Standing valves should be mechanically prevented from being blown out from their seating nipples by the great pressure differentials occurring during the after flow period of the intermittent cycle.
- A properly sized bleed valve or bleed port installed on the dip tube is desirable to vent formation and injection gas from the chamber. Bleed holes of $3/_{32}$ in. or $1/_8$ in size are sufficient for low producers but bleed (differential) valves are required for wells with high formation GLRs and/or high cycle frequencies. For insert chamber installations in gassy wells, an annulus vent valve must also be used to vent formation gas from below the packer into the tubing string.
- If possible, dip tubes should be of the same diameter as the tubing because gas breakthrough can significantly increase when the liquid slug is transferred from the dip tube of a greater diameter.
- Wire line retrievable equipment are desirable to reduce the operating cost.

- The bottom unlading valve should be set immediately above the operating (chamber) valve to ensure proper unloading of the well. Its opening pressure should be at least 50 psi higher than of the steering of operating valve.
- The top of the chamber should never be placed above the maximum dynamic liquid level. This is because the space between the liquid level and the top of the chamber is filled with injection gas in every cycle which increases the gas requirements and also the time needed for pressure blow-down. The latter requirement severely limits the cycle frequently and production rate.

4.21.5 *Design*

Since chamber lifts are intermittent gas lift installations, all the necessary calculations and design procedures follow those required for conventional cases. *The design of the unloading valve string, the selection and operation of surface injection controls, the calculation gas injection requirements, etc.* The most important differences between a conventional and chamber lift installation are that (a) well fluids accumulate in the chamber, and (b) gas is initially injected at the top of the chamber. Implications of these conditions necessitate the knowledge of the length of the chamber, which is discussed as follows.

Determination of Chamber Length: Figure 4.48 on Page 218 illustrates a two-chamber installation immediately before the opening of the operating valve (on the left-hand side) and just before gas injection into the tubing takes place (on the right-hand side). If the chamber has completely filled up, then a liquid column of a length equal to the Chamber Length (CL) is standing in the chamber and the tubing string. After the chamber's liquid content is completely from a balance of the respective volumes and can be expressed using the chamber's geometrical data as,

$$H = CL \left(1 + \frac{C_a}{C_t} \right)$$

Where CL = chamber length, ft
C_a = capacity of the chamber annulus, bbl/ft
C_t = capacity of the tubing above the chamber, bb/ft.

The height of the liquid column can also be calculated from a balance of the pressures at the instant the annulus liquid level drops to the bottom of the chamber, if the pressure at the bottom of the liquid slug is known:

$$H = \frac{p_i - p_t}{\text{grad}_1}$$

where,

p_i = design tubing load, i.e. the pressure at the bottom of the liquid column, psi

p_t = sum of wellhead and gas column pressures at top of liquid column, psi
$grad_l$ = static gradient of accumulated liquid, psi/ft

The length of the chamber:

$$CL = \frac{p_i - p_t}{(1 - R_c)\, grad_1}$$

where, $R_c = C_a/C_t$ = ratio of annular and tubing capacities.

The length of the chamber heavily depends on the design tubing load P_i which is selected by the design as to ensure an efficient lifting of the liquid slug to the surface. Liquid fallback minimum if the tubing load equals 60–75% of the available injection pressure. Since injection pressure identification the operating (chamber) valve's opening pressure, the proper value of tubing load is found as follows:

$$0.60\, P_{io} < p_i < 0.75\, p_{io}$$

where, p_{io} = the chamber valve's opening injection pressure at its setting depth, psi.

The previous formula can only be used if the chamber fills up completely and its top is located at the dynamic liquid level. Similar formulas for other possible chamber configurations (insert chamber with a dip tube were developed by Winkler and Camp).

Surface design: The design of a chamber installation closely follows the procedure for conventional intermittent installations. A complete design for choke control, including the unloading valves is given here, following the procedure required for using a constant surface closing pressure of unloading valves. Although it is developed for a two-packer chamber installation other installation types can be handled similarly.

Spacing of the unloading valves is identical to the procedure described in Steps 1–8 of the previous section rest of the design calculation is given as follows.

1. Select the setting depths of the two packers: the lower packer should be set at the depth of the upper packer's setting depth is less by an assumed *CL*.
2. Delete all unloading valves below the assumed depth of the upper packer.
3. Set the operating chamber valve to the depth of the upper packer and its surface closing pressure 50 psi less than that of the unloading valves. Calculate the appropriate downhole closing pressure p_{ic}.
4. Calculate the tubing pressure at the chamber valve when the valve is closed.

$$P_t = WHP + Lgrad_g$$

Where,
  WHP = wellhead pressure, psi
     L = setting depth of the chamber valve, ft.
  $Grad_g$ = gas gradient, psi/ft.

5. Assume the opening pressure P_{io} of the chamber valve at depth.

6. Find the intermittent cycle's gas requirement based on the chamber valves opening pressure P_{io} and its setting depth.

7. The pressure differential Δp necessary to store the required gas volume in the annulus is found. Since a choke is used to control the surface gas injection into the well, an operating valve with a spread equal to this pressure differential should be selected.

8. The valves opening pressure is calculated as the sum of its closing pressure at depth and pressure differential just found,

$$P_{io} = p_{ic} + \Delta p$$

9. Compare the assumed and calculated opening pressures and repeat steps 5–8 until the two values are in agreement.

10. The required port size of the chamber valve is found by solving the valve opening equation for the ratio of valve areas, R,

$$R = \frac{p_{io} - p_{ic}}{p_{io} - p_t}$$

11. The pilot operated gas lift valve with the nearest R ratio is selected from manufacturer's data. The port size selected is for the control part only; the main port can be selected independently so as to ensure a proper instantaneous gas injection rate through the valve.

12. The actual opening pressure at depth of the chamber valve is checked with the valve opening equation as follows:

$$P_{io} = \frac{P_{ic}}{1-R} - p_t \frac{R}{1-R}$$

13. Select proper tubing load p_i corresponding to the valve opening pressure p_{io}.

14. Calculate the capacities of the chamber annulus and that of the tubing; find their ratio by the formulas given as follows:

$$C_a = 9.71 \times 10^{-4} (ID^2_{cb} - OD^2_t)$$

$$C_t = 9.71 \times 10^{-4} ID^2_{cb}$$

$$R_c = \frac{C_a}{C_t}$$

Where,

ID_{ch} = chamber inside diameter, in
ID_{cb} = tubing inside diameter, in
OD_t = tubing outside diameter, in

15. Calculate the required length of the chamber.
16. Find the depth of the chamber valve from the *CL* just calculated. If the calculated value is not close end then one assumed in Step 3, repeat the design with the calculated valve depth. Otherwise set the unloading valve two tubing joints (about 60 ft.) higher than the chamber valve and continue.
17. Make sure that the bottom unloading valve doesn't open when the liquid slug passes for this, as tubing pressure opposite valve P_t when the bottom of the liquid slug has risen above the valve setting.
18. Assume the opening pressure of the bottom unloading valve P_{io} and find the required R value equation (g).
19. Select an unloading gas lift valve with the nearest R ratio and calculate the actual opening pressure. Compare this value with the assumed one and repeat steps 18–49 until the two values.
20. Select the type and choke sizes for the unloading valves.
21. The flowing temperature at each valve setting depth T_i is found.
22. The dome charge pressure p_d of the valve at their setting depths are calculated from the valve opening as follows.
23. The dome charge pressure at surface conditions p'_d are calculated at a charging temperature of 60 nitrogen gas charge.
24. Finally, TRO pressure are found form the surface dome charge pressure using the

$$TRO = \frac{P'_d}{1-R}$$

CHAPTER 5

Electrical Submersible Pump

5.1 Introduction

Electrical Submersible Pump (ESP) are used as an artificial lift techniques for oil well when inflow of fluid to well bore decreases. They are a versatile form of pumping, especially where high production rates are envisaged. They are very suitable for high PI, low GOR, and medium bottom-hole pressure well condition.

The Electrical Submersible Pump (ESP) is basically a high volume mode of lift system. ESP is a relatively efficient system of artificial lift, and under certain conditions even more efficient that a beam pump, with lower lifting costs and a broader range of production rates and depths. A typical ESP installation is given in Figure 5.1a. Surface and Sub-surface set-ups are shown in Figure 5.1b and Figure 5.1c respectively. ESP, in some situations, can provide the maximum possible draw-down by bringing annulus level to the top of the perforations.The ESP operates over a wide range of depths

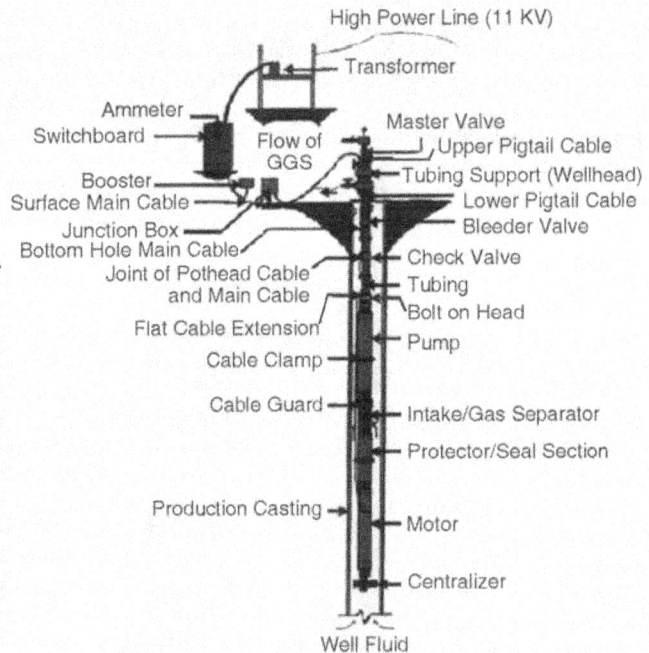

Fig. 5.1(a): Electrical Submersible Pump

and volumes. The maximum depth is 12000 ft (3650 meters). The minimum capacity of ESP is known to be around 200 BPD and the maximum capacity is as high as 90,000 BPD. However, environmental variables i.e. free gas, temperature, viscosity, depth, sand and paraffin can severely limit the pump performance. Excessive free gas results in motor load fluctuations and cavitation leading to reduced run life and reliability. Temperature may limit application because of limitations of the thrust bearing, epoxy encapsulations, insulation, and elastomers. Viscosity increases cause a reduction in the head the pump system can generate, leading to an increased number of pump stages and motor horse-power. Depth limitations are due to the burst pressure ratings of the components such as the pump housing and the thrust bearing. Sand and paraffin problems lead to component wear and choking conditions inside the pump.

The ESP is extremely suitable for a very low viscosity liquid. This pump is also used to pump high viscosity fluids and can operate in, gassy wells and high temperature wells. The prime mover of the

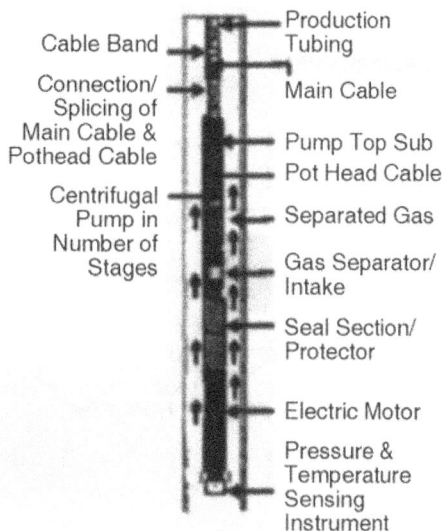

Fig. 5.1(b): Subsurface Assembly of ESP

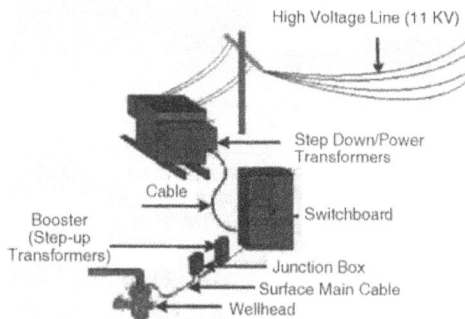

Fig. 5.1(c): Surface Set-up for ESP

electrical submersible pump is the downhole motor coupled directly with the pump.

The pump consists of several pump stages, each consisting of an impeller connected to the drive shaft and a diffuser that directs the flow of fluid from one stage to next. The number of stages required is determined by the lift and volume of fluid. Sizes vary from less than 3.5″ to 10″ in diameter, and from 40 to 344 in. in length. The motor is a 3-phase, squirrel cage induction motor varying from 10 to 750 Hp at 60 H2, ranging from 3.75″ to 7.25″ diameter. Voltage requirements vary from 420 to 4200 Volts at 60 Hz.

The seal section serves to separate the well fluids present in the pump from the motor oil in the motor. The fluids must be kept from entering the motor, regardless of differential pressure. Also, this seal must accommodate expansion of the motor oil due to heating. The electric cable supplies electric energy to the down-hole motor, and must therefore be capable of operating in fluids at high temperature and pressures, and deliver maximum electric currents efficiently.

The motor controller serves to energize the submersible motor, sense such conditions as motor overload, well pump-off, etc. as well as shutdown or startup in response to pressure switches, tank levels or remote commands. They are available in conventional electro-mechanical and solid state devices. Conventional controllers give a fixed speed, fixed rate pump. This limitation can be overcome with a variable-speed controller where the frequency of electric current is varied, thereby changing the rpm of the motor and the resulting produced volumes. This drive allows changes to be made whenever a well changes volume, pressure, GOR, or water cut, as well as in wells where PI is not accurately known. The transformer simply changes the voltage of the distribution system to voltage required by the ESP system. Under normal operating conditions, the operating life of ESP can be expected from 1.5 to 5 years, with some units operating even over five years. With the recent improvement of ESP metallurgy and cable technology, some manufacturers claim that ESP run life is even more than five years under normal operating conditions. One of the main reasons of failure of ESP is the breakdown of insulation at downhole either in the cable, cable joint, motor, etc.

When examining the use of ESPs, there are four main considerations, all of which interlinked with each other:

1. Determining and optimizing the well performance with associated pump, motor, cable type and protector.
2. Choosing a method of deploying and recovering the pump-tubing.
3. Venting of annulus gas and the requirement and selection of a gas separation method and other monitoring systems like bottom-hole pressure transmitter.
4. Setting depth of pumps depends upon PI of the well.

5.1.1 *ESP Well Performance*

The starting point for well performance is combining the tubing in-take performance with the reservoir inflow performance. Almost all pump designs are now done with help of software packages, but a graphical approach demonstrates the principles. If the pump is going to be run with a packer or run with the suction pressure above the bubble point, then all of the gas flow through the pump and tubing and conventional tubing, performance curves are used. If gas flows up the annulus then this will reduce the GOR of the tubing flow which needs to be accounted for.

The downhole rate will be higher by the Formation Volume Factor (FVF).

The pump performance can be derived either experimentally or predicted by mathematically equation based on the geometry of the pump stage. In practice, the ESP vendors usually provide the pumps performance curves with pump being specific to the casing size. The pumps curves are provided for software programs as a polynomial equation for ease of calculations.

The motor must also be sized to match the same casing inside diameter as for the pump. A given motor will be available in a range of power and voltage combinations. Running the motor below its maximum current rating will reduce operating temperatures and prolong the life of the motor.

The Power Factor (PF) is the ratio of real power to apparent power. Because of inductance of the motor, there is a lag between the voltage and the current; some of the energy taken from the electrical supply is stored and transmitted back to the supply, later in the cycle. A low PF, therefore needs a larger current and this will lead to larger cable energy losses. In case, the PF is 0.85, so the apparent power is 308 kVA,

$$\text{Power Factor} = \frac{\text{Real Power (kW)}}{\text{Apparent Power (kVA)}}$$

From the apparent power, the cable current can be calculated. The cable current ratings will be higher than the current 'used' by the motor. From the cable current, the voltage drop along the cable can be calculated by reference to a voltage drop chart. The thicker the cable, the lower the cable loss will be required for the cable. When a motor starts, the starting current will be 6/7 times the continuous running current the cable losses will, for a short period, be very high, and if the cable is too thin or long, this might be enough to prevent the motor from starting. Cable sizes reference conductor diameters, cross-sectional areas or the American Wire Gauge (AWG). This rather obtuse measurement (at least to non Americans) refers to the number of drawing operations needed to produce a given size of wire. The conversation to conductor diameter (in inches) from AWG is:

$$D = 0.005(92^{(36\text{-AWG})/39})$$

Up to now, the pump has been treated as rotating at a fixed frequency, although the motor speed varies due to slippage. The engineer can also deliberately vary the rotational speed by varying the electrical frequency. Speed controllers, called Variable-Frequency Drivers (VFDs) or Variable-Speed Drives (VSDs), are used to vary the electrical frequency. They use solid-state electronics that first convert the input Alternating Current (AC) coming in at 50 or 60 Hz (depending on location) to Direct Current (DC).

Gas Handling

Generally, allowing gas to enter a conventional ESP pump is detrimental to performance and reliability. There are three main methods mitigating these problems:

1. Pump should be set at deeper depth or operate at low enough rates such that the pump suction pressure is above the bubble point pressure.
2. Provide the separate the gas out line before it enters the pump and produce the gas separately, usually via the annulus of the casing.
3. Pump does not have to be changed to handle gas.

Gas in solution in the oil at the pump inlet is not a problem and generally beneficial (reduced viscosity and, further up the well, reduced tubing pressure drop). If there is no free gas at the first stage of the pump, there will be no free gas through any of the following stages, as the pressures will be higher. The gas that affects the pump is the free gas, expressed as the Gas Void Fraction (GVF); this is the volumetric fraction of the gas in the fluid. Through knowledge of PVT, it can be calculated from basic PVT data. Assuming no slippage of the phases through the pump, the GOR in terms of in situ conditions is,

$$GOR_{(rcf/rcf)} = (R_{sb}-R_s)\left(\frac{B_g}{B_0}\right)\frac{1}{5.6146}$$

Where R_{sb} is the solution GOR at the bubbles point (*scf/stb*), R_s is the solution GOR at the bubbles point (*scf/stb*), R_s is the solution GOR at the pressure being considered (*scf/stb*), *Bg* and B_0 are the gas and oil FVFs.

The GVF can then be worked out from this ratio:

$$GVF = \frac{GOR_{rcf/rcf}}{1+GOR^{rcf/rcf}}$$

Water can also be incorporated into the equation and results in a lower gas volume fraction.

5.1.2 *An ESP System*

Electric Submersible Pumping (ESP) Systems (refer Figure 5.1d) incorporate an electric motor and centrifugal pump unit run on a production string and connected back to the surface control mechanism and transformer via an electric power cable.

The downhole components are normally suspended from the production tubing above the perforations. In a conventional system the motor and downhole monitoring tool are located on the bottom of the work string. Above the motor is the seal section, the intake or gas separator, and finally the pump, which is connected to the production tubing by the discharge head, which is basically a crossover from the ESP flange to the tubing thread. A specially designed flat cable (motor lead extension) connects to the motor at the upper pothead section, and is then spliced to a round or flat power cable above the ESP.

This cable is then banded or clamped to the tubing all the way to the tubing hanger, where an electrical connection is made. A surface cable is then connected from this point to a junction box and finally to the Switchboard or VFD. The Switchboard/VFD receives power from a utility grid or generator. When the ESP is started it draws the required voltage from this source. This powers the motor and thus generates pump rotation. As the fluid comes into the well through the perforations it passes by the motor and into the pump. This fluid flowing past the motor brings about a natural cooling process, critical for long and efficient motor life. The fluid then enters the intake and is taken into the pump. Each stage (impeller/diffuser combination) adds pressure or head to the fluid at a given rate. The fluid will build up enough pressure as it reaches the top of the pump to lift it to the surface and into the separator or flowline.

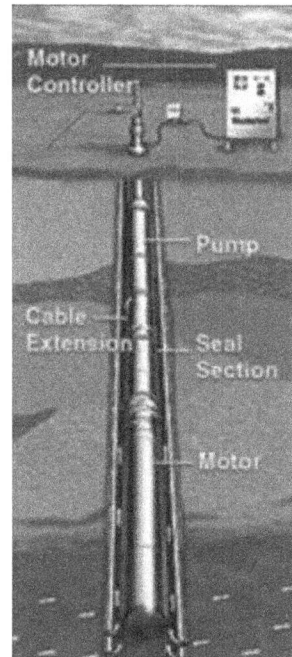

Fig. 5.1(d)
(Courtesy: Weathered Ford)

5.2 Applications

If we hark back a few decades from now, we find that ESP had application in lifting water from water well and thereafter ESP was used to produce an oil well with high water cut. Perhaps the first version of ESP was brought out in the name of REDA. The full form of REDA is 'R' stands for Roto, 'E' for Electro, 'D' for Dynamo, and 'A' for Arutunoff after the, name of a Russian Scientist who had first patented the pump for lifting water from under the ice covered Alaska region. Many offshore and onshore wells are currently being produced by ESP, especially where wells are high producers. Companies like M/s. REDA, Schulumbeger, Centrilift Saker Hughes, TRICO, Wood group etc. manufacture electrical submersible pumps.

There are many applications where an ESP will provide the most efficient cost-per-barrel lifting method. The following applications are typical areas where ESP technology would provide best-in-class service:

- High volume lift requirements (100–30,000 BPD)
- Deep, hot and/or deviated wells
- Water floods or high water-cut wells
- Limited offshore surface footprint.

VSD/Monitoring Applications

- Depleting reservoir
- Uncertain reservoir characteristics

- Well testing operations
- Well management.

5.3 Surface and Sub-Surface Components of ESP

Electrical submersible pumps consist of various equipment and their allied parts. The equipment can be broadly segregated as surface and downhole components.

The downhole components are: Electric motor, Protector, Pump intake/gas separator, Multistage centrifugal pump, Pressure Sensing Instrument (PSI), Pothead extension of power cable, Power cable, Centralizers, Cable bands, Check valve, Bleeder valve (drain valve), Pump in take connection pig tail penetrator of packer.

The surface components are: Wellhead, Mini-mandrel, Upper pig tail, Surface cable, Junction box, Booster, Switch board, Power transformer, VSD.

5.4 Downhole Components

5.4.1 *Electric Motor*

The electrical submersible pump motor is of two-pole three-phase squirrel cage induction type one. These motors operate at a nominal speed of 3500 rpm on 60 Hertz cycle and 2915 rpm on 50 Hertz cycle. These motors are filled with highly refined mineral oil. This highly refined mineral oil provides the necessary dielectric strength as well as a good thermal conductivity which prevent motor from getting overheated and thereby damaged. This mineral oil by virtue of its quality also serves as a lubricant for the bearings installed around the shaft inside the motor.

In general the motor, in any electrical submersible pump assembly, is attached to the bottom-most part of the assembly. The intake part is placed above the motor. Therefore, when the motor is in operation the well fluid first passes on over the exterior of the motor Figure 5.2 and thereafter it enters the intake section of the pump. In doing so, the fluid carries the heat generated by motor operation and thereby keeps the motor relatively cool to perform the motor operation within the recommended temperature range. In some cases, the motors are placed above the pump assembly and in that case, some sort of shroud is provided to divert the liquid so that the well fluid flows past the motor housing before entering intake section. In some cases the pumped fluid is routed through the exterior of motor for transferring heat from the motor to the moving fluid up the surface.

The motor normally consists of low carbon steel housing with brass and steel laminations placed inside. The motor shaft material is carbon steel or high strength steel. The steel laminations are aligned with the rotor section whereas the brass laminations are aligned with the radial sleeve bearings. The squirrel cage rotor is made up of one or more sections depending on motor horse power and length of the motor. In the case of motor stator it is winded as a single unit in a fixed housing.

Heat generated by the motor is dissipated in the well fluid as it flows by the motor housing.

The precision steel stator laminations focus the magnetic forces on the rotors to reduce the energy loss. The stator windings have been enhanced with increased copper fill and a superior high temperature insulation system. There is an expanded oil reservoir in the base for extra cooling capability is provided.

The standard motor has thrust bearing which is a type of fixed pad and whose purpose is to support the thrust load of rotor stator as well as to keep the rotor shaft aligned vertically with the stator's magnetic field.

Motors are manufactured with different diameters as more conveniently named by different series to suit the various physical dimensions of the well i.e., the minimum I.D. of the casing. The smallest diameter motor is of 375 series for 4 1/2 inch cased-hole. The other series are 456, 540, 738 etc. The horse power of the motor ranges from 6.3 H.P. to even more than 1000 H.P. (Refer Figure 5.2(a) Reda Motor, courtesy M/S Reda Co.).

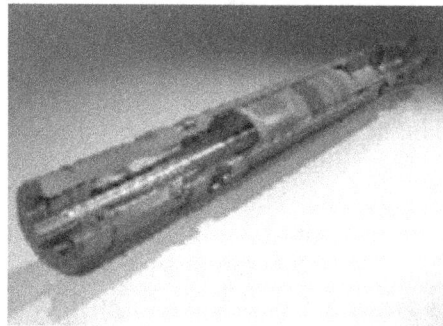

Fig. 5.2(a): REDA Motor
(Courtsey: REDA Company)

Generally length of the motor ranges from a few feet, say 5–8 feet, to even more than 100 feet. When large H.P. motors are required, two or more number of motors are coupled, this total combination of motors are called TANDEM motor. Generally 375 series motors with 50 Hz, tandem configuration of motor is required only when the motor H.P. exceeds 25 H.P. Most of the wells in Assam oil fields of ONGC are being operated on single housing motor of around 18.3 H.P. and 22 H.P. with only a few have high H.P. motors which are of tandem configuration (two single motors are joined).

5.4.1.1 *Submersible Electric Motor*

The submersible electric motor is the prime mover of an electric submersible pumping system. It is supplied with electrical power from surface via the ESP cable and converts this into meaningful mechanical work.

Fig. 5.2(b)

Global statistics show that the motor is among the most likely components to cause premature failure of an ESP system. As such, most of the ESP companies are dedicated to ensure that only the best material and technology is used in motors and application engineers ensure that the appropriate selections are made for the particular downhole conditions of each well.

5.4.2 *Protector*

The very name of the protector implies that it protects the motor (Refer. Figure 5.3a). That is the reason protector is connected just above the motor. During operation of the motor, the highly refined oil inside the motor gets heated up and owing to that, the internal pressure of motor gets increased. Again during the idle period of motor, the oil inside the motor remains cool and due to this, a low pressure is created inside the motor. If this is allowed to continue, then motor housing may burst or collapse because of the differential pressure across the wall of it. The protector acts as a breathing element of the motor. During running of the motor, the protector breathes out, meaning, it releases some motor oil through the protector in the well. Again during the idle hour, because of the low pressure inside the motor, the motor inhales the well fluid inside it through the protector. In this way, it maintains a pressure balance inside and outside the motor/protector by keeping the differential pressure to a minimum.

If the well fluid is allowed a straight entry from the protector to motor then in no time the motor oil gets displaced by well fluid due to gravity segregation which entails a complete break down of the motor insulation. This is reason the protector operates utilizing the labyrinth path principle. This is accomplished by allowing the well fluid and motor oil to communicate through labyrinth tube paths in several U-tube fashion where each U-tube is enclosed in sealed chamber. There is also another way of preventing the well fluid to have a direct access to the motor by providing a bag or balloon inside the protector housing. The bag/balloon collapses during the running of the motor and expands during the idle moments of motor. This design is termed as "positive seal" design. (Refer Figure 5.3b, Reda Modular Protector, which is a combination of labyrinth and positive seal, courtesy M/s Reda Co.). The protector also houses pump thrust bearing to carry the axial thrust developed by the pump.

The labyrinth type of protector can be of two chamber, four chambers, six chambers, or eight chamber type as manufactured by different companies as per their patented design. Again number of chambers means the equal number of seals of the rotating shaft and the housings. For example two chambers means two seals. It has been found by experience that even two chamber labyrinth protector is sufficient to prevent the well fluid from making an entry in the protector but whenever the breakdown of motor insulation has resulted, it has been found, in most of the cases, the well fluid has entered the motor, through the leaking seals. That is why we prefer 4 chamber protector over 2 chamber protector. Since 4 chamber protector is not the common product of all ESP manufacturers, two number of 2 chamber protectors are coupled to make 2 + 2 i.e., 4 chamber protector. In many North sea offshore wells 4 chamber protector applications have been in use.

5.4.3 *Pump Intake/Gas Separator (Refer Figures 5.4a and 5.4b)*

The pump intake is connected in bolt-on-fashion to the lower side of the pump section of the electrical submersible pump and to the top of the protector. In other words this pump intake is connected between the protector and the pump section. This provides a path for the fluid to enter into the pump. Very often, the straight intake section is replaced by the other forms of intake sections called gas separator for separating out the free gas from the liquid before the liquid enter the pump. The free gas is routed up through the annulus to ultimately get discharged in the flow line. The non-return valve installed after the valve of the annulus of wellhead prevents the gas/flow line fluid from flowing back into the annulus. In India all the ESP wells, gas separator intake is used rather than straight pump intake. Gas separators are broadly categorized into two types. The first one is the poor-boy type gas separator where the fluid bends 180 degrees i.e., from upward direction of flow to downward direction. In the process

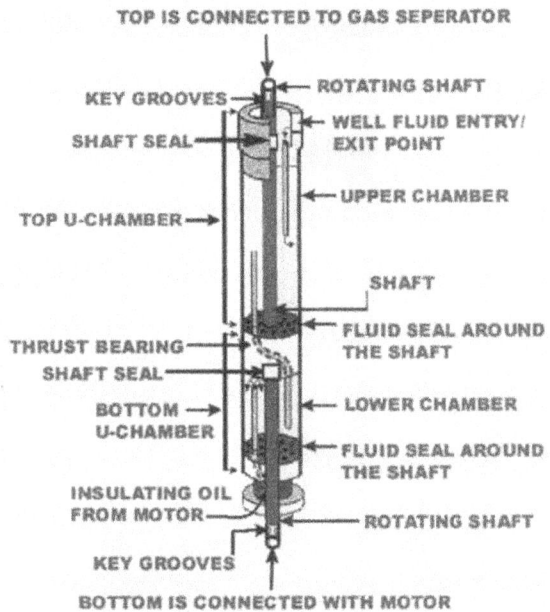

Fig. 5.3(a): A Typical Labyrinth Path Type Protector or Seal Chamber

Fig. 5.3(b): REDA Modular Protector (Combination of Labyrinth and Positive Seal) [Courtsey—REDA Company]

free gas separates out and the liquid enters in the pump through inner pull tube of the gas separator. The separated gas finds a way out to the surface through the annulus. This type of separator is also called static or reverse flow separator.

Some companies like M/s Schlumberger REDA employ one inverted impeller just below the pull tube of the static type gas separator. This inverted impeller owing to its inverted operation, pressurizes the fluid to some extent and in the process, if at all some free gas is present, it gets dissolved into the liquid which then moves up into the pull tube. Most of the wells in India are equipped with this type of separator.

The second category is the rotary type separator. (Refer Figure 5.4b, Reda Rotary Gas Separator, Courtesy: M/s REDA Co.) This rotary type separator by its rotary centrifugal motion separates out the gas and liquid. This centrifugal action keeps the denser fluid to the periphery and allows the lighter fluid like gas to rise from the center of the rotary gas separator through the path of flow divider/cross-over section into the annulus, finally the separated gas to be discharged into the flow line.

Rotary gas separator has some distinct advantages over the reverse flow type. Due to centrifugal action, separation of liquid from free gas is more effective. Secondly, remaining free gas in the form of minute bubbles can be dispersed all through the liquid medium and make the liquid less dense. (This gas is other than the free gas which is at the center of the rotary separator). This less dense liquid finally enters the pump and increases its efficiency. But in some cases, this rotary gas separator has not proved effective.

As the liquid rotates, it can create an unbalanced lateral thrust and shaft vibration, which can accelerate seal failure in the protector. Many rotary gas separators are in use in Indian oil fields located in Assam.

Example: Intakes/Gas Separators

Many companies offer several intake configurations to provide operators the flexibility to customize ESP systems to specific well conditions. Depending on well conditions, it can be in the form of a simple intake adapter with inlet-holes or a more complex gas separation equipment.

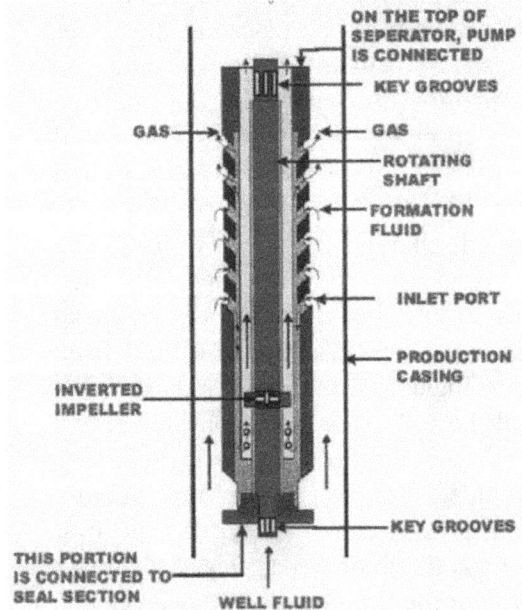

Fig. 5.4(a): A Typical Reverse Flow Gas Separator

Intake: A standard intake can be used when the well is producing above the fluid bubble point pressure, or if a maximum of 10 % free gas is being produced.

Gas Separators: It has long been recognized in the ESP industry that free gas entering the pump causes significant problems, such as gas locking, lower bearing lubrication, decreased efficiency, few companies continues to develop equipment to cope with the problem of gas, and can currently offer the following different separatism systems.

Vortex Gas Separator: The vortex gas separator is used for a higher efficiency separation system. Based on well-known fluid separation principles the vortex separator has been improved significantly to provide effective higher rate of gas separation based on reduced vibration, high specification bearings, lower H.P. requirements, and improved hydro-dynamic efficiency.

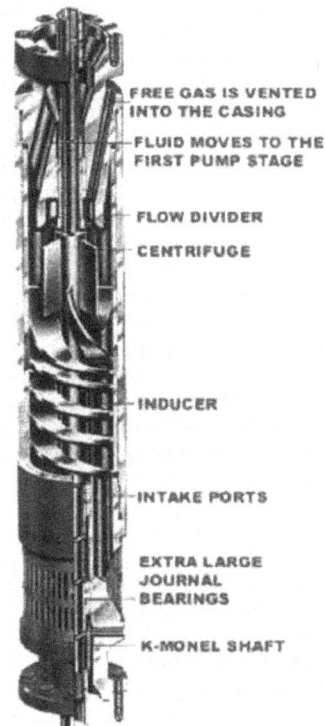

Fig. 5.4(b): REDA Rotary Gas Seperator (Courtesy: REDA Company)

Tandem Vortex Gas Separator: The tandem system was an obvious development, offering gas separation facility for high PI wells, where the single system efficiency would be compromised at the higher rates.

5.4.4 *Multistage Centrifugal Pump*

Electrical submersible pumps are centrifugal pumps in a multi-stage fashion (Refer 5.5a, Reda high efficiency pump, courtesy M/s Reda Co.). Obviously because of the physical parameters of the well (inside diameter of the casing), diameter of the pump is very much restricted and therefore, its design are different from surface centrifugal pumps. The OD and the type of impeller design determine the rate of fluid production. Whereas, the number of stages where each stage consists of one impeller and diffuser are governed by the requirement of the head of fluid to be lifted to the surface against the given tubing head back pressure.

The principle of ESP operation to accomplish its job simply follows the principles of Physics. As the liquid under positive head enters into the eye of the impeller, the liquid flows out laterally by the rotating impeller. As the impeller rotates, it imparts a rotary motion to the fluid thereby considerably increasing its (Figure 5.5a) kinetic energy. Then the diffuser with its expanding area changes the high velocity energy

into low velocity energy i.e. trans-
formation to pressure energy, before
the fluid is re-directed into the eye of
the next impeller. In this way pressure
energy continuously gets built up in a
particular ratio in each successive
impeller as per the design of impeller-
diffuser.

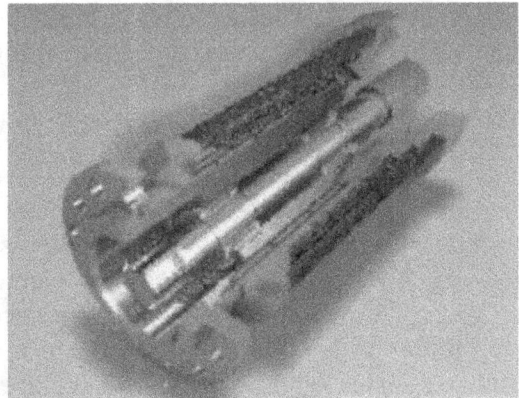

Fig. 5.4(c)

Two types of setting of the impeller
diffusers are in vogue. One is floating
or balanced type where the impeller
floats up and down a little and axially
along the shaft. Floating impeller means
impeller is firmly fitted with the shaft and shaft moves up and down a little.
Depending on the flow rate, the impeller either sits on the down thrust pad or
touches the up-thrust pad or freely floats in between them. Most of the centrifugal
pumps are of floating or balanced types especially those for deeper wells and for
low, moderate and moderately high fluid volume. Therefore when a pump is

operating at greater than designed
flow rate, it may induce an excessive
up-thrust which results in excessive
friction between the up-thrust pad and
impeller. On the other hand when a
pump is operating at less than designed
rate it can create an excessive down-
thrust due to the friction k) between the
impeller and down-thrust pad. This is
precisely the reason why a centrifugal
pump should be operated within a
recommended capacity range where
frictional force is minimum. This

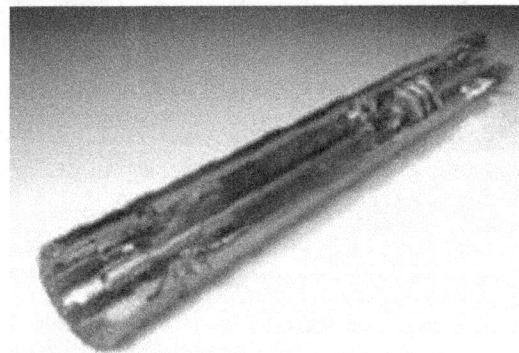

Fig. 5.4(d)

recommended capacity range is available from the pump performance curve as
supplied by the manufacturer (Refer. Figure 5.5b, Head capacity and pump
efficiency curve of a typical ESP).

The other type is the fixed impeller type pump. It is used for pumping very high
volume of liquid. In this type, impeller is fixed to the shaft and the shaft cannot
move up and down axially. The impeller also does not sit on the diffuser pad.

In some earlier cases, electrical submersible pump companies sometimes used to
prefer combination pump where a certain number of the stages at its bottom is of
floater design and the remaining stages at the top are of fixed type. Now a days, this

type of combination of floating and fixed impeller pumps are not normally being used.

During lowering of pump, pumps of different housing lengths are joined in series as per the requirement of the total head to be generated. The lengths of Pump housings are normally available from as low as 2.1 feet to 14.8 feet or more. Each stage of the submersible pump handles the same volume of fluid, therefore the total stages are only linked with total head generation. The pump stages are available in different groups called housings, where one housing houses a number of stages like 54, 74, 99, 151 stage's etc. Two or more housings are connected to create the necessary stages as per the requirement of well. So far, the maximum pump size available is of around 400 stages, as per the information available from M/s Schlumberger.

Metallurgy: The pump housings are normally seamless, heavy walled, low carbon-steel tubing in order to withstand the normal operating pressure of the pump. The outside diameter of this housing ranges from 3.38 inches to even 11.25 inches

NEW STAGE DESIGNS INCREASE OPERATING EFFICIENCIES UPTO 6%

Fig. 5.5(a): REDA High Efficiency· pump (Courtsey: REDA Company)

(designated as per the OD of the housing). Stages are manufactured with the materials which can provide optimum performance as well as resist the corrosion and erosion. Generally K-Monel shafts are provided as a standard shaft material. The impellers normally are Nickel, Ryton and Bronze. Diffusers are generally made of Ni, which provides hardness whereas, bronze imparts ductility.

5.4.5 *Pressure Sensing Instrument*

This pressure sensing instrument is coupled below the motor. It is composed of a surface read-out unit and a surface detachable downhole pressure and temperature sensing instrument. The surface unit is connected to the downhole sensor through the motor windings and through the same power cable which is used to operate the pump. It records the pressure and temperature of the fluid at the pump depth.

5.4.6 *Pothead Extension Power Cable*

Pothead extension power cable is used to connect the motor with the main cable. One end of the pothead is joined with the main cable and the other end is joined to the motor head. While connecting with the motor head, there should be a

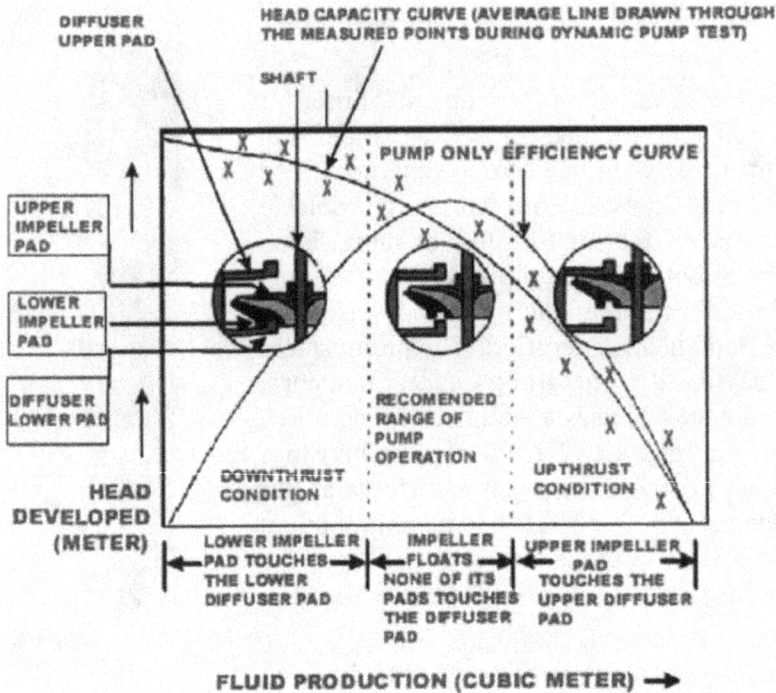

Fig. 5.5(b): Drawal of Head Capacity and Pump Efficiency Curve of a Typical ESP

compatibility of the motor head joining section with the pothead cable joining section. There are two types of pothead extension power cable available in the market.

- Plug-in type pothead.
- Tape-in type pothead.

The plug-in type pothead is similar to a three pin plug with necessary 'O'-ring fitted for fluid seal. This is inserted into motor three pin-hole and then rigidly bolted with the motor body. This type of connection has some disadvantages:

1. When the plug-in pothead is used a number of times, it looses its proper fitting with the motor body.
2. Due to more rigidity of this type of cable-motor connection there is always a possibility that a hair crack in the pothead just above the plug-in point may accidentally develop. This generally goes unnoticed and once the tubings are lowered into the well it (the damaged pothead) comes in contact with the well fluid and thus break in insulation results.

Tape-in type pothead is a better proposition and for this motor head should also have the compatibility for this type of connection. This connection is similar to a connection between two cables. The flexible wires inside the motor are taken out first with the help of pliers and then necessary splicing job is carried out to connect the pothead extension cable with the motor cable. Finally the flange fitted with the pothead extension cable is fitted with the 'O'-ring on the motor body for fluid seal and rigidly fitted with bolts. Because of the very flexible nature of connection with this tape in type system there is absolutely no room for having any break in insulation, once the proper splicing job is ensured.

When electrical submersible pumps were introduced in India as early as in 1975–76, only the plug-in pothead connections were being used as per the recommendations of the concerned manufacturers.

Subsequently with experience it was realized that tape-in type was a better option. Numbers of pot-head insulation failures were reduced drastically to a bare minimum.

These pot-head connections are normally available in lengths of 40 feet, 50 feet and 60 feet so that connecting section of the main and pot-head cable remains above the pump assembly. Normally, the pot-head cable being connected, just above the motor, is subjected to less conducting capacity than the main cable. For example the main cable is of # 4 AWG (American Wire Gauge) whereas the pot-head extension cable is of # 6 AWG. Primarily this is because of economic reasons, if the lower capacity cable is technically permissible.

Pot-head extension cable is a very sensitive part of the total ESP system. ONGC has made a principle to procure pot-head extension cables in small packages with sufficient shock absorbing cushions. Also, the pot-head extension connection with the main cable is made at the well site after the main cable passes over the hanging ESP pulley over the well head.

5.4.7 *Monitoring System*

As operators move to deeper and more demanding reservoirs the requirement for downhole operating data and analysis has never been more critical. Monitoring downhole conditions and pump performance is finally being recognized as the largest sole mechanism for maximizing the full potential of the reservoir.

The Weatherford Downhole Monitoring System will help operators realize this full potential by improving the ultimate reservoir recovery and extending the life of the equipment.

Power Cable: Power cable is the means through which power is supplied from the surface to the downhole motor. The cable has been standardized by AWG (American Wire Gauge) standards. In this standard, sizes of conductor range from # 1 AWG to # 6 AWG. #1

Fig. 5.5(c)

AWG signifies a thicker conductor. As we approach from # 1 to # 6 AWG, the conductor sizes become thinner and thinner. That means its current carrying capacity becomes less and less. The total range meets all electrical submersible motor amperage requirements. Almost all cables use copper as conductor, however there are few where the conductors are of aluminum. Although aluminum is cheaper than copper, their current carrying capacity is much less than that of the latter. If the aluminum cable is used in place of copper cable for a similar requirement of electrical submersible pump then cable diameter will become bigger and it may not physically permit running of cable into the well. In such cases the pump and tubing diameter have to be reduced and therefore copper conductor is always used.

Power cable is made up with three separate conductors separated from each other by proper insulating material. Each conductor is meant for each power phase where three phase supply is present. The composition of the insulation with proper thickness which determines the cable's resistance to current leakage as well as to prevent permeation by well fluid especially gas are most important aspects of the cable. Very few electrical cable manufacturers make as per these standards suitable for use in electrical submersible pump operation in oil wells.

All ESP cables are armored, like galvanized armor, which prevents the cable from getting physically damaged. During lowering and pulling out jobs of ESPS, it is obvious that cable suffers severe abrasion by the rubbing of casing-tubing wall. Since bare cable (i.e., without any armor) cannot sustain this abrasion, armored cable is used.

The cable configuration is of two types:

- Round cable;
- Flat or parallel cable.
  (Refer Figure 5.6(a) for Reda Round and Flat cables).

Fig. 5.6(a): REDA Round and Flat Cable (Courtesy: REDA Company)

Round Cable: Round cable, as the name implies, is round in shape. The three conductors, each enclosed by insulation and sheathing material are placed side by side at 120° to each other and then finally all three are covered with insulation and sheathing materials. On the exterior of it, armour is provided. It therefore forms a round shape. To make a positive fluid seal at the wellhead with Hercules make wellhead, round shape cable is preferable. But, sometimes overall diameter of the cable and tubing coupling becomes more in case of round cable than that of flat cable. More so, round cable, because of its less surface area in contact with the tubing, the tight gripping of cable with the tubing is often difficult. The frequency of failure of clamps holding round cable is more. As a result, a portion of cable gets accumulated at one place and prevents normal retrieval operation and finally calls for complicated fishing jobs.

Flat or Parallel Cable: Flat or parallel cable as the name suggests, are those where all the three conductors are placed side by side and parallel to each other and the cable looks flat. Like in round cable, each conductor is enclosed by insulating and sheathing materials and finally the whole cable is enclosed by insulating and sheathing materials followed by metallic armor like galvanized armor.

Fig. 5.6(b)

With the use of parallel cable, the running in of ESP becomes comparatively easy simply because of more rigid gripping with the tubing owing to the flat surface of the cable and less chance of damage during work over operation.

However, this type of cable always poses a problem with the "Hercules" type of wellhead. Since flat cable type stuffing box in its same configuration, as such, cannot pass through the rubber packing of the wellhead, the individual conductors with their usual sheathing material get separated and are made to pass through the rubber packing. This does not ensure a fool-proof sealing. However with modified, high pressure wellhead like "seaboard" wellhead, this problem has been overcome.

Example—Weatherford Cables

Various companies offer a complete range of cables suitable for all ESP applications. Pump cable selection requires careful consideration in order to meet anticipated mechanical, temperature and chemical stresses. The correct selection is critical to the efficiency and reliability of the entire system.

All the cables are available in either flat or round configurations. The flat configuration is available when the standard round cable cannot fit in the well due to limited space between the down-hole string and the well casing.

All ESP cable are manufactured to all applicable API and IEEE standards and under ISO 9000.

5.4.8 *Centralizers*

At least one centralizer can be installed below the motor in the pump seat assembly and another above the top of the pump assembly. This can keep the motor in the central portion of the casing for better cooling effect. Also, many more centralizers can be placed in the entire length of tubing string, which will greatly minimize the dragging effect of cable between the tubing and casing wall.

5.4.9 *Cable Bands*

Cable bands, though a very small item, is one of the vital components of ESP system. Cable band attaches the cable rigidly with the tubing with the help of cable tensioner hand machine and clamping tool. A number of cable bands are required to rigidly grip the cable with the tubing during the lowering of ESP. About 3 to 4 cable bands are required for one stand (2 for single tubing). In this way whole cable weight is distributed equally to the total cable clamps used (Refer Figure 5.7). In case, one clamp breaks, the lower clamp has to bear the cable weight of two segments above it which make it susceptible to give way and then, automatically, the successive lower cable bands give way under the increasing cable load above them. This leads to the coiling of cable, while tubings are pulled out during pulling out of ESP. As the tubings are being retrieved leaving cable behind, this may ultimately create jamming of tubing and even create a situation where it is impossible to retrieve anything from well and virtually one has to abandon the well. So proper number and quality of cable band must be used to prevent such occurrence. Also the clamping tool must be in perfect condition. Time to time it is necessary to replace this tool by a new one. These cable bands are of one-time use only. During re-run of ESP again new cable clamps are required.

In the Motor-Protector-Separator-Pump section cable bands are protected from getting damaged because of friction with

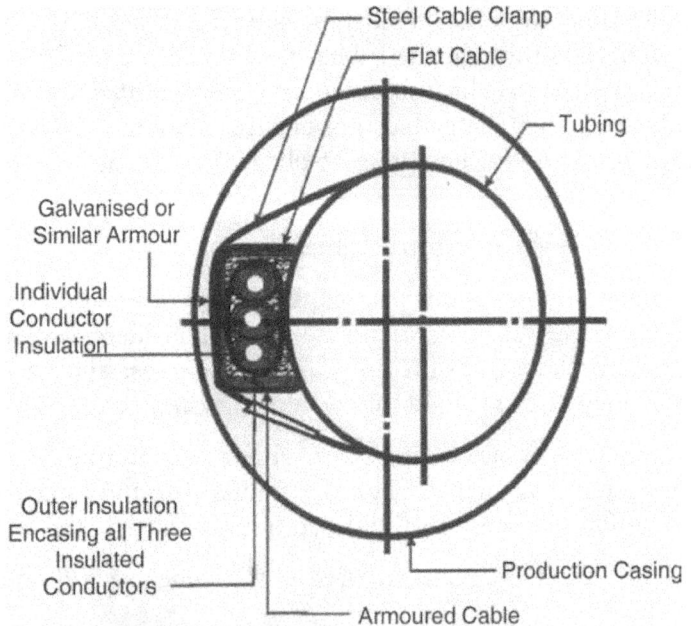

Fig. 5.7: ESP Downhole Flat Cable (In Clamped on Position)

the casing, by GI guard channel. In the rest portion up to the top of the well, no such guard is necessary.

5.4.10 *Check Valve*

Check valve is a non-return valve installed in the tubing just above the pump. It is a flapper disc type valve and it allows the flow from bottom to surface and does not allow the liquid to run down the well through it. It always helps to keep the tubing full with the liquid and does not allow the liquid to run down through the pump during the idle condition of the pump. Running down of liquid may sometimes cause the impeller to rotate in opposite direction and if the pump starts at that moment, it draws a sudden huge current, which can damage the electrical components of down-hole equipment resulting in the insulation failure and thereby costly workover job.

5.4.11 *Bleeder Valve*

Bleeder valve (or Drain valve) is installed above the check valve. It is used to drain out liquid from the tubing during pulling out job. Before ESP is pulled out the drain valve is broken by dropping a heavy rod from the top. If it is not installed, only the wet tubings will be retrieved and on opening the tubing, oil/water will be splashed on the derrick floor making it (floor) very slippery and create difficulty for the persons to work there. Besides this, it may create condition for blow-out of the well.

This bleeder valve has got a disadvantage too. if during mechanical scraping operation of tubing, scraping wire gets snapped and sinker bar falls, it breaks the bleeder valves nipple and there will be no alternative other than to pull out the tubing for changing the bleeder valve, which means a costly workover job.

5.4.12 *Pump Top Substitute*

Pump top sub is a connecting substitute to connect the top of the tubing. Its lower portion is a flange to flange connection with the top portion being box-threaded to connect with the pin end of tubing.

5.4.13 *Lower Pig Tail*

Lower pig tail is a small length of main cable with one end spliced with the main cable just before the wellhead (as and when the running-in is completed) and the other end being connected (coupled) with the electrical mini-mandrel is installed in a specially drilled-hole in tubing hanger by the side of the tubing connection. This type of lower pig tail—mini-mandrel connection is meant for the "seaboard" or equivalent types of wellhead.

5.5 ESPs Design

Like most down-hole tools in the oil field ESPs, are also classified by their outside diameter (from 3.5 to 10.0 inch). The number of stages to be used in particular outside diameter sized pumps is determined by the volumetric flow rate and the lift (height) required. Thus, the length of a pump three-phase (AC), squirrel cage, induction type. They can vary from 10 to 750 hp at 60 H_z (and range from $3^{3/4}$ to $7^{1/4}$ in diameter). Their voltage requirements vary from 420–4,200 V.

The seal system (the protector) separates the well fluids from the electric motor lubrication fluids and the electric wiring. The electric controller (surface) serves to energize the ESP, sensing such conditions as overload, well pump-off, short in cable, and so on. It also shuts down or starts up in response to down-hole pressure switches, tank levels, conventional electromechanical or solid-state devices. Conventional electromechanical controllers give affixed speed, fixed flow rate pumping. To overcome this limitation, the variable speed controller has been developed (solid state). These controllers allow the frequency of the electric current to vary. This results in vairiation in speed (rpm) and, thus flow rate. Such a device allows changes to be made (on the fly) whenever a well changes volume (static level), pressure, GLR, or WOR. It also allows flexibility for operations in wells where the PI is not well known. The transformer (at surface) changes the voltage of the distribution system to a voltage required by the ESP systems.

Unlike positive-displacement pumps, centrifugal pumps do not displace a fixed amount of fluid but create a relatively constant amount of fluid but create a relatively constant amount of pressure increase to the flow systems. The output flow rate depends on backpressure. The pressure increase is usually expressed as pumping head, the equivalent height of freshwater that the pressure differential can support (pumps are tested with freshwater by the manufacture). In U.S. field, the pumping head is expressed as,

$$H = \frac{p}{0.433}$$

Where,

h = pumping head, ft
p = pump pressure differential, psi.

As the volumetric throughput increases, the pumping head of a centrifugal pump decreases and power slightly increases. However, there exits an optimal range of flow rate where the pump efficiency is maximal. A typical ESP characteristic chart is shown in Figure 5.8.

- Free gas in oil
- Temperature at depth

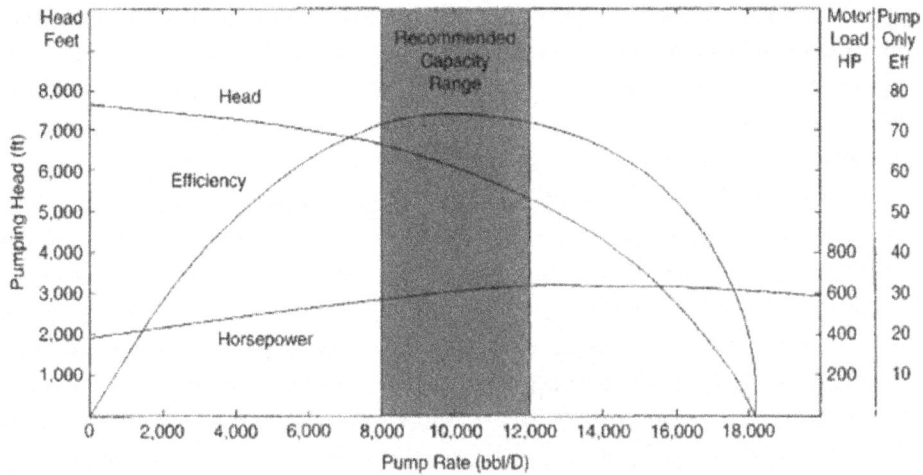

Fig. 5.8: A Typical Characteristic Chart for a 100 Stage ESP

- Viscosity of oil
- Sand content fluid
- Paraffin content of fluid.

Excessive free gas result in pump cavitation that leads to motor fluctuations that ultimately reduces run life and reliability. High temperature at depth will limit the life of the thrust bearing, the epoxy encapsulations (of electronics, etc.), insulation, and elastomers. Increased viscosity of the fluid to be pumped reduce the total head that the pump system can generate, which leads to an increased number of pumps stages and increased horsepower requirements. Sand and paraffin content in the fluid will lead to wear and choking conditions inside the pump.

5.6 Application

The following factors are important in designing ESP application:
- PI of the well
- Casing and tubing sizes
- Static liquid level.

ESPs are usually used for high PI wells. More and more ESP applications are found in offshore wells. The outside diameter of the ESP down-hole equipment is determined by the Inside Diameter (ID) of the borehole. There must be clearance around the outside of the pump down-hole equipment to allow the free flow of oil/water to the pump intake. The desired flow rate and tubing size will the determine the Total Dynamic Head (TDH) requirement for head immediately above the pump (in the tubing). This is converted to feet of head (or meters of head). This TDH is usually given in water equivalent thius, TDH = static column of fluid (net) head + friction loss head = back-pressure head.

The following procedure can be used for selecting an ESP:

1. Starting from well Inflow Performance Relationship (IPR), determine a desirable liquid production rate q_{LD}. Then select a pump size from the manufacture's specification that has a minimum delivering flow rate q_{LP} that is $q_{LP} > q_{LD}$

2. From the *IPR*, determine the flowing bottom-hole pressure p_{wf} at the pump-delivering flow rate q_{Lp} not the q_{LD}.

3. Assuming zero casing pressure and neglecting gas weight in the annulus, calculate the minimum pump depth by,

$$D_{pump} = D - \frac{P_{wf} - P_{suction}}{0.433 \, Y_L}$$

Where

D_{pump} = minimum pump depth, ft
D = depth of production interval, ft
P_{wf} = flowing bottom-hole pressure, pisa
$P_{suction}$ = required suction pressure of pump, 150–300 psi
Y_L = specific gravity of production fluid, 1.0 for fresh water.

4. Determine the required pump discharge pressure based on well head pressure, tubing size, flow rate q_{Lp}, and fluid properities. This can be carried out quickly using the computer spreadsheet Hagedorn Brown Correlation. Xls.

5. Calculate the required pump pressure differential $\Delta p = d_{ischarge} - S_{uction}$ and then required pumping head.

6. From the manufacture's pump characteristics curve, read pump head per stage. Then calculate the required number of stage.

7. Determine the total power required for the pump by multiplying the power per stage by the number of stages.

Example Problem

A 10,000 ft. deep well produces 32° API oil with GOR 50 scf/stb and zero water cut through a 3-in. (2.992-in. ID) tubing in a 7-in casing. The oil has aformation volume factor of 1.25 and average viscosity of 5 cp. Gas-specific gravity is 0.7. The surface and bottom-hole temperatures are 70°F and 170°F, respectively. The IPR of the well can be describe by the Vogel model with a reservoir pressure 4,350 psia and AOF 15,000 stb/day. If the well is to be 8,000 stb/day against a flowing wellhead pressure of 100 psia, determine the required specifications for an ESP for this application. Assumed the minimum pump suction pressure is 200 psia.

Solution

1. Required liquid throughput at pump is,
    $Q_{LD} = (1.25)(8,000) = 10,000$ bbl/day
    Select an ESP that delivers liquid flow rate $q_{LP} = q_{LD} = 10,000$ bb l/day in the neighborhood of its maximum efficiency (Figure 5.8).
2. Well IPR gives,

$$P_{wfd} = 0.\ 125\ p \sqrt{81 - 80 \left(\frac{qLd}{qmax} \right) - 1}$$

$$= 0.125(4,350) \left[\sqrt{\left(81 - 80 \frac{8000}{1500} \right) - 1} \right]$$

$$= 2,823$$

3. The minimum pump depth is,

$$D_{pump} = D - \frac{P_{wf} - P_{suction}}{0.433\ Y_L}$$

$$= 10,000 - \frac{2,823 - 200}{0.433(0.865)}$$

$$= 2,997\text{ft.}$$

    Use pump depth of $10,000 - 200 = 9,800$ fit. The pump suction pressure is,
    $P_{suction} = 2,823 - 0.433(0.865)(10,000 - 9,800)$
    $= 2,748$ psia.
4. Computer spreadsheet Hagedorn Brown Correlations.xls gives the required pump discharge pressure of 3.728 psia.
5. The required pump pressure differential is,

$$\Delta_p = P_{discharge} - P_{suction}$$
$$= 3,728 - 2,748$$
$$= 980 \text{ psi.}$$

    The required pumping head is,

$$h = \frac{\Delta p}{0.433} = \frac{980}{0.433}$$
$$= 2,263 \text{ feet of freshwater.}$$

6. At throughput 10,000 bbl/day, Figure 5.8 gives a pumping head of 6,000 ft for the 100 stage pump, which yield 60 ft pumping head per stage. The required number of stages is $(2,263)/(60) = 38$ stages.
7. At throughput 100 stage pump of 600 H.P., which yields 6 H.P./stage. The required power for a 38 stage pump is then $(6)(38) = 226$ H.P.

5.7 Surface Components

5.7.1 *Wellhead*

When the final run-in of the tubing is completed, it is required to cap the well properly so that only tubing and cable or the tubing and mini-mandrel are protruding out of the surface. On the tubing, X-mass tree is fitted and protruded cable or mandrel is connected to the surface cable. So, the wellhead is to provide a perfect seal around the tubing and power cable and keeps the tubing hanging by its tubing hanger.

There are numerous types of wellheads available in the market. Broadly, ESP wellhead can be categorized into two types:

(a) The wellhead through which the sub-surface power cable protrudes at the surface: This type of wellhead is having necessary rubber packings for providing, leak proof sealing around the cable. "Hercules" make wellhead is one such type. Wellhead is always required to be specified for parallel cable or round cable as well as for the required AWG specification. This type of wellhead, can withstand a pressure of around 1500 psi (100 kg/cm^2). Produced fluid leakage from such type of wellhead through cable-rubber packing seal section is a common problem. Once the wellhead leaks, it is sometimes required to subdue the well and new wellhead packings are installed.

(b) The wellhead through which mini-mandrel protrudes at the surface: Here the main subsurface cable is joined with one end of the lower pig tail and the other end of the lower pig tail is coupled with the mandrel. Similar type of pig tail (upper pig tail) connection is there at the surface. Mini-mandrel is screwed on the wellhead with necessary "O"-ring. "Seaboard" make wellhead is similar kind of wellhead, This type of wellhead can withstand much higher pressure of around 3000 psi (200 kg/cm^2).

Both type of wellheads i.e. 'Hercules' and 'Seaboard' were used in India. Now-a-days "Seaboard" type is being preferred for onshore operation.

5.7.2 *Mini–Mandrel*

Three copper conductor of the required size and around them very good non conducting solid material in a cylindrical shape form mini-mandrel. It is screwed into the slot of seaboard type well-head. Rubber 'O' ring seal is provided for effective fluid seal. Upper and lower pigtails are coupled with the two ends of the mini-mandrel.

5.7.3 *Upper Pig Tail*

The upper pig tail is similar to the lower pig tail. One end is connected to the surface cable and the other end is coupled with the top of the mini-mandrel fitted at the wellhead.

5.7.4 *Surface Cable*

Surface cable is similar to the power cable. Approximately 100 m or so length of cable is laid on the surface to connect the wellhead to the switch board. It is advisable to pass the surface cable through a tubing so that the cable can be protected from any kind of physical damage.

5.7.5 *Junction Box*

Junction box is required especially when low pressure wellhead is used. It is a junction point of well cable and surface cable located at a safe distance from the wellhead. It is a well ventilated box. In case any well gas migrates through power cable at the surface, it gets vented at junction box. This prevents gas getting vented in switchboard which can result in an unsafe condition to work and potential fire hazard. (Refer Figure 5.9(a)).

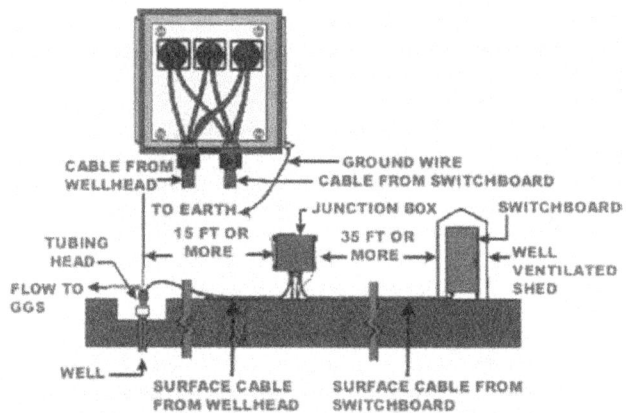

Fig. 5.9(a): Junction Box shall be well Ventilated and Properly Grounded Junction Box Installation

Though there is no scope of gas escape system to lay the cable Junction box to the switchboard. Migrating from wellhead to the mini-mandrel, it is always Junction box and then from Junction box to the switchboard.

5.7.6 *Booster*

Booster is required to boost the surface voltage according to the requirement of rated down-hole voltage. It is connected preferably in between the junction box and switch board.

5.7.7 *Switch Board*

The standard switch boards are weatherproof, but not flameproof. They are available in different ranges of voltage say from about 440 volts to about 4900 volts. Also the selection criteria depends upon other factors like amperage, horse power etc.

It has many features like recording ammeter, fused disconnects, overload protection, signal lights, timers for intermittent auto start etc. In case of under-load operation due to incoming of less fluid in the pump, which causes motor to draw less current, the controller shuts down the unit automatically. However with a selected time delay apparatus, say from 30 min. to 2 hour the unit can be automatically restarted. Similar kind of auto re-start is not provided in case the pump is stopped due to overload operation. Overload shutdown has to be manually restarted, only after the necessary verification of faults causing overloading.

Switch board and booster compressor are housed in a well ventilated room at a safe distance away from the wellhead. The room or switchboard house has its floor padded with adequate thickness of rubber padding. The switchboard house is also properly earthed.

5.7.7.1 *Switchboard Control Panels*

The Switchboard control panel is designed to protect the pumping system from malfunctions and premature failures by positive shutdowns based on pre-programmed alarm parameters, and preventing re-start until it is deemed safe. These control panels can be simple units with push button magnetic contacts and overload protection to more complex assemblies with fused disconnects, recording ammeters, underload and overload protection signal lights, timer for automatic restarts and instruments for automatic remote control.

Fig. 5.9(b)

Fig. 5.9(c)

Should be designed Switchboard to interface seamlessly with performance setting motor controller technology.

All enclosures are NEMA 3R environmentally rated, suitable for even the most arduous outdoor conditions. The enclosure has three isolated compartments for added safety.

Switchboards are available in the market in following four standard sizes:

600 V, 1500 V, 2500 V and 5000 V.

5.7.7.2 *Variable Frequency Drive*

The Variable Frequency Drive (VFD) is an extremely powerful tool in ESP operations. The essential principle of varying the motor rotational speed above and below that dictated by line frequency to improve performance has developed into a myriad of complex functions and features.

VFD technology has much superior automation. This knowledge has more recently been focused on meeting the unique needs of an ESP drive system, and the result is a best in class flexible system which is tailor made for ESP applications while maintaining the focus on providing near unity power factor. The advanced pump control strategies reduce the energy cost-per-barrel of fluid produced, providing significant power and operating cost savings.

As well as the usual drive features of soft starter, electrical parameter monitoring, harmonic reduction etc, the VFD also features a ground breaking data extrapolation system which uses known well, fluid and motor data to monitor critical downhole parameters, ideal for situations where a downhole sensor may not be practical. The non-traditional ESP troubleshooting unit of amps the some system additionally uses nodal intelligence to monitor the system performance at all critical zones, be it fluid level, intake pressure or motor torque. This system ensures that when downhole conditions or pump performance changes the operator will know in real time, and warnings, alarms and shutdowns can be designed for best protection available to the ESP for the life of the well.

Standard low and mid range VFD's are available up to 600 KVA. Larger drives up to 1200 KVA are available for higher horsepower application.

5.7.8 *Power Transformer*

Standard power transformer i.e. step down transformer say from 11 kV to 420/440 V is available with different kW range.

Other vital components of ESP are different types of Electrical tapes for splicing the cable, copper-sleeve for cable conductor to conductor connection, different sizes of rubberized "O"-ring, galvanized armour etc., are required during the installation of pump.

5.8 Standard Performance Curves

The standard performance curves are the most important graphs for ESP design (Refer Figures 5.10, 5.11). For every type of ESP, in its dynamic flow condition standard performance charts are drawn. The abscissa (horizontal axis) indicates the capacity of pumping in bbls/day or m³/day and the ordinate (vertical axis) indicates liquid head to be generated, brake H.P. and efficiency of ESP. For a small type of pump with different RPM, the standard performance chart will be different. So for every pump performance chart, RPM is mentioned.

The head capacity is plotted with the head either in feet or in metres. For simplistic approach fresh water of density 1 gm/cc has been used to generate the performance curve by the pump manufacturing companies.

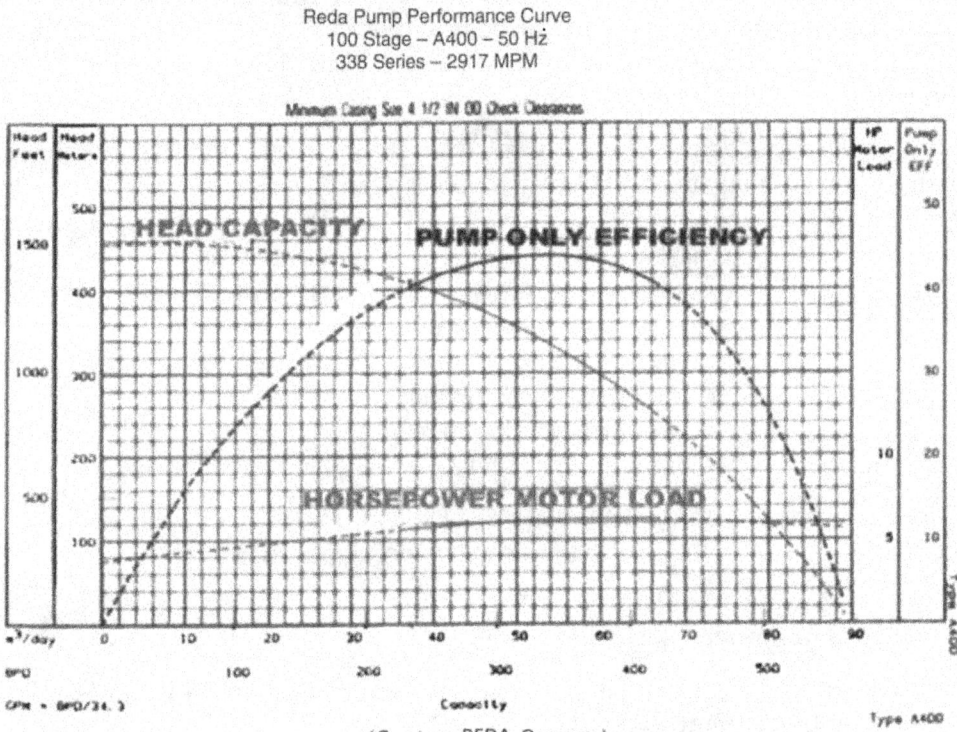

(*Courtsey*: REDA Company)

Fig. 5.10

Reda Pump Performance Curve
100 Stage – A400 – 50 Hz
338 Series – 2917 MPM

Minimum Casing Size 4 1/2 IN OD Check Clearances

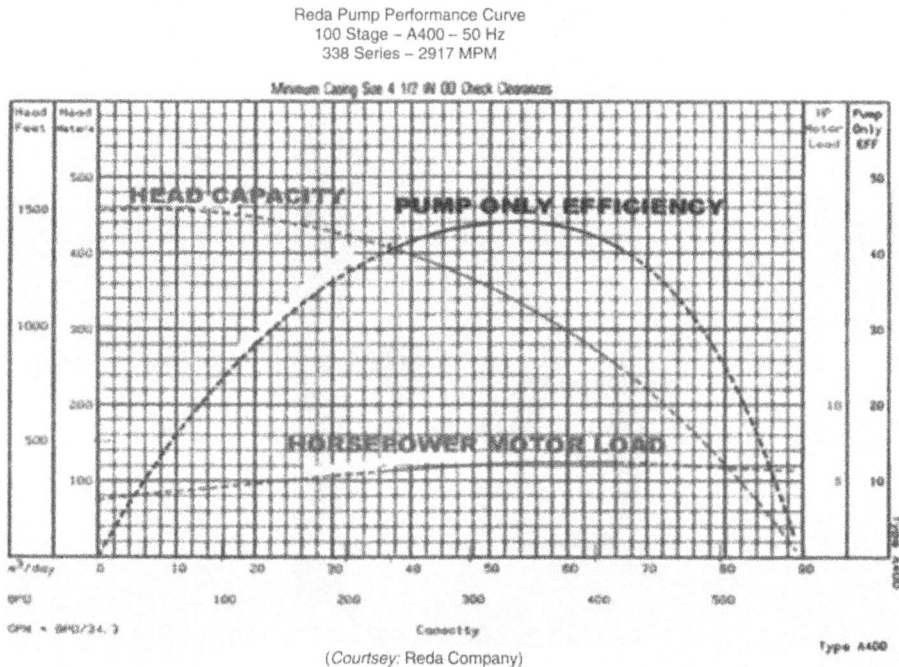

(*Courtsey*: Reda Company)

Fig. 5.11

Also, the performance curve is plotted either with 100 stages of pump or with single stage, as such, some companies prefer the former one and some the latter.

In the pump performance curve, at very low rate or almost zero rate, the head capacity to be developed by the pump is maximum and as the pumping volume increases the head capacity decreases and at one point of pumping the head capacity is zero. It means there will not be any lifting of liquid in the tubing beyond that volume. Keeping an eye on the pump efficiency, every manufacturer has drawn a maximum and minimum range in each performance curve, as such all ESPs are supposed to operate within this range. The space between the maximum and minimum lines is called the recommended range.

As an example, say pump is required to pump 80 m³/day. The type of pump suitable with RPM (as per Indian condition, it is 2915 RPM) and casing size is to be chosen where 80 m³/day falls somewhere in the middle of the recommended range. Let the pump performance chart be made for 100 stages of pump. So the total head is to be developed to the extent of 800 mts.

So, now 80 m³/day is marked on the abscissa. From there a vertical line is drawn which cuts the head capacity curve and from there a line is drawn to horizontally cut the ordinate. Let the value at the ordinate be 400 m of head. It means each stage can develop 4 meters of head (i.e. 400 m is divided by 100 stages). Therefore, to develop 800 meters, number of pump stages required = 800 m/4 m per stage = 200 stages. So, in this way required number of stages can be calculated.

Therefore, in order to find the motor capacity i.e. horse-power of motor = H.P./ Stage × Total stages of pump × Specific gravity of fluid.

It is always advisable to mark the highest B.H.P. from the graph. Say maximum H.P. is marked as 7.65, then it is better to consider 7.7 H.P. (i.e. the next whole number after decimal). Therefore, H.P. = 7.7/1 00 × 200 × 1.05 = 16.17 H.P. (Here kill fluid Sp. Gr is taken as 1.05).

Now say, no motor of that rating (H.P.) is available. So, always, it is advisable to choose the motor of H.P. just higher than the calculated one, say 18 H.P.

5.9 Total Dynamic Head

Total Dynamic Head (TDH), written in short as TDH, is a very common concept in calculating the total stages of a centrifugal pump. This includes:

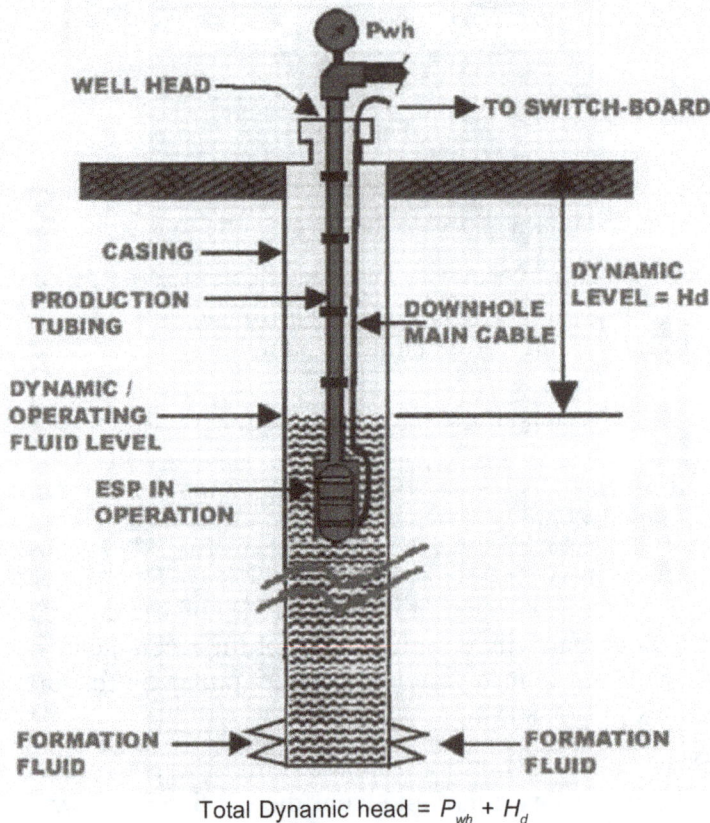

Total Dynamic head = $P_{wh} + H_d$

Fig. 5.12: Calculation of Dynamic Head for ESP

(i) The friction losses in the tubing and surface flow line,

(ii) Tubing pressure against which pumping is to be done,

(iii) The difference in elevation of the dynamic level and the surface,

(iv) Any losses due to valves etc. in the flow-line.

By taking into account only the elevation and tubing pressure and neglecting all other factors, TDH can be written as:

TDH = Tubing pressure [in terms of equivalent fluid (liquid height) + Dynamic level as measured from top i.e., from the surface, (Refer Figure 5.12).

For example, Say tubing pressure = 10 kg/cm² = 100 m of water and dynamic level from the surface = 600 m, then, TDH = 100 + 600 = 700 m.

5.10 Troubleshooting

Whenever any problem of ESP operation occurs, it is required to be properly identified, as some rectification of the surface equipment can overcome the problem. In this manner costly work-over jobs can be saved. So, in this respect, it is required to generate sufficient well data regarding rate of oil production, water cut, run-life of ESP unit, dynamic and static fluid loads, GOR/GLR, pump setting depth, sand cut/ corrosive fluid, how many times ESP has been serviced and reasons thereof, reservoir pressure, reservoir drive mechanism, rate of fall of static pressure, bottom-hole temperature, electrical power supply details i.e. voltage, current, voltage fluctuation frequency of power cuts and other related details.

These data should be recollected regularly, and particularly more frequently in the beginning i.e. just after commissioning or re-commissioning of the ESP unit (say at first, every day and then in every 3 to 4 days, then say after a month and like that). One of the very important source of information, when troubleshooting an ESP installation, is the recording ammeter.

The recording ammeter continuously records the amperage drawn by ESP, either it can be put on operation daily or periodically. It is already inbuilt in the switch board system. Different probable recordings of ammeter charts have been discussed in the following section.

5.10.1 *Typical Ampere Charts of ESP*

Normal and Smooth Operation (Figure 5.13a): It shows the ideal operating conditions. The chart draws a smooth symmetrical amperage curve at or near the name-plate amperage. The producing rate and dynamic head are steady and possibly can vary by approximately 5 %, which is negligible. One spike is shown here, which is normal the occurrence during the start-up of the motor.

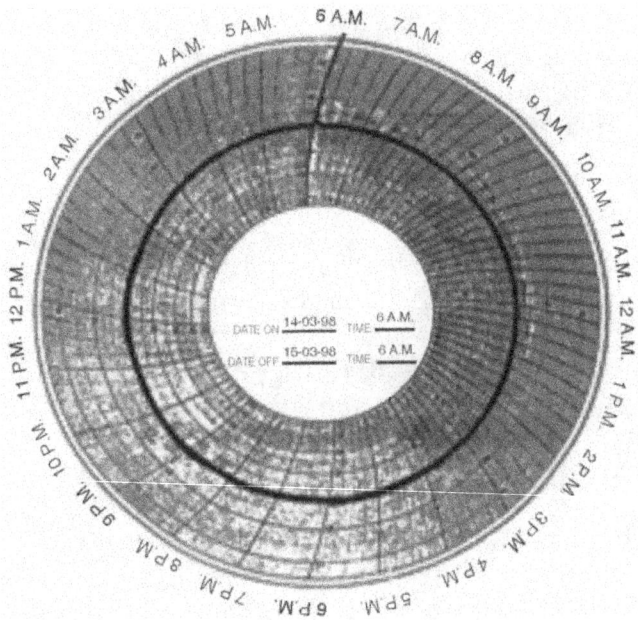

Fig. 5.13a: Pump is in normal and smooth operation, since pumping operation continues with almost negligible fluctuation in amperage.

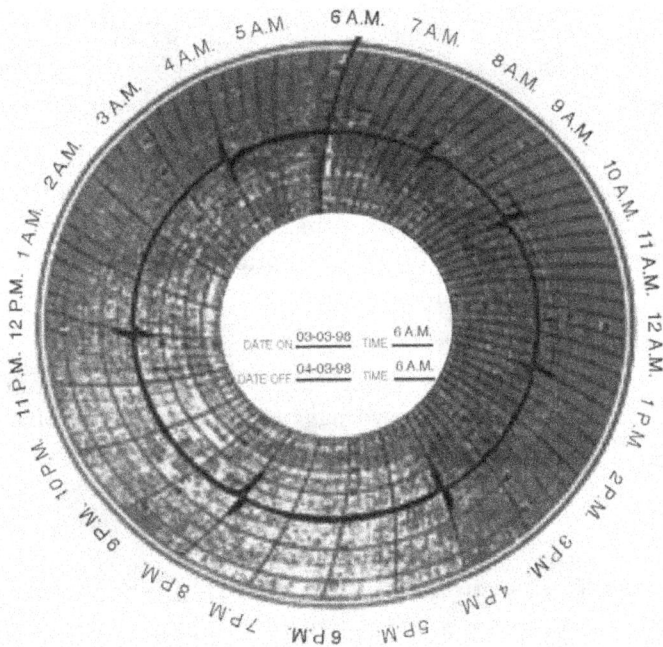

Fig. 5.13b: Pump is operating as normal. However, spikes are noticed, which are presumably due to voltage fluctuation. As a resua current fluctuates as spikes, where range of spikes is limited within the overload and underload settings.

Normal Operation with Frequent Spikes (Figure 5.13b): The graph depicting the spikes from time to time indicate the power fluctuations. This may arise due to various reasons like:

(a) If the primary power supply voltage at source fluctuates, the amperage naturally fluctuates in an attempt to retain constant horsepower output. This type of fluctuation is commonly seen.

(b) Periodic heavy drain of power in adjacent areas may cause such spikes, especially when there is a common transformer for more than one ESP. In most of the wells in India, separate transformer is provided for each well, barring a few cluster wells, for avoiding this problem.

(c) These spikes can also be observed during an electrical disturbance such as lightning/storm, since fluctuation of voltage can be witnessed during such eventualities. Here, the range of spikes is contained within the overload and under-load settings.

Fluid Pump Off Conditions/Fast Lowering of Annulus Level (Figure 5.13c): This chart shows intermittent operation of the pump. The fluid pump off condition occurs when the pumping unit's capacity is larger than well intake capacity. As a result the continuous drop in amperage is observed. When the amperage goes below the under-load setting, switchboard automatically switches off the power. Immediately then re-start timer starts working and after a pre-fixed pause (time delay here is set at 2 hour, which is maximum.) the switchboard automatically switch on the power to ESP.

Pump Off/Gas Lock (Figure 5.13d): It shows the typical chart of a pump which has gas locked and consequently had an automatic shutdown. The well is being pumped out intermittently with the help of timer device, where the well remains shutdown for 1/2 hr. to 2 hr. (since its automatic trip). In each idle period, the fill up of the fluid in the annulus takes place. So, initially, production rate and amperage are more and then amperage comes down to its normal rate/flow value with more or less designed production rate. Thereafter a decrease in amperage takes place, when the fluid level falls below the desired level. Finally, before the pump gets underloaded, the erratic amperage, pointing to the cyclic loading of free gas and liquid slug in the pump is observed.

Insufficient Pause Time (Figure 5.13e): Sometimes, it happens when sufficient annulus pressure build up does not take place during the idle period. This occurs when the P.I. of the well is very poor. This is also a case of pump-off phenomenon.

Frequent Short Duration Cycling (Figure 5.13f): Here the running time of pump is very brief and therefore it has created more cycles. This type of situation arises due to various reasons like faulty adjustment of amperage, very poor P.I. of the well, which is much less than the pump capacity, excessive flowline pressure etc.

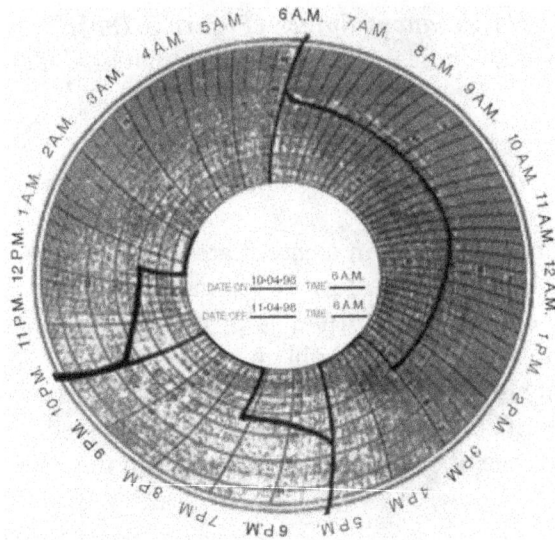

Fig. 5.13c: This is pump off condition of the pump. The discharge of the pump > Fluid intake from the wellbore. As a result continuous drop in amperage is observed and as the amperage drop below the underload setting, switchboard automatically switches off the power. Immediately then re-start timer start working and after a pre-fixed pause time the switchboard automatically switches on the power to ESP.

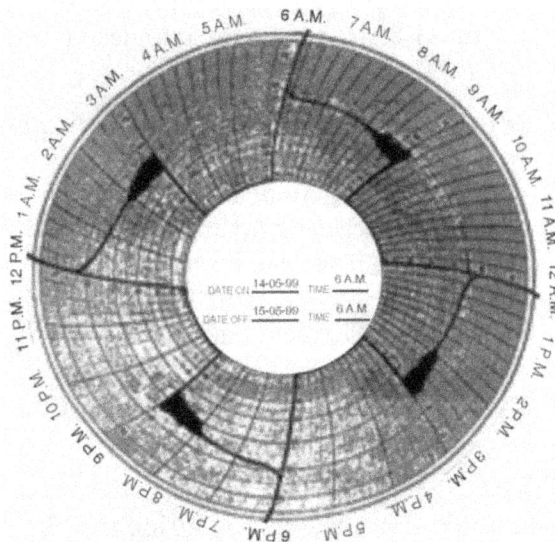

Fig. 5.13d: This is a case of pump off/gas lock, due to more capacity of pump discharge than the inflow capacity of well, fluid level in the annulus goes down continuously, with generation of more and more free gas in the pump. Finally the current goes below the underload setting and pump stops. Thereafter the auto restarting unit starts operating and pump gets restarted after a predetermined pause. Very frequent current fluctuations prior to shut-down indicates gas locking phenomena.

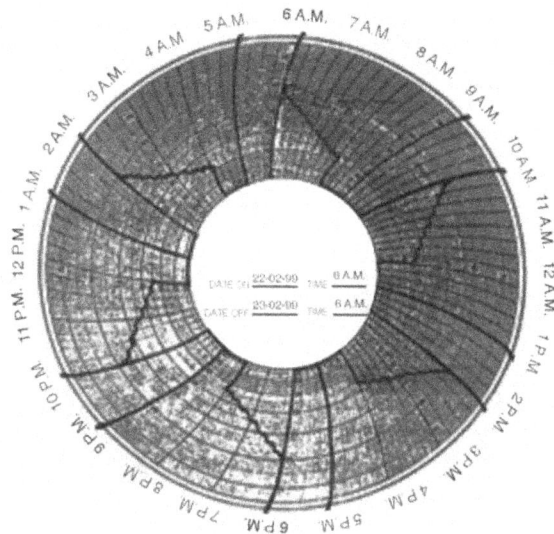

Fig. 5.13e: This indicates insufficient pause time before the pump is restarted. This is possibly due to very low influx into the wellbore and as such length of the pause time should be more. This is also a case of pump-off phenomena.

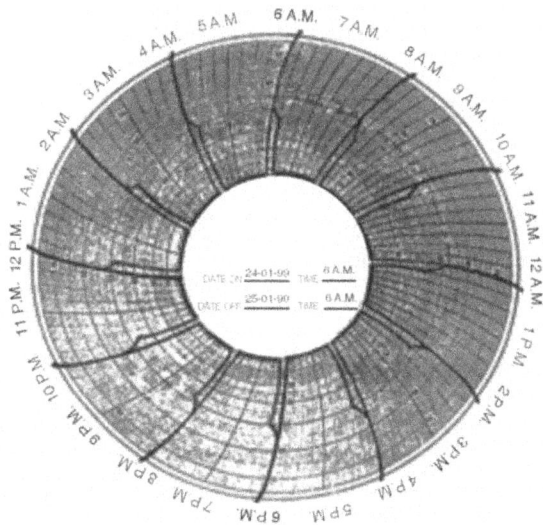

Fig. 5.13f: This indicates a large number of times of operation and shutdown of pump. The possible reasons may be:

1. If pump is of very high capacity and inflow into the wellbore is very less, then this phenomena can occur, however, in practice. Except at the initial stages of pump commissioning. This is remote.
2. If flowline pressure is excessively high, it will lead to early shoutdown of pump due to underload phenomena, after a pre-fixed pause, the automatic restarter starts. This condition shortens the operating life of pump and hence should be avoided.

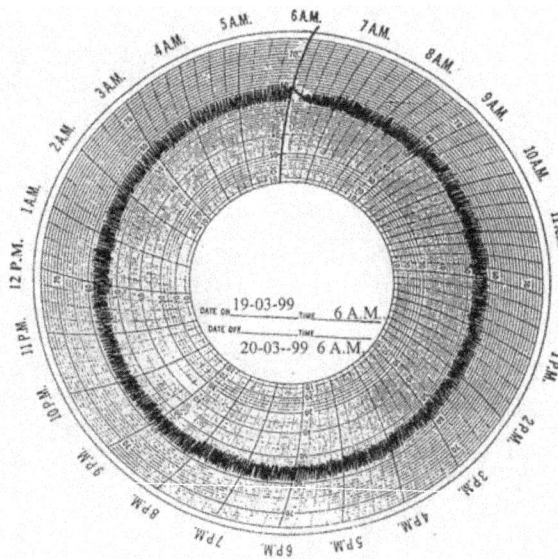

Fig. 5.13g: It is a normal/high GOR pump operation, provided there is no voltage fluctuation to that extent. The fluctuation indicates the loading and off-loading of pump very frequently due to interference of gas.

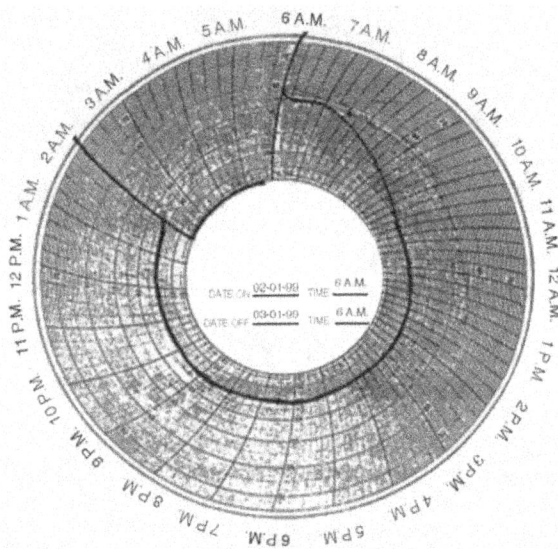

Fig. 5.13h: This is case of pump off/dry run of pump. This case arises when underload setting is not done. Due to relatively low intake of formation fluid into the well-bore, amperage gradually drops and then maintains a very low value all through the pump operation. There is no outflow of fluid at the surface.

This resulted in abnormal heating of the motor/cable and thereby damage the insulation. This condition must be avoided.

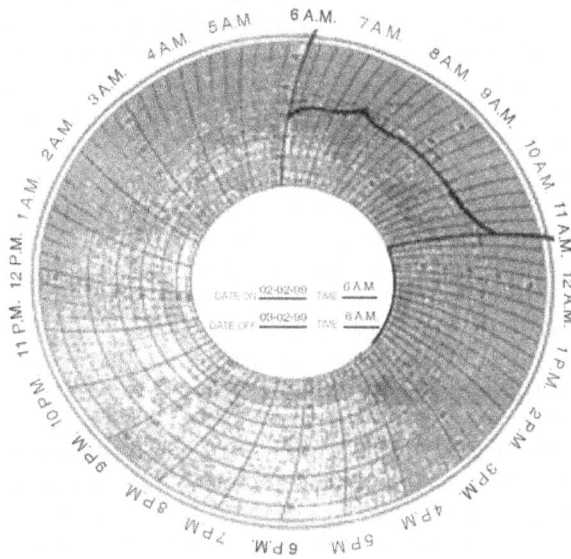

Fig. 5.13i: This is case of overload. The current goes up gradually and finally when it crosses the overload setting mark, the pump stops. Overloading phenomena may arise due to power fluctuation or due to mechanical/fluid problem. As per convention of the manufacturer, after overload shout-down, pump will not get automatically restarted. This needs to be checked by electrical and production engineers for locating the actual fault. Once the fault is located and rectified, pump is then run manually.

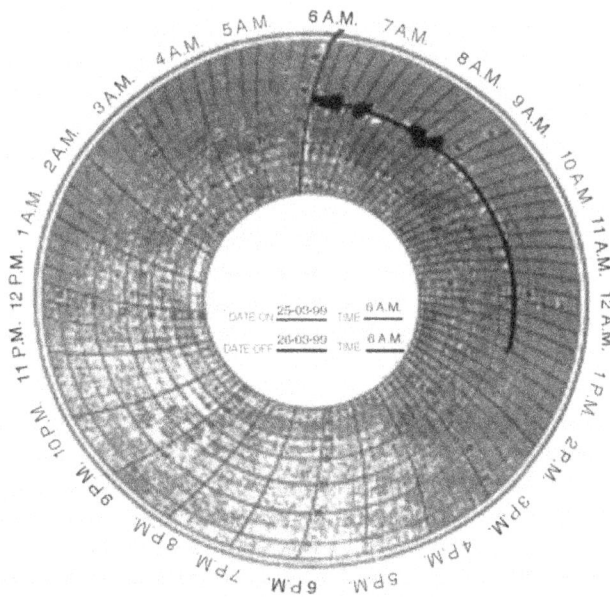

Fig. 5.13j: This is case of viscous/emulsified fluid entry in the pump. Once the viscous and emulsified fluid are pumped out, pump is in normal operation.

This type of situation fails to maintain proper cooling of motor which may result in early breakdown of motor insulation.

Normal/High GOR Pump Operation (Figure 5.13g): Interference of free gas is noticed here. Possible cause of this is the emulsified/heavier fluid with free gas trapped in it. The fluctuation indicates very frequent loading and off-loading of pump.

Pump Off/Dry Run (Figure 5.13h): This chart indicates that after a normal start-up, a decline in amperage has taken place. The fluid production is very less. After a prolonged period of idle operation, the unit shows overloading and stops. One of the possible reasons is that underload relay system in the switchboard which is either not set properly or not working. This condition is very serious as it may lead to breakdown of motor/cable insulation.

Overload Condition (Figure 5.13i): Overload condition of the pump is another very common phenomenon. It may be due to emulsion or very viscous liquid at the pump, mechanical problem of pump or because of disorder of power supply. Just after the well is worked over this type off phenomena has been observed mainly because of emulsion problem due to the mixing of technical water with incoming gas-free oil from the formation. Low voltage also makes the current high leading to overloading. Overloading due to the mechanical problem arises when there is excessive drag of impeller with the diffuser pads, shaft is not rotating freely or some bearing is not working properly.

Other possible causes of overloading are lightning, motor overheating, etc.

Once the pump is overloaded, it will not get automatically started. When it occurs, pump has to be started again only manually and if the same overloading phenomenon repeats, it is mandatory to check it thoroughly from the electrical aspects and with necessary rectification, if any, before the pump is restarted.

Viscous/Emulsified Fluid Entry (Figure 513j): The chart shows that during the initial starting period, there appears to be current fluctuations, followed by drawing of normal current during the pumping operation. This type of situation is also encountered after workover operations. It is regarded as the cleaning operation with the help of pump for pumping out muddy water or brine, high viscous oil-water emulsion, etc. Once these are pumped out pumping operation becomes automatically normal. So electric motor coupled with ESP should have adequate capacity to overcome this type of situation.

5.11 Design of ESP

To design a downhole pump and motor, one must have knowledge of Productivity Index (PI) of the well. The desired operating conditions will determine the depth at which the pump is run to ensure submergence below the pumping fluid level as well as the size of the pump and motor. The desired flow rate and tubing size will determine the Total Dynamic Head (TDH) requirements for the pump.

5.11.1 *ESP Design Example 1*

Given Data

Pumping fluid at	: 4000 ft
Well depth	: 7418 ft
Separator pressure	: 50 psi
Frictional pressure loss in tubing	: 100 psi
Specific gravity of fluid being pumped is	: 0.85 (0.37 psi/ft)
Desired production rate	: 1268 b/d
Casing size	: 5½″

Calculated TDH $= (7418 - 4000) + (150 \times 2.31)$
$= 3762$ ft of head, water equivalent

Table 5.1: Pump Selection Table

Pump capacity, b/d	377	548	925	1165	1645	2090	2365	2742	3427
Peak efficiency 50 Hz,b/d	300	500	733	1000	1333	1583	1792	2250	2792
Optimum range, 50 Hz,b/d	152–410	373–608	550–900	770–1167	937–1708	1042–1958	1250–2205	1458–2917	1833–3625

Table 5.2: Motor Specifications for 5½″, 20 ppf and Larger Casing Sizes

Sizes, hp		Volts/Amps		Length, m	Weight, Kg
60 Hz	*50 Hz*	*60 Hz*	*50 Hz*		
60	50	735/51	612/51	5.13	381
60	50	840/44	700/44	5.13	381
60	50	945/40	787/40	5.13	381
60	50	1270/30	1058/30	5.13	381

From Table 5.1, 1165 b/d pump would be appropriate choice since it has the highest efficiency for desired production rate.

Based on this choice, we can select a typical pump performance chart (refer Fig. 5.14). At 1268 b/d, the chart indicates a head capacity of this pump as 18.5 ft/stage,

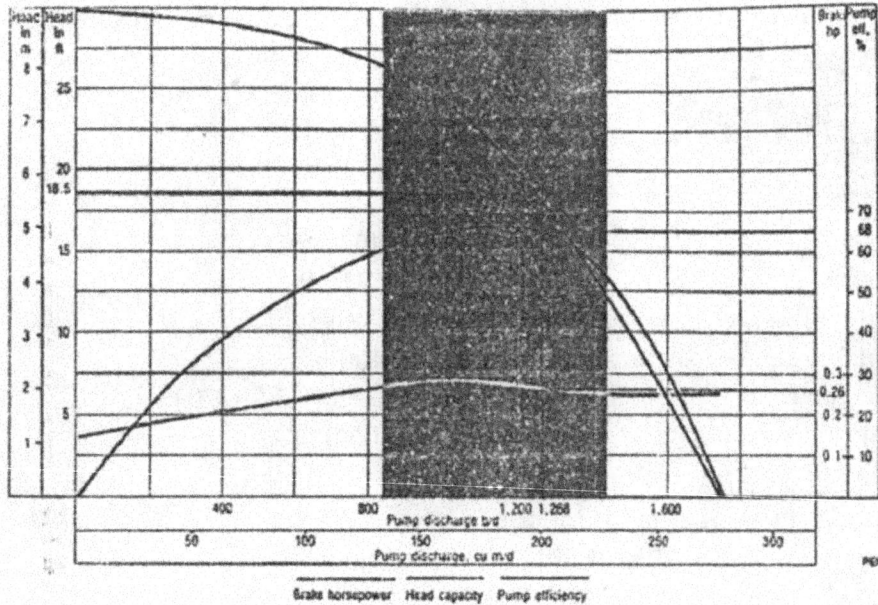

Fig. 5.14: Typical Performance Curve for a 34 gpm, 4 in. Pump
in 5½″ Casing, Lifting Water at 3475 rpm

0.26 hp/stage and pump efficiency of 68%. Thus, total stages = 3762/18.5 = 203 stages; H.P. required = 203 × 0.26 = 52.8 H.P. From Table 5.2, a 60 hp, 1270 V, 30A motor is selected for this well.

Tubing	: 2 7/8 inch
Casing size	: 5 ½ inch (17–20 ppf)
Reservoir pressure	: 205 kg/cm² (2915 psi)
Flowing Bottom-hole pressure	: 180 kg/cm² (2560 psi)
Wellhead pressure	: 7 kg/cm² (100 psi)
Water cut	: 60%
GOR	: 45 m³/m³ (252 SCF/b)
GLR	: 17 m³/m³ (96 SCF/b)
Design Liquid rate (at stock tank)	: 35 M/D (225 b/d)
Degree API of oil	: 35° API (Sp. Gr. = 0.8489)
Specific gravity of gas (Air = 1)	: 0.65
Specific gravity of water	: 1.05
Bottom-hole Temperature	: 70°C (158°F)
Wellhead temperature	: 30°C (86°F)

Bubble point pressure	:	80 kg/cm^2 (1137 psi)
Electric supply system	:	400/440 V: 50 Hz
Well profile	:	S-shaped (Build-up and Build-down Profile) and from 1000 mts depth from surface it is vertical.
Formation volume factor (B$_0$)	:	1.15

The following are the step-wise calculations:

Step 1: Size of Pump

From the catalogue of ESP manufacture the best suited pump primarily with respect to its OD and capacity is to be selected. Let the available ESP is of REDA make.

Since casing size is 5½″, at the first instance, 400/450 series REDA pump/ protector as applicable in 5½″, is considered (Reference: REDA catalogue). Now, maximum OD of REDA pump set with cable, cable guard and cable clamp in position is required to be checked with 5½″; 20 ppf casing (that is minimum ID of casing).

(i) OD of 450 series protector = 114.3 mm
(ii) Thickness of Armoured cable of 6 AWG of parallel shape = 12.3 mm

Thickness of cable guard and cable = 2.0 mm (approx.)

Total = Max. OD of READ Pump (400/450 series) = 128.6 mm

I.D. of 5½″, 20 ppf casing = 121.4 mm

Drift diameter of 5½″, 20 ppf casing = 118.2 mm

Since drift diameter of 5½″, 20 ppf casing is less than Max. O.D. of REDA pump, it is required to find out pump of one size lower.

The next lower size is of 338/325 series pump/protector as applicable in 4½″ casing,

(i) OD 338/325 series pump/protector = 85.85 mm
    (Thickness off Armoured cable (6 AWG, parallel) = 12.30 mm
(ii) Thickness of cable guard and cable clamp = 2.00 mm
    Total = Max. OD of 338/325 series Pump/protector = 100.15 mm
    Therefore, clearance between minimum casing I.D. and max. Pump O.D.
        = (118.20–100.15) × ½
        = 18.05 × ½ = 9.0 mm

From REDA catalogue, the compataible pump/protector set of 338/325 series is selected which is to be coupled with 375 series motor OD = 3.75 inch = 95.25 mm).

Step 2: Static and Dynamic Level

Considering the datum level at 2500 Mts., and with specific gravity of water as 1.05, the fluid level at static condition = $2050 \times 1/1.05$ = 1952 Mts. So, static fluid level from the surface = 2500–1952 = 548 Mts.

The fluid level at flowing condition (that is, dynamic condition)

$$= 1800 \times 1/10.5 = 1715 \text{ Mts.}$$

Therefore, dynamic level from the surface = 2500–1715 = 785 Mts.

Step 3: Location of Pump Depth

The pump has to be located below the dynamic level in the well. Also, to minimize the interference of free gas, the pump, if possible, can be located in deeper depth.

 (i) Dynamic level from surface = 785 Mts.
(ii) Bubble point pressure of 80 kg/cm^2,

   which is equivalent to = 762 Mts.

Total = 1547 Mts.

Therefore, location of Pump = 1600 Mts. form surface.

Step 4: Fluid Volume in the Pump (Q)

$$\begin{aligned}Q \ &= 35 \text{ m}^3/d \times B_0 = 35 \times 1.15 \\ &= 40 \text{ m}^3/d\end{aligned}$$

Step 5: Pump Selection

A 400 pump is selected from performance curve as supplied by the manufacture for 50 Hz supply and 338 series pump with the desired fluid production rate of 40 m^3/d lies in the recommended range for operating the pump on the accepted efficiency level.

Step 6: Pump Stage Calculation

From the performance curve, 100 stages develop 400 Mts. of head.

Therefore, 1 stage develop 400/100 = 4 Mts. of head.

Now, total head required, that is Total Dynamic Head (TDH) will be.

TDH = Dynamic level from surface + Fluid friction in the tubing + Tubing
         head pressure
    = 785 Mts. + Negligible + 70 Mts. = 855 Mts.

Therefore, total stages of pump required $= \dfrac{855 \text{ Mts.}}{4 \text{ Mts./stage}} = 214 \text{ stages.}$

From the catalogue of the manufacturer, the number of stages and housings have been selected, so that total stages of pump is slightly more or equal to 214 stages.

2 numbers of housing each having 81 stages and 1 number having 60 stages have been selected.

So, total stages $= ((2 \times 81) + 60) = 162 + 60 = 222$ stages.

Step 7: Motor Horsepower Requirement

From performance curve, max. H.P. $= 6.0$ H.P./100 stages

The nearest whole number $= 6.0$ H.P./100 stages $= 0.06$ H.P./stage

So, total H.P. requirement $=$ H.P./stage $\times$ Number of Stages $\times$ Specific Gravity of Water
$= 0.06 \times 222 \times 1.05 = 13.98$ H.P.

From catalogue, 375 series motor has to be selected, which has H.P. either equal to this value or next higher value.

H.P. Motor selected $= 16.3$ H.P.

It is always advisable to choose a motor with low amperage rating, provided its voltage rating is not very excessive. So, from two categories of 16.3 H.P., 50 Hertz motors.

That is from, 16.3 H.P.: 238 V: 38 A; 50 Hertz

and 16.3 H.P.; 323 V; 25 A; 50 Hertz.

Step 8: Main Cable Selection

From manufactures catalogue "Redelene" type (can work up to 205°F, where B.H.T. is 158°F) flat cable and 4 AWG (considering cost and voltage drop factor) has been considered.

Step 9: Surface Voltage Calculation

From cable voltage chart, supplied by the manufacturer.

Voltage drop $\qquad = 11$ volt/1000 ft.

Total cable length = Subsurface cable length + Surface cable length
= 1600 Mts. + 100 Mts. (say)
= 1700 Mts. = 5576 ft
= 5600 ft

So, the total voltage drop $= \dfrac{11\,V}{1000\,ft} \times 5600\,ft$

$= 61.6\,V$

Therefore, voltage required at the surface = name plate voltage + Total voltage drop

$= 323 + 61.6 = 384.6\,V = 385\,V$

Step 10: Calculation of KVA (KW) Requirement of Power Transformer (Step Down Transformer)

$$KVA = \frac{(Required\ Surface\ Voltage) \times (Name\ Plate\ Amps.) \times (1.73)}{1000} + 2.5\%$$

$$= \frac{285 \times 25 \times 1.73}{1000} + 2.5\% = 16.65 + 2.5\%$$

$$= 16.65 + \frac{16.65 \times 2.5}{100} = 16.65 + 0.42$$

$$= 17.07 = 18\ KVA$$

Since the power transformer of 18 KVA is not normally available. The next size available is 25 KVA So, power transformer of 25 KVA (25 kW) is selected.

Step 11: Selection of Switch Board

From the manufactures catalogue, the switch board of the following type is selected depending on max. volt, H.P. and max. full load amps.

Switchboard is class DFH-2, type 72, size 2 max, volt 600, H.P. 25 and max. full load amp. 50.

Step 12: List of Suitable Designed Pump, its Components and Miscellaneous Accessories

All pump components must be compatible to each other.

Pump : A-400 : 222 stages : 2 Nos. of Housing each of 81 stages (338 series) 1 No. of Housing of 60 stages.

Motor : 375 series : 16.3 H.P.; 323 V; 25 A; 50 Heartz; Tape-in type

Protector	: 325 series: labyrinth type
Intake section	: Reverse flow gas separator
Pot-head Cable	: 6 AWG: Redelene flat galvanized: 50 ft in length
Main Cable	: 4 AWG: Redelene flat galvanized; around 1700 Mts. (5600 ft) in length wound 1700 Mts
Wellhead	: "Seaboard" wellhead or equivalent for 5½″ casing: $2\frac{7}{8}″$ tubing with necessary fitting like upper and lower pigtails, mini-mandrel etc.
Switchboard	: Class DFH-2, type 72, size 2, max, Volt 600, H.P. 25 max. full load amp. 50.
Power Transformer	: 25 KVA capacity with step-down voltage from 11 kV power transmission line 400/420 Volt (standard industrial voltage).
Accessories	: Junction box, Pulley and its arrangement for lowering/ pulling of ESP in and out of the well, sufficient quantity of high quality insulating oil ammeter recording charts full splicing kits, sufficient number of cable bands, clamping of cable band tool-set, check valve, bleeder valve with nipples, pump discharge head, Pot-head extension guards or channels, centralizers, necessary mechanical handling tools and necessary electrical instruments.

5.12 Recent Advances in Electrical Submersible Pumping

5.12.1 *Sub-Sea ESP Completions*

Reda Production Systems, a Schlumberger Company, Bartlesville, Oklahoma, helped form a JIP project funded by the European Commission and six oil operators for a review of nonviable offshore field developments. Subsea applications of ESPs have turned marginal or unprofitable fields into profitable producers, Figure 5.15. The greatest distance for a subsea ESP step out from the host power supply is in excess of 15 km (24 mi). The total number of ESPs in subsea applications has reached more than 25 successful completions. Reda has supplied 100 % of the subsea ESP systems now installed.

5.12.2 *High-Performance Gas Separator*

Reda Production Systems introduced the second in a new line of high-performance Vortex Gas Separators, the 538 Series VGSA S70-150. This newly designed gas separator offers greater efficiency in higher-flowrate applications. Improved hydraulics, coupled with Reda's compliant-mount radial bearing system, provides

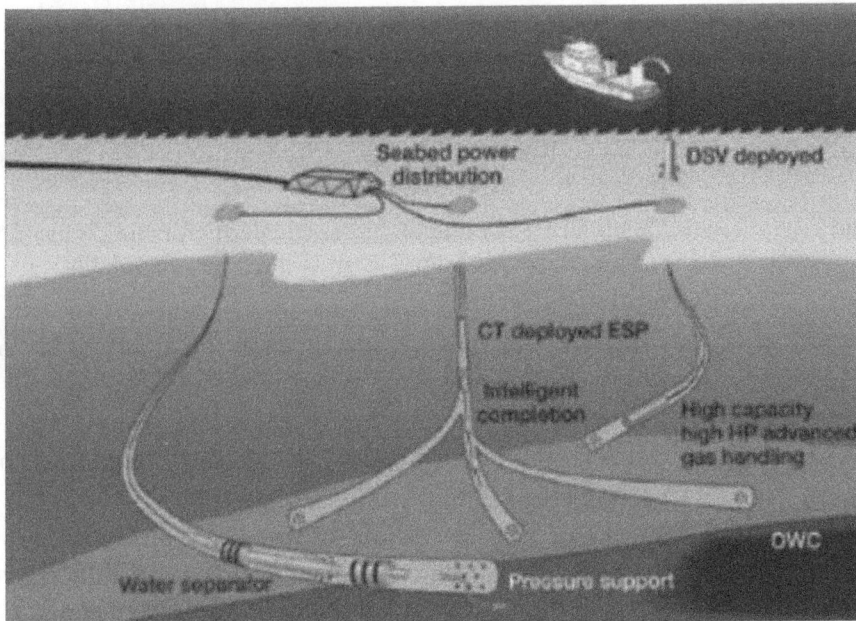

Fig. 5.15: Sub-Sea Applications of Electrical Submersible Pumps

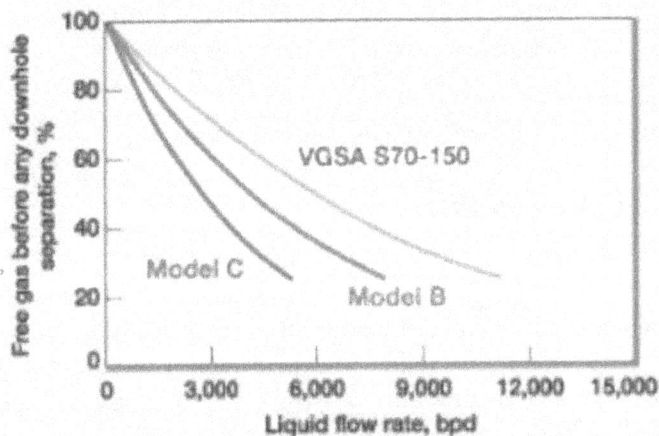

Fig. 5.16

high reliability. Figure 5.16 shows a performance comparison between the 538 Series system and other gas separators for liquid flows up to 12,000 bpd. Performance tests, on a fluid that contains 15% free gas after total separation, show how new vortex gas separator is more efficient at high flow rates. For example, at 10,000 bpd, the VGSA can handle fluid with 30% of the total 15% free gas before downhole separation; Model B can only handle this much gas at 7,000 bpd.

5.12.3 *Downhole Gauge*

Schlumberger developed the PumpWatcher Sapphire four-channel downhole gauge for ESP systems. The tool measures downhole conditions in real time using a patented method of digital transmission. The high-precision gauge provides continuous pressure and temperature measurements at the pump intake and discharge head, Figure 5.17.

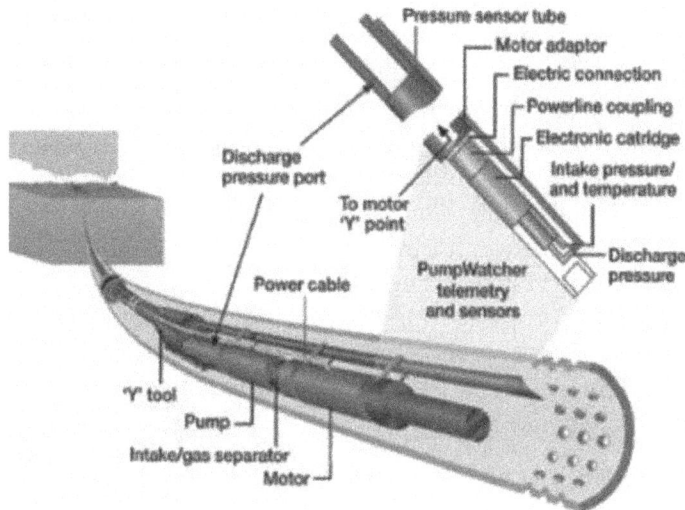

Fig. 5.17: Pump Watcher Four-Channel, Downhole Gauge

Measurement of pressure produced by the pump is used to indicate flow, pump efficiency and pump wear. Monitoring motor temperature prevents operating the pump in adverse conditions. Data is transmitted on the ESP power cable, eliminating need for an "I" wire. This feature simplifies installation, improves reliability and reduces total system cost. Digital telemetry is not affected by electrical noise.

5.12.4 *Surface Readout for Downhole Instrumentation*

Reda Production Systems has developed a new surface unit to operate with its line of downhole monitoring equipment. The WB2 provides power to the gauges, demodulates the signals, stores and conveniently delivers data. The optional SD1 Smart Display is used to read measured parameters directly at the wellsite. The small, robust display can be secured in place or carried to multiple installations.

The rugged WB2 is totally enclosed in a NEMA enclosure, Figure 5.18, and is rated for operation between 55°C and 75°C (67°F and 167°F) ambient temperature. All data connections are isolated from supply voltage by solid shields to increase safety. There are four analog outputs, eight digital outputs and connections for Modbus to provide local and remote control—Modbus 232 and 485 protocols are supported. A communications port readily connects a laptop or similar PC device.

5.12.5 *Downhole Power Cable*

Wood Group ESP of Oklahoma City, Oklahoma, has introduced a newly designed downhole power cable. Tradenamed Powerline, the cables are designed with the toughest oilwell environments in mind. Three standard cables are currently available: the 205, 300 and 450. The 205 has an operating temperature range of –5 to 96°C and is available in voltage ratings of 3, 4 or 5 kV. It has a polypropylene insulation, a nitrile rubber jacket and choices of armor. Figure 5.19 represents typical product lines with round and flat versions.

Powerline 300 is an EPDM-insulated cable with a rubber or EPDM jacket. It is designed for moderate-to-harsh environments and operating temperatures below 149°C. The 450 is the choice for extremely harsh environments, with EPDM insulation and a lead jacket. The operating range of temperatures for this cable is 40 to 232°C; and it is configured in parallel only.

5.12.6 *High-Efficiency Motors*

Wood Group ESP has released its E4 and E5 series high-efficiency motors for 5½- and 7 in. casing, respectively. Incorporating newly designed stator and rotor laminations, these motors operate at higher efficiencies and produce more horsepower per unit length than standard motors. Figure 5.20 represents a cross-section of the E5 series.

Fig. 5.18: Surface Readout Unit to Monitor Downhole Instruments/Equipment

Fig. 5.19: Downhole Power Cables in Round and Flat Configurations

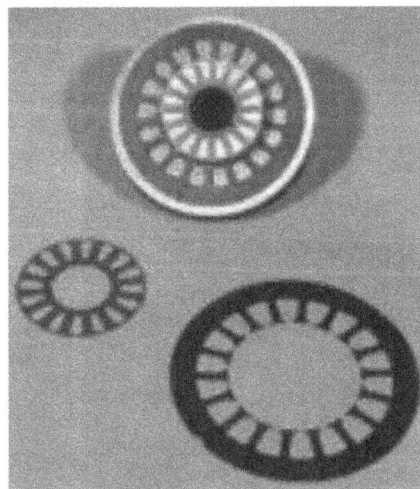

Fig. 5.20: Cross-Section of New E5 Series, High-Efficiency Motor

Fully compatible with the company's current product lines, the E-series motors offer an all-steel, open-slot stator and "teardrop" rotor bars, both of which will result in increased operating speed and efficiency. The E4 offers hp ranges from 25 to 150 hp in a choice of voltages. The E5 series ranges from 35 to 315 hp.

5.12.7 *Labyrinth Seal*

Also new from Wood Group ESP is a labyrinth seal section. The 98L seal has a totally re-designed thrust bearing and runner. The runner's thickness has been increased to a load capacity of 5,500 lb, and a tilting-pad bearing surface is utilized. To reduce influence of well fluid on the thrust bearing, it has been moved to the bottom of the seal section. This allows for two mechanical seals above the thrust bearing rather than the one conventionally found on most labyrinth seals. Application guidelines are available.

5.12.8 *Skid–Mounted System*

Centrilift, a division of Baker Hughes, Clare-more, Oklahoma, now offers deserdrive, skid-mounted systems that allow platform or field operators to install ESP Variable Speed Drive (VSD) equipment in arduous, hot desert environments. The VSD skid system is fully portable and can be operated outdoors without need for an air-conditioned switchroom building, Figure 5.21. With the package, installation operators can locate the drive equipment outdoors close to the wellhead, in hot ambient locations, without need for expensive air-conditioned switchrooms.

Fig. 5.21: Skid-Mounted System for VSD Equipment in Harsh-Climate Areas

5.12.9 *Containment System*

Centrilift's drive-containment system allows platform or rig operators to install ESP variable speed drive equipment in hazardous-rated areas, Figure 5.22. Before this development, drive systems had to be located in nonhazardous areas for safe operation. The system is especially suited

Fig. 5.22: Drive Containment System

for projects with a short duration or on long-term projects that do not have sufficient safe-area space available for location of standard VSD equipment. With the system, installation operators can locate the drive equipment close to the wellhead in Hazardous Zone 1 or Zone 2 rated areas.

5.12.10 *Simulator Screen*

Since Centrilift's release of Version 1.1 in 1995, AutographPC has gone through several major upgrades each year. The most exciting feature yet is called the Simulator option, which adds the time variable to sizing to view well dynamics operation, in conjunction with the pump operating point, Figure 5.23. The system can accelerate or slow time, start/stop the pump or change fluid characteristics like watercut—all while simultaneously recording changes in parameters like pressure, flowrate, amps and others. This virtual simulation enables users to match active well performance to predicted performance, plus use this information to optimize production and equipment run life.

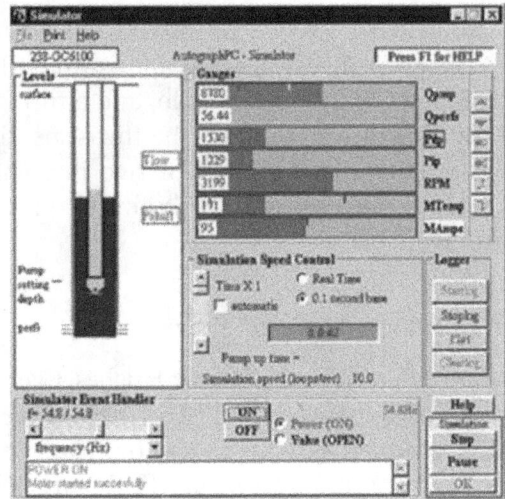

Fig. 5.23: Simulator Option for Autograph PC ESP-Control System

The scope includes well performance and sizing of the centrifugal pump, motor, seal, cable and controller based on customer-provided well data and production requirements. The program can also size tapered pumps, ESPCPs and SubSep (downhole oil/water-separation system). Besides sizing, Autograph PC can also compute and check bending-stress levels that occur when passing equipment through doglegs, with the included ASAP (A Stress Analysis Program) routines.

5.12.11 *Dual–Booster Pump System*

Centrilift has broken new ground for customers in Ecuador by completing some of the world's most difficult and unique oilwell pump systems. The company has been behind several new developments, including the world's first double- and triple-booster oilwell pump completions, Figure 5.24. Eight of these unusual booster pump systems were installed in Ecuador in 1999.

Ecuador was also the site where the company completed the industry's first "separate-formation" completion, in which two pumps are placed in the same-hole with each producing oil from a different formation. Even though the pumps are situated in the same well, fluid from each formation remains isolated to the surface.

5.12.12 *Large-Motor Development*

Centrilift has expanded its 725 motor series (which will fit in 9–5/8-in. casing) to include motors up to 2,000 hp that can produce over 60,000 bpd in oilfield conditions. With casing up to 36 in., these pumps can lift 6,700 gpm for mine dewatering. The 725 series motors will operate in wells with BHTs up to 400°F. With these new systems, operators can now drill deeper and produce more in smaller-diameter wells with higher-horsepower motors even in high-temperature environments.

Surge protection devices. MVC, Inc., Amarillo, Texas, is a manufacturer of surge protection devices designed for voltages of 40 VDC to 4,160 VAC applications. The MV Series (120 to 480 V applications) offers one of the highest degrees of protection against line noise transients, Figure 5.25. They have been tested through UL labs and hold a UL 1449 2nd Edition listing. They carry a full five-year, unconditional warranty. The WS Series (1,500 to 4,160 V applications) has been specifically designed for medium-voltage motor and pump applications. These units have replaceable modules and considerably extend motor, pump and control life.

The ICP Series was specifically designed for individual-circuit protection such as Programmable-Logic Control (PLC) and solid-state controls. These units are built in either parallel or series configuration, according to requirements of the customer. The ICP also has a remote alarm, which allows the user to monitor Surge-Protection Devices (SPDs) from a computer terminal located miles away. All MVC surge-protection devices are externally fused for safety and have failure indicators.

5.12.13 *Cable Insulation Protection*

Years of experience has caused the industry to adopt a "floating" electric power connection for extending ESP

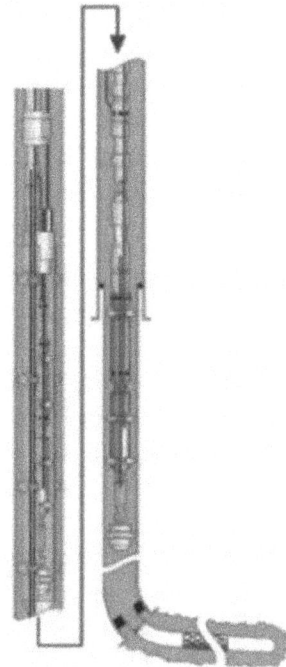

Fig. 5.24: Dual Boost ESP Pumping System

Fig. 5.25: Line Surge Protector

run life. This connection allows continued operation with one phase shorted to ground. Total failure occurs when a second phase shorts to ground in onshare operation. Many ESP failures are still associated with thunderstorms or power outages. With no limit to the peak insulation voltage, there is nothing to limit insulation degradation. Such degradation and, ultimately, failure relate to partial discharges that cause carbon tracks or "trees" to grow in the insulation between phase wires and ground.

To solve these problems, PM&D Engineering, Inc., in Bluffton, Indiana, has introduced a Transient-Voltage Surge Suppressor (TVSS), called Pro-MoDr, designed specifically for floating, constant-frequency to ESP installations. Mounted in the junction box to minimize ground-wire inductance, Figure 5.26, it provides full line-to-line and line-to-ground protection with UL 1449-listed surge-suppression devices. A three-lamp circuit indicates a line-to-ground short. By limiting the peak voltage that can be applied across cable and motor insulation, the system greatly reduces partial discharges, thereby increasing the life of ESP insulation.

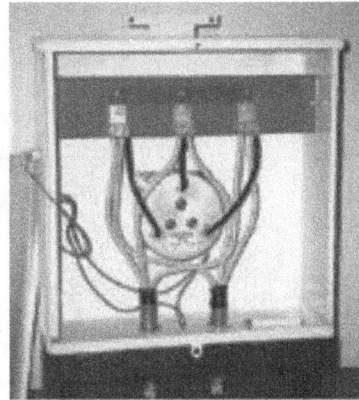

Fig. 5.26: Transient-Voltage Surge Suppressor for Cable-Insulation Protection

5.12.14 *Reeled ESP*

ReELIFT offered by Innovative Engineering, Aberdeen, is a reeled completion system that utilizes ESPs as the means of lifting liquids. It comprises of the required surface and subsurface assemblies—with the exception of the ESP. There are two system series: "C" is used with conventional pumping technology—motor at the bottom, Figure 5.27. "I" is configured for wells that require inverted pumps—motor at the top of the ESP assembly. Both series utilize the electrical power cable placed in the internal of the CT.

According to the supplier, the system:

- Reduces cost and minimizes downtime by avoiding rig use and allowing faster installation/recovery.
- Extends run life/improves reliability by isolating electrical parts from production fluids.
- Minimizes formation damage by allowing live-well deployment and retrieval.

Allows well intervention without retrieving the completion.

It can be used with conventional, as well as inverted, pumping to allow flexibility in pump/supplier choice. It incorporates subsurface flow control for well safety and reservoir isolation. It provides a release/circulating system for safe well circulation/control, disconnect and effective completion retrieval.

The surface assembly is a compact production spool connected to the cable/coiled-tubing assembly which houses the cable. The subsurface assembly comprises elements from the bottom of the CT string to the pump assembly, depending on the configuration selected. For the two main completion system options, the surface and CT assemblies are common and almost identical.

Series C systems are designed for wells requiring packer-type completions using conventional ESP technology. This configuration has the cable inside the CT string. A packer-type completion allows the ESP to discharge fluids through the packer and crossover into the annular space using a flow diverter located at the top of the packer. A hydraulic, surface-operated flow control is located at the pump discharge for isolation and well control. The cable is connected to the surface facility using conventional connectors.

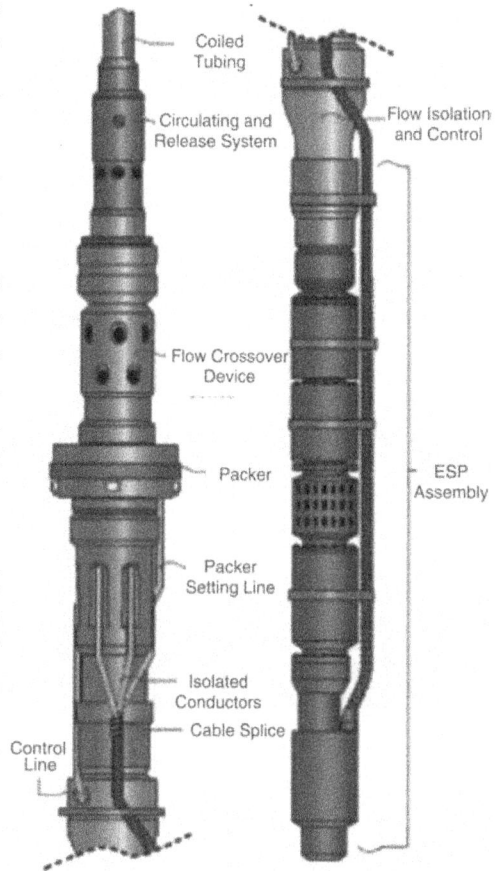

Fig. 5.27: Type "C" Reeled ESP Completion System

The Series I is designed for inverted (bottom-intake) ESPs. The subsurface assembly comprises cable-termination and isolation devices at the top of the pump assembly (top of motor also) and a release/circulation device, which allows well neutralization and release (disconnect) from the ESP assembly. The subsurface assembly incorporates a pump discharge sub based on packer technology. This flow diverter, placed between pump and motor (or seal), provides isolation between suction and discharge sides of the pump assembly and allows flow to cross from pump internals into the annular space.

5.12.15 *High-Volume ESP Gas Separator*

Multi-phase fluid wells historically have been challenging for ESP electrical submersible pump systems and, consequently, a constant goal for the industry is to push the limits of gas handling technology. Baker Hughes Centrilift, Claremore, Oklahoma, a leader in gas handling technology, had designed high-volume rotary/vortex downhole gas separator, GasMaster, to maximize production from wells with significant free gas content.

The new separator incorporates dramatic design changes compared to standard gas separators, improving total flow rate through the separator into the pump while providing better separation efficiency. Centrilift had designed a high gas volumes, older separator which could not provide enough pressure increment to overcome flow losses within the unit, and the inducer design choked at higher rates.

Based on this research, the company realized that a new approach to gas separator pressure charging was required, which led to development of the High-Angle Vane Auger Design (HAVA), incorporating the optimum vane angle for efficiently handling free gas and larger two-phase fluid volumes.

Further testing indicated that improvement to the intake/base, discharge head/diverter and spider bearing was needed to optimize fluid path geometry throughout the gas separator. Various configurations were tested in the high-pressure gas loop test facility in Claremore until designers were satisfied that this goal was reached with the new system's component parts. Improved production rates and higher separation efficiencies obtained confirmed the success of these designs.

The new separator is available in both rotary chamber or vortex options and high/low flow options to specifically fit each well application. The gas separator also incorporates a range of abrasion protection levels that can be customized to specific well conditions.

5.12.16 *Downhole Monitoring Tool*

Baker Hughes-Centrilift is developing a full suite of downhole monitoring gauges, another important tool used with optimization services. In 2004, the company introduced the GCS-Centinel downhole monitoring system, designed specifically for use with ESP equipment, Figure 5.28. The gauge tool provides data on pump intake pressure, and intake and motor winding temperatures, along with system diagnostics. The digital data transmission system eliminates potential interference and resulting inaccurate information or even loss of downhole data, which can occur with traditional analog data transmission systems.

Fig. 5.28: Downhole Monitoring System Designed for ESP Equipment Use

The new system's surface electronics panel receives signals from the surface inductor panel and displays data through the graphics display panel standard within the GCS family of products. The surface inductor panel, or choke panel, is the interface between the ESP power system and the surface electronics equipment, filtering the digital communication signal from the AC power line.

The company is continuing to add tools and expand system monitoring and production control capabilities. The new tool, combined with the GCS Electrospeed II variable speed drive and the GCS vortex motor controller, allows active real-time production monitoring and control, as well as ESP performance monitoring. Predictive analysis of equipment performance is now within reach, as are the use of adaptive control strategies geared to maximize system performance linked to real-time technical and commercial drivers.

5.12.17 *Coalbed Methane Dewatering Pump Motor*

Coalbed Methane (CBM) production continues to grow in the US and is garnering more attention worldwide. However, a constant challenge for CBM operators is economically dewatering the coal seams to release gas. Baker Hughes-Centrilift has developed a new motor with an integral seal specifically for CBM low-cost applications.

The 450 CBM motor is a two-pole, slimline design for 5 ½ in. casing, Figure 5.29. Traditional oilfield technology ESP systems are economically challenging in CBM applications, but water-well systems require larger-diameter equipment for the same horsepower. The new motor's integral seal design significantly reduces motor/seal costs while offering higher operating parameters and a more robust design vs. water-well pumping systems. This new motor has an operating temperature rating of 212°F (100°C).

Fig. 5.29: Slimline Design ESP Motor for Coalbed Methane Well Dewatering

It can deliver up to 30 H.P. in a 4 ½ in. diameter design. This allows the unit to be run in 5 ½ in. casing vs. 7 in. casing necessary with water-well pumping systems for comparable horsepower, thus providing operators significant drilling and completion cost savings.

The innovative integral seal design is a single chamber bag seal section attached to the motor, minimizing the number of necessary components and reducing cost and installation time. Additional technical features include: 1) the high-load, self-aligning thrust bearing for the motor, which carries rotor stack and thrust from the pump; and 2) the premium face mechanical seal, designed to handle coal fines in the fluid.

5.12.18 *ESP Optimization*

Baker Hughes Centrilift had a range of optimization services to offer technologies and processes necessary to tailor solutions for a specific well or field performance improvement. This will integrate a broad range of expertise and technologies to achieve the lowest possible total lifting cost, making a sustained contribution to present and future net asset value.

A key tool supporting the ESPeXpert Optimization Services is the new ESPeXpert software tool powered by Centrilift's industry leading Autograph PC sizing and simulation software. In-house experts can use this new tool to automatically match downhole/surface performance measurements, using both data provided by the SCADA system and equipment sizing information. It also provides a built-in, rules-based diagnostics engine.

This new functionality enables early identification and prediction of challenges or anomalies in the well, or the ESP, to enable performance intervention.

This optimization services also will offer remote monitoring and control, providing next-generation low-power, wireless technology via radio, satellite or cellular communications.

5.12.19 *Smart System Controller*

Wood Group ESP, Inc., Midland, Texas, has developed a new bottomhole sensor that does much more than just giving pressure readings. This breakthrough in down-hole monitoring has broken barriers and presented numerous methods of assisting operators in maintaining their submersible equipment in more productive and cost-effective ways. This Smart Guard sensor and surface interface system, Figure 5.30, is a reliable, accurate and cost-effective way of putting the operator in control of submersible equipment. Some of the improved capabilities include:

- Pressure accuracy, 0.1% psi/bar
- Temperature accuracy, +/– 1.8°F (+/– 1.0°C)
- Vibration accuracy, 1%
- Current leakage detection
- Temperature shut-down capabilities
- Discharge temperature/pressure
- Flow rates, and
- 30-day memory.

A few of the many applications include: 1) ability to eliminate cycling in CO_2 applications and wells with very high GLRs; 2) ability to access accurate, reliable information for sizing/troubleshooting; 3) operating from the pump intake as opposed to relying on the surface controller; and 4) temperature shut-down capabilities, an intricate feature as operators venture into deeper, hotter production zones.

These and many more features are assisting operators in solving problems, analyzing failures and extending running times in areas once thought impossible.

ESPs to replace rod pumps. Wood Group ESP, Inc. has designed a low-volume pump stage. These pumps are with higher efficiencies, wider vane designs, beefed-up thrust washer areas and lower operating ranges, Figure 5.31. Efficiencies of these new stages are much higher than the old technology still being used in today's market. In many cases, a typical 5,000 ft design can require 100 less stages to lift the same Total Dynamic Head (TDH) as the old stage designs.

The wider vane designs have been accepted by operators faced with the challenges of trying to maintain small-volume pumps in CO_2 floods, and producers with high GLRs. This breakthrough has greatly reduced gas locking problems previously associated with narrow vane clearances and fluid paths. The thrust-washer areas have been greatly increased to extend run times in both ideal operating conditions and harsh environments.

Finally, there is an option for operators with low-volume applications that suffer from deviated wellbores or wells that are just too deep to achieve long run times with a conventional rod pump. This issue has been resolved with development of the TD 150. This stage has an operating range of 80 bpd to 280 bpd. Further, the new ESP's operating cost is now competitive with that of today's rod pumping systems.

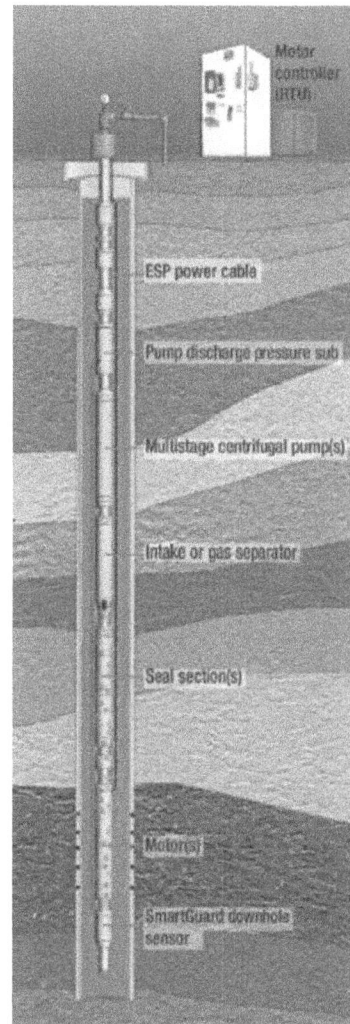

Fig. 5.30: Downhole Sensor and Surface Interface System to Control Submersible Systems

Fig. 5.31: Lower-Volume Pump Stages. New Wider Vane Design, Right, Reduces ESP Stages

5.12.20 *Lifting Service*

The Axia Lifting Service from Schlumberger, Sugar Land, Texas, is an end-to-end offering to improve production and reduce lifting costs in wells and fields with operating ESPs. The service uses well/field information in combination with real-time ESP operational data for surveillance/control of submersible pump operation, and broader evaluation of field-wide production improvement opportunities, Figure 5.32. The results are more efficient field operations, reduced lifting system costs and increased production.

This new lifting service is based on reliable REDA-ESPs and Phoenix downhole monitoring systems. The service is completed by expertise from the Schlumberger Data and Consulting Services team, the ESP Watcher surveillance/control service with integration of the LiftPro well optimization process. The Axia service seamlessly integrates downhole equipment with expert analysis through secure, remote connectivity to provide operators a simple interface to their own processes. Operators can now rely on the new lifting service to improve their own multi-well, field and reservoir management workflows, while increasing their personnel's productivity.

Broad-based lifting service evaluates/controls field and downhole operations for overall ESP optimization.

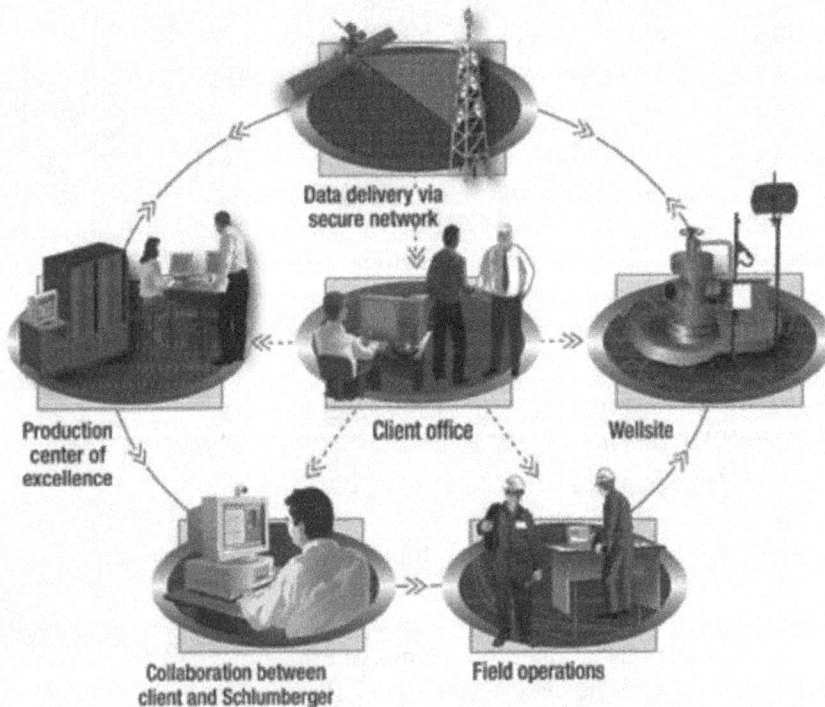

Fig. 5.32

5.12.21 *Low–Profile Cable Protector*

The Cannon Services, Stafford, Texas, cable protector has proven to be an ideal solution for securing, supporting and protecting both round and flat ESP cable configurations in downhole. The protectors are available for single and dual completions and have been used as deep as 18,500 ft on single ESP completions, and 12,000 ft on dual ESP completions. Company engineers regularly create custom designs and configurations to accommodate unique cable bundles and other special requirements such as round-to-round, flat-to-flat and round-to-flat cable splice protectors.

Fig. 5.33: Low-Profile ESP Cable Protector

The new low-profile protector, Figure 5.33, is designed with a minimal running profile to protect ESP cables where running clearances are critical, such as casing liners or casing patches. The standard and low-profile protectors can be installed and removed quickly, and they are reusable. All protectors are extremely durable and are manufactured from a variety of materials including low-carbon and stainless steel.

5.12.22 *High Temperatures*

Baker Hughes Centrilift, Claremore, Oklahoma, has developed an extreme temperature system for Steam Assisted Gravity Drainage (SAGD) tar sand applications ESP protect, Figure 5.34. Production from tar-sands in Canada is growing dra-

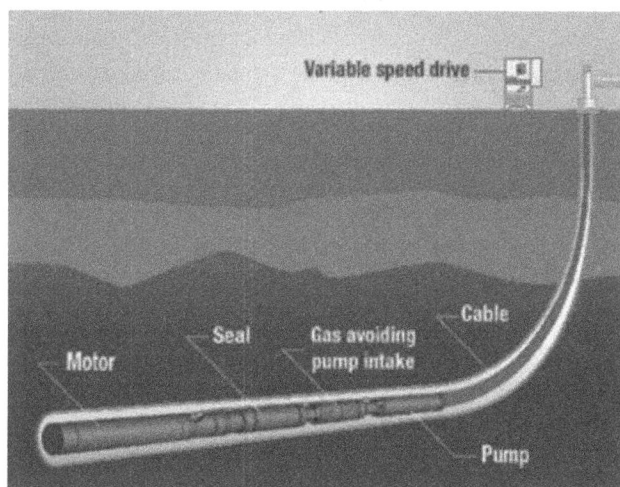

Fig. 5.34

matically; however, SAGD systems—two directionally drilled, stacked horizontal wellbores with the top for steam injection, the bottom for producing melted tar— severely challenge conventional ESP technology, due to extreme temperatures and temperature cycling.

The extreme temperature ESP system is designed for applications such as SAGD projects to recover tar sand oil.

To overcome these challenges, Centrilift minimized system elastomers; strengthened electrical connections and insulation; designed components for axial and radial thermal growth; and allowed for extra oil expansion. The SAGD environment test facility, a high-temperature, closed loop, allows for 18-day cyclic temperature testing of the ESP system to: fluid temperatures exceeding 500°F (260°C), pressures to 1,000 psi (6,900 kPa), and maximum flow to 27,800 bpd (50 l/s).

A key consideration in the extreme-temperature design is reducing overall rig time, to achieve earliest production. The seal section and motor are pre-assembled, prior to shipment, saving time during installation. The extreme-temperature, plug-in pothead vs. a tape-in design, minimizes field splice time during installation.

Primary design features of the system include, for the: 1) Motor—special internal metallurgy to withstand demanding downhole conditions; 2) Seal section—early testing indicated need for improved thrust/journal bearings, shaft seals and expansion chambers. Due to the elastomer, the seal is the most challenging design component for extreme temperatures; 3) Pump—thermal cycling characteristics of SAGD operations required special design considerations, including metallurgies/coatings to extend run life in abrasive/corrosive conditions. Additional mechanical design was required to handle stage compression issues specific to SAGD applications; and 4) Motor lead extensions—SAGD testing indicated the design must mitigate thermal expansion; and a robust, severe-duty, 4-bolt plug-in pothead, combined with motor-lead cable featuring individually encapsulated phases inside capillary tubing, was developed.

5.12.23 *ESP Monitoring and Automation*

Baker Hughes Centrilift has introduced a new downhole sensor designed to provide a broader range of measurements aimed at optimizing ESP-system run life and enhancing production. The WellLIFT* sensor, Figure 5.35, measures seven downhole parameters, including: intake pressure; fluid temperature; motor-winding temperature; gauge electronics temperature and vibration on both X and Y axis; and discharge pressure and temperature. The system also tracks 17 surface/diagnostic parameters, which allows high-level well management. Such costs can be further reduced by remote well monitoring and control through SCADA systems and via the Internet.

Downhole WellLIFT sensor is designed for a broader range of measurements.

5.12.24 *Combined with WellLINK, it Helps Optimize Run Life*

Centrilift also offers WellLINK, a comprehensive well-data distribution, retrieval and analysis service that provides data acquisition, display, control and analysis of ESP surface/downhole systems. Well data can be accessed online with a standard web browser using the service.

WellLINK features the following capabilities: 1) ESP Global* is a low-earth-orbit communication device for remote data acquisition and ESP system management. It works over any distance or terrain, providing two-way communication via one or more devices; 2) ESP Vision* web-based monitoring system can be used with or without a SCADA system. Without SCADA system, it provides an end to end, remotely hosted system accessible through a secure web log-in, offering the advantage of SCADA without the support/maintenance. With SCADA system, it provides real-time data collection/characterization via the internet or private network connection.

And ESPExpert* is a real-time data analysis tool used exclusively by Centrilift. It merges well data with AutographPC* proprietary ESP system sizing/simulation software, to help understand downhole conditions and compare those to original system sizing, allowing experts to change well parameters or ESP performance to match well data via a user-friendly graphical interface.

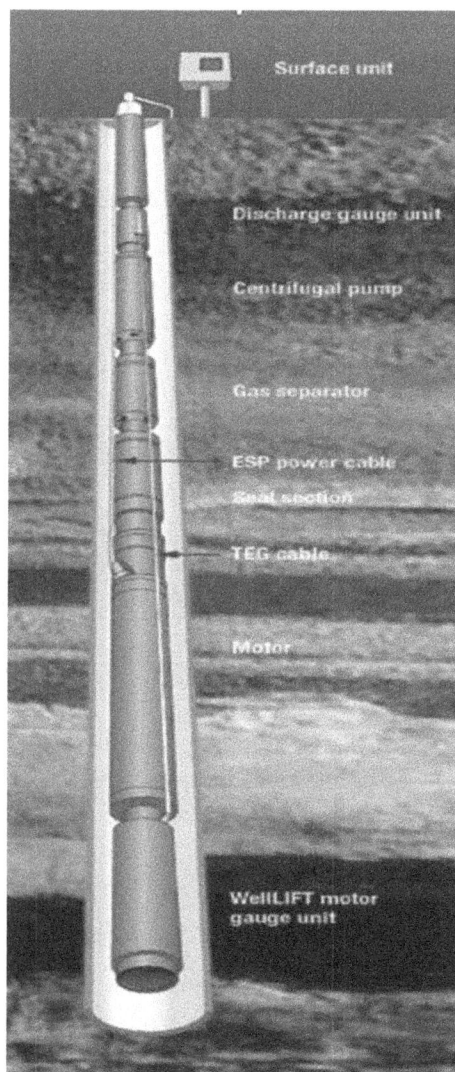

Fig. 5.35

Low Flow Hydraulic Diaphragm ESP. SmithLift, LLC, Provo, Utah, a division of Smith International, Inc., has developed and field proven the Hydraulic Diaphragm Electric Submersible Pump (HDESP). The pump is powered by a three phase, triple insulated electric motor and uses two hose-like "diaphragms" to positively displace formation fluid into the production tubing, Figure 5.36. The pump was designed for low-production-rate (< 200 bpd) oil, gas and coal bed methane applications with depths less than 2,500 ft. This pump has undergone a thorough development and testing program since its introduction in World Oil May 2004, "What's new in artificial lift," Part 2. The pump offers operators numerous advantages over rod pumps, PCPs and centrifugal pumps.

Operators have reported production gains due to higher drawdown while avoiding problems associated with cavitation and gas locking. As the pumping mechanism is completely isolated from the well fluid, the pump can handle relatively more solids than conventional beam pumps and centrifugal ESPs that use small vane openings for low liquid volumes. Operators also report other advantages, such as: ease of installation due to the pump's compact size (3¾ in. OD, 8 ft long, weighing less than 120 lb); a single integrated pump and high-efficiency electric motor; reduced electric power costs; pump-off tolerance; and decreased overall operating costs.

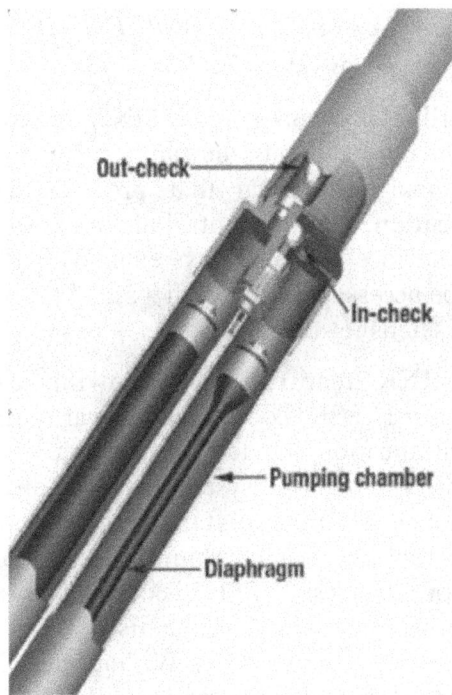

Fig. 5.36: Dual Hose-Like Diaphragm, Electric Powered Downhole Hydraulic ESP

5.12.25 *Artificial Lift Downhole Monitoring*

The Phoenix Select* monitoring tool from Schlumberger is the newest addition to Phoenix* artificial lift downhole monitoring systems. This cost-effective system features the latest generation of downhole sensors for in-depth diagnostics and analysis in wells and fields on artificial lift. By accurately identifying operating problems or anomalies in real time, it facilitates effective intervention strategies, protects lift systems, improves well integrity, and helps optimize production and maximize ultimate recovery.

The downhole systems can be used to monitor ESPs, beam (rod) pumps, PCPs, and gas lift systems. When monitoring an ESP, the sensor is connected to the pump motor; and the system transmits data to surface via the pump cable, Figure 5.37. Monitoring options are: 1) Lite: Basic ESP monitoring and protection; 2) Standard: Adds discharge pressure, temperature and vibration for improved pump protection and performance analysis; 3) Advanced: Complete monitoring of both pump discharge and intake conditions; and 4) Reservoir: Extends pressure, temperature and vibration measurements to the sand face. This flexibility allows operators to select the appropriate level of monitoring for any artificial lift system.

Traditional monitoring applications require data sampled at higher rates; the new system adapts sampling rate for key downhole parameters so that sufficient data is available. Trip and alarm relays for all monitored parameters are field programmable to match individual reservoir well and operating conditions, thereby reducing false alarms. This reduces compatibility issues by using a single data acquisition/communications platform for all monitored wells. The combination of advanced transducer technology, state-of-the-art microelectronic components and digital telemetry ensures reliable and accurate information.

Phoenix* monitoring systems are fully compatible with other monitoring/control technology, enabling an integrated system comprising: a sensor unit, an integrated surface panel or universal site controller, a surface choke assembly, software for manual data retrieval, and an optional portable data collector. They are SCADA-ready, with a Modbus remote terminal-protocol port and RS-232 and RS-485 ports for continuous data output. Downhole and field components can be further integrated with the espWatcher* surveillance and control system for: real-time, remote data acquisition; alarms and alerts via satellite; remote pump startup and speed control; and remote resolution of a variety of pump problems.

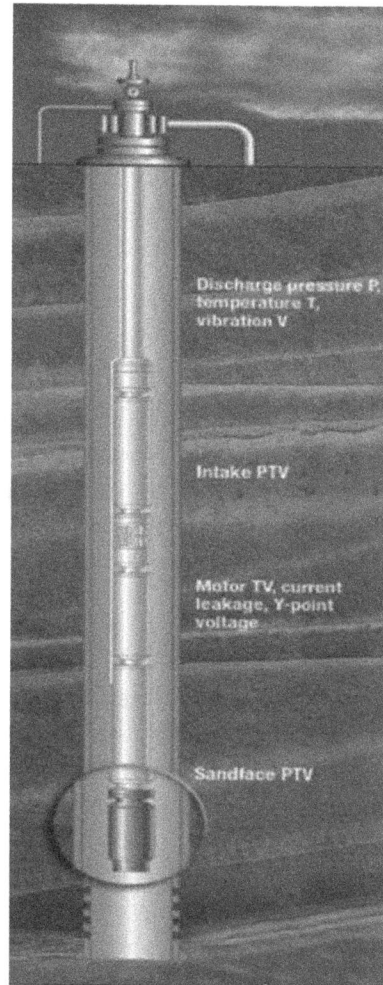

Fig. 5.37: Artificial lift Downhole Monitoring System for ESP

5.12.26 *Through-Tubing Conveyed ESP*

Baker Hughes Centrilift, Claremore, Oklahoma, has introduced the Centrilift-Thru-Tubing Conveyed (TTC) ESP system which provides an innovative approach to economic well completions. The system is designed to install and pull the submersible pump assembly without a conventional workover rig.

Pump installation and removal can be accomplished using a variety of methods, including: wireline, slickline, electric line, coiled tubing, small-diameter tubing and sucker rods. Systems are operating in 4½ in. tubing, with larger system deployment capabilities under development. TTC systems use standard Centrilift ESPs.

The new technology has numerous benefits, including: reduced workover costs on pump-only failures; reduced cycles in and out of wellbores give increased power cable longevity; economical pump replacement in response to real-time changes in well dynamics; and less-intrusive workover procedures, ideal for environmentally sensitive areas. The system uses standard Centrilift components.

Applications that can benefit include: small offshore platforms without permanent rigs; remote locations with logistical problems for workovers; environmentally sensitive areas or urban well sites; and any high-cost workover environment.

Wells with production strings 4½ in. or larger are candidates. With average workover costs on Alaska's North Slope running between $140,000 and $240,000 per day, fields have high workover rates. The first unit was installed in a well in a North Slope field in late 2001. The project, which initially deployed TTC technology in 4½ in. tubing, resulted in cost savings of at least $100,000 per pump change-out.

5.12.27 *Variable Speed Drive*

Their VSP & VSD are same different company had given different name which is introduces by GCS Electro Speed-II, the only Variable Speed Drive (VSD) in the industry to offer a choice of six-step or filtered Pulse Wave Width (PDM) output waveforms. The patented filter controller protects the motor by either switching to six-step mode or shutting down the unit if the filter fails. On the input side, the system offers standard 6 or 12 pulse converter input options. Drives with higher pulse count converters can be special ordered to meet IEEE 519 Standards for increased harmonic reduction.

The VSD has a range of 66 kVA to 2,000 kVA. It is designed for easy setup/operation, with advanced features that reduce the number of circuit boards and enhance drive reliability/versatility. The system's enclosures can be customized to meet the harshest environmental demands, Figure 5.38.

Fig. 5.38: Variable Speed Drive Enclosure

The new VSD is a fourth-generation drive representing over 25 years of ESP-specific experience. Surface controllers are an integral component of ESP systems. VSDs offer many advantages over constant-speed controllers by allowing operators to vary ESP operating parameters to: control motor speed to lower motor temperature; improve gas handling; control well drawdown; adjust ESPs to changing well conditions; and decrease start-up stresses.

Positive-seal shroud hanger: Motor shrouds and shroud hangers are commonly used to set ESP systems below well perforations to maximize well drawdown and/or to minimize gas interference in the pump, Figure 5.39. The shroud orients wellbore fluid around the downhole motor, thus helping to cool it.

Conventional shroud hangers often create multiple problems, including: misdirected fluid flow which creates motor heating; excessive heat causes scale buildup between motor and shroud; optimum water production and well drawdown are not achieved due to scaling; poor gas separation occurs due to fluid/gas leakage through the shroud hanger; and gas production can be below expectations. Further,

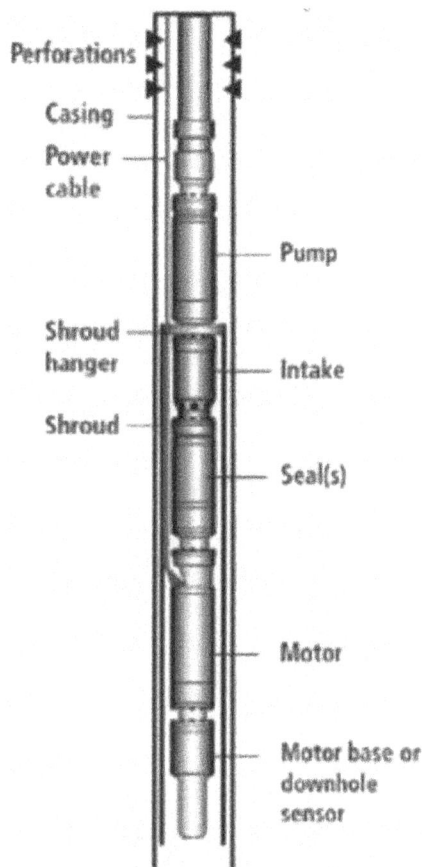

Fig. 5.39: ESP System and Shroud Installed Below Perforations

wells can continually shut down due to high temperature affecting motor performance; and scale-restricted fluid flow can create additional heat and cause eventual motor failure. Downhole re-circulation systems have been only marginally successful in solving these problems.

Testing performed at Wood Group ESP, Oklahoma City, determined that wellbore fluids and gas were leaking through the conventional shroud hanger, and leakage increased at very high rates, as flow between shroud and motor became restricted. Based on these findings, a new intake body and shroud were designed.

The new, patented-design, positive-seal shroud hanger: efficiently forced wellbore fluid past the motor; reduced motor temperature 50°F to 30°F; reduced scaling; allowed for more accurate equipment sizing, which in some cases lowered the HP requirement by 50%; provided better gas separation; increased drawdown efficiencies; and increased equipment run life and production.

Downhole sensor/surface interface: Wood Group ESP has introduced the new SmartGuard V and SmartGuard VI downhole sensor and surface interface systems for ESP systems, Figure 5.40. These enhanced models offer expanded monitoring of vibration and leakage.

SmartGuard monitoring systems enable operators to reliably and accurately monitor the condition of their ESP systems, and track wellbore parameters. All of the systems monitor wellbore or pump intake pressure, wellbore or intake fluid temperature, and pump motor oil or winding temperature. The new models also measure vibration of the pump system and current leakage, which allows cable degradation to be determined. The SmartGuard VI model includes the measurement of pump discharge pressure. This system is suitable for all major brands of ESPs.

The sensors have been used successfully as part of an integrated control system that automatically cycles wells off and on based on predetermined parameters. The system is programmable to sample data continually with selectable recording capabilities.

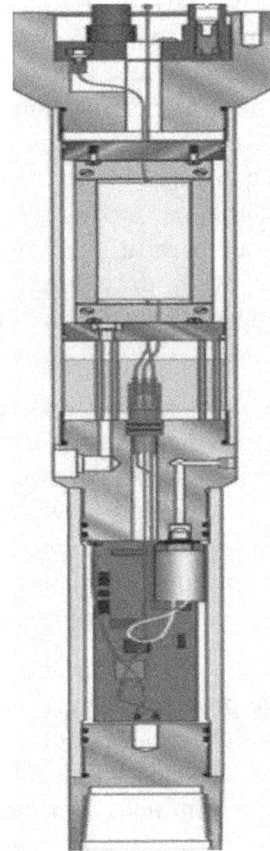

Fig. 5.40: SmartGuard Sensor System as Mounted to Bottom of Motors

5.12.28 *Downhole Desander System*

ESP systems specialist, Pumptools, headquartered in Aberdeen, has launched its SandCat system, which separates sand from the produced fluid in sandy oil wells and prevents it from entering the ESP, thus reducing destructive consequences. The system is a downhole centrifugal sand separator connected below the ESP that separates sand from produced fluid, allowing non-abrasive fluid to the ESP, Figure 5.41. Costly well preparation methods such as sandscreens and gravel packs are avoided, as it can also be retrofitted to any ESP well.

The sand management system includes unique calculation software to predict separation efficiency on a well-by-well basis. The supplier provides this service for all SandCat applications to determine the appropriate system spec. Available for flowrates up to 10,000 bpd, the tool comes in sizes suitable for casing sizes from $5\frac{1}{2}$ in. to $9\frac{5}{8}$ in., and this is engineered for each installation. Unique software enables the well engineer to calculate efficiency required of the system on a well-by-well basis, and predicts how much sand is being separated and, therefore, allows the operator to decide on the best sand management option.

The tool separates the sand before it enters the ESP, leaving it downhole and collected either in a tailpipe, or dumped into the rat-hole below the perforations, need to be removed in a later work-over operation. Pumptools offers four different deploy-ment methods to optimize sand separation management.

5.12.29 *ESP Motor Protector*

Advanced motor protector systems from Schlumberger, Houston, are designed to handle challenging environments. They are available in different versions to provide a fit-for-purpose solution for sand and solids production, corrosive fluids and high-temperature environments.

The new design features an optional metal bellows to replace the elastomeric bag, Figure 5.42. Using of metal bellows extends temperature operating limits far beyond those of the bag-type protector. In the case of gassy wells, the bellows also prevents gas from migrating through the rubber bag and

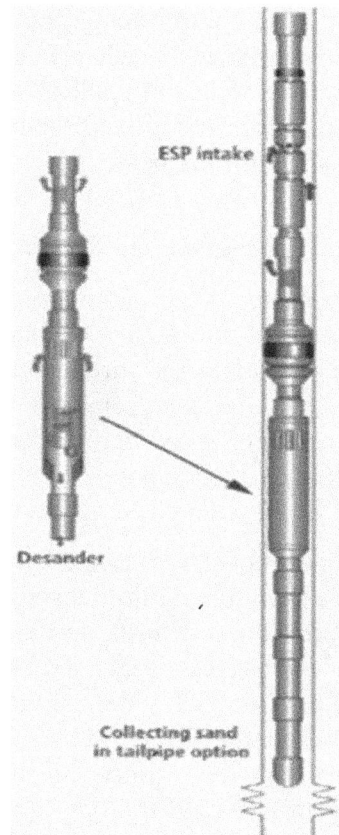

Fig. 5.41: SandCat Desanding System Run below the ESP, Dumping Separated Sand into a Tail Pipe

displacing the motor oil. The bellows is resistant to H_2S and impermeable to gas and, thus, is more suitable for wells in harsh environments, i.e., high temperature and H_2S, as well as high gas content.

The advanced protector has special features for wells producing sand or solids, including a novel protector head design. This allows falling sand or solids from the protector intake to drop back into the wellbore while protecting the top shaft seal from abrasion damage. It also utilizes new bearing types to improve shaft stability and reduce abrasive wear.

If temperature, gas content and corrosive chemicals are not a concern, the metal bellows can be replaced with the traditional elastomeric-bag of labyrinth-type design options, while still incorporating advanced options to handle abrasive environments.

With a maximum temperature rating of 425°F, the high-temperature, advanced protector has a proven record in steamflood injection applications at its maximum operating limits. These wells would have never been produced with ESPs at maximum rates before development of this new protector.

5.12.30 *Gas-Handling System*

Schlumberger's Poseidon gas-handling system is a multiphase, axial flow device installed below the pump to handle up to 75% free gas through the pump. The system can be installed above a gas separator where gas can be vented into the casing, or it can be installed above a standard intake, if all produced gas must go through the pump. It can also be used in subsea wells and wells with nonvented packers.

The axial flow stages prime the main pump and push the gas-liquid flowstream into the stages. Gas volume is reduced by compression. Lab tests and field installations have shown that it can successfully operate in the ultrahigh-gas-cut well with gas volume fractions up to 75%, far exceeding conventional systems' 40% to 45%. Figure 5.43 shows before and after ammeter charts documenting improved pump performance with the new system's installation.

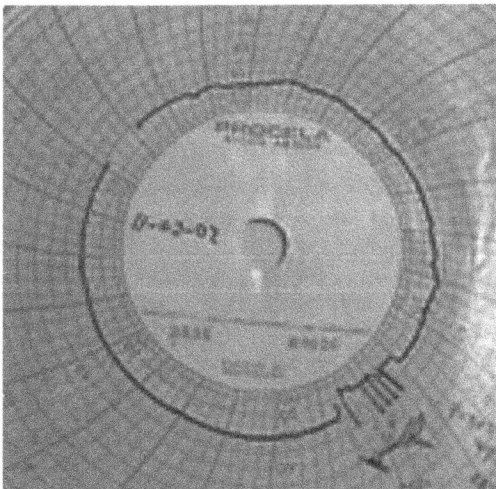

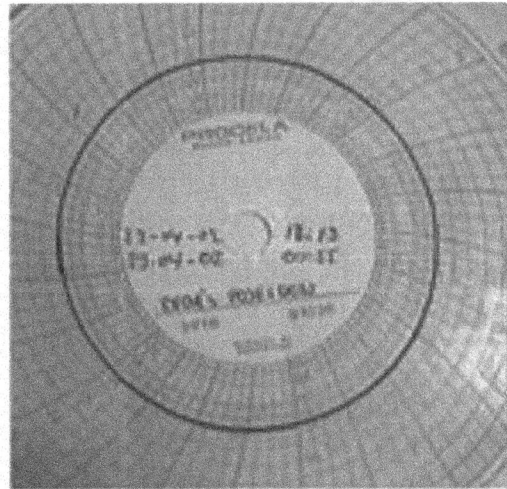

Bellows   Bag   Labyrinth

Fig. 5.42: Advanced ESP Motor Protector System

Fig. 5.43: Ammeter Charts before (A) and after (B) Installation of Axial Flow Gas Handling System below the ESP

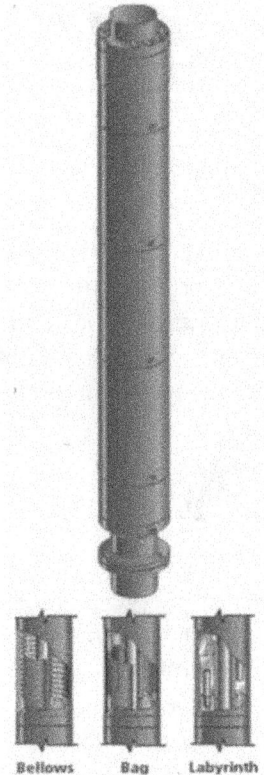

The first field installation was in Colombia in a field operated by HOCOL, one of the Mimir Companies. Well SF-75 is in a block relying on water-alternate-gas injection at fixed intervals to improve secondary oil recovery. The injection program causes frequent high variations in the produced gas. The system was operating with a gas-handling device, but production significantly dropped with the increase of gas. After change-out to the Poseidon system, high and steady production rates were achieved at higher gas volumes. The system improved pump efficiency in handling free gas. The well's production increased by more than 40%, and the new installation managed free gas rates as high as 60% in the pump intake.

5.12.31 *ESP By-Pass Tool*

There has been a need for a downhole Electrical Submersible Pumps. ESPs are prone to have relatively high failure rates due to the unavoidable use of electrical and moving part components in hostile downhole environments.

ESP By-pass Tool allows through-tubing reservoir/production data instrument running below the pump.

1. Determining and optimizing the well performance with associated pump, motor, cable selection and protector.
2. Choosing a method of deploying and recovering the pump-tubing.
3. Venting of annulus gas and the requirement and selection of a gas separation method and other monitoring systems like bottom-hole pressure transmitter.
4. Setting depth of pumps depends upon influx of the well.

ESP companion tool by which BHP, BHT and production logging data could be obtained in both producing and static conditions. Monarch Engineering of Gardena, California, designed a bypass tool, through which reservoir data could be obtained. This tool, in its original design, is available from Oilwell Survey Testing Systems (OSTS), Santa Fe Springs, California, Figure 5.44. The tool's key features are its: simplicity; friendly wireline operation; attractive cost; and trouble-free long life.

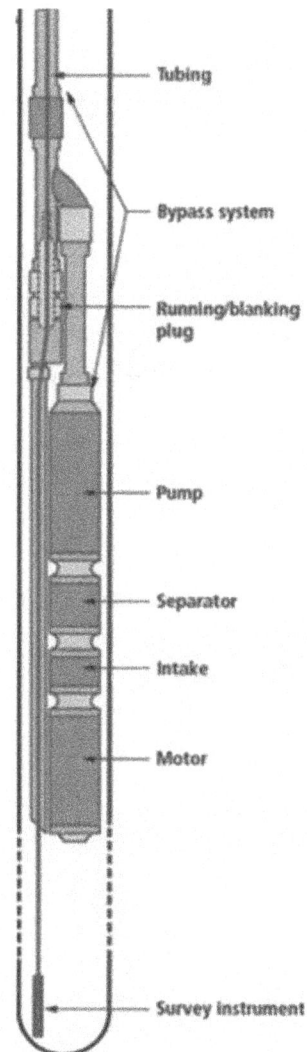

Fig. 5.44

The tool body is fabricated/machined from plain-end mechanical tubing. The side arm for hanging the ESP is certified welded to the main body. The instrument tubing is drawnover mechanical tubing in range 2 lengths with a flush joint thread. It is hung (pin end up) from an oval swivel connection just below the blanking plug seat. This connection permits orientation of the instrument tubing to fit the OD of various series of ESPs.

The blanking/running plugs are 1-3/4 in. or 2-1/4 in. The seat for these plugs will pass 1-11/16 in. logging tools. The tool is available in 2-3/8, 2-7/8, 3½ and 4½ in. sizes, with 1-3/4 or 2¼ in. blanking plug. It is available in stainless steel.

5.12.32 *Diaphragm-Type ESP*

Smith Lift, LLC, Provo, Utah, a subsidiary of Smith International, Inc., is introducing a new submersible pump line that has been in development and testing for the past three years. The pump is similar in size to conventional submersible centrifugal pumps, but uses a patented, true-positive-displacement diaphragm pumping system to improve efficiency, handle solids and pump mixtures of gas and liquids.

The pump, known as the Hydraulic Diaphragm ESP (HDESP) provides a downhole electric-driven hydraulic system that powers a hose-like "diaphragm" to positively displace formation fluid into the production string (Figure 5.45). The current pump design is 3–3/4 in. in diameter, 10 ft long and weighs less than 150 lb. These are designed to produce 50 bpd to 400 bpd from depths of up to 6,000 ft.

The new ESP has been used to move a wide variety of fluids, from heavy oil to fresh water, under a wide variety of fluid conditions, including up to 2 % sand/coal fines, gassy fluids and high H_2S/CO_2. The pump is constructed of stainless steel and is installed similarly to a conventional submersible centrifugal pump.

Chemical treatments, such as acidizing and paraffin removing can be accomplished by the operator with the pump in place. The pump typically consumes about one-third of the power required to drive a conventional rod or centrifugal pump.

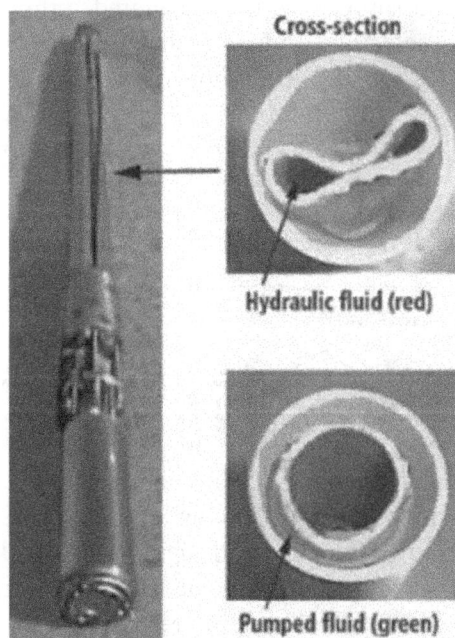

Cross-section

Hydraulic fluid (red)

Pumped fluid (green)

Fig. 5.45: Submersible Electric/Hydraulic Diaphragm Pump Cross-Section

5.12.33 *Multi-Vane Pump*

Centrilift, a division of Baker Hughes, Claremore, Oklahoma, has introduced the MVP multi-vane pump. The new ESP design features a unique, patent-pending impeller that can lift more fluid or reduce horsepower in wells with high gas-to-liquid ratios. The pump handles higher free-gas-content fluids, prevents gas locking and maximizes lift compared to conventional ESP systems.

The improved multi-phase-fluid dynamic design improves performance by producing more free gas through the pump, utilizing the greater number of vanes, Figure 5.46. The free gas then lightens the fluid column in the production tubing, reducing Total Dynamic Head (TDH) and required horsepower. Therefore, it can lift more fluid with the same horsepower, or the same amount of fluid with reduced horsepower. In either case, the result is lower power cost per volume of fluid lifted. The pump can be used alone, or employed as a charge pump with a standard ESP. It can also be used in combination with rotary gas separators for high gas applications to reduce intake gas to acceptable levels.

The system can be sized using Centrilift's Autograph PC sizing and simulating software. MVP performance was measured during development, with extensive testing in the on-site gas test loop facility. In an oil well in the western US, with a GOR greater than 16,000, the system was added to an existing ESP. Well drawdown improved and oil production increased over 75 %, 32 bopd from 18 bopd. At the same time, gas production nearly doubled.

ESP for high-viscosity fluids. VIPER, introduced by Centrilift, is the first and only ESP specifically designed to produce high-viscosity fluids. It features a unique, patent-pending pump stage design with radically different vanes that improve fluid shear and reduce surface friction, Figure 5.47. As a result of this breakthrough technology, the pump optimizes production in wells with high-viscosity fluids. The

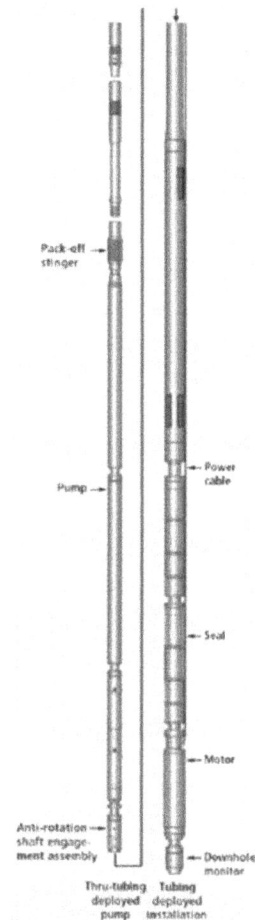

Fig. 5.46(a): Through-Tubing Deployed ESP Pumping System

Fig. 5.46(b): Multi-vane Impeller Design Delivers More Head and Increased Production

design reduces power requirements and maximizes lift, compared to conventional ESP systems. It employs an inducer that improves fluid shear at the pump intake, and larger vane openings in the impellers to reduce internal friction. The result is higher total dynamic head, improved efficiency, and a lower horsepower requirement.

The technology gives operators a choice: it can lift more fluid with the same horsepower, or lift the same amount of fluid with reduced horsepower. In either case, the result is lower power cost per volume of fluid lifted. The system is designed to significantly improve ESP performance in high-viscosity fluids. As viscosity increases, performance improves, compared to conventional ESP technology. Performance of the ESP in adjustable-fluid-viscosity applications was measured in Centrilift's test loop facility, allowing design optimization.

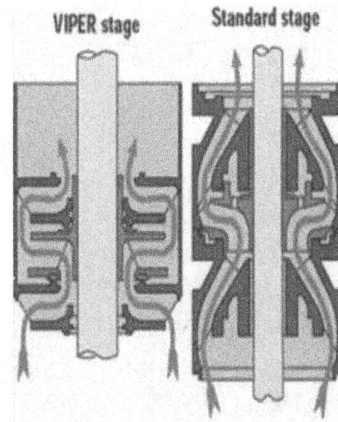

Fig. 5.47: The VIPER Stage Improves Fluid Shear with a Shorter Fluid Path

5.12.34 *Wider-Vane-Opening Pumps*

Wood Group ESP, Inc. (WGESP), Oklahoma City, Oklahoma, had designed two new, higher-efficiency, wider-vane-opening pumps that have recently been incorporated into its 400 series, the TD150 and TD460. In a commitment to continual improvement, the company has replaced one of its high-usage pumps with more-efficient stages. A new model TD460 offers wider-vane-opening stages which can help produce lower-gravity fluid, handle more gas and reduce the effects of scaling, Figure 5.48. The ESP's stages have highest efficiency and head per stage in its class in the industry.

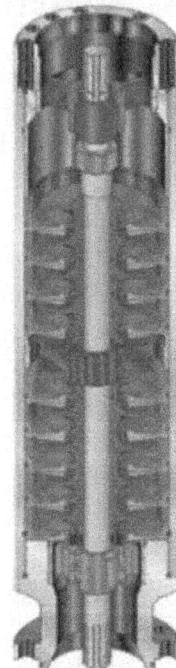

Fig. 5.48: Cutaway of New TD460 pump Shows Wider-Vane-Opening Stages

Horsepower requirements for model TD460 stages are about 15–22 % less, compared to existing stages offered by different manufacturers in the 450 bpd range. Due to its high head coefficient, it requires fewer stages to generate the same amount of lift. Development of TD460 was preceded by TD150 stages, which also had impeller to wider-vane-opening stages. Together with TD150, TD300 and TD460, WGESP offers a large choice for wider-vane-opening stages for 400 series pumps.

5.12.35 *Downhole Sensor with Surface Interface*

Wood Group ESP, Inc., had designed the Smart-Guard, a downhole sensor and surface interface system for reliable, accurate, real-time monitoring of ESPs and reservoir conditions, Figure 5.49. Operators should expect reliable information about equipment performance and vital downhole conditions. Now, this new family of monitoring systems can reliably/accurately track and retrieve real-time information related to critical performance parameters. Output control loads that respond to selected inputs are also available. This can lead to lower operating costs and longer run times.

Employing the latest state-of-the-art transducer technology and advanced linear calibrations, this new-generation monitoring system surpasses conventional pressure and temperature sensors. Engineers designed the sensor for oilfield-rugged with no moving parts to wear out. It is resistant to vibration and has been field-proven. The strain-gauge technology used for pressure monitoring is robust and accurate. Wellbore, motor oil and motor winding temperatures are measured via platinum Resistance Temperature Detectors (RTDs) and Type J thermocouples. The technology improves the accuracy/resolution of pressure/temperature data with 0.1% pressure accuracy and 0.6% downhole temperature accuracy. The system is programmable to sample data continuously with selectable recording capabilities.

5.12.36 *New Sensor Technology*

Schlumberger has developed new sensor technology for coiled tubing—deployed ESPs. To further enhance the benefits of Elutrical Coil, the patented technology utilizing power cable inside the coil tubing, the company now offers real-time sensor data as part of the system. Like standard ESP monitors, the sensor transmits real-time data to

Fig. 5.49: Cross-Section of Smart Guard Downhole Sensor to Measure Critical ESP Performance Data

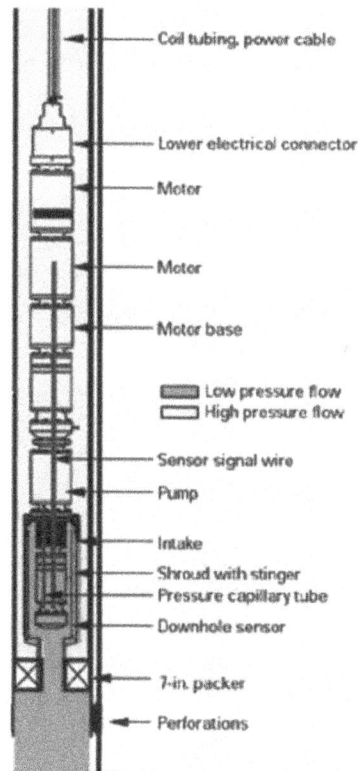

Coil tubing, power cable

Lower electrical connector

Motor

Motor

Motor base

▨ Low pressure flow
▢ High pressure flow

Sensor signal wire

Pump

Intake

Shroud with stinger

Pressure capillary tube

Downhole sensor

7-in. packer

Perforations

Fig. 5.50: Downhole Sensor Sends Data to Surface through Power Cable Inside Coiled Tubing

surface through the power cable, Figure 5.50. Data such as intake pressure/temperature, discharge pressure, motor winding temperature, vibration and current leakage data are collected.

5.12.37 *CT-deployed ESP*

In many applications, it is not possible to flow production up the coiled tubing due to the high flowrates—up to 30,000 bpd in 7-in. casing—produced by some ESPs. Now, Schlumberger provides the patented CrossFlow system, coiled tubing—deployed ESP, for many difficult applications, Figure 5.51. With the system, high-flowrate wells can be produced, as the system does not have fluids produced up the coiled tubing. For wells with high volumes of free gas, the system can incorporate the company's high-flowrate gas separaors.

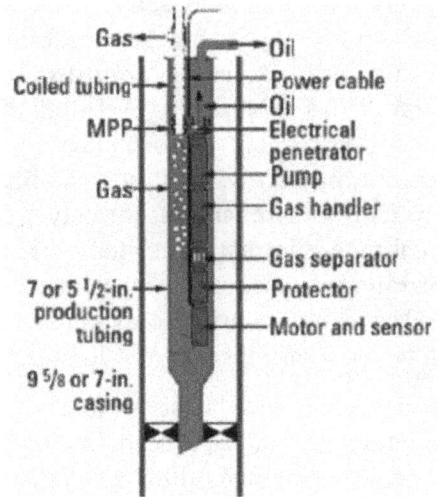

Fig. 5.51

High viscosity oils can reportedly be produced more easily with the system and the number of pump stages are lowered by not pumping high-viscosity oil and emulsions up the CT. Access to the reservoir, and well treatment, is possible without retrieving the CT-deployed ESP. And the spoolable gas lift system, can now be added to the CT-deployed ESP, providing back-up in case of ESP failure, or used in conjunction with the ESP to lower H.P. and fluid-lifting requirements. Schematic of CrossFlow system for CT-deployed ESPs.

5.12.38 *Eliminating Harmonic Problems*

With the release of Sine Wave Drive, Schlumberger provides technology that can eliminate harmonic problems on both the load side and line side of ESP applications. The new drive combines a built-in filter with advanced pulsed-width, modulation-VSD technology to provide near-sinusoidal current and voltage waveforms to the ESP motor, Figure 5.52.

The filter does not require any application-specific tuning or adjustment,

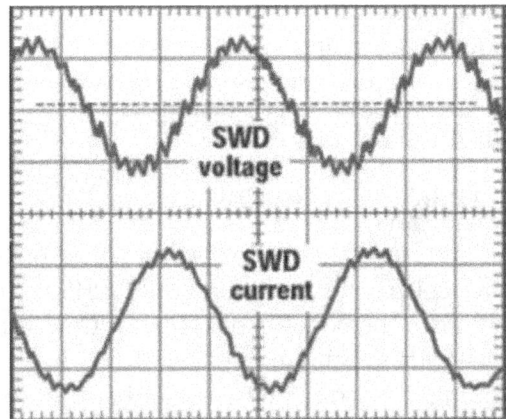

Fig. 5.52

making the system applicable for land-based VSD operations, which are commonly moved to multiple locations. Offshore operators will be pleased with the minimized footprint achieved by incorporating the filter section in the base of the system and not adding additional footprint or ancillary devices. Load-side harmonics and power issues are addressed with diode technology on the drive input. Use of diode front ends allow for line-side filters to effectively eliminate harmonics over the various operating frequency ranges desired in ESP operations. New Sin Wave Drive provides near sinusoidal current/voltage waveforms to ESP motor.

5.12.39 *Pump-Monitoring System*

Weatherford Artificial Lift Systems, Houston, has expanded its ESP range, adding a combined downhole ESP sensor and switchboard/variable speed drive motor controller. This rugged, downhole ESP sensor is attached at the rig site to the ESP motor and communicates well pressure/temperature and motor-winding temperature, vibration (two planes) and insulation quality to the surface using the ESP power cable, Figure 5.53.

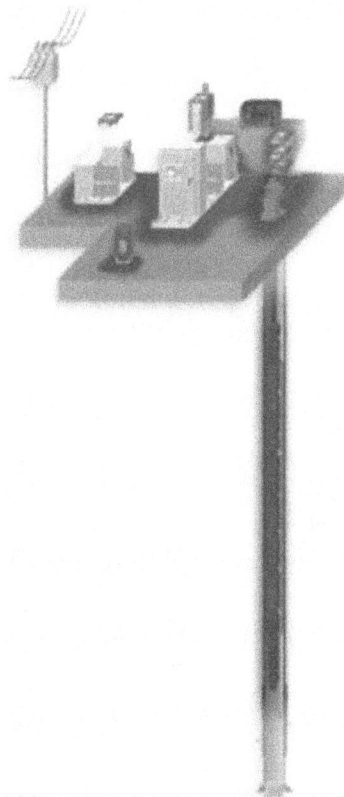

At the surface, all high-voltage equipment is integrated within an existing switchboard, transformer or the Weatherford control system. Fitted to the switchboard or variable speed drive, the New Weatherford Artificial Lift controller (a component of the system) for the first time combines all functions of a motor control, data logging unit and human machine interface for both downhole sensor data and motor running parameters. No separate motor controller is required.

The sensor is designed for in-country field re-dress and upgrade. Field upgrades may be optionally factory fitted. Available upgrades include discharge-pressure measurement and reservoir quality monitoring using a quartz sensor for well pressure. The surface

Fig. 5.53

controller is a complete Modbus-based unit with a flexible architecture to allow unlimited analog and digital input/output. PC and PocketPC software is supplied. In addition, by using the company's management software, this system handles all the data received and allows for easy utilization of the logged data. Downhole ESP sensor and surface controller system.

5.12.40 *ESP for Coalbed Gas Wells*

The coalbed methane gas industry has attempted to use ESPs to deliquify CBM gas wells, with limited success. Weatherford determined that a new approach was needed that: offered extended run times, was cost effective, repairable and fully automated. As ESP pumps held the most potential for CBM gas applications, the company focused on adapting its technology for these unique situations and created the patented CBM-ESP pump. The pump works in deep or shallow wells, with low or high water volumes, over a wide range of pressures/temperatures.

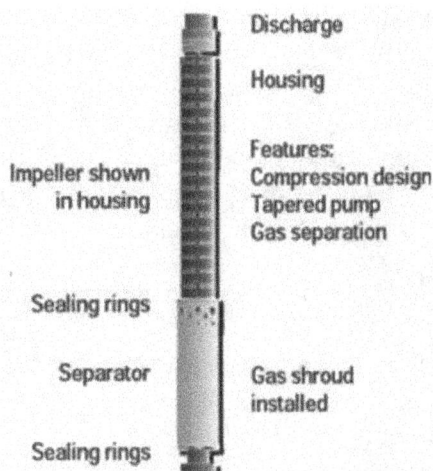

Fig. 5.54

Three main operating challenges are: handling large and varying amounts of gases, dealing with abrasive coal fines, and rapid water decline accompanied by rapid gas increase. When pumping begins, these problems must either be diverted from, or handled by, the pump. A special effort was made to design a submersible pump that was easy to repair.

The new pump can be manufactured in many different types of gas-shrouding configurations to handle gassy fluids. These shrouds divert gas from around the pump and allow it to flow up the wellbore, Figure 5.54. The tapered design compresses gas that enters the pump so that it passes into the liquid phase and is produced. The system uses compression stages that help with low-fluid-producing wells. And the pump is built with hardened shafts and support bearings that are tolerant of abrasive particles—unique intake screens capture/divert many of these fines. Principal components of coalbed methane liquid removal ESP system.

5.12.41 *ESP Testing/Reconditioning*

Independent Testing Services of Midland, Texas, offers submersible equipment testing services to production operators in the Permian basin to help lower submersible pumping systems costs. For pumps, the service provides cleanup, acidizing and chemical cleanup. Once cleaned, each pump section is computer tested utilizing a horizontal test bench, Figure 5.55. Test results are compared to the manufacturer's performance curve. Over 65 % of the used pumps test within satisfactory limits and require no expensive repairs.

For motors, a series of tests determine whether motor insulation is dry and suitable for rerun. And a series of tests on the protector/seal section—including a load test on the thrust bearing and pressure tests across the mechanical seals-to be done. ITS provides an inventory stocking system for production companies to store their used equipment at no charge, along with providing adapters to enable operators to mix and match different manufacturers' equipment. This allows operators to maximize the use of surplus inventory and lower the cost of equipment to be utilized. Minor repair on all types of equipment is also provided. Facility for testing/reconditioning ESP equipment.

Fig. 5.55

5.12.42 *New Motor/Capacitor Combination*

A new tool for engineers is projected to lower electrical costs. Electrical power is a major economic factor in mature oil fields and is often the greatest single cost for upstream operations. Low VA-efficiency motors, characterized by low power factor and low efficiency, contribute to high electrical costs. NEMA D, high-slip, high-torque motors characteristically used in beam pump applications exhibit low efficiency and low power factor. Annual operating costs for these motors can be 50 times their installed cost. Thus, transmission, reception and power conversion inefficiency adds up to the high electrical cost of production.

Load-based solutions can significantly reduce electrical costs. However, frequent phase-leg losses and variably cyclic beam-pump loads are generally incompatible with load-based shunt capacitors. PACCAP, Inc., of Odessa, Texas, is developing a new set of load-based solutions. It has developed a new capacitor implementation. These durable, continuous duty electrolytic capacitors provide AC capacitor size reduction. They are suitable for series applications. Series capacitors, Figure 5.56, provide voltage support, and power factor correction. They allow and require optimal electrical motor designs. Financial payback is targeted at one year, but will vary with electricity cost, utility billing practice, annual run time and motor designs.

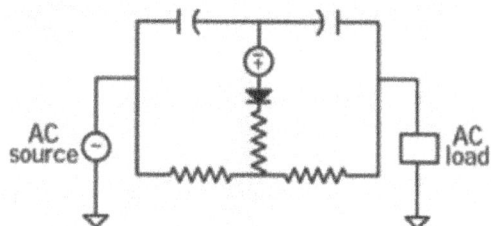

Fig. 5.56: Biased Anti-series Capacitors

CHAPTER 6

Hydraulic Pumping

6.1 Introduction

Hydraulic pumping systems (Figure 6.1) employs surface pumps to supply energy to a downhole cyclone and pump assembly. The main manufacturers for hydraulic pumps include Kobe, Guiberson, National Oilmaster, KSB and Wier (for turbine system).

There are three types of hydraulic pumping systems *i.e.,* piston pump, jet pump and hydraulic turbine (weir). Hydraulic pumping has a wide variety of applications but is particularly attractive as an alternative to rod pumping in deep or crooked wells. In the moderate to high volume range where formation gas liquid ratio is moderate to poor, hydraulic pumps are an alternative to gas lift or electrical submersible pumps.

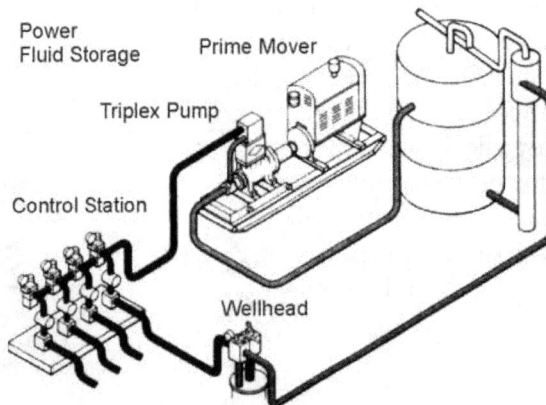

Fig. 6.1: Surface Facilities for Hydraulic Pumping
(*Source:* After a Trico Document, Composite Catalog 1990–1991)

Some of the specific advantages of hydraulic pumping system:

- Can be used where gas supply is not available for gas lifting.
- The pump can be pulled from hole by simply reversing the flow of power fluid and pumping it to surface. No service rig is required.
- Compared to sucker rod pumps, the hydraulic pumping system is capable of lifting a larger daily volume from a greater producing depth.
- Displacement rates can be controlled more easily since piston and jet pumps are interchangeable without pulling the completion.
- The system can be used in crooked, deviated or horizontal wells.
- Chemical treatment for paraffin and corrosion control can be easily added with power fluid.
- Hydraulic pump can be used to produce any desired draw-down whereas gas lift is un-economic for very low bottom hole pressures.

The major disadvantage is the need for at least two conduits from surface facility to the bottom of the well, and high capital and maintenance costs on the high pressure power fluid supply system.

6.2 Power Fluid

To achieve efficient hydraulic pump installation, clean and solid free power fluid is to be maintained. Surface facilities must include cyclone cleaners as well as separators and pumping equipment.

The power fluid can be either oil or water. Oil reduces maintenance costs because of its built-in lubricity and is generally easier to treat. However, there is a higher risk of flash fire with high pressure oil and stock and make up costs are high. Water requires less surface power because of its density, but introduces higher corrosion risks and therefore treatment costs, especially in an open power fluid system. The low bulk modulus of water causes more severe pressure pulses, and can contribute to fatigue failure of pump components. Subsurface pumps are sensitive to viscosity and lubricating qualities of the power fluid. Because water has practically no lubricating ability at bottom hole temperatures, it can, if not adequately treated, contribute to shorter pump life. Leakage of power fluid past the various sliding surfaces in the pump is a function of viscosity and is greater with water than crude oil.

Quality of power fluid is critical *i.e.,* solids should be less than 15 ppm, particle less than 15 µm and salt less than 12 lb/1000 bbls. Cyclones and/or centrifuges are generally used to clean the power fluid.

6.3 Down-Hole Completion for Hydraulic Pumping

6.3.1 *Single String Configuration*

Most common is the casing free pump in which tubing is used for power fluid supply and production and spent fluid is lifted up the annulus. Packer is required to isolate the production interval. Sometimes the casing must be isolated from corrosion in which case two tubing strings are used. Free pumps are run in the power conduit and can be pumped in and out of the well for servicing by simply reversing the flow. The ease with which Bottom Hole Assembly (BHA) can be recovered, inspected and replaced, and the interchangeability of jet and piston pumps with well conditions, make this type of hydraulic pump particularity attractive. However, this option comes at a cost of reduced pump capacity.

Conventional or fixed installations have the pump assembly run on the tubing. If the pump requires the service or replacement, the entire completion must be pulled out.

6.4 Surface Pumps

The surface pumps can be driven by electrical or any type of internal combustion engine. They are designed especially of power fluid service. To generate required pump pressure, the pumps should be positive displacement triplex type.

Metal plungers and liners, and ball type valves are used for oil power fluid. For water power fluid, plungers and liners with packings are usually used due to decreased lubricity and viscosity of water. Relief valves, pressure gauges and safety switches must be installed and properly serviced to provide hazard free operation.

The discharge from the relief and safety valves must be routed back to the main separator. This prevents flash gas from being drawn into the pump.

Pulsation in the system caused by opening and closing valves may be reduced with a pulsation dampener. The higher bulk modulus of water makes a pulsation dampener mandatory in a water power fluid system.

6.5 Other Facilities

As shown in Figure 6.1, a single well installation would include power fluid storage, cyclone separator for fluid cleaning, a spill over system for produced fluid, a flow control valve, a power and return fluid monitoring system. Full field systems will require manifolding to test and bulk separators for power fluid supply to each well. Also high pressure power fluid supply lines have to be laid to the individual well heads. These normally operate at 3000 to 5000 psi.

The power fluid tank should be sized to ensure a maximum vertical velocity of 1 ft/hr at the expected power oil withdrawal rate. This allows for settling of particulate matter and provides sufficient capacity for small power fluid losses.

6.5.1 *Surface Facilities for Hydraulic Pumping System*

(a) *Produced well fluid:* When power fluid is the well fluid, separation of the produced fluids at the surface is done with a three phase separator, which also acts as a reservoir for the surface power fluid pump. The power fluid from the separator is cleaned of solids by means of cyclone separators, before it is taken into the suction of high pressure power fluid pump. A centrifugal pump is used to supply the power fluid to the inlet of the cyclone. The pressurized power fluid is then metered and sent to the well. Oil, gas and part of the water is taken to the GGS for further separation. Chemicals (corrosion, scale or paraffin inhibitor) suction side of the power fluid pump, if required. A typical scheme of the surface facilities used with is shown in Figure 6.1.

(b) *Water as power fluid:* Condition of well base and can well be dosed into the fluid as power fluid suitable conditions, external water source, such as available high pressure water injection water can also be used as power fluid. In this case, the separation of power fluid at the well site is avoided as the same can be done at the GGS.

6.6 Piston Pumps

The hydraulic piston pump is a closely coupled reciprocating engine and pump. Essentially, this is a competitor to the rod pump. It is the most effective deep (> 8000 ft) lift system with units operating down to 18000 ft. It is highly attractive for crooked or deviated wells. It is not a high rate lift system.

6.7 Jet Pumps

An oil well jet pump is a ventur-type device where high pressure fluid is caused to accelerate to a high velocity and thereby create a low pressure area into which reservoir fluids will flow. The pump has no moving parts. This is accomplished by momentum transfer of fluid through an ejector nozzle, throat and diffuser assembly. As the high pressure fluid comes out from the nozzle and enters in the throat or mixing tube, it is converted into a high velocity and low pressure fluid jet. Thus the surrounding well fluid, having comparatively higher fluid pressure and having access to the throat chamber will be sucked in the throat. On passing from throat to diffuser, the mixture of well fluid and power fluid loses velocity and consequently acquires equivalent discharge pressure to a value sufficient enough to lift the total fluid to the surface. A typical downhole jet pump is shown in Figure 6.2. A closer look at the effective mechanism is shown in Figure 6.3. These pumps require a high pressure fluid pump on the surface, and this fluid is circulated down the well, typically the tubing. The fluid flows through jet causing a low pressure region, and the power fluid - produced fluid mixture is brought to the surface by a second conduit, typically the casing annulus. Once installed, the pump jet can be changed simply reversing the circulation and circulating the pump up the tubing to the surface for repair. A replacement is pumped into place.

Jet pumps can produce high volumes and can handle free gas very well. However, they are not as efficient as positive displacement pump, thus leading to higher surface horsepower requirements. This type of pump is very useful in certain situations, for example, where high production rates are desired. Also, locations where beam pumping units, for whatever reasons, can not be used, such as populated areas, offshore platforms, where gas lift is not available etc., are all installations for this pump. Deviated wells are candidates as well if gas lift is not possible.

Three flow configurations are possible for jet pump installation. The first is the standard circulation type (Figure 6.4) where the power fluid is pumped through the production tubing (hence called power fluid tubing or PFT) and the production is

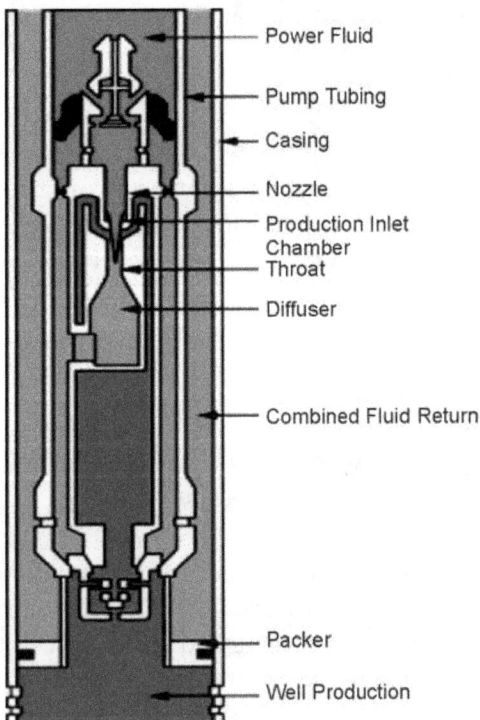

Fig. 6.2

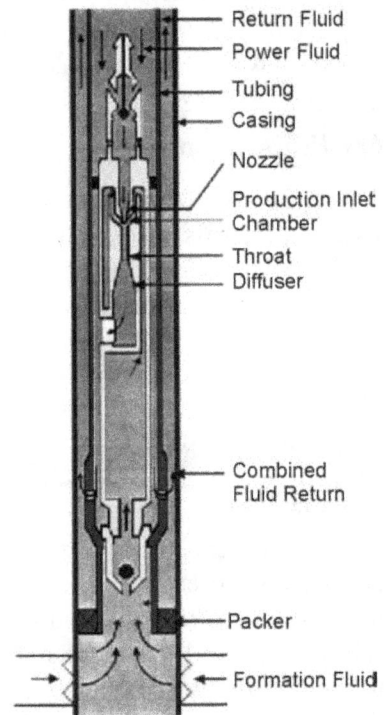

Fig. 6.3: Schematic of Jet Pump (Free type – standard circulation)

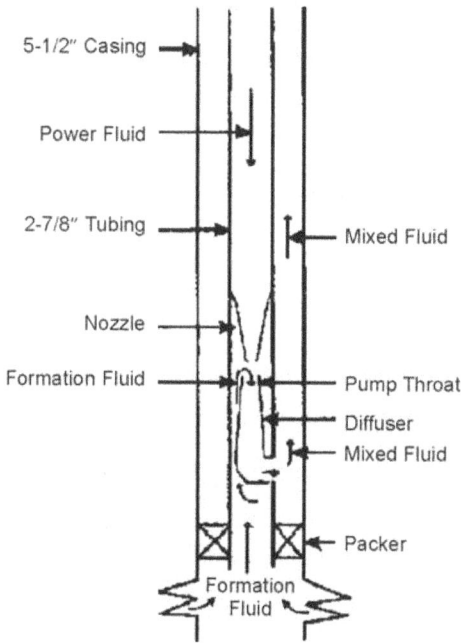

Fig. 6.4: A Schematic Diagram of Standard Circulation Configuration with Jet Pump in Power Fluid Tubing

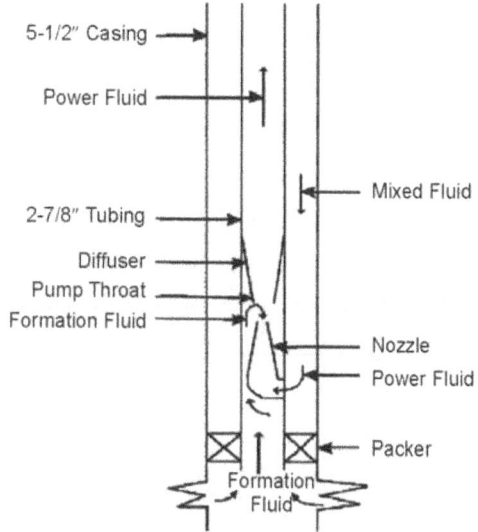

Fig. 6.5: A Schematic Diagram of Reverse Circulation Configuration with Jet Pump in Return Tubing

taken through the tubing-casing annulus. The second one is the reverse circulation type (Figure 6.5), where the power fluid is pumped through the tubing-casing annulus and the production is taken through the tubing. The third configuration is the parallel tubing completion type (Figure 6.6) where the power fluid is pumped through one tubing (power fluid tubing or PFT) and the production is taken through the other string (return tubing or RT). The size of the return flow tubing is normally bigger than the power fluid tubing since only power fluid flows through power fluid tubing whereas, power fluid plus produced fluid flow through return flow tubing. All the three configurations will require

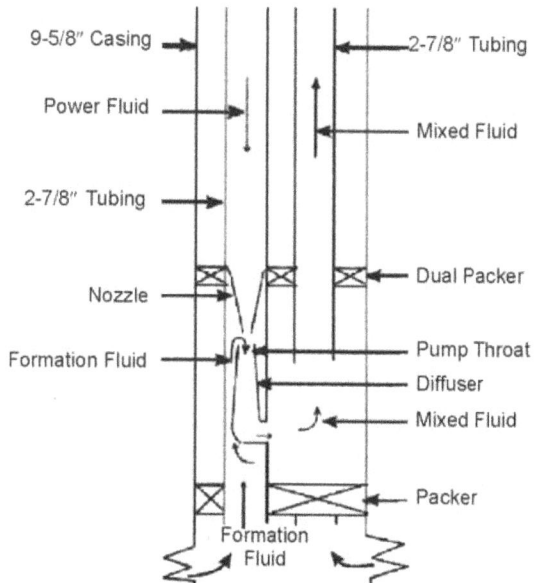

Fig. 6.6: A Schematic Diagram of Parallel Tubing Configuration with Jet Pump in Power Fluid Tubing

seating of jet pump in the bottom hole assembly (landing nipple) which is required to be lowered along with the tubing at pre-determined depth. Normally, the pump is lowered and retrieved by standard wireline techniques. Only the standard circulation type enables the pump to be retrieved by pumping the power fluid through the annulus (which is opposite to the normal flow direction of power fluid). In this case, wireline operations are not required and the pump is known as **"Free pump"**.

6.7.1 *Jet Pump Design*

Since fluid density, gas and viscosity are the variables needed in the calculations to size the jet pump, the calculations are complex and require iterative solutions. Here, computer enters the picture. Any service company in the jet pump business can furnish characteristics plots of the form shown in Figure 6.7. This plot is for a particular pump size and well since the PI of that well is needed as well as tubing length and diameter, fluid type, and fluid properties. However, once this plot has been generated, an engineer can predict the power fluid and surface pressure require-ments for a desired production rate. Since pump size is known, the intake pressure generated is predicted as well. Since PI must be known to design any lifting mechanism, the well's performance will follow the PI line overlaid on the graph. Note, that cavitation zone is also shown, above which the pump cannot operate. That is, if one attempts to pump a very high rate of power fluid at a high surface pressure, the pump will cease pumping reservoir fluid due to cavitation.

Nozzle and throat sizes determine flow rates while the ratios of their flow areas determine the trade off between produced head and flow rate. For example, if a throat is selected such that the area of the nozzle is 60 % of the throat area, this will result into relatively high head, low flow pumping operation. There is a comparatively small area around the jet for well fluids to enter in the throat, leading to lower production rates compared to the power fluid rate, and with the energy of the nozzle is being transferred to a small amount of production, high heads will be developed.

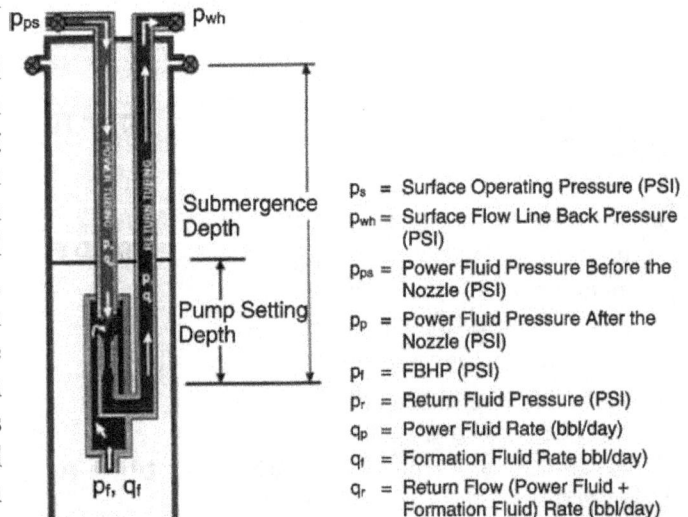

p_s = Surface Operating Pressure (PSI)
p_{wh} = Surface Flow Line Back Pressure (PSI)
p_{ps} = Power Fluid Pressure Before the Nozzle (PSI)
p_p = Power Fluid Pressure After the Nozzle (PSI)
p_t = FBHP (PSI)
p_r = Return Fluid Pressure (PSI)
q_p = Power Fluid Rate (bbl/day)
q_f = Formation Fluid Rate bbl/day)
q_r = Return Flow (Power Fluid + Formation Fluid) Rate (bbl/day)

Fig. 6.7: Sizing Calculations of Downhole Jet Pump

Such a combination of nozzle throat sizes of the jet pump is suited to deep wells with high lifts. Conversely, if a throat is selected such that the area of the nozzle is only 20% of the throat area, more production rate is possible, but since the nozzle energy is being transferred to a large amount of production compared to the power fluid rate, lower heads will be developed.

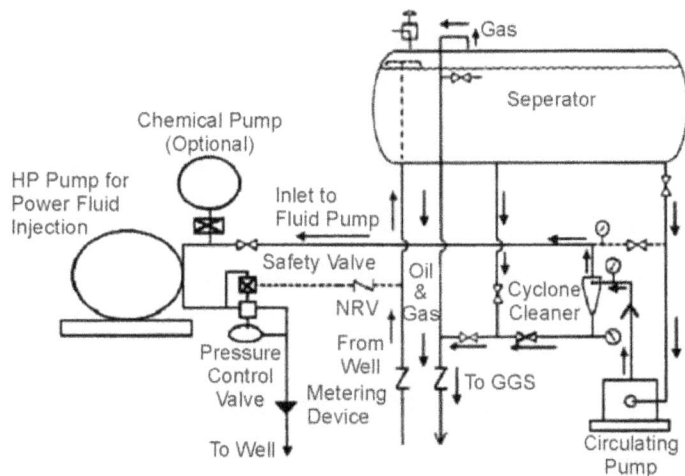

Fig. 6.8: Surface Facility for Jet Pump

Shallow wells with low lifts are candidates for such a combination of nozzle and throat sizes of the jet pump. A large number of combinations of nozzle and throat areas are possible to match different production rates and head requirements. Attempts to produce small amounts of well fluids as compared to the power fluid rate with a nozzle throat ratio of 0.2 will be inefficient due to losses as a result of high turbulent mixing between the high velocity jet and the low rate (slow moving) of production. Conversely, attempts to produce at high rates with a nozzle-throat ratio of 0.6 will be inefficient due to losses resulting from high friction as the produced fluid moves rapidly through the relatively small throat area. Selection of required ratio involves a trade-off between these two extreme losses viz. losses due to turbulent mixing and losses due to friction.

It is also required to be ensured that cavitations must not occur in the pump. The throat and nozzle flow areas define an annular flow passage at the entrance of the throat. The smaller this area, the higher the velocity of a given amount of produced fluid passing through it. The pressure loss of the fluid in the annular passage is proportional to the square of the velocity and eventually may reach the vapor pressure of the fluid at high velocities. This low pressure will cause vapor cavities to form, a process called **cavitation**. This results in choked flow in the throat, and then, no more production is possible at that pump intake pressure, even if the power fluid rate and its pressure are increased. Subsequent collapse of the vapor cavities happens as pressure is built up in the pump diffuser. This may cause erosion due to implosion of the vapor cavities which is known as *cavitation damage.* Thus, for a given rate of production and pump intake pressure, there will be a minimum annular flow area required to avoid cavitation problem.

The pump is defined by the nozzle and throat sizes. A given nozzle number coupled with same throat number will always give the same ratio of nozzle area to throat area. This is designated as 'A' ratio. Successively larger throat number when matched with a given nozzle number will give the B, C, D and E ratio. Sometimes "A-" ratio is also used, which will have the throat just smaller than that of 'A' ratio pump for any nozzle. The standard nozzle to throat area ratios, and their designations as provided by the reputed manufacturer of hydraulic Jet pump M/s. KOBE have been given here under:

Pump Ratio Designation	*Nozzle Area to Throat Area Ratio*
A	0.410
B	0.328
C	0.262
D	0.210
E	0.168

The other manufacturers of hydraulic jet pumps may have slightly different pump ratios from the ones as given above by M/s. KOBE. The most commonly employed area ratios fall between 0.235 and 0.40. Area ratios greater than 0.4 are sometimes used in very deep wells, or when the power fluid pressures are low. Area ratios less than 0.235 are used in shallow wells or when very low bottom hole pressures require a large annular flow passage to avoid cavitation. Thus the higher ratio pumps are suitable for low production rates and high heads, while the lower ratios are suitable for high production rates and low heads.

The design steps are summarized as below:

Step 1: Determine the pump minimum annular area needed to avoid cavitation. This is done with equation:

$$ASM = (AT - AN) \qquad \qquad \qquad \dots (1)$$

or

$$ASM = \{(q_f/691) \times \sqrt{(G_s/P_f)}\} + q_f \times (1-WC) \times GOR/(24650 \times P_f) \dots (2)$$

Where, AT   = Flow area of the throat, inch²
        AN   = Flow area of the nozzle, inch²
        GS   = Gradient of produced fluid, psi/ft
        P_f = Producing bottom hole pressure, psia
        q_f = Production flow rate from formation fluid, b/d
        GOR = Producing gas-oil-ratio, scf/bbl.

A nozzle and throat combination which has an annular area greater than ASM would then be chosen from a list of standard nozzle throat sizes indicated in Table 1.

Step 2: This is trial and error calculation for each power fluid pressure (surface operating pressure). The iterative nature is due to the fact that flow rates in the tubing and casing annulus (or another tubing string, depending on actual configuration of the well) are required to calculate frictional pressure losses, but are not known until final pressures are known. The power fluid rate is determined by nozzle equation:

$$q_p = 832 \times AN\sqrt{P_p - P_p)/GP}$$

Where, P_p = Pressure at the nozzle entrance, psia
P_f = Producing bottom hole pressure, psia
GP = Gradient of power fluid, psi/ft
q_p = Power fluid rate through nozzle, b/d.

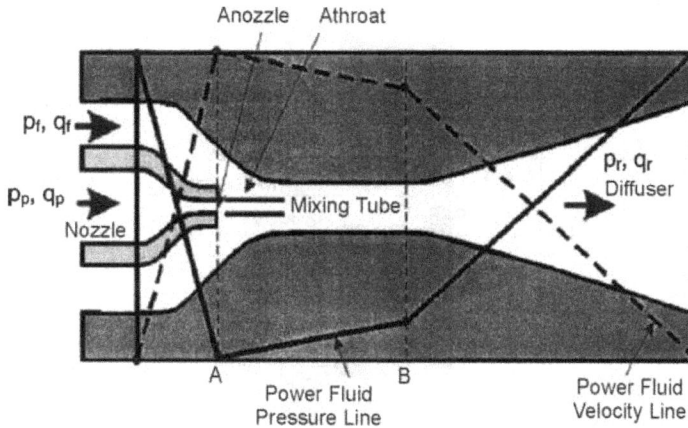

Fig. 6.9: Jet Pump Working Principle

The produced fluid is the sum of the power fluid and reservoir production. This combination may require use of two phase flowing gradients. As we can see, this procedure becomes very complex and is not efficient for each operating engineer to oil master attempt, especially since computer programs are already available from major vendors (National, Guiberson and Kobe).

Example: A typical well has following data.

PI: 2.65 b/d/psi;	Expected liquid rate: 2000 b/d; WC: 98%	
Sp. gr.: 1.03;	Depth of well: 7000 ft;	Tubing: 2-3/8"
Casing: 7";	GOR: 1250 scf/bbl;	SBHP: 2250 psi
FTHP: 100 psi;	Oil API: 40°API	

The flowing bottom hole pressure, P_f = 2250 – (2000/2.65) = 1495 psia.

The calculated value of ASM by equation (2) is = 0.05134 inch^2

Table 1: Different Combination of Nozzle and Throat (mixing tube) Possible

Nozzle		Throat		ASM (1)	ASM1 – ASM2	Status
No.	AN	No.	AT	(= AT – AN)		
DD	0.0016	7	0.053093	0.05149	0.00015	'OK'
DD	0.0016	8	0.066052	0.06445	0.01311	'OK'
CC	0.0028	8	0.066052	0.06325	0.01191	'OK'
BB	0.0038	8	0.066052	0.06225	0.01091	'OK'
A	0.0055	8	0.066052	0.06055	0.00921	'OK'
B	0.0095	8	0.066052	0.05655	0.00521	'OK'
C	0.0123	8	0.066052	0.05375	0.00241	'OK'
DD	0.0016	11	0.11946	0.11786	0.06652	'OK'
CC	0.0028	11	0.11946	0.11666	0.06532	'OK'
BB	0.0038	11	0.11946	0.11566	0.06432	'OK'
A	0.0055	11	0.11946	0.11396	0.06262	'OK'
B	0.0095	11	0.11946	0.10996	0.05862	'OK'
C	0.0123	11	0.11946	0.10716	0.05582	'OK'
D	0.0177	11	0.11946	0.10176	0.05042	'OK'
E	0.02405	11	0.11946	0.09541	0.04407	'OK'
F	0.03141	11	0.11946	0.08805	0.03671	'OK'
G	0.04524	11	0.11946	0.07422	0.02288	'OK'
H	0.06605	11	0.11946	0.05341	0.00207	'OK'

Note: For Nozzle No. 12 to 14, all combinations are OK (non-cavitational).

A typical jet pump sizing program of Guiberson, the calculations for 500 b/d. Method of sizing of Downhole Hydraulic Jet Pump is adopted by down hole jet pump manufacturers. In order to select the appropriate pump design, they have drawn head ratio and efficiency Vs flow ratio curves. Dr. Brown in his book "Artificial Lift Methods" has provided the procedures for sizing of downhole jet pump in the similar fashion. The computer program (excel sheet) evaluates cavitation by different method and calculates pressures losses. For each design production rate at specific surface operating pressure, the pump size (nozzle and throat) are calculated to avoid cavitation.

Table 2: The Power Fluid Rates for Different Pump Intake Pressures

Surface Operating Press, psi	Power Fluid (b/d) Requirement for q_f, b/d (P_f, psi) →			
	500 (2061)	1000 (1873)	1500 (1684)	2000 (1495)
2000	929	3898	–	–
3000	610	1531	4038	–
4000	464	1128	2040	3512

The final design calculations done for 2000 b/d indicate a power fluid rate of 3512 b/d (refer Table 2). The nozzle area is calculated as 0.038 inch² which is close to nozzle code "G" having area as 0.045239 inch². Optimum area ratio is 0.3 which results in throat area (AT) of 0.045239/0.3 = 0.15, which is close to mixing tube code of "12".

6.7.2 *Merits of Jet Pump*

- *Handles solids:* The wear and tear is less in the jet pump, because primarily the pump has no moving parts. However, high velocity power fluid as the fluid ejects through the nozzle can cause erosion of pump parts. Therefore to minimize the erosion of pump parts, the abrasion resistant material like tungsten carbide is used in the construction of nozzles and mixing tubes.

- *Handling corrosive fluids:* The simple construction of the jet pump allows the use of corrosion resistant alloys, thus corrosive fluids too can be handled.

- *Use in crooked holes:* Short length of the pump allows it to pass through the tight spots created by highly deviated well bore profile. Since only the high pressure fluid flows through the tubing and there is no tubing movement as such, tubing wear does not arise.

- *High volume pump:* Jet pumps can handle high volumes of well fluid.

- *Adaptability to sliding sleeves:* The size of jet pump is easily adaptable to sliding sleeves.

- *Handles gas:* The simple mechanical design of jet pump with no moving parts enables it to produce gassy well fluids with no damage to the pump. However, the volumetric efficiency of the pump goes down with the increase in free gas content at the pump intake.

- *Adaptability to variation in well production rates:* The jet pump can be adjusted to the varying production rates by changing the power fluid rate to the pump. Higher production rates can be achieved by increasing power fluid rate, provided the well is capable of giving the higher rates of production.

- *Suitability to low gravity crude oils:* Mixing of power fluid (water or light oils) with formation fluid, especially in case of heavy crude, provides viscosity blending, which results in lower viscosity of the overall mixture. This in turn reduces the frictional pressure losses in production string as well as makes the handling of produced fluid at surface easier.

- *Saves work-over cost:* A "Free pump" that can be circulated in and out of the well without the necessity of a work-over rig under normal circumstances, reduces reinstallation/repair costs to a great extent. For the case of "Free Pump", wire-line job is also not required.

6.7.3 *Demerits of Jet Pump*

- The jet pump requires higher pump intake pressure than other conventional pumps to avoid cavitation. A minimum of 20% submergence is required at the pump intake under dynamic condition.
- The mechanical efficiency is low in the jet pumps as compared to the positive displacement pumps, because of higher surface pump horsepower requirements.
- The power fluid should be clean and free from solids.
- The fluid handling capacity of surface facilities (including the separation facilities) needs to be increased to handle the increased fluid volumes (*i.e.,* produced fluid plus power fluid).
- Parallel string/concentric tubing completions may often be required for jet pump application, when flow through casing-tubing annulus is not desired. Parallel tubing completion increases the completion cost. Use of diesel or other similar liquids as power fluid increases operating costs.

Coiled tubing deployed jet pump: This is a type of jet pump installation which does not require deployment of work over rig to pull out the existing tubing. This involves the attachment of the bottom hole assembly of pump either with 1-1/4″ or 1-1/2″ coiled tubing and running-in of the same as a unit in the existing production tubing. The pump can be operated as a "free" pump, that is, it can be circulated in and out of the coiled tubing, thus this avoids utilization of a service rig as required for total maintenance of conventional down hole jet pump. The pump can also be set and retrieved with the help of standard wireline tools. The power fluid can be pumped down the well through the coiled tubing and the production can be taken through the annulus of coiled tubing and production tubing. Though it is a standard circulation type of jet pump the larger casing or production casing is not subjected to any fluid flow and associated pressures. The completion would, however, require a deeper setting of SCSSV which is to be placed below the jet pump with 2-7/8″ or 3-1/2″ tubing. A miniature hydraulic fluid line as usually will trace the tubing O.D from SCSSV to its surface control set-up for actuation of SCSSV. Coil tubing deployed jet pump reduces the size of the jet pump and hence the production capacity of the pump is reduced. However M/s. Trico, a reputed manufacturer of hydraulic jet pump claims that the tubing jet pump can handle upto 1000 bbl/d formation ideal pumping conditions. Trico-manufactured coiled fluid in 2-7/8″ tubing under the rig-less servicing job of the jet pump makes it a cost-effective option. The entire hydraulic surface package can be custom-designed to suit the space requirement and the existing equipments/facilities.

Critical observations on hydraulic jet pumps: Hydraulic jet pumps are extremely useful to produce oil from very deep well, where all other Artificial Lift Systems do not perform properly. Hydraulic jet pumps can be very economically utilized to produce oil from marginal and isolated offshore oil fields/wells. Hydraulic jet pump has also found its applicability for de-watering to produce coal-bed methane gas. Hydraulic jet pump appears to be very effective in deep and highly deviated well like S-profile well. Since hydraulic jet pump is very sensitive to surface pressure of return fluid (*i.e.*, tubing pressure or back pressure of the well), this lift can be combined with continuous gas lift for significant reduction in horse power and power fluid requirement. However along with sensitivity analysis, trial field application is required to establish all the benefits of this type of combined lift in the same well.

6.8 Recent Advances in Hydraulic Pumping

6.8.1 *Hydraulic, Long-Stroke Pumping System*

DynaPump, Inc. of Northridge, California, has introduced an automated, long-stroke, computerized surface pumping system that has solved basic problems associated with artificial hydraulic lift systems such as heat, power consumption and reliability. Those solutions allow for a design that offers a very long stroke in every pump model and a wide range of flow capacity within those models. Where these advantages are applicable, this lifting system can increase production and lower production costs.

The pumping system consists of two main components, pumping and power units (See: Rosman, A. and Nofal, M., "Computer controlled pump unit cuts power, increases output," *World Oil*, November 1996, pp. 53–56). The pumping unit stands over the tree and comprises a gas reservoir, a hydraulic cylinder and a pulley system. The power unit drives the pumping unit and consists of a computer control system with radio modem, solid-state electronics, motor controllers and hydraulic pumps. The unit comes in seven different models, with the largest having a maximum rod-load capacity of 80,000 lb and a 360 inch stroke. The power unit comes in different hp models up to 200 hp, and is matched to the pumping unit depending on dynamics of each well.

Fig. 6.10: Hydraulic, long-stroke pumping system comprises two main components, hydraulic piston pumping unit over tree and computer-controlled power unit. pulley system doubles piston movement for up to 360-inch. stroke

This pumping system is designed for longevity, reliability and ease of maintenance with the use of solid-state electronics. Other advantages include low acquisition costs, light weight/portability, easy/fast installation, automatic diagnosis of well operations/pump operations, automatic flow control, long stroke, differential speed up and down, less installed-power requirements, less power consumption and remote computer control. Some of these pump systems have exceeded the world record for total flow for hydraulic pumping systems and are capable of producing in excess of 10,000 bpd.

6.8.2 *Ultra-Long-Stroke, Rod-Pumping Improvements*

Rod pumping successfully in space-restricted areas (such as offshore platforms) has traditionally been very difficult; and pumping in deviated wells has been expensive due to severe-wear conditions. With the help of RPS Canada, Hydraulic Rod Pumps, Int'l. (HRPI), Foothills Ranch, California, has installed a tubing rotator on the world's only low-profile, ultra-long-stroke, sucker rod-pumping unit. The proprietary design of this pumping unit (in use since 1987) allows installation of sucker rod pumps in height and space-restricted locations, such as offshore platforms, town lots and multi-well cellars.

The tubing rotator was added to complement the ultra-long, 336 inch stroke length—this combination can extend tubing/rod life. In addition to tubing rotation, the rods are rotated by tubing string rotation-induced torque. This is achieved by allowing the integral polished rod to rotate freely inside the hydraulic cylinder, as the hydraulic fluid that lifts the piston also acts as a thrust bearing.

The current installation, however, does not ensure that the rods are being rotated, as it only allows the rods to rotate if tubing-induced torque friction is present. To improve on this design, a free-sliding linear guide is being developed, and will be available for future installations, to ensure rod-string rotation at the same speed as the tubing.

The low-profile, long-stroke cylinder design is submerged inside the tubing string to eliminate the vertical profile. To make room for the "subsurface" cylinder, tubing is hung at the surface with a single casing joint, which is sized to allow ample clearance for the cylinder inside the tubing string. The bottom of the casing joint is attached to the top of the existing tubing string with a casing-by-tubing crossover.

The rod string is then installed, and the cylinder is hung from the top cylinder head via an API-type flange, while the entire cylinder body is submerged below the flange inside the tubing string (casing joint). The cylinder assembly includes an integral polished rod, a high-pressure stuffing box and internal flow passages for produced fluid outlets, so no other equipment is required.

Since the cylinder is larger in diameter than a rod string, standard tubing rotators are not compatible with low-profile cylinder design. RPS Canada was approached by HRPI and Stocker Resources (Los Angeles) to design a large-diameter tubing rotator that would adapt to the HRPI subsurface-cylinder wellhead design. The first such rotator went into service in June 1999, and operated trouble free.

6.8.3 *Tubing Drain for Hydraulic Pump*

Hydraulic tubing drains have been used for many years as a method of communicating between tubing and annulus. Weatherford Artificial Lift, Houston, recently adapted the drain sleeve to a hydraulic artificial-lift, bottomhole assembly (BHA), resulting in an enhanced downhole tool and process that saves time and money.

The S Drain BHA sleeve is available on equipment for 2-3/8 and 2-7/8 inch tubing. The sleeve has been adapted to BHAs for both jet and hydraulic reciprocating pumps. It is a simple design, that is easy to install and operate. It can be shifted with hydraulic surface equipment normally installed at hydraulic-lift locations. It has a slim design for maximum annulus flow area. And the BHA has large fluid-discharge ports to reduce turbulence and fluid erosion.

Currently, when a new well which is expected to free flow initially is completed, a packer is set and tubing is run. The well is perforated and allowed to free flow for as long as it will. After the well stops flowing, the tubing is pulled and a new completion string is designed & run. This sequence requires the tubing to be run twice. A BHA with blanking tool could be run in Step 1, however this limits access through the completion when work below the packer is required.

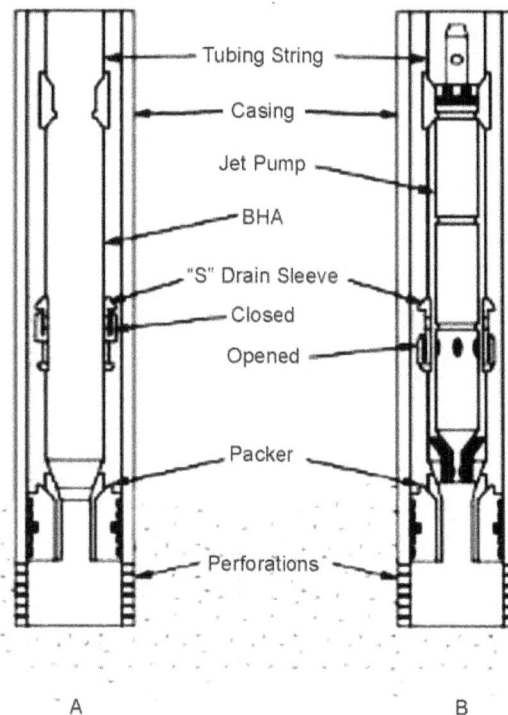

Fig. 6.11: Tubing drain for hydraulic pumping system. S drain assembly can be run in new well, A. When artificial lift is needed, drain sleeve is opened and hydraulic pump is run, B

If a S Drain BHA is used in a new well that is expected to free flow initially, a packer is set and tubing with BHA is run, Figure 6.11. With the drain in the closed position, tools can be run through the BHA, and the well can be perforated. Once the well stops flowing, the standing valve is dropped and pressure is applied to the tubing string, causing the pins to shear and the drain sleeve to shift, opening the tubing string to the annulus. The hydraulic pump is then circulated into the BHA, and pumping begins.

This installation technique does not require the tubing to be pulled nor any type of wireline operation to shift the sleeve. All additional operations are done with fluid flow and pressure. The savings, therefore, are in cost of pulling the tubing and increased production due to equipment availability.

6.8.4 *Multiplex–Pump Plunger Seal*

Weatherford Artificial Lift Systems, Houston, is introducing a multiplex-pump seal improvement for high-pressure, positive-displacement pumps used in the surface hydraulic power fluid unit for downhole hydraulic pumping systems.

Most packing systems used today have one primary seal that is exposed to the product being pumped and separates the product from the atmosphere. When the seal leaks, the product can go on the ground or into the air; the Hydro-Balanced Packing System is designed to control the product with a barrier fluid that is pre-chosen.

The technology can best be described as having a primary seal that is exposed to a barrier fluid and a pressure transmitter that is used as a secondary seal to separate product from barrier fluid. When the primary seal leaks, an environmentally friendly barrier fluid is the only thing exposed to the atmosphere. Figure 6.12 shows a pump plunger stuffing box with the transmitter/piston, barrier fluid and the primary seal arrangement.

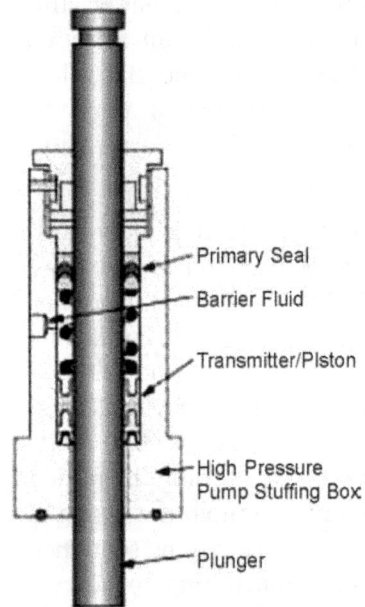

Fig. 6.12: Pump plunger stuffing box with transmitter/piston barrier fluid and primary seal arrangement

6.8.5 *Pneumatic Lift*

Two field-proven systems that utilize natural gas supplied either from the well annulus itself, or compressed gas from a local unit or a multiwell system, are described here. Both systems utilize a vertical surface cylinder and piston-lift concept.

6.8.6 *Pneumatic Pumping System*

The McCoy Pneumatic Rod Pumping System (MPRPS) offered by Permian Production Equip., Inc., Midland, Texas, lifts the rod string and strokes the downhole pump by utilizing natural gas energy available in the field. The MPRPS is operated with pressurized gas from a source such as a compressor discharge line, trunk line, another well or perhaps from the well to be lifted itself.

The unit incorporates a piston in a cylinder. The piston rod acts as the polished rod of the well and passes through the stuffing box where it connects to the well's rod string. It operates solely by gas pressure and does not require any type of external energy such as a motor.

The system takes supply gas through a pneumatic motor valve into the chamber under the piston. When pressure under the piston exceeds weight of rods and fluid, the piston travels up the cylinder bore. At the top

Fig. 6.13: McCoy Pneumatic rod pumping system uses natural gas energy available in the field. Piston rod acts as polished rod, going directly through stuffing box to sucker rod string

of the stroke, pressure under the piston passes through a sensing port and shifts the pneumatic motor valve to the exhaust position.

The weight of the rods forces the piston back to the start position, where pressure holding the pneumatic motor valve in the exhaust position is relieved. This action allows the spring-loaded valves to shift back to the supply line, which starts a new stroke. The weight of the rods forces used gas into the well's flowline, which returns it to the production facility along with the well's production.

6.8.7 *Pneumatic Pump Jack*

Maranatha Industries, Farmington, New Mexico, has designed a pneumatic pump jack, called Pneulift, as a niche product to address certain environmental concerns and lower operating costs in artificial-lift situations. The unit is available in stroke lengths from 40 to 78 inch and is intended for use to a maximum depth of about

8,000 ft Stroke speed and length are easily field adjusted to accommodate specific fluid-removal requirements on a well-by-well basis.

Fig. 6.14: Pneulift Pneumatic Pump Jack, Available in 40 to 78-in. Stroke for Pumping to 8,000 ft, Operates on Well Annular Gas or Compressed Gas

Pneumatic pump units are designed to operate in a constantly pressured environment, with power being applied on both up- and downstroke. By maintaining a state of pressure equilibrium, dynamometer tests show extremely well-balanced rod string harmonics, virtually unaffected by stroke speed or length. The unit operates free of any external energy source and provides quiet lift operations. Since no motor is required, fuel use is eliminated, and all gas used is exhausted into the sales line. There are no atmospheric emissions. The cylinders are powered by annular gas, well site compression or a centralized compression site. By varying stroke length and speed, the unit is capable of lifting up to 150 bpd.

Operating pressures required to power the unit depend on well depth, downhole pump size and sales line pressure. Typical operating pressures are in the range of 25 to 175 psi. A solar-powered, telemetry-compatible automation package is currently under development to allow intermittent operation and address pump-off situations in low-rate wells.

6.8.8 *Planned Subsea Jet Pump*

Weatherford, Houston, has designed and manufactured a special wireline-installed 7-in. subsea jet pump, currently installed for offshore operator, Lundin, in January 2007. The operator is drilling the well about 100 miles off the coast of Tunisia in

820 ft of water depth. The subsea completion will include 11-3/4 inch casing and 7 inch tubing. The well is expected to produce in the range of 20,000 bpd.

Lundin recognized up front that cost for intervention after the well is put on production would be prohibitive. Therefore, it commissioned a study by another company to determine the best method of artificial lift. That study determined jet pumping to be the best choice. Weatherford was then contacted by Lundin and, in response, proposed a 7 inch wireline jet pump. While a 5 inch pump would handle expected production, the larger size was selected to minimize fluid velocities inside the pump for the longest run time. The special type of jet pump, as illustrated in Figure 6.15 was designed and manufactured for installation as noted.

The initial power-fluid rate of the wireline-installed pump in the 7 inch tubing will be about 13,500 to 15,000 bpd, depending on reservoir response. Lundin plans to use water, but that is open to change. Initial injection pressure will be about 3,500 psi and will be limited to 4,800 psi as watercut increases; initial watercut is expected to be 1.0%. This is believed to be the first subsea jet pump installation. And a 7 inch jet pump is rare. The only other one known to exist was in Oman long ago, reportedly producing 50,000 bpd. The expected 20,000 bpd will likely decline over time, with increasing watercut and the injection pressure limit of 4,800 psi. The gas oil ratio is only 25:1, so gas is not an issue. If it were, the throat would be sized to accommodate any free gas at the pump.

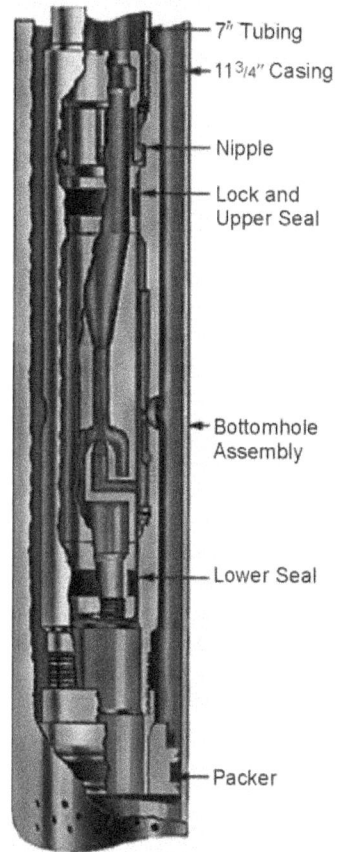

Fig. 6.15: Basic Design of Wireline-installed Subsea Jet Pump for 7 inch Tubing

CHAPTER 7

Progressive Cavity Pump

7.1 Introduction

The progressive cavity pump system (Figure 7.1) is composed of rotary pump, sucker rods and a rotary power transmission unit. The reciprocating motion of conventional beam units is replaced by purely rotary motion.

The progressing cavity pump has been in use as a fluid transfer pump for many years in various industrial applications. For the last several years the progressing cavity pump is being used as a method of artificial lift in oil wells. The use of progressing cavity pump, as a means of artificial lift has one distinct advantage over other conventional artificial lift methods in that it is perhaps the most efficient method in lifting of very high viscous crude oil from shallow wells. Through years of research and development in PCP design, the production capacity and lift efficiency of PCPS are increasing their horizon to cover a wide range of areas. For example, the progressing cavity pumps have now the ability to pump out abrasive fluids. Their applications are also extended in other types of fluids. With various and improved elastomer materials available, a wide range of well fluids can be handled efficiently using the PCP. The low initial investment, ease of installation, minimal maintenance and high volumetric efficiency are some of the other advantages of the PCP.

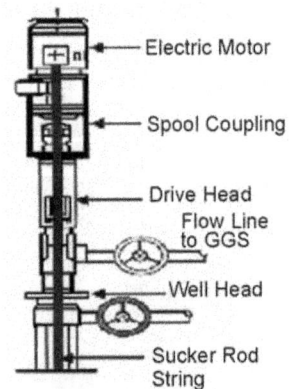

Fig. 7.1: Surface Setup for a Progressive Cavity Pump

With the improved materials of pump construction, the Progressing cavity pump has found its adaptability to a wide range of well conditions. The rotor suspended by the rod string is the only moving part of the down hole PCP. It is a single external helix with a round cross-section. It is made of high strength steel, precision machined with chrome plating for abrasion resistance. The stator, connected to the tubing string, is a double internal helix inside of which is lined with synthetic elastomer (factory moulded). The elastomer lining of the standard stator is made using a Buna N elastomers which is best suited for oil, gas and water applications. Other stator elastomers available are 'High Nitrile' and 'Nysar'. High Nitrile elastomer is used for fluids containing higher percentage of aromatics, while Nysar is used for fluids containing hydrogen sulfide at elevated temperatures.

7.2 Principle

The progressing cavity pump consists of a single helical or spiral system (rotor) which rotates inside a stationary elastomer-lined double helical or spiral system (stator) of the same minor diameter and twice the pitch length. The rotor is lowered down hole with the help of sucker rods. The sucker rods are rotated with the help of electric motor (prime mover). Thus the rotary motion is transmitted to the rotor of the down hole pump. Some PCP manufacturers have recently developed rodless PCP using down hole electric motor which with a speed gear arrangement is directly coupled to PCP. The schematic of progressive cavity pump is shown in Figure 7.2 and 7.3. The movement of the rotor inside the stator is actually a combination of two movements: A rotation around its own axis, a rotation in opposite direction of its own axis around the axis of the stator. Because of this second type of movement, it is also sometimes referred to as "eccentric screw pump".

As the rotor rotates eccentrically within the stator, a series of sealed cavities are formed 80 degrees apart which progress almost pulsation-free from the suction to the discharge end of the pump, that is, from bottom end to top end of the pump. As one cavity diminishes, another is created at the same rate resulting in a constant non-pulsating linear flow. The total cross-sectional area of the cavities remains the same regardless of the position of the rotor in the stator. The progressing cavity pump overcomes pressure because it has a complete seal line between the rotor and stator for each cavity. The pressure capabilities in the pump are based on the number of stages and the number of times the elastomeric seal lines are repeated.

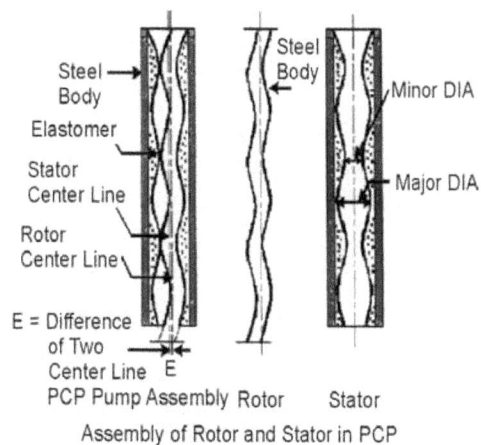

Fig. 7.2: Sucker Rod Pumping System

The minimum length required for the pump to create effective pumping action is the pitch length of the stator. One pitch length forms a stage of the pump. Each additional pitch length results in additional stages. Normally a stage is designed and manufactured to be 1.1 to 1.5 times the pitch length of the stator. The reason for this is to ensure a proper seal between the rotor and the stator to achieve the desired pressure increase per stage. By increasing the number of seal lines or stages the pressure build-up capability of the pump is increased, allowing it to pump from deeper depths. As the pump is of the positive displacement type, the head capability is independent of the speed *i.e.,* high lifting pressures can be generated even at low speed. As pressure increases for the same number of stages and speed, the flow rate decreases as this increases slip. All rotary positive displacement pumps experience some slippage. The amount of slippage depends on a number of factors, which are as follows:

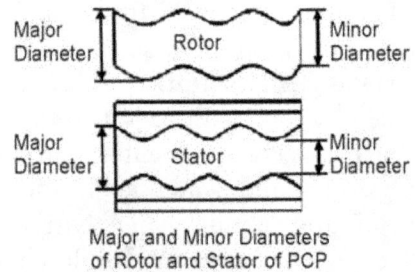

Major and Minor Diameters
of Rotor and Stator of PCP

Fig. 7.3: Sucker Rod Pumping System

- Differential pressure between suction and discharge *i.e.,* total head
- Number of stages
- Degree of compression fit between rotor and stator
- Viscosity of production fluid
- Temperature at pump level.

Slip is however independent of speed. Slippage means a loss of efficiency, but at the same time it ensures lubrication of pump.

7.3 System Components

7.3.1 *Progressive Cavity Pump*

The pump consists of an elastomer stator and a stainless (or chrome plated) steel rotor. The stator (pump body) is connected to the turbine string. It is a double internal helix shape precision-molded synthetic elastomer bonded to a steel tube. The pitch length is double that of the rotor. The rotor is precision-machined into an external helix with a round cross-section.

When the stator and rotor are fitted together a chain of individual cavities are formed between the matching surfaces. As the rotor turns within the stator, the cavities progress from the inlet (bottom) to the outlet (top) of the pump. The cavity size depends on the rotor pitch length, diameter and offset which vary from one model to another. The areas where the rotor and stator contact provides a fluid seal to provide positive displacement. Similar curves can be obtained from all suppliers (Griffin, Corod, Robins and Meyers, ProCav).

Typical operating speeds are:

Light Oil	:	300–500 RPM
Medium Oil	:	200–400 RPM
Heavy Oil	:	10–200 RPM

The maximum speed should also be restricted in the presence of abrasive sand:

Slight Sandy	:	400 RPM
Moderate Sandy	:	300 RPM
Very Sandy	:	100 RPM

The pumps are tailored to the application by selecting a model type based on the volume capacity requirements. The net lift in terms of head, fluid friction and back pressure is then determined. The total number of stages, which can vary from 4 to 18, determines the maximum differential pressure drop which the pump can withstand without excessive slippage. It is generally not recommended to exceed 100 psi/stage to prevent excessive wear and fatigue of the stator material. Although larger, higher capacity stators are possible, they increase the risk of landing and misalignment problems and may therefore have shorter run lives in severe operating conditions. The suppliers QA/QC program is extremely important in this regard, especially when units are operated near the limits.

Fluids with gravities ranging from 5 to 45 degrees API containing water, paraffin, sand or gypsum has been pumped. Fluids with sand contents as high as 85 % can be handled with minimal wear.

The Buna-N elastomer (standard stator material) is not suited to hydrogen sulphide and high temperature environments (> 176°F) which has been experienced during steam stimulation. The elastomer swells causing increase of friction between the rotor and stator which further increases the temperature, eventually leading to failure. Recent developments of high temperature chemical resistant stator materials such as Nysar have shown promise in overcoming these problems and increasing the operating range to 356°F temperature.

7.3.2 *Power Transmission Unit*

The power transmission unit supplies torque to rotate the rod string and carries the thrust loads caused by the pump and string. Drive can be provided by:

(a) Hydraulic Drive on the wellhead.

(b) Direct belt drive via a torque limiting hub
- Gas engine
- Electric motor
- Hydraulic motor.
(c) Direct wellhead mounted electric motor.

Although somewhat expensive, the hydraulic drives are generally the most popular because of the ease of speed adjustment.

Maximum supply torque is limited by safety devices. With hydraulically powered units a bypass is installed to sense and relieve excess pressure slowly so as to avoid tubing or rod back-off. On belt drive units, protection is obtained from a torque limiting hub. Approximate horsepower requirements can be estimated as:

$$HHP = \frac{P \times Q}{58766}$$

$$But \ HHP = \frac{T \times S}{5252}$$

$$Therefore, \ T = \frac{P \times Q}{11.25 \ S}$$

$$BHP = \frac{HHP}{0.75}$$

HHP = Hydraulic Horse Power
BHP = Brake Horse Power of Transmission System
P   = Pressure (psi)
Q   = Production (bbl/day)
T   = Torque (ft.lbs)
S   = Speed (RPM).

However, torque and horsepower requirements are difficult to predict because of frictional influences of sand and heavy oil. Field experience or manufacturers advice should be used when sizing equipment.

Charlynn model 104–1024 hydraulic motors are commonly used which have a stall torque of 512 ft.lbs shows typical power fluid pressure to rod torque relationships for some of the commonly used units.

7.3.3 *Rods and Tubing*

It is normally recommended that Class D rods of at least 19 mm (3/4″) be used with progressive cavity pumps. Severe pumping conditions often require larger and stronger rods with sizes up to 25.4 mm (1″) being commonly used. A few operators insist on using Class C rods, in which case torques must be carefully controlled to avoid torsional failure.

Clockwise rotating Moyno pumps result in anti-clockwise torque on the tubing. This has resulted in numerous instances of backed-off tubing. Tubing should be made up to the "optimum" torque, or even slightly greater. Tubing that is only made up to minimum torque will probably back-off.

Recent publications have further downgraded the maximum rod torques, previously recommended based on experimental results from a test rig built by M/s Corod. The values are actual yield points and therefore should not be exceeded during operation. Torque requirements are greatest at the surface and may be higher than expected, if very viscous oil or emulsion is pumped due to the rotational drag force of the oil. The effects of low winter ambient temperatures on flowline back-pressure should also be considered.

A conventional 32 mm (1–1/4″) polished rod is used with a rotary stuffing box mounted on the flow tee. It is essential that projection of the polished rod through the drive is limited to less than 2 ft, so that any back-lash resulting from a rod break does not cause a whiplash action that might injure personnel or damage equipment.

There is some debate of the pull back necessary to properly space the pump to compensate for rod stretch due to fluid load and length decreases due to increasing tension. A minimum pull back from the tag position of 8″ is recommended by all manufacturers. Additional pull back for fluid load is generally 2″ to 20″ depending on unit size and depth. Some companies also use an additional depth related safety factor to adjust for elastic and buckling uncertainties involved in establishing the proper tag position at zero set down weight. However, this is better handled by repeated space out measurements on both pick-up and set-down.

7.4 Principle

The progressing cavity pump theoretically it is pulsation free pump has a constant cross-sectional flow area with a constant velocity. This results in constant quantity of flow by pump, which is calculated as: $Q = AV$.

The cross-sectional area can easily be determined by calculating the cross-sectional area of the opening and subtracting the same by cross-sectional area of the rotor. By obtaining the areas of the circle and rectangle which make up the cross-sectional area, the cavity area can be determined.

A_{ROT} = area of a circle = $(\pi \times D_{ROT}{}^2)/4$

A_{STA} = area of a circle + Area of rectangle = $(\pi \times D_{ROT}{}^2)/4 + (E \times D_{ROT})$

A_{CAV} = $A_{STA} - A_{ROT}$ = $4 \times E \times D_{ROT}$

D_{ROT} = Diameter of Rotar

D_{STA} = Diameter of Stator

This shows that the area of the cavity depends upon the size of the rotor diameter and eccentricity. The length of the cavity is determined by the pitch of the stator. The pitch length of the stator determines the velocity of the fluid moving through the pump. For each rotation of the rotor, the fluid moves one pitch length of the stator. The longer the pitch length, the higher is the velocity of fluid through the pump.

Velocity of fluid is given by $V = P_s N$

Where P_s = stator pitch length, N = number of revolutions

The flow formula, $Q = A \times V = 4 \times E \times D_{ROT} \times P_s \times N$

7.5 Pump Life

Although the manufacturers claim an average service life of 2 years in Alberta, Canada based on pump replacements, frequency, however in some operaters report that a service frequency of 4–6 months is more common in Heavy Oil. Not all workovers necessitate pump changes (e.g. sand bailing, parted rods, parted tubing, etc.) but where this is required a complete new pump unit is needed ($4,000 – $6,000). On average, service costs are 66% higher than for a conventional insert pump. Where the workover frequency with a conventional pump is less than once per year, it is probably not worth considering the progressive cavity pump. Based on this type of analysis, several operators use the progressive cavity pump to 'clean-up' heavy oil wells initially when sand production is high and subsequently install a conventional pump, leaving the progressive cavity pumps on problem wells. Moreover, conventional rod pumps often have a larger lift capacity, if sand production is not too severe.

The manufacturers' claim, with some justification, that the performance of the progressive cavity pumps is under-estimated by many engineers, since they are predominantly used in problem conditions. Moreover, some of the non-pump related workovers are the result of operators not following manufacturers recommendations:

- Inadequate cleaning of pipe and tubing
- Running damaged threads
- Not using heavy enough rods (normally 22 mm (7/8″) are needed)
- Not using Grade D rods
- Ignoring over torque limits.

Under favorable conditions, some progressive cavity pumps have been running for over 4 years without a workover. Some companies are using the progressive cavity pump in light oil applications (e.g., at Valhalla).

7.6 Design of Progressive Cavity Pump—Rotor

A PCP rotor is a precision-machined, single external threaded helical gear made of high-strength steel and coated with hard chrome plating to protect against abrasion and minimize friction. Its cross section is a circle with the diameter of d at any place. The centers of all the cross sections are on a helical line which has an eccentricity 'e' with the rotors axis.

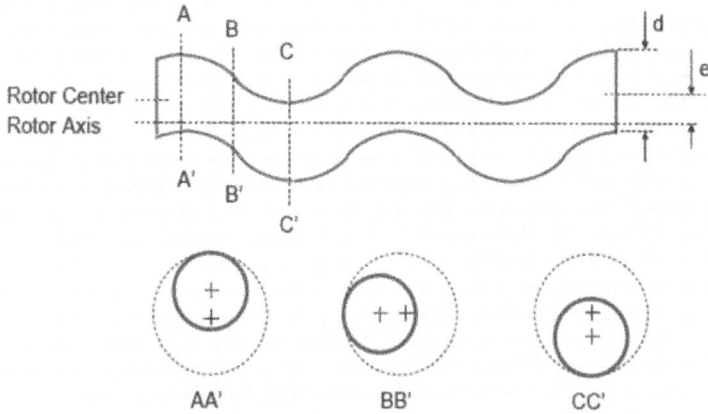

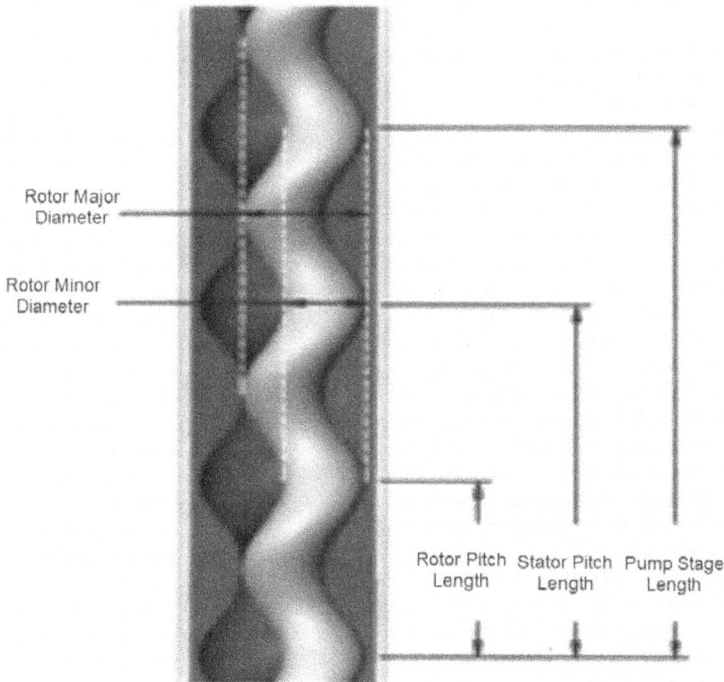

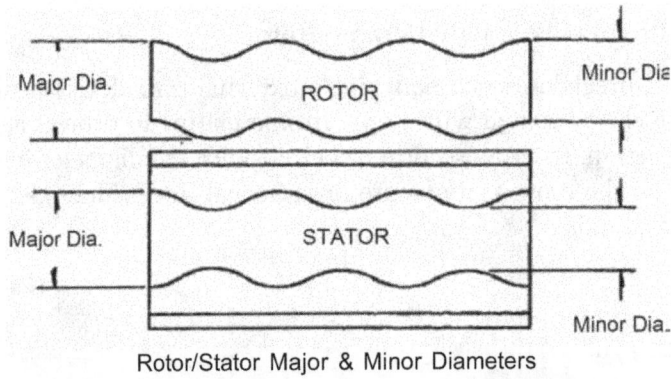

Rotor/Stator Major & Minor Diameters

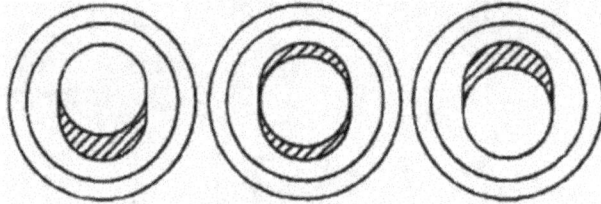

Constant cavity Cross sectional Area

7.6.1 *Progressive Cavity Pump—Stator*

The stator is a double internal helical gear and has the same minor diameter as the rotor. The stator has twice pitch length as that of the rotor.

A cross section of stator at any place is a long circle and can be described as two half circles of diameter 'd' departed by a distance 4e. The stator has the same cross sectional shapes along its axis but with different angles.

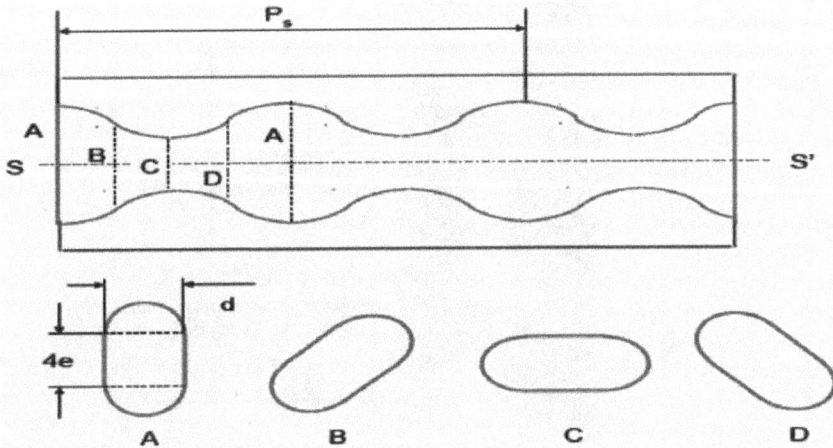

7.6.1.1 *Flow Rate Derivation*

The cross sectional area of rotor and stator are:

$$A_{rotor} = \pi/4d^2$$
$$A_{stator} = \pi/4d^2 + 4ed$$

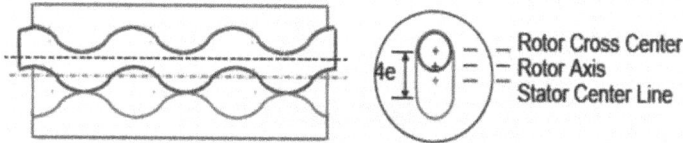

Rotor Cross Center
Rotor Axis
Stator Center Line

After setting a rotor in a stator, the rotor axis is not coincident with the stator line. In addition to rotating around its axis, the rotor rotates eccentrically around the stator center line with the same eccentricity 'e'. The cross sectional area at any place of a PCP reduces to a rectangle with width d and length 4e.

7.6.1.2 *Flow Rate Derivation*

The rotor divides stator chamber in two crescent sections at any place. The two sections belong to 2 cavities and their areas change with the rotation of rotor. The two cavities are 180° apart and work alternatively as suction and discharge. The total area of the two sections is constant at any place along the pump, and it is the fluid flow area.

$$A_f = 4ed$$

The length of the cavity is the pitch length of the stator, P_s. The 180° departed two cavities move one pitch when the rotor turns 360°. Therefore, a PCP will move fluid $4edP_s$ per rotation.

For rotational speed n, cavity moving speed along stator center line is:

$$V = nP_s \qquad \qquad \text{... (7.1)}$$

So the total flow rate $Q_t = A_f * v = 4ednP_s$.

7.6.1.3 *Actual Production and Slip Rate*

In short to get a high production rate, one can use a larger pump or make a pump rotate faster. To overcome high differential pressure, one can use a longer pump or add more stages.

The slip q_s, varies with the structure of the PCP and the differential pressure on it. Also, it doesn't change with pump rotational speed.

$$Q_a = Q_t - q_s \qquad \qquad \text{... (7.2)}$$

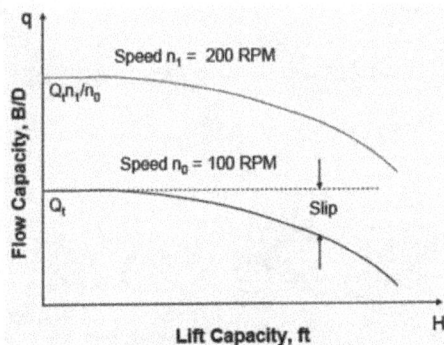

7.6.2 *Rotational Speed Design*

Actual production rates depend upon the slip, which is a function of differential pressure across the PCP, and the differential pressure is determined by the actual production rate.

The pressure profile consists of pressures at well perforation place (p_{wf}), at pump intake (p_i), pump discharge (pd) and at well surface (p_{wh}). The differential pressure across a PCP is the between points B & A. For different production rate, the profile and differential pressures are different

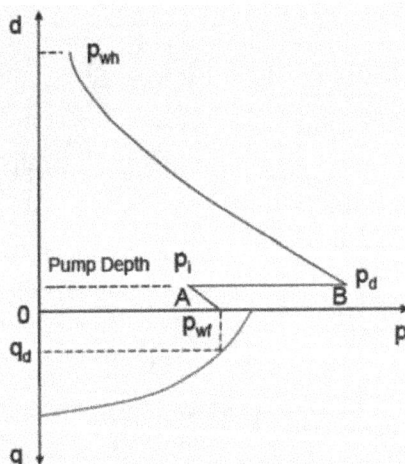

For a desired production rate q_d, one can get FBHP p_{wf} from the IPR of the reservoir. Then from the p_{wf} and reservoir temperature at bottom hole, one can calculate P&T at any place from bottom hole to pump intake using Multiphase flow correlations.

The fluid properties like water cut, gas oil ratio, bubble point pressure, densities of oil and gas, and water are known variables for calculation. In addition to P&T, in-situ fluid density, liquid rate, and free gas are also calculated.

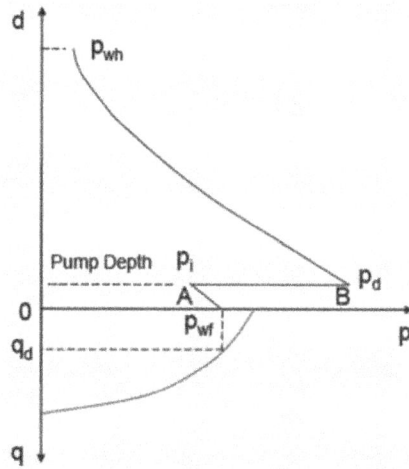

The outflow performance curve is calculated from wellhead down to the pump discharge by using Multiphase flow correlations. The Pressures, temperatures, fluid rates and densities are also calculated.

The outflow performance curve is calculated from wellhead down to the pump discharge by using Multiphase flow correlations. The Pressures, temperatures, fluid rates and densities are also calculated.

Once differential pressure across the PCP has been determined, one can convert it to head by dividing it by the average fluid density between pump intake and pump discharge. The calculated head and fluid rate at pump intake are used to design the rotational speed of the pump.

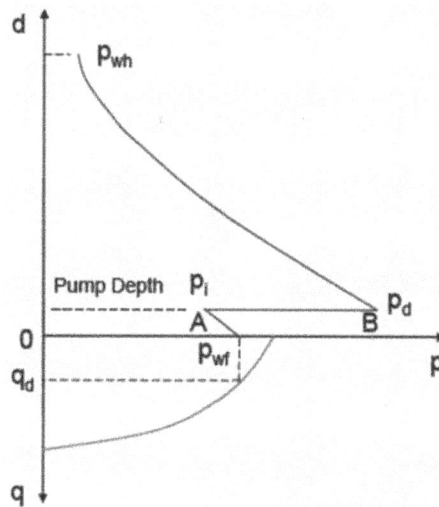

7.6.3 *Graphical Method*

Draw a vertical line from the calculated head H_a, horizontal line from the total fluid rate q_a, find the intersection point 'A'.

Move the performance curve at 100 rpm until it matches with point A.

Find the theoretical flow rate Q_m from the new curve at zero differential pressure. The solution speed will be,

$$n = 100 \times Q_m/Q_t \qquad \qquad \dots (7.3)$$

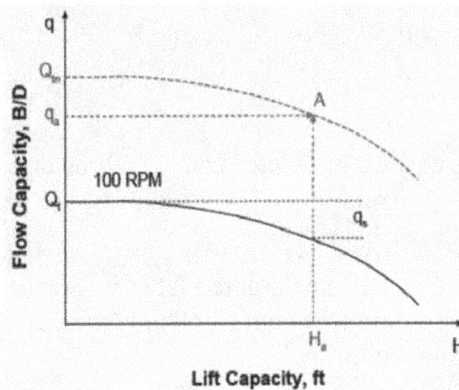

7.6.4 *Production Rate Design*

The pump depth is used as a node and then calculate the inflow and outflow curves for the node.

Assume a series surface liquid flow rates (STB/d) and calculate their corresponding pressures at pump intake from reservoir IPR curve and Multiphase flow correlations.

Construct the inflow curve using the liquid flow rates and calculated pump intake pressures.

From wellhead P&T, calculate the pressures at pump discharge for a given surface liquid rates and construct the outflow curve. The vented free gas needs to be taken away from the fluid stream for the outflow calculation.

The inflow and outflow curves can be used to construct a well system curve at pump depth

7.6.5 *Production Rate Design*

For any flow rate q_i, the differential pressure is Δp_i. Select a series of inflow rates, calculate their corresponding differential pressures and convert them to heads by using corresponding fluid densities. Calculate the corresponding total rates at pump intake. Draw a curve of the heads vs. total flow rates to obtain well system curve.

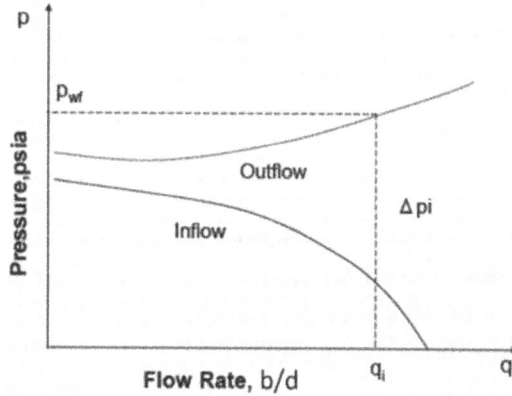

7.6.5 *Production Rate Design*

The system curve gives the required head for a given total flow rate at pump intake. Also plotted are pump performance curves. The intersections of the pump performance curves and well system curves are the production rates at the rotational speeds.

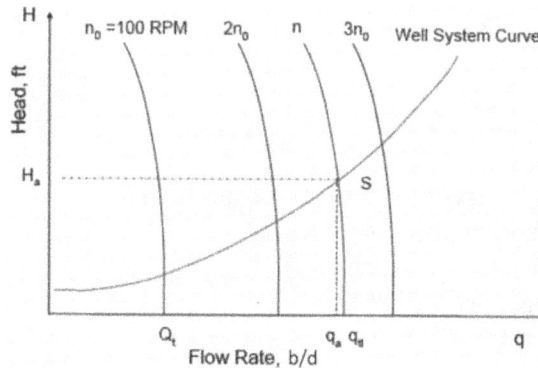

7.6.6 *Viscosity Effect on Slip Rate*

The slip rate is controlled by pump type, pump size, clearance between the rotor and stator, differential pressure and fluid viscosity.

For a given pump and differential pressure, the higher the viscosity of pumped fluid is, the lower the slip rate. This is due to the resistance to flow for a viscous fluid. Viscosity increases the difficulty of flowing. Therefore, pumping viscous fluid has higher volumetric efficiency than water.

$$\frac{q_s - \mu_2}{q_s - \mu_1} = \sqrt{\frac{\mu_1}{\mu_2}} \qquad\qquad\qquad \dots (7.4)$$

Viscosity varies with Fluid pressure & temperature. It also depends upon fluid shear rate for Non Newtonian fluids. The shear rate in PCP is controlled by rotational speed.

7.6.7 Cavity Filling

As the rotor of the PCP rotates, a cavity opens and well fluid enters the cavity. Until the rotor closes the cavity and pushes the fluid, the pump does not do any work to the fluid. The rotor rotation just creates void for fluid entrance. The fluid is not pulled into the void by the PCP, but is pushed into by pressure at the pump intake.

To ensure quiet and efficient operation, the cavity should be filled completely with the well fluid before the rotor closes. Fluid flow into a cavity depends on the fluid viscosity, size and shape of the opening and the pressure at the pump intake.

7.6.8 Progressive Cavity pump

The critical intake pressure for Newtonian fluid is given by:

$$P_{in} = \frac{1}{8.04\ E6} \frac{l_s}{d_t{}^3} \mu\ (nQ_t - q_s\mu) \qquad\qquad \dots (7.5)$$

Where,

   D   = diameter of rotor
   L_s = spiral length of a cavity in one stator pitch (in)
   T   = average thickness of the cavity (in)
   P_{in} = critical pump intake pressure (psi)
   n   = pump rotational speed (RPM)
   Q_t = designed flow rate (BPD)
   q_s = Slip rate (BPD)

Progressive Cavity pump

Where,

$$l_s = (P_s^2 + (\pi(d + 4e))^2)^{1/2} \ \text{ and } \ t = \frac{4eP_s}{\pi l_s} \qquad\qquad \dots (7.6)$$

Below this critical intake pressure, cavity cannot be completely filled. The critical fluid level above the pump can be calculated from the pump intake pressure, casing pressure, pump depth, net zero flow correlation.

For a PCP well, there is a rotational speed at which fluid fills the cavity,

$$n = \left(\frac{8.04\,E6dt^3\,p_{in}}{l_s\mu} + q_{s\_\mu}) \right) / Q_t \qquad \text{... (7.7)}$$

For Bingham and Power law Non-Newtonian fluids, the flow correlations are

$$\frac{dp}{dL} = \frac{12\mu_p}{wt^3}\left(q_a + \frac{wt^2}{4\mu_p}\tau_y \right) \text{ and } \frac{dp}{dL} = K\left(\frac{2^{1/n}(4+2/n)}{wt^{(2+1/n)}}q_a \right)^n \qquad \text{... (7.8)}$$

Where,
  μ = Plastic viscosity (poise)
  τ = Yield Point (dynes/cm^2) in Bingham Model
  K = Consistency Index (dyne-sec/cm^2)
  n = Power law exponent in power law model.

7.7 Merits of Progressive Cavity Pump (PCP)

The main attractions of the progressive cavity pump installations are:
- Low Capita; Cost (approximately 60% of an equivalent conventional pump installation).
- The Ability to handle very high sand cuts (up to 85%).
- Lower Power Consumption Per Barrel.
- The Ability to Increase Production from Heavy Oil Wells by Avoiding Rod Falling Problems.
- Reduced Wax Problems.
- Convenient for Temporary Well Tests and Well Clean-up.
- Increased Efficiency due to Elimination of Rod Stretch.
- Small and Lightweight Surface Equipment.
- *Special Applications:* Hot oil recycle pumps can be fitted to a progressive cavity pump system. They are used to transfer a specific amount of heated production oil from the lease tank, down the annulus to the pump. By controlling the discharge rate of the recycle and bottomhole pump, the fluid velocity in the tubing can be increased to a level that is higher than the sand settling velocity thereby reducing sand-in-problems. In addition, recycled oil can be used to carry wax inhibitors and other chemicals and to thermally inhibit wax deposition.

7.8 Demerits of PCP

The main problems with the progressing cavity pump are:
- High service costs.
- Reduced capability for handling hot fluids.

- Limited total dynamic head.
- Difficult to lift from marginal wells with high GOR's, gas separation may be required.
- Limited production capacity.
- Possibility of backed-off tubing occurrence.
- Gas: Progressive cavity pumps must not be allowed to gas lock by completely pumping the well off or by failure to vent the annuals. A continuous flow of a least 10% volume of the liquid capacity is required to prevent rapid wear and torque problems. The pumps rate should therefore be adjusted to ensure that the pump is always fully submerged.
- Sand and Water Production: Sand settlement is a problem in the lower viscosity fluids and especially as water cut increases. Sand settlement rates can be calculated using Stoke's Law. Shut down operations always involve sand settlement risks, it is therefore often necessary to lift the rotor out of the stator for a flush-by operation before restarting a sandy well.
- High water cuts also result in more apparent pump wear because of the reduced lubricity and higher slip tendency.
- Stalling and Start-up Torque: High start-up torques are sometimes experienced due to increased static friction between the stator and rotor. This can often be overcome, if the rods can be lifted. Some designers are therefore considering the use of wellhead assemblies with built-in lifting jacks.
- Frequent stalling may be indicative of:
  - A worn pump
  - Sand settlement
  - Waxing or scale
  - Poor pump space-out
  - Excessive viscosities
  - Stretched rods.

7.9 Recent Advances in Progressing Cavity Pumping (PCP)

7.9.1 *PCP Stator Redesign*

A more "stable" stator design called uniform elastomer or even wall stator technology, is being developed and tested by Weatherford Artificial Lift Systems, Lloyd Minister, Alberta, Canada. This upgrade to conventional stators is showing promise in improving PCP durability/reliability/flexibility.

The initial pump design being tested produces 60 bpd/100 rpm, with a pressure rating of 1,800 psi. Two other designs for higher-volume applications are scheduled for field tests. Currently, few are being undergoing long field trail testing in heavy oil applications.

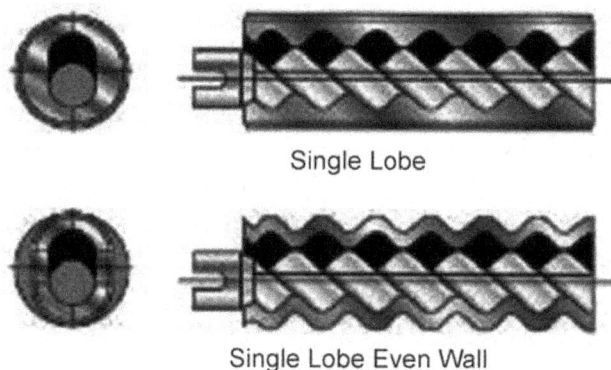

Single Lobe

Single Lobe Even Wall

Fig. 7.4: Uniform Elastomer-Thickness Wall, Compared to Conventional PCP Elastomer Design

In a conventional PC stator, the elastomer is injected into heavy-walled steel tubing, forming a double internal helix. For years, this design was considered the only economical way to produce stators. However, heat can build up in the widest portions of the helix, possibly leading to early burnout. The newer, uniform-thickness elastomer stators have numerous advantages, including:

- Improved heat dissipation—the elastomer runs cooler, leading to improved mechanical properties and fewer failures due to stress or wear of the stator elastomer minors.
- Uniform elastomer swell—Having a uniform elastomer thickness means the elastomer also swells in a uniform fashion. Properly sizing the rotor for aggressive applications is, therefore, much simpler.

7.9.2 *Wider Applicability*

Applications that once pushed the envelope for PC pumps may now be within reach. These include wells with higher concentrations of aromatics, CO_2 and H_2S, or higher temperatures.

And the system has a higher pressure rating. Conventional PCPs rely on interference between rotor and stator to create a seal. Uniform-thickness stators are more consistent, providing a better rotor/stator fit with less interference. This enables the pump to handle more pressure.

7.9.3 *Hi-Torque Sucker Rods for PCPs*

Weatherford is currently developing a hi-torque sucker rod for progressing-cavity pumping systems. The new product, called Corod DER 8.5, represents the latest technology enhancement to continuous sucker rods, which are unique to Weatherford Artificial Lift Systems.

The DER 8.5 continuous sucker rod is 1–5/32 inch in diameter and can withstand 1,400 ft-lb of torque and operate in 2–7/8 inch tubing. It provides the capability to produce more fluid from deeper depths without increase of tubing size. This results in cost savings, since the operator does not have to incur additional capital spending for larger tubing. The added-size feature of the sucker rod also enhances wear characteristics of continuous sucker rods in crooked, deviated and horizontal wells. Judging from current field test results, it is expected that DER 8.5 will be available for use shortly.

7.9.4 *Surface-Drive Head/Hydraulic Power Skid*

The new progressing cavity pump-surface-drive head and hydraulic power skid system from M/s R&M Energy Systems, Houston, was engineered and developed for use in conjunction with Moyno Down-Hole Pumps for artificial-lift applications in oil/gas production. As part of the Ultra-Drive family of surface drives, the DHH system includes a low-profile, surface-mounted drive head and an accompanying electric motor or gas-driven engine and hydraulic power skid for unlimited, variable-speed control with automatic backspin braking. The system's low-profile design is more desirable than bulky, conventional-style pumping units. It is also optimally suited for remote, non-electrified areas due to its ability to utilize a gas-driven engine

Fig. 7.5: Surface-drive head for downhole PCP. Accompanying motor or engine and hydraulic power/control skid are not shown

powered by natural gas produced at the well-head. It also features easy field installation, simplified maintenance and an improved stuffing box to prevent leaks.

The surface-drive head includes the following specifications:

- Hollow shaft
- 1,250 ft-lb maximum rod torque
- 65 HP maximum power
- Flanged or pin wellhead connection
- 1–1/4 inch polished rod size.

Further, the system operates on a 500 rpm maximum polished rod speed, depending on the hydraulic pump/motor, and can handle a 33,000-lb (maximum) axial load.

7.9.5 *Leak-Free Stuffing Box*

R&M Energy Systems has also developed a new stuffing box for use on progressive cavity pump drive heads. The Ultra-Drive Enviro-Stuffing Box can be used with all

Moyno surface drives, as well as other drive-head brands. The system incorporates special "memory" type braided rope packing that provides an excellent seal while never losing its original shape. The unit requires lower maintenance than other types of environmental seal units and is easy to install and service. The Ultra-Drive box is an external environmental catch chamber. This chamber can either re-direct the leaked well fluid to a safe location or provide 100% containment. The sealing unit is completely detachable, allowing wellhead pressure being maintained while the mechanical portion of the top drive removed for servicing. An optional Anti-Pollution Stuffing Box Adapter (APA) leak-detection and switching device is also available for maximum safety against costly stuffing box spills.

Fig. 7.6: Leak-free stuffing box for moyno or other drive heads features braided rope packing and external environmental leak catch chamber

7.9.6 *Pump Controller*

Users of PCP can maximize run life of pumps and minimize lift costs by closely monitoring the performance of the pump and adjusting parameters that the controller uses. Offered by Case Services, Inc. of Houston, the system provides users with an easy-to-use interface that displays pump curves and trends such as rpm, surface tubing pressure, gross liquid rate, surface temperature and tubing pressure. With that information, and electrical usage history, the user can accurately assess a problem in the PCP and even predict future problems. The user is provided with current status of the well(s), including customizable alarms, fault history, torque and speed. The module is part of the csLIFT suite, so it is integrated with existing features of the suite, such as well-test information and alarm callouts.

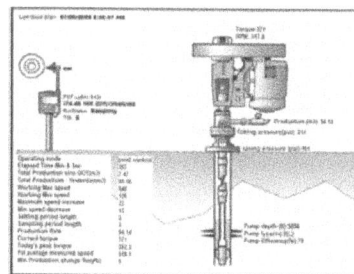

Fig. 7.7: PCP controller interface displays pump curves and trends such as rpm, tubing pressure, liquid rate

7.9.7 *Low Horsepower Drivehead PCP*

Baker Hughes Centrilift has expanded its drivehead series to include the **LIFTEQ LT30E** model for lower horsepower requirements. The system is designed to effectively meet the needs of coalbed methane dewatering operations, as well as low-volume conventional oil well production.

The compact design has no externally mounted hoses, pumps, belts or other fragile components that can be damaged during installation or operation. An internalized braking system is protected from effects of outdoor exposure, enhancing durability and trouble-free/maintenance-free backspin control. The braking system is self-regulating, with no manual adjustments required to regulate recoil speed.

Fig. 7.8: New drivehead for progressing cavity pump expands the LIFTEQ series, for lower horsepower requirements, and features a positive brake system without ball bearings

The effectiveness of many traditional braking systems that include a set of ball bearings for brake engagement, are subject to varying oil viscosity, which can negatively affect brake operation. The new model uses a positive brake system that eliminates need for ball bearings and, as a result, minimizes the influence of changing oil viscosity within the drive head. A stand-pipe design replaces bottom seal requirements, eliminating the possibility of lubricating oil leakage. A stuffing box and a rotating mechanical seal are both available to provide a seal around the polished rod.

7.9.8 *PCP for Heavy Oil Production*

Heavy oil has been produced using progressing-cavity pumps since the late 1980s. The accumulation of field experience and incremental product developments over time have enabled engineers to continually upgrade the technology. Contributing to this trend, Baker Hughes Centrilift has added the newly developed 110-D-2600 PCP (1.1 bbl/rpm @ 2,600 psi @ 100 % efficiency) to the LIFTEQ PCP product line specifically for heavy oil production applications.

The various combinations of conditions such as high viscosity, sand production, high flow losses and low BHPs commonly associated with heavy oil production require a design that balances the impact of all these conditions. The large 3.75-in. OD tube of the 110-D-2600 provides a greater cavity cross-sectional area which enhances viscous inflow. The internal geometry decreases fluid velocity in the stator to reduce erosive effects of sand and abrasives.

The pump geometry also provides a larger rotor diameter, capable of sustaining torque-induced stresses common to this application. The shorter rotor pitch length enables both a greater pump differential pressure over a comparably short pump and more effective movement of solids through the pump.

The new pump is manufactured with Centrilift's field-proven LT 2000 elastomer. In heavy oil applications, this has operated with minimal swell, and has proven resistant to abrasives. In addition, the mechanical strength of the elastomer can provide sustained operation at high pressure.

7.9.9 *Insertable High–Volume PCP*

Weatherford took a good idea and made it even better with the Arrowhead insertable progressing-cavity pump. Insertable PCPs reduce downtime during running and pulling phases by requiring removal of only the rods, rather than the tubing string. This new patented design reduces pump assembly length by as much as one-half to two-thirds.

As a result, this PCP is capable of high lift and large volumes. Pumps with volumes of up to 5,000 bpd are available, depending on tubing size and operating conditions. Pump lifts as high as 5,250 ft have been achieved. In addition, the shorter pump configuration also simplifies shipping and handling.

The new PCP is positioned with a single-pump seating nipple and a no-turn tool does not require a tubing interface device. This simplified arrangement eliminates the need to space out tubing to accommodate specific pump lengths. As a result, different sizes of insertable pumps can be run into the same tubing string over well life without having to pull the tubing. The simplified assembly is easier to run and is more reliable than dual-attach-point systems. It is also likely to get stuck less by sand/debris.

The enhanced tool can essentially be used in applications where a conventional PCP is used, as long as the tubing housing the insertable PCP fits inside the casing. Significant savings can be realized in reduced servicing costs by using the pump in wells where change-outs are frequent. In remote locations where a complete service rig comes at a premium cost, the new pump allows servicing by less expensive means with a flush-by or other rod-pulling device.

Weatherford recently used the new system in an application in Canada.

The original pump in the horizontal oil well was pulled after nearly 17 months in use. A flush-by unit performed the workover to replace the insert pump, using the new PCP, which took about 3 hrs. An operation of this type with a standard PCP takes about 8 hrs. A flush-by unit usually cost about US $141/hr, compared to $266/hr for a conventional service rig, as required for change-out of a standard PCP. After the change-out, the new pump operated at 91 % efficiency.

The new tool effectively enhances a pump-system workover operation with a simple, yet innovative approach. Insertable PCPs are currently applicable inside 3-1/2, 4-1/2 and 5-1/2 inch tubing and larger, if requested. Additional installations are in place and operating in Canada, Venezuela and off-shore of West Africa.

7.9.10 *Variable Speed Drive*

R&M Energy Systems has launched a new software product, the **Guardian** variable speed drive (VSD), an innovative techno-logy advancement that offers substantial production/cost-reduction benefits to end users. The new VSD is an effective digital solution that provides versatile production control and informative monitoring of downhole PCP systems. A combination of precise torque and speed control in a single durable package is the key to maximizing production and increasing energy efficiency, while providing total protection of PCP systems to ensure equipment longevity.

The proprietary software incorporates a complete model of the total PCP system. This program provides continuous feedback

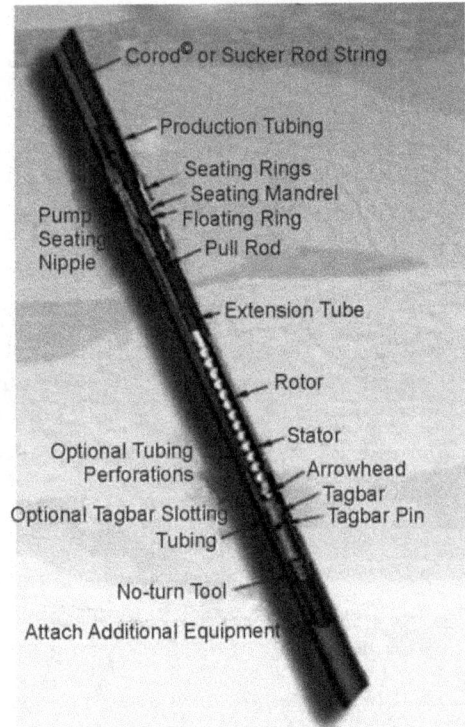

Fig. 7.9: Insertable High-Volume PCP Cuts Running/Pulling Time and Reduces Pump Assembly Length by up to Two-Thirds

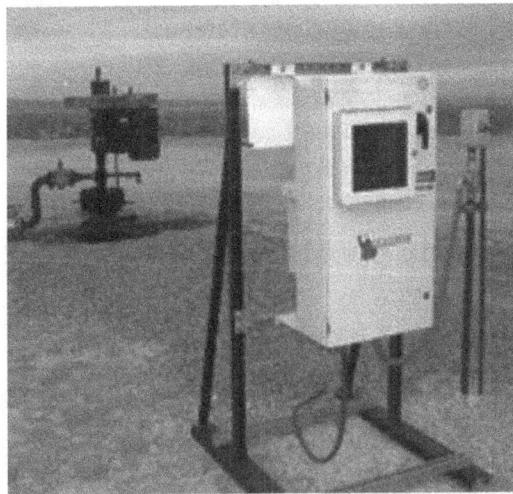

Fig. 7.10: Variable-Speed Drive Features Cost-Saving Benefits through Digital Solution Combining Precise Torque/Speed Control

of torque, speed, pump efficiency, fluid rate and estimated downhole pressure. This feedback, measured against operator set points, allows the system to provide a highly effective control method for protecting downhole equipment and maximizing production.

The Guardian VSD:

- Monitors the PCP, motor, rod speed and torque
- Offers a pump flow monitor and production accumulator
- Provides embedded data in the software for rod string, tubing, reservoir and PCP model
- Monitors and controls fluid level, and
- Provides remote monitoring by network servers.

Fluid level over the pump intake is controlled with manual speed adjustments or automatically by the system software. The new pump-off control capability maximizes well production for any given inflow characteristic, and a power optimizer reduces electric utility cost for any inflow rate.

7.9.11 *PCP Drive System Dewaters Gas Wells*

Hydraulic Energy Products, Inc., (Applied Energy Products, AEP, spinoff), Denver, Colorado, has developed a three-component drive system to automatically de-water gas wells, especially coalbed methane wells. The complete system is self-monitoring and regulating, with signal capability to alert the operator if drive speed goes above

or below pre-determined set points. Variable speed with manual control is achieved by applying the Top Head Drive and the Hydraulic Power Skid only.

The direct Top Head Drive hydraulic, motor-driven unit from HEP is offered in four torque ranges and utilizes the conventional friction lock block to connect to the 1-1/4 inch diameter polished rod. The largest drive motor is capable of supporting string and water column weights to 5,000 ft depth, while delivering 2,000 bwpd; or 4,000 bwpd from 2,500 ft. The maintenance-free drive operates through a rotary union, replacing the conventional stuffing box.

Fig. 7.11: The Three-component Hydraulic Top Drive PCP System Monitors Output to Automatically Dewater Gas Wells

Complete control of "back-spin" is achieved in conjunction with the HEP hydraulic power skid, and in the two larger sizes, a fail-safe brake provides release or dissipation of stored energy. The drive can be mounted on a threaded adaptor, to the flow tee, or secured on the standard API 8-bolt flange. The drive is Hydraulic Power Skid, presently offered in two sizes: 12 inch^3 and 26 inch^3/rev. At present, there are 10 drive heads running, with initial units exceeding two years without service or maintenance.

The HEP, is offered with natural gas powered 2.5- or 4.0 HP engines or 460 volts electric motors up to 100 HP. The HEP design incorporates complete secondary containment of all liquids and a full enclosure to prevent water buildup. The variable volume hydraulic pump is electro-proportionally controlled, with a feedback loop to assure accuracy and volumetric/operating pressure efficiency. The onboard computer optimizes fuel mix and governs operating speed. It also monitors engine oil pressure, radiator temperature, hydraulic reservoir liquid level and temperature. The skid can reverse rotation of the Top Head Drive, with limited torque and speed, to dissipate stored energy, flush the PCP, or rapidly equalize fluid levels. The constant-speed electric motor drive system provides variable Top Head Drive speeds hydraulically and eliminates the need for a Variable Frequency Drive (VFD).

Fig. 7.12: Power Skid with Gas Engine or Electric Motor-Driven Hydraulic Motor Dirves a Downhole PCP, with Computerized Control

7.9.12 *Liquid Level Monitor and Controller*

The proprietary controller from Applied Energy Products monitors, calibrates and alters drive speed of the downhole PCP—maintaining the desired liquid level, with calculations based on physical data and formation flowrate, for the specific well. Recalculation frequency can be altered to meet fluctuating well conditions. The Liquid Level Controller (LLC) has two separate computers. The first gathers information or observes formation water flow increase or decrease into the well casing, while the second analyzes and generates the appropriate signal, causing the hydraulic pump to increase or decrease flow to the drive motor, changing the speed of the downhole PCP. The LLC maintains the liquid level between lowest perforation, or any user defined level, and the PCP inlet, thereby maximizing gas production.

The controller offers a manual mode for setting a desired speed, and an automatic mode to remove guesswork. A signal is generated that may be transmitted by a SCADA system - if the operating range (high or low) is exceeded - with a detailed recording of system performance. The LLC, being the last component introduced, has been installed on operating wells and the concept proven; however, it has not yet been released to the general market.

7.9.13 *Experimental Simulation Loop*

C-FER Technologies of Edmonton, Canada, recently commissioned an experimental loop to simulate downhole conditions for pumping systems operating in Steam-Assisted Gravity Drainage (SAGD) applications. Pumping equipment of up to 80 ft (24.4 m) in length can be tested at maximum discharge pressures, intake temperatures and flowrates of 1,100 psig (7,580 kPag), 200°C (400°F) and 5,050 bpd (800 m³/d), respectively.

The flow loop has multiphase capabilities (oil, water, air) and is fully instrumented to allow real-time measurements of all key variables (pressures, temperatures, flowrates, pump speed, torque, etc.). This makes the loop an ideal tool to ascertain performance of new artificial lift systems under controlled conditions before actual field tests are undertaken. Testing of four pumping systems has already been completed under a joint industry project. Services offered by C-FER include design, execution and data analysis of pump testing programs, for both operating companies and equipment manufacturers.

Fig. 7.13: An Experimental Loop for Downhole Pumping Systems Allows Testing of New Artificial Lift Systems under Controlled Conditions

7.9.14 *PCP Software Program*

A new version of PC-Pump (2.67), a well-known proprietary software program for interactive design and evaluation of PCPs, has been recently released by C-FER Technologies. Upgrades incorporated in this latest version include: updated equipment databases from vendors (pumps, rods, drive heads, etc.); improved wellbore temperature gradient and inflow performance calculations; and an improved batch-run comparison mode to facilitate assessments over a range of operating conditions.

7.9.15 *Insertable PCP System*

The new Moyno Insertable Progressing Cavity (IPC) Pump System from R&M Energy Systems, Houston, is engineered to allow downhole PCP installation and retrieval without having to pull the production tubing. This considerably reduces time/pulling costs. In addition, it provides high operational reliability/flexibility, reducing overall life cycle costs compared to other artificial lift systems. These systems can handle a wide range of conditions, including low- and high-viscosity fluids and solids transport. They have been engineered for strong performance in oilfield and coalbed methane dewatering applications.

Other benefits of the IPC pump systems include: 1) unique hold-down devices that provide tubing-pump annulus sealing at pump intake, and accurate installation at seating assembly; 2) easy rotor-tag spacing without retrieving pump from the locked position; 3) system allows flushing operations without disengaging pump from seated position;

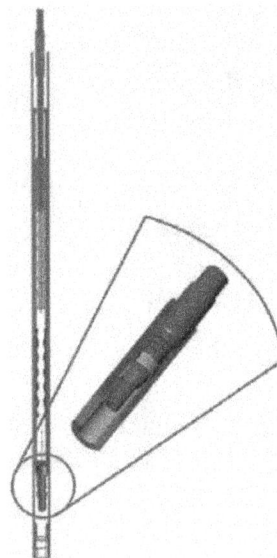

Fig. 7.14: Insertable PCP System can be Installed in 2-7/8 inch to 5 inch Strings without Puling the Tubing

4) the landing nipple is the only system component attached to the tubing string at the bottom of the hole; and 5) the system is ideal for quick, inexpensive pump changes, allowing the user to match well conditions in terms of differential pressure and/or flow without retrieving the tubing string.

The new IPC pumps can be installed in 2-7/8 inch to 5-1/2 inch diameter tubing. In the interest of installation time, it is recommended that all production tubing and components, through which the insertable must pass, be drifted to prevent any problems. No special tools are required for installation.

7.9.16 *Heavy Duty Drive Head*

The new Moyno Ultra-Drive Model EX1 Surface Drive Head from R&M Energy Systems is designed for use with Moyno downhole PCPs in applications that require larger pump sizes and/or extraction from deeper wells. This new unit offers rugged construction and improved operational features for heavy-duty downhole PCP applications. It offers easy installation, safe/consistent recoil control during shut-down, and improved dynamic wellhead sealing capability.

The system's motor support frame is designed to accept dual electric motors in horsepower ranges from 50 to 150 (maximum 300 hp with both motors installed) with maximum polished rod speed of 600 rpm and operational polished rod torque of 4,000 ft.lb. Weight of rod string and rotary PC pump thrust load is supported by the axial thrust bearing in the heavy-duty gearbox.

Fig. 7.15: Heavy Duty PCP Drive Head Designed for Applications Requiring Larger Pump Sizes and/or Production from Deeper Wells

Dynamic load rating of the standard bearing is 80,369 lb, based on a CA90 rating. The gearbox assembly contains a spiral bevel gear set, directly linked to a hydraulic motor drive. In the backspin mode, normally resulting from release of energy stored in the well at shutdown, this hydraulic motor drive retards rotational speed for effective, safe control.

The low profile design simplifies field mounting and installation, and a removable sealing system makes it versatile and ensures easy maintenance. It can accommodate polished rod sizes between 1-1/2 inch and 2-1/16 inch.

7.9.17 *Drive Head Seal for PCPs*

The new Rotary Seal Unit (RSU) from Baker Hughes Centrilift, Claremore, Oklahoma, is the latest component to complement the LIFTEQ LT Drivehead Series for PCP Systems. It is a unique sealing system rated for operation to 550°F (288°C), with redundant leak prevention features, and is approved for use with LIFTEQ LT30E, LT50E, LT100E, LT150E and LT300E drive heads.

The new seal utilizes multiple lip seals as a primary seal, and a secondary seal area that includes conventional Teflon and graphite rope-style packing. The system provides an early warning in advance of a primary seal failure and, as such, the secondary seal can be easily engaged to avoid an unscheduled pumping system shutdown. The secondary sealing system operates as a conventional stuffing box and can continue to be utilized until maintenance can be scheduled, thus avoiding costly additional shut downs.

An added benefit of the RSU is reduced polished rod wear. The primary seals are designed so that, while in operation, polished rod wear is reduced or eliminated. The Centrilift RSU can also be adapted to other manufacturers' products.

7.9.17 *High-Torque Rod and Coupling*

Weatherford, Edmonton, Alberta, Canada, has developed larger PCPs and top drives to meet industry's artificial lift needs. As these two components increase in size, torque on the sucker rod used to join the top drive and PC pump has increased as well. Traditionally, the approach has been to make a larger sucker rod; however, this restricts fluid flow and increases both string weight and torque, requiring larger, more expensive top drives.

To address these applications, Weatherford's proprietary ultra high-strength sucker rod proves a viable solution. The EL rod is capable of handling high torques of new top drives and PC pumps. To further enhance the rod's high torque rating, the high-torque (Hi-T) coupling was developed. This coupling has pushed the new rod's torque rating, with a built-in service factor

Fig. 7.16: Rotary Drive Head Seal for PCP Systems

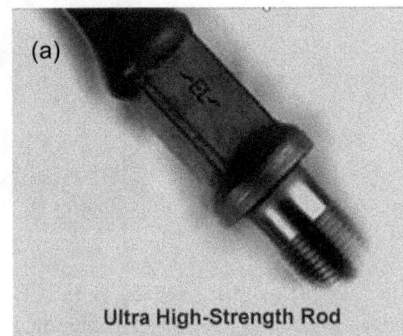

(a)

Ultra High-Strength Rod

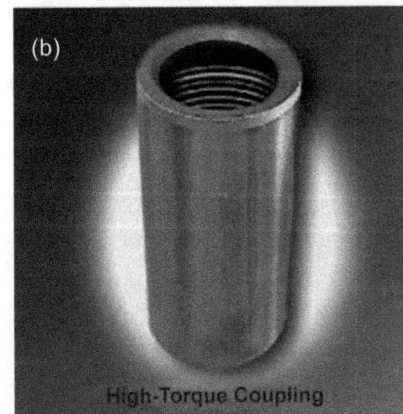

(b)

High-Torque Coupling

Fig. 7.17: (a) High Strength PCP Sucker Rod, and (b) High-Torque Rod Coupling

beyond industry's traditional thresholds. Because of this built-in safety factor, there is no need to decrease the new rod's published torque rating as is done with conventional rods.

The ultra-high torque ratings of the new rod and coupling typically allow the operator to drop down one size, e.g., 7/8 inch, vs. 1 inch conventional. This gives a lighter string weight, larger flow areas and lower torques. In addition, the new rod can also have centralizers when required, unlike larger sucker rods.

Fig. 7.18: Model CV1 Downhole PCP Drive Head

7.9.18 *Surface Drive Head*

R&M Energy Systems, a unit of Robbins & Myers, Inc., Houston, Texas, has introduced the new Moyno Ultra-Drive Model-CV1 downhole pump drive head. This is the latest development in the Ultra-Drive "C" series drive heads designed for medium-duty, PCP applications. Significant advancements of the new design include:

- Sealing versatility that allows the end user to utilize various stuffing box configurations to meet sealing needs of a broad range of well conditions, as well as maintenance preferences and environmental requirements. The Ultra-Guard Sealing System is available as an option, offering leak-free operation, minimal maintenance and extended shaft/seal life.
- With reduced overall height, the new drive head is a more compact, lower-profile design—an important consideration when clearance is an issue, but also in terms of head stability.
- Improved balance: Overall drive head stability is enhanced because overall unit weight is centralized more directly over the wellhead. The result is improved weight distribution for drive head balance and overall stability.

Electrically driven, the drive head offers rugged construction, and it features polished rod speeds to 600 rpm and a 1,600 ft-lb torque rating. It can accommodate either 1-1/4 or 1-1/2 inch polished rods.

7.9.19 *Tubing Rotators*

R&M Energy Systems have introduced new models for its RODEC line of tubing rotators. Identified as the RODEC RII Series Tubing Rotators, these units provide greater application versatility for a broader range of wellheads than was previously available. A new, modular design allows the rotators to adapt to any wellhead configuration, including threaded cap as well as flanged wellheads. The new product line includes three types of systems: 1) the threaded-cap-type model in which the main body is the same for all configurations; 2) the adapter flange model that is studded down for attachment to any existing API tubing head; and 3) the Flow T/BOP model that combines the new rotator design with a production flow tee and a rod BOP. This module can be added to any rotator of the new line.

Additional features and benefits include: retrofit of any well with an existing tubing head; virtually required no maintenance/ monitoring; units are ideal for all down-hole pump types and they maximize

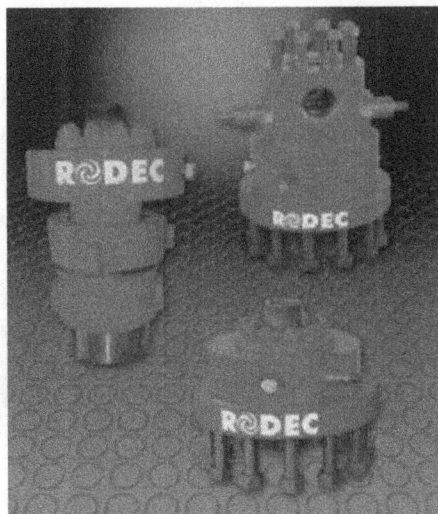

Fig. 7.19: New Models of Tubing Rotators for PCP or Beam Pumping

production and extend tubing life. The rotators provide a cost-effective solution for evenly distributing wear about the internal circumference of production tubing. Field usage shows that RODEC rotators extend tubing life by 6 to 10 times, reducing operating costs and downtime for maintenance.

7.9.20 *Drive-Heads for Rod-Driven PCPs*

The primary purposes of drive-heads are to support a prime mover, provide a seal to contain well fluids, support the sucker rod string, provide safe control of rod recoil and to transfer energy from the prime mover to the rod string. Centrilift now markets the LIFTEQ line of drive-heads for rod-driven PCPs (RDPCPs). The series is available in various configurations that support up to 300 hp. Safety features include hydraulic back-spin control that automatically regulates the speed of sucker rod recoil. Five models are offered for ratings of 50 to 300 hp. The LT100E, LT150E and LT300E share the same frame and can be con-figured in the field to match changing production requirements. Figure 7.20 illustrates the LT150E model.

Fig. 7.20: LT150E Drivehead for Rod-driven PCPs

7.9.21 *Variable Speed Drives for PCPs*

With the Electrospeed GCS and Electrospeed PC, Centrilift now supplies a complete line of variable speed drives (VSDs) to meet the needs of all LIFTEQ RDPCP and ESPCP systems. The PC model is designed primarily for RDPCP systems up to 125 hp. It features direct torque control (DTC), a motor control that allows control of both speed and torque without pulse encoder feedback. The GCS model will control systems of 100 hp and greater. The GCS features the graphic control system for ease of use. The VFDs will shut down the system when conditions develop may damage the pump. They operate with surface or downhole instrumentation. If neither of these is connected, the driver will operate based on motor torque.

Fig. 7.21: The Electrospeed GCS Features a Graphic Control System for Ease of Use

7.9.22 *Doppler–Effect Fluid Level Finder*

Wilson, Houston, Texas, a business unit of Smith International, is introducing a revolutionary, patented fluid-level control system called the FM1100, in conjunction with the Wilson VFD1000. This system will give the operator the ability to monitor and control the fluid level in many wells on a real-time basis without need for any downhole sensors. It will use multiple sensors highlighted by doppler-effect radar to measure empty space in the well annulus. A frequency generator creates doppler-frequency waves in the annulus; the wave harmonics change form as the fluid level changes. A system of highly sensitive sensors at the surface recognizes the harmonic changes and sends the data to a processor, which creates a fluid column height on a real-time basis.

This system can be utilized with standard rod pump, PCP and ESP systems. In standard rod pump applications, it can be utilized to set shut-down and start-up fluid levels. In both PCPs and ESPs, the system can be utilized with the VFD to maintain a constant fluid level by changing the motor speed. As a reservoir management tool, it can monitor fluid level changes in producers as injection rates change in supporting injection wells. The system will also monitor flowrate, temperature, pressure, amperage, and actually optimize the fluid level to maximize production.

Fig. 7.22: Fluid Level Finder Using Droppler-Effect Radar

7.9.23 *Stuffing Box Sealing System*

The new Ultra-Guard Sealing System from
R&M Energy Systems, Houston, a unit of
Robbins and Myers, Inc., successfully
addresses environmental and maintenance
problems caused by stuffing box leakage in
the top drives of PCP artificial lift systems.
Designed and developed specifically for
harsh, abrasive environments, including high
pressures and speeds, the system utilizes a
unique hydrodynamic principle to provide
seal-to-shaft separation via a lubricant film,
virtually eliminating seal and shaft wear

Fig. 7.23: Sealing System for Stuffing Boxes on PCP Top Drives

while excluding abrasives. Dual independent seal chambers permit continued leak-free operation even if one of the rotary seals fails.

Available in 1-1/4 and 1-1/2 inch polished rod sizes, the system is compatible with the entire line of Moyno Ultra-Drive drive heads and can easily be adapted to fit most other drive heads. As additional features and benefits, the system:

- Positively excludes abrasives from the bearing system
- Operates under transient conditions (frequent starts/stops, vibration, pressure/temperature changes)
- Is compact and self-contained.

Further, the system has proven reliable in the presence of abrasives; it is suitable for high-pressure/high-speed combinations; and it is easily installed and maintained.

7.9.24 *Medium/Light Duty Drive Heads*

R&M Energy Systems offers the new Moyno Ultra-Drive Model CD1 downhole pump drive head. This electrically driven system offers rugged construction and numerous operational features that make the "C" series drive heads attractive for medium-duty PCP applications. The system features polished rod speeds ranging to 600 rpm and a 1,600 ft. lb torque rating on its recoil braking system.

The low profile design of the unit simplifies field mounting and installation, and a removable sealing system permits flexibility and easy servicing. It can accommodate either 1-1/4 or 1-1/2 inch polished rods. Additional features and benefits include: 1) side lifting lugs and multiholed lifting bracket facilitate lifts; 2) single-point belt tensioning system; 3) a low profile for use with overhead irrigation systems; 4) its hinged belt guard allows unobstructed access; and 5) increased thrust bearing capability assures long service life.

For light-duty PCP applications, R&M offers the Moyno Ultra-Drive Model AD1 drive head, designed for use with Moyno downhole PCPs in oil-well production and gas-well dewatering applications. This electrically driven drive head offers rugged construction and numerous operational features. It offers polished rod speeds ranging up to 600 rpm and a 460 ft. lb torque rating on its recoil braking system.

Fig. 7.24: PCP Model CD1 Drive Head for Medium-duty Applications

The hollow-shaft, lightweight, compact design of the AD1 unit simplifies field mounting and installation and can accommodate 1-1/4 inch polished rods. Additional features and benefits are the same as those listed for the Model CD1 drive head.

7.9.25 *Composite PCP Elements*

The newest technological advancement in the PCP industry has been use of composites in the manufacture of PCP elements. This patented technology is being developed by G-PEX, Tulsa, Oklahoma, with manufacturing in Pearland, Texas (Houston). In this early development stage, the stator is made of a hard composite material and is placed in a steel tube jacket. The composite material offers similar wear characteristics to metal, but makes the stator the longer lasting element.

The rotor is made of steel and coated with an even thickness of a soft and durable polyurethane. The urethane offers increased wear resistance and mechanical properties over conventional elastomers and, when placed on the rotor, offers the advantage of the wear element being located on the end of the sucker rod string rather than on the end of the tubing string.

The composite PCP becomes a highly durable system that incorporates the emerging even-thickness-elastomer technology, which has been under development in conventional stators for years and has been proven to offer increased mechanical properties. The even thickness allows more consistent thermal expansion and chemical swell within the pump, enhancing performance predictability in most applications. The method of processing the coated rotor allows for use of more advanced materials, which resolves many compatibility issues realized with conventional nitrile-based elastomers.

With the elastomer placed on the rotor, the wear part is much less time-consuming to retrieve/replace, which will cut workover costs by more than 50%. The composite stator offers increased wear and corrosion resistance, allowing it to remain in the well longer. This PCP technology advancement is currently being thoroughly lab and field tested. The next development phase will be all-composite stators and rotors. Applications are being sought to help prove this emerging technology.

CHAPTER 8

Artificial Lift Optimization and SCADA System

8.1 Introduction

In offshore operation numerous control systems are employed for automation and effective management of oil and gas production. The most comprehensive and sophisticated out of these is SCADA which stands for "**S**upervisory **C**ontrol **A**nd **D**ata **A**cquisition". Inopportunely SCADA has received less attention for corporate monitoring than corporate network, making them especially vulnerable when it comes to maximizing usage in optimization of offshore production system.

In Artificial Lift optimization, it is fundamental to protecting base production in oil fields worldwide. With real-time wellhead surveillance and automation data available on a SCADA system, the engineer can quickly identify well problems and opportunities to improve gas lift performance. A continuous gas lift automation and wellhead surveillance system was installed on offshore field can have a minimum 5 % increase in oil production.

The gas lift automation is a solution to the daily struggled of gathering enough data to understand field production. Automation equipment and software were put together in a combination that allows engineers and operators to measure and monitor well information and control each well's gas lift rate. With automation, each well can be monitored in real-time from the engineer's laptop computer anywhere in the world. What previously required three or more days of analysis can accomplish in real time on a SCADA from a dial-up or network connection.

Before installing the automation system production optimization and field management was very challenging because of the nature in which gas lifted wells use to flow, the number of gas lifted wells, compression constraints for low pressure gas and the overbooked well testing schedule. Gas lift optimization process has been gaining momentum with more structured decision flow chart which allows overcoming

Business Objective	Technical Objective	Monitoring Regime	Control Regime
• To increase production of field through gas lift optimization • To reduce unplanned deferment	Optimise gas lift gas allocation and distribution by using identified suite of PI based applications to do following: • Carry out well optimization work online to supply lift gas to individual well at optimum level for max. performance	• Gas volume monitoring at mother platform and onshore office • Gas balance for gas lift distribution • Lift gas distribution monitor to well platform • Well monitoring at mother platform & well platform • Trunk line pressure monitoring at mother platforms & well platform	• Set-point control on gas lift valve to individual gas lifted wells for optimizing gas lift injection rate

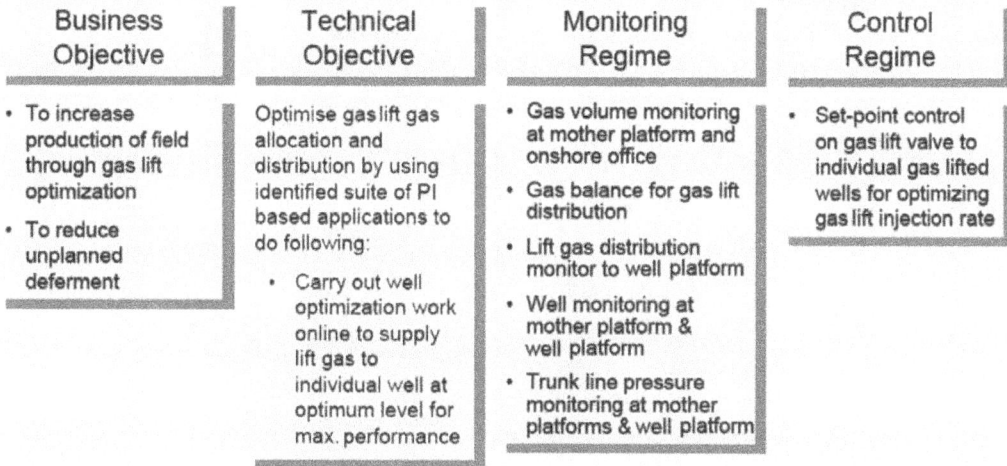

Fig. 8.1: Gas Lift Automation Process

existing issues and clear management objectives. Figure 8.1 illustrates the basic flow chart behind gas lift automation process.

One of the critical issues while addressing gas lift optimization process is gas dehydration. In offshore pipe line usually gets condensation effect, due to variation in temperature and quality of gas dehydration. This causes condensation formation in the flow lines. To overcome on this issue flow line either has to be over size and regular pigging needs to be performed during routine maintenance. This has a great impact on offshore gas lift optimization process. Efficiency of flow lines provides better system performance including gas lift valves.

8.2 Offshore Gas Lift Optimization

Offshore gas lift optimization is implemented using Continuous gas-lift method, which is the most common artificial lift method employed in offshore operations. The major problem most operators encounter with continuous gas-lift is maintaining an optimum gas injection rate into each well. Since injection gas is limited in offshore facilities, each well cannot use its optimum gas injection rate. In the past, this optimum gas injection rate was assigned to the well that would yield the maximum oil production. Today, a unique point on a gas lift performance curve, as seen on Figure 8.2 "Optimum Gas Injection Rate", has been recognized where the cost of the additional injection gas is greater than the additional profit that will be made from increased oil production.

In a typical continuous gas-lift installation, a standard adjustable choke controls the gas injection rate at each individual well. The injection rate often varies because of fluctuations in the gas-lift supply pressure. Supply pressure fluctuates due to compressor down-time, equipment maintenance, and increases in other wells' injection rates. Historically, these fluctuations were stabilized by adjusting operating conditions such as gas injection choke, on a trial-and-error basis. This approach to stabilizing injection gas requires increased manpower and results in production loss.

The primary goal of the gas-lift optimization system is to inject less gas to the less productive wells but continue to inject the optimum rate to the most

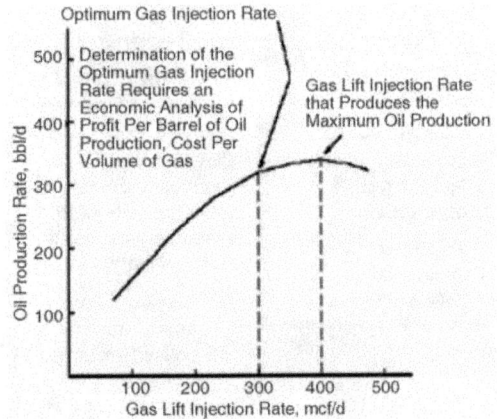

Fig. 8.2: Optimum Gas Injection Rate

productive wells when supply gas becomes limited (1). In order to accomplish this goal, the optimization system must also allow engineers to observe live data from the field. Then, the engineers can understand how to improve well performances in the field. The optimization system has four main tools that work together to provide the overall benefit:

8.2.1 *Constant Gas–Lift Injection Rate*

The computer constantly measures gas lift rates and adjusts the injection choke according to the set point. This prevents tubing head pressure fluctuations from affecting the injection rate. Maintaining a constant injection rate decreases the amount of slugging in wells and therefore, decreases instability in production processes.

8.2.2 *Real Time Wellhead Surveillance*

This tool allows the engineer to see temperature and pressure data from individual wells. This replaces two pen chart recorders that are read manually by field operators.

8.2.3 *Optimization Well Testing*

In order to determine the optimum gas lift injection rate on each well, an optimization well test must be run on each well. The computer runs the test and plots fluid flow versus gas lift rate. There are various intelligent automatic testing system with multiphase capability now being run by E&P companies.

8.2.4 *Data to Desktop*

The last tool transmits all the automation data back to the office and stores it in a database. The engineers can access this data and customize the interface to display any information from the optimization system.

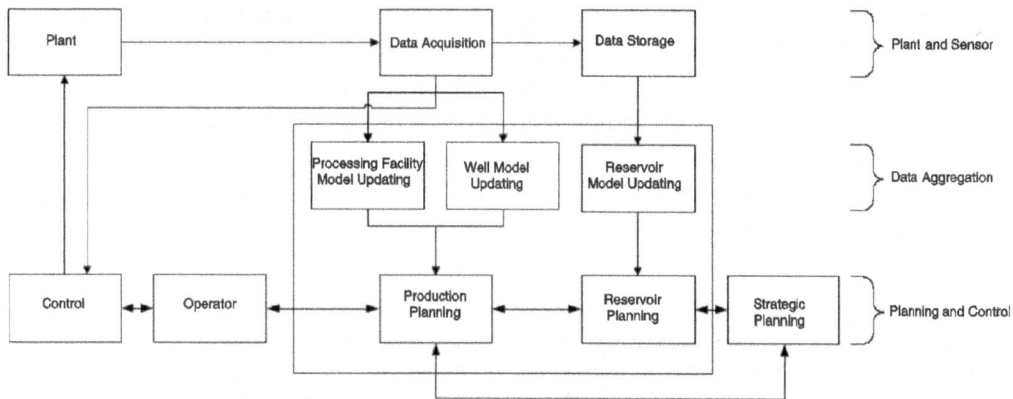

Fig. 8.3: Offshore Gas Lift Optimization Network

Flowchart above displays typical offshore gas lift optimization network, which provides clear picture about flow path for data and its application for further use.

8.3 Manual Surveillance and Optimization

The well surveillance and optimization efforts in the field to adding the automation system is performed using data gathered completely by hand. A systematic diagram is given at Figure 8.4 flow of total gas network system. A daily report in electronic format is e-mailed from the field every morning, to the engineers in the office containing field data including the following information from the previous day:

- Flow line pressure and temperature
- Casing pressure
- Gas lift rate, measured twice a day
- Well tests performed and completed
- Field production rates etc.

Too often, operators have little or no information on pressure and temperatures at the point of gas injection, and little control or flexibility in altering injection rates as production variables change. In fast moving offshore fields, where field and well conditions can fluctuate dramatically, this is a significant disadvantage.

Furthermore, with traditional gas lift, the primary method of gas injection is the "side pocket mandrel" configured

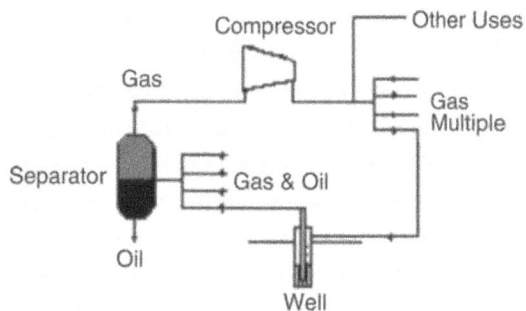

Fig. 8.4

in well completions system where wire line interventions are used to change the operating valve when injection rate changes are necessary. This can be a long, cumbersome process that can damage existing infrastructure (if the wire snaps, for example) and halt production while a new side pocket mandrel unit is installed. Here, the dangers of well instability also can come to the fore, with artificial gas lift actually increasing the probability of dramatic flow fluctuations and unpredictable surges in liquid and gas production rates.

8.4 Digitization of Artificial Gas Lift Process

There are various solutions available to address gas lift automation process. However one solution is a digital artificial lift system that addresses many of these issues. There are various commercial products available in market to address this issue, one of well-known as APOLLO (Commercial Name), it is based around binary actuation technology (BAT), which also has applications for the automotive, manufacturing, and life sciences industries. Central to the technology is a low energy pulse control which signals to switch an actuator between two stable positions to digitally operate a valve. Particular benefits include high switching speed and low power consumption.

With help of automated system all process can be integrated to synchronized so that system can be utilized in optimum manner. A illustrative Figure 8.5 is provided below of SCADA system run in offshore environment with wireless radio frequency link for better understanding.

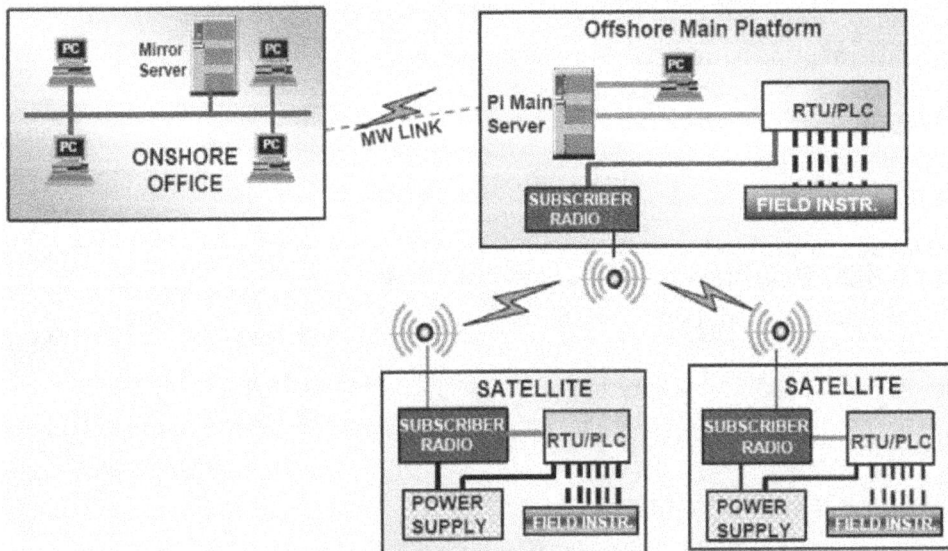

Fig. 8.5: SCADA System Run in Offshore Environment

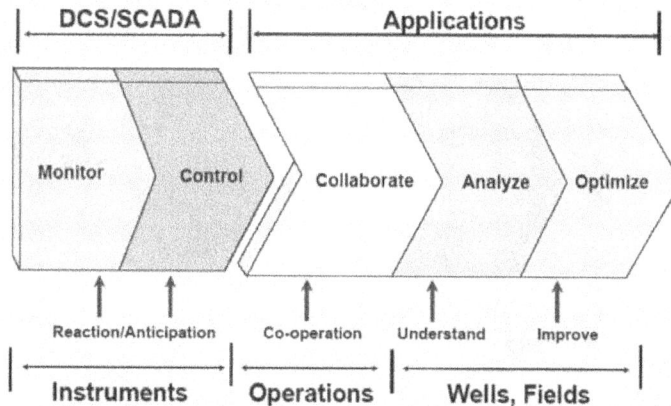

Fig. 8.6: Digital AL and SCADA System Process Cycle

This technology now is customized for artificial gas lift with a seemingly simple but technically advanced concept with implications for offshore artificial lift system. The actuator is coupled to a gas flow control valve to enable that valve to be opened or closed remotely, eliminating the need for side pocket mandrel units and wireline intervention. The series of digitally operated valves then enable the real-time setting of injection rates, with the technology able to be fitted straight into the tubing, so it can be deployed more flexibly—in fish-hook wells or where there are highly deviated sections, for example.

Offshore operators will be able to vary injection rates in real-time without wireline or slickline intervention, meaning no lost uptime, continuous production, and reduced risk. Furthermore, as opposed to conventional gas lift where the operator has no information about operating conditions at the point of injection, the live information allows operators to optimize extraction conditions, minimize gas usage across the reservoir, enhance oil recovery, and protect the wells from instability.

In short, digital artificial lift provides above ground control over the downhole gas lifting process and eliminates the need for well intervention. The ability to remotely alter conditions without intervention should enable operators to address the performance gap between topside and subsea wells. The ability to monitor remotely downhole conditions as well as to alter injection conditions without the need to touch the wellhead increases operational flexibility. Digital AL and SCADA system process cycle is illustrated in Figure 8.6.

8.5 System Architecture

Following Figure 8.7 is the architecture of SCADA with screen shot illustration.

- Well model (Well flow software)
- Trouble Shouting

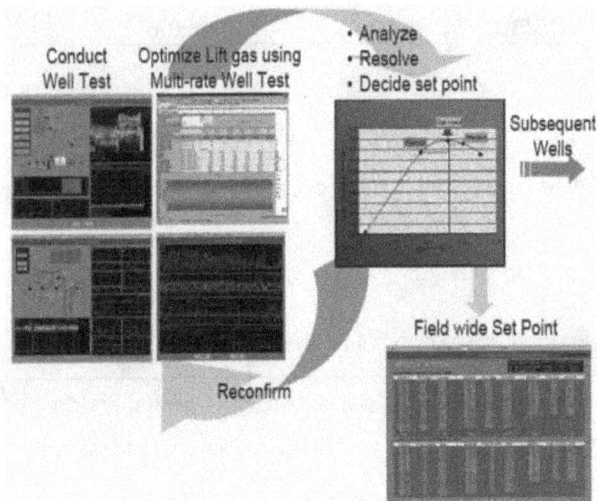

Fig. 8.7: Architecture of SCADA

- Gas separation and quality control
- Gas compression
- Gas distribution network (Pipsim or equivalent software)
- Gas metering Surveillance (Integrated Asst. Management software)
- Data access requirements such as gathering data simultaneously from multiple types of data sources
- Visualization/reporting requirements for time series charts (Real-time or historical) as well as design/analysis charts for gas lift diagnostic and performance plots
- Communication requirements for emails, no notifications, alarms and instant messages
- Integration requirements to work with web servers for posting reports, analytical applications by supporting various "protocols" (COM/DCOM, TCP/IP, API, etc.).

Surveillance of gas-lift wells were implemented in several offshore fields. Each platform can improve operations and increase production over other. In order to take advantage of any existing gas injection pressure at process platform, the real-time measurements from intelligent system and basic inferred data are made readily available through SCADA. Total gas lift pipe line network distribute gas to multi platforms through integrated asset management software of Schlumberger, it is a more complex Pipsim or equivalent simulation process in which well-by-well modelling, troubleshooting and optimization can be integrated with well flow software. Perhaps the most benefit will be gained by the results of the workflows implemented during this process. Once the well models are built and value is realized from well-by-well optimization, the next level of integrations is to model the field as a whole and incorporate any constraints in the gas injection or production system to improve the total fields efficiency and productivity. This approach, in fact, can easily be extended to other types of artificial lift wells. Further improvements such as implementation of artificial intelligence or expert systems, integration with reservoir modelling and real time operation can be rolled into the system at later stages. In offshore proper pipe line network planning for gas distribution is the ultimate success of a surveillance system.

8.6 ESP SCADA System

8.6.1 *Introduction*

A real time technology gains momentum in the oil industry, more wells are equipped with permanent gauges. SCADA systems and data historians also are becoming the norm as reduced operating cost and increased recovery factor value is demonstrated consistently. More than 11,000 ESPs have been fitted with gauges over the past six years and more than 3,000 of these wells have remote monitoring capability using a SCADA system.

With help of downhole measurements, it is possible to plot flow rate, flow pressures, and reservoir pressure accurately, which enable engineers to see production changes as low as approximately 10 bbls/day. This granularity cannot be achieved using monthly well tests, particularly when production fluctuates rapidly.

Basic function on which ESP-SCADA works is based on power absorbed by the pump is equal to that generated by the motor. On one hand, pump power is a function of its pressure differential, flow rate, and pump efficiency. On the other hand, motor power is a function of downhole voltage, current, power factor, and motor efficiency. Because equilibrium exists between both pump power and motor power, it is possible to solve for the unknown, which is the flow rate through the pump. With the ability to calculate instantaneous pump flow rate, engineers can use data to observe trends that are virtually invisible to those depending on quarterly or monthly well tests. Once calibrated, trends can be used to monitor well performance. Because they are taken in real time and at high resolution, they are valid in both transient and steady state conditions, unlike those of NODAL analysis, and also can capture transients associated with slugging effects. The qualitative nature of the trend is accurate, even if pump efficiency has been compromised through pump wear. Further, by using the value measured from the previous well test, trend data can be calibrated, providing instantaneous flow rate over the interval between tests.

8.6.2 *SCADA System Architecture*

The acronym SCADA stands for **S**upervisory **C**ontrol **A**nd **D**ata **A**cquisition. In reality, the primary purpose of SCADA is to monitor, control and alarm plant or regional operating systems from a central location. While override control is possible, it is infrequently utilized; however control set points are quite regularly changed by SCADA.

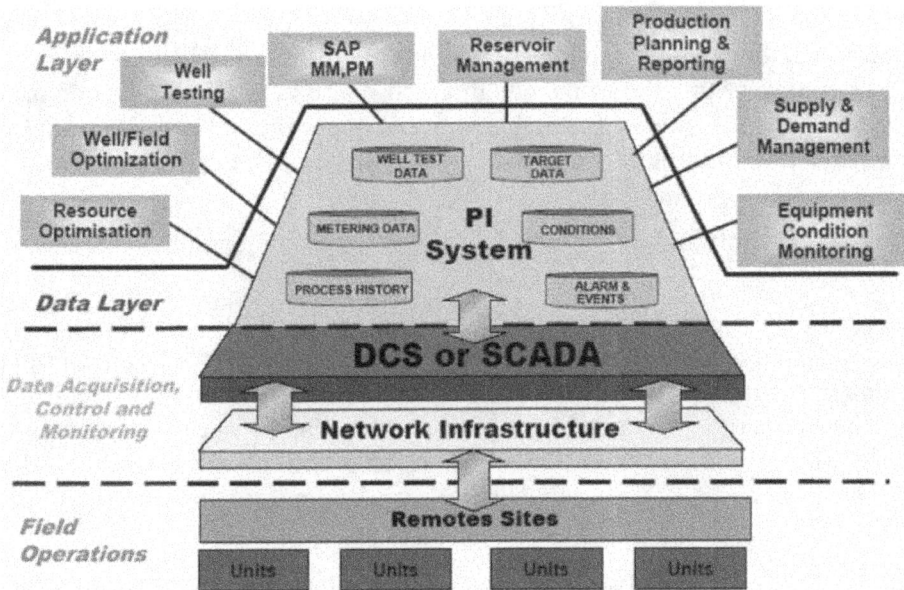

Fig. 8.8: SCADA System Architecture Application

There are three main elements to a SCADA system, various RTU's (Remote Telemetry Units), communications and an HMI (Human Machine Interface). Each RTU effectively collects information at a site, while communications bring that information from the various plant or regional RTU sites to a central location, and occasionally returns instructions to the RTU. The HMI displays this information in an easily understood graphics form, archives the data received, transmits alarms

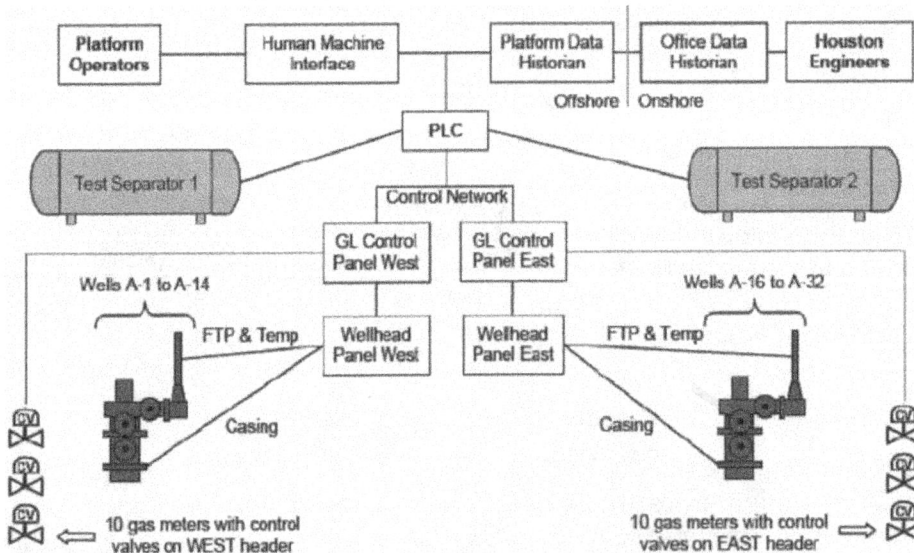

Fig. 8.9: Offshore Functions of SCADA System

and permits operator control as required communication within a plant will be by data cable, wire or fibre-optic, while regional systems most commonly utilize radio. The HMI is essentially a PC system running powerful graphic and alarm software programs.

There are five phases to creating a functional SCADA system:

Phase 1: The DESIGN of the system architecture. This includes the all-important communication system, and with a regional system utilizing radio communication often involves a radio path survey. Also involved will be any site instrumentation that is not presently in existence, but will be required to monitor desired parameters.

Phase 2: The SUPPLY of RTU, communication and HMI equipment, the latter consisting of a PC system and the necessary powerful graphic and alarm software programs.

Phase 3: The PROGRAMMING of the communication equipment and the powerful HMI graphic and alarm software programs.

Phase 4: The INSTALLATION of the communication equipment and the PC system. The former task is typically much more involved.

Phase 5: The COMMISSIONING of the system, during which communication and HMI programming problems are solved, the system is proven to the client, operator training and system documentation is provided.

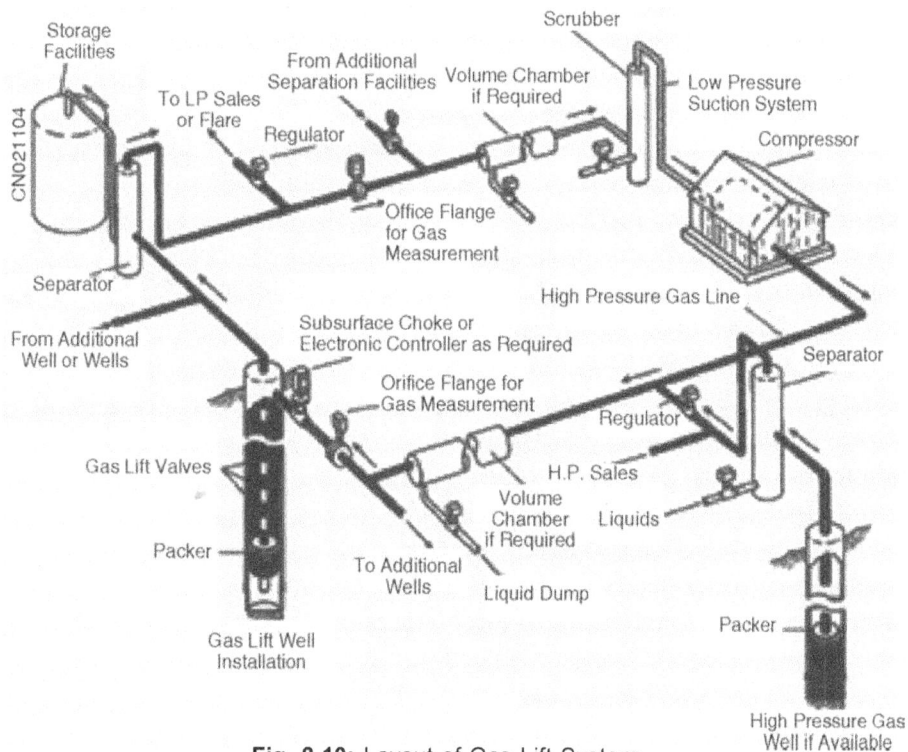

Fig. 8.10: Layout of Gas Lift System

The ESP-SCADA system consists of a universal site controller which can monitor multiple well and operating data points—electrical system data, information from external analog or digital devices, data measured by the downhole monitoring system, and remote commands. Users can program alarm and trip settings locally or remotely. In its basic configuration, the device is a fixed-speed motor controller and data acquisition device. It can be easily adapted for use as a data acquisition and communication hub or VSD controller by adding as many as four expansion cards. Each expansion card is plug-and-play with card control menus automatically added to the VSD front panel screen. Firmware upgrades are typically performed at the wellsite using a laptop computer connected to the serial port on the front of the controller. Typically the upgrades are completed in less than 1 minute. The ESP SCADA device can accommodate up to four analog and six digital input channels and provide two analog and four digital output channels, each individually configurable. For remote monitoring and control, the ESP-SCADA system can be connected to other SCADA system and/or the ESP watcher surveillance and control system—in parallel, if required.

8.6.3 *Advantage of Process Automation*

The main objectives of a automation system are to increase the well intervention time for offshore wells. Loses collaboration one of the means of reaching the strategic objectives of the real-time gas lift surveillance system is to improve the capability to predict the performance of the gas lift system. Benefits of automated SCADA system as follow:

- Gains in Production
- Collaboration between Asset Planning, Well Surveillance and Operations (because they use the same data and system accessible from Desktop)
- Capture and keep the history of the lift cycle of each Wells (CHP, FLP, FTHP, GLI, etc.)—to be used for analysis
- Well-head and Well Test data is all online and easy to access. (E.g. Faster turn-around of test results)
- Historical data to perform analysis, modelling and perform optimization.

Automation industry has brought significant step forward for Gas Lift Optimization in offshore production process and facilities. This process involves some major CAPEX however the Return on Investment (ROI) varies depending on the level of implementation (virtual metering only, abnormalcy monitoring, production optimization, etc.) and varies based on particular situation. Even in recent years some small producers are being benefited in production increase as high as two fold. In verdict the gas lift optimization benefits are proven, as they are the largest and most tangible when it comes to spending v/s revenue.

Appendix A

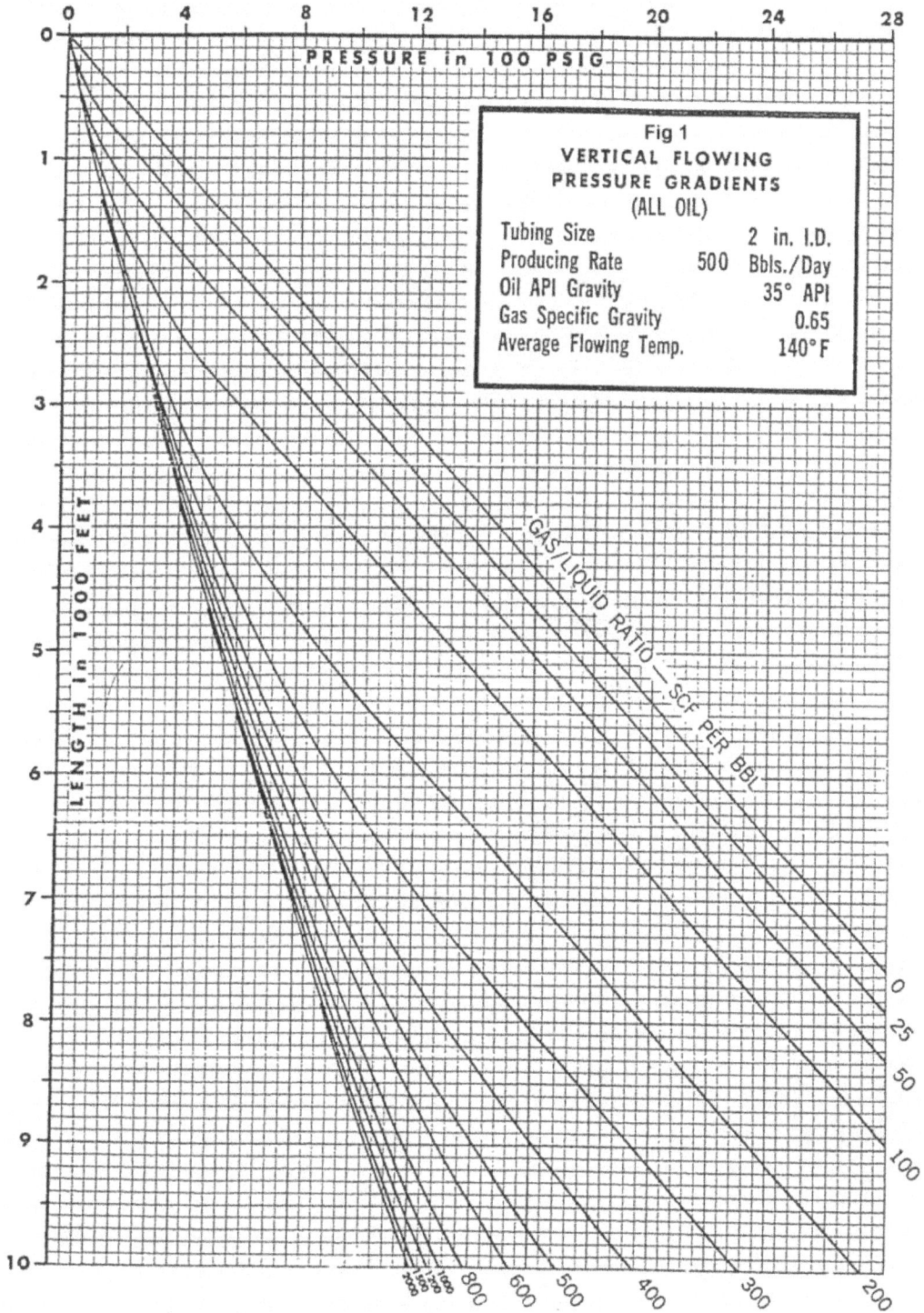

Fig 1
VERTICAL FLOWING
PRESSURE GRADIENTS
(ALL OIL)

Tubing Size		2 in. I.D.
Producing Rate	500	Bbls./Day
Oil API Gravity		35° API
Gas Specific Gravity		0.65
Average Flowing Temp.		140°F

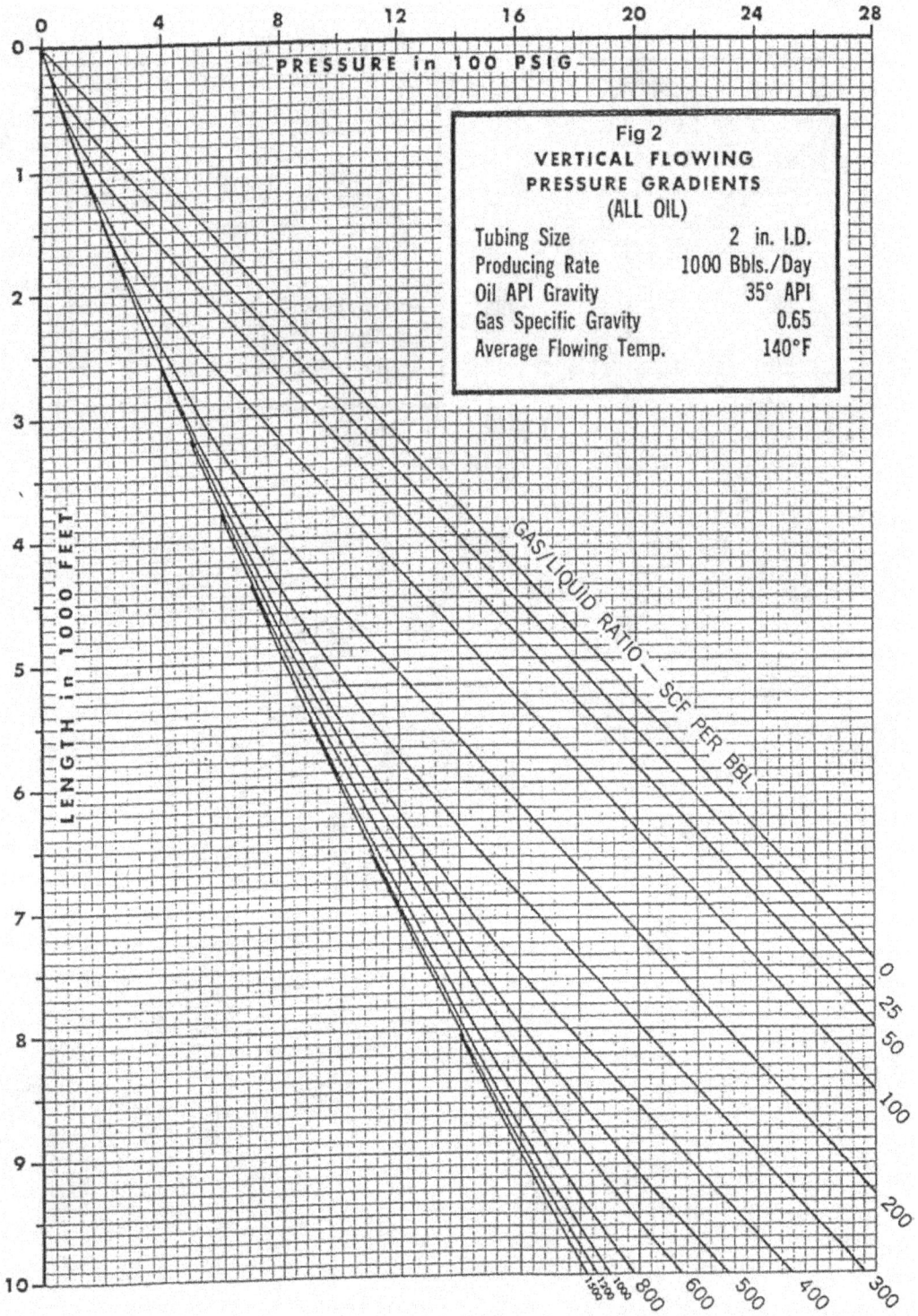

Fig 2
VERTICAL FLOWING
PRESSURE GRADIENTS
(ALL OIL)

Tubing Size	2 in. I.D.
Producing Rate	1000 Bbls./Day
Oil API Gravity	35° API
Gas Specific Gravity	0.65
Average Flowing Temp.	140°F

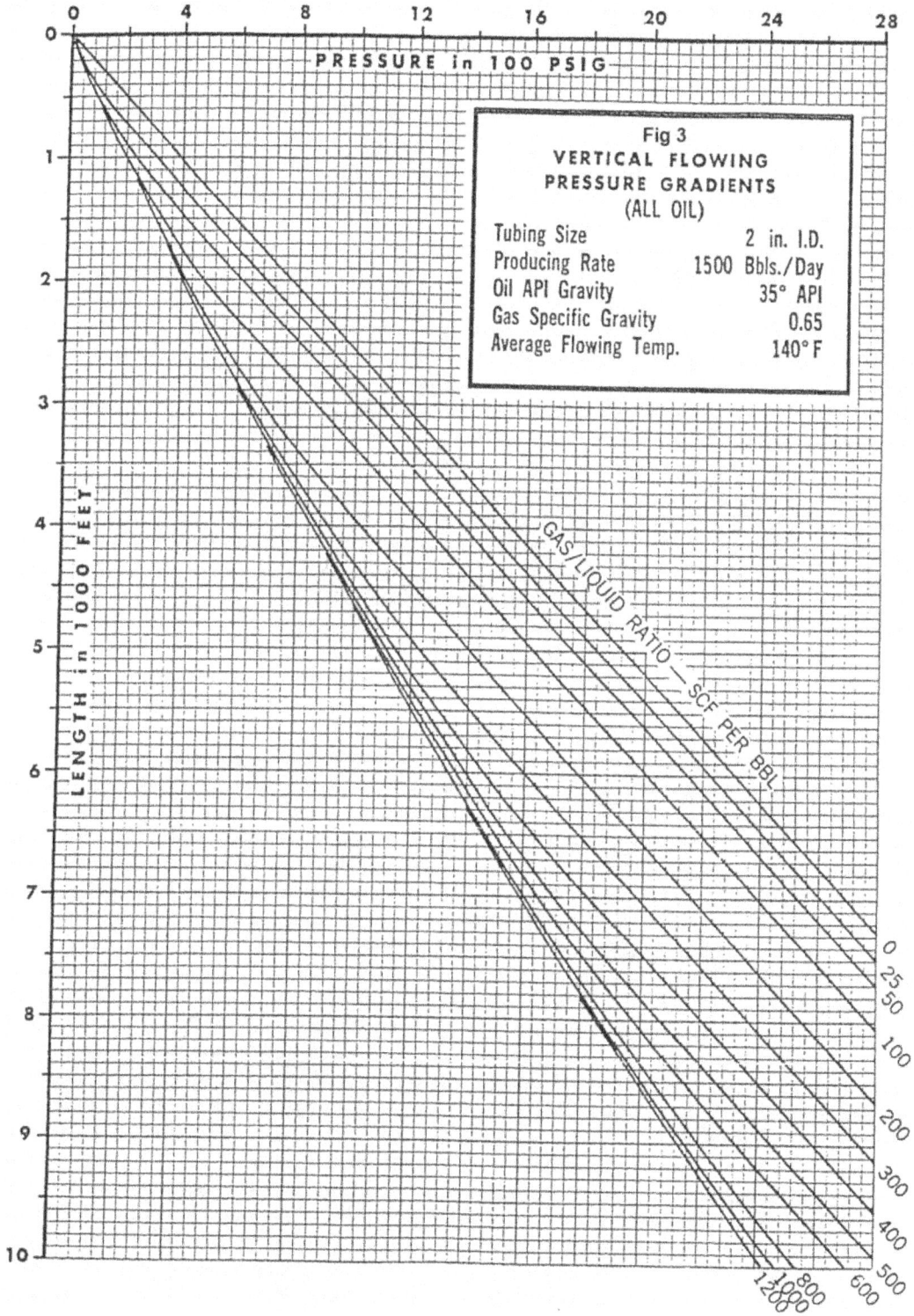

Fig 3
VERTICAL FLOWING
PRESSURE GRADIENTS
(ALL OIL)

Tubing Size	2 in. I.D.
Producing Rate	1500 Bbls./Day
Oil API Gravity	35° API
Gas Specific Gravity	0.65
Average Flowing Temp.	140° F

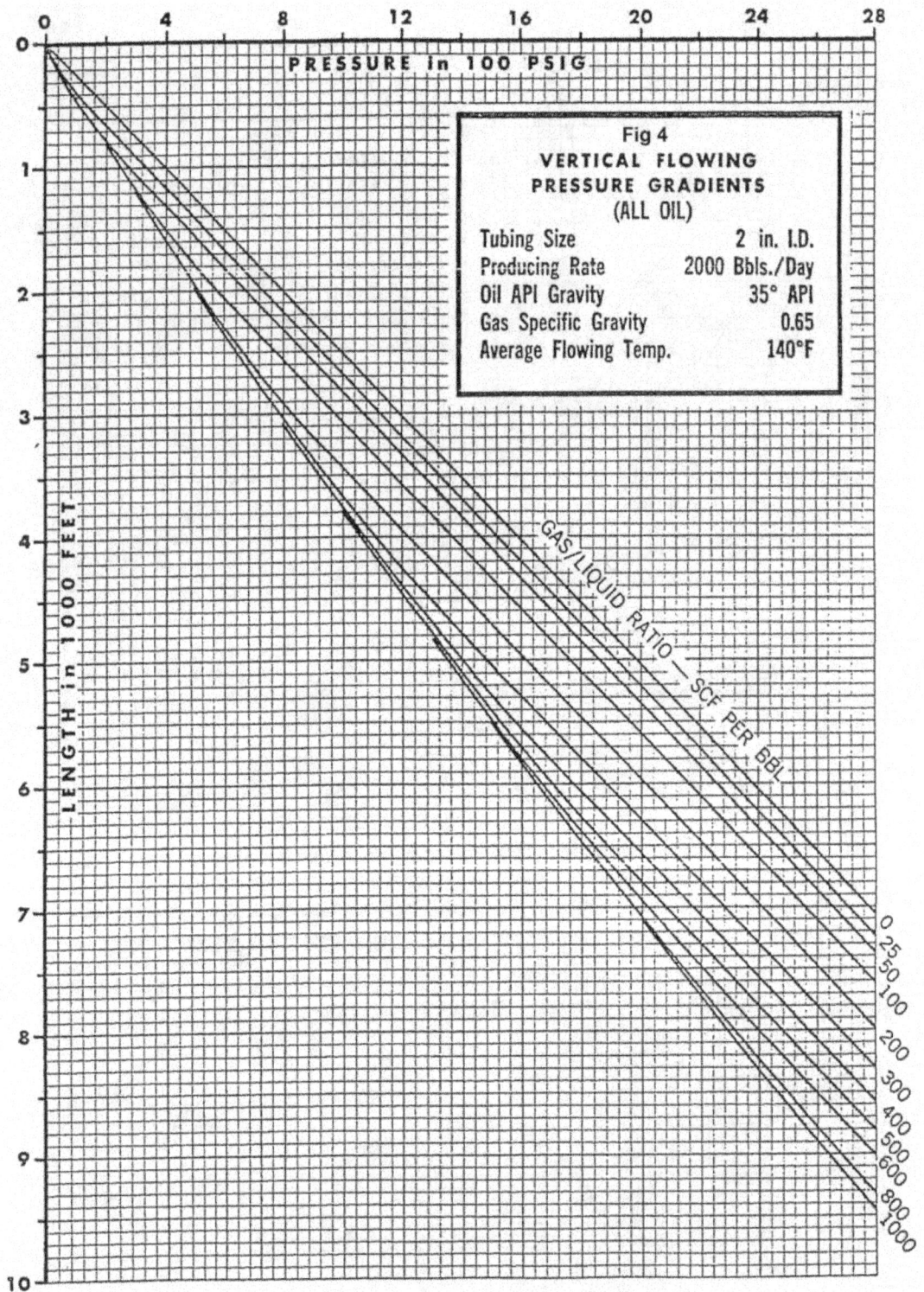

Fig 4
VERTICAL FLOWING
PRESSURE GRADIENTS
(ALL OIL)

Tubing Size	2 in. I.D.
Producing Rate	2000 Bbls./Day
Oil API Gravity	35° API
Gas Specific Gravity	0.65
Average Flowing Temp.	140°F

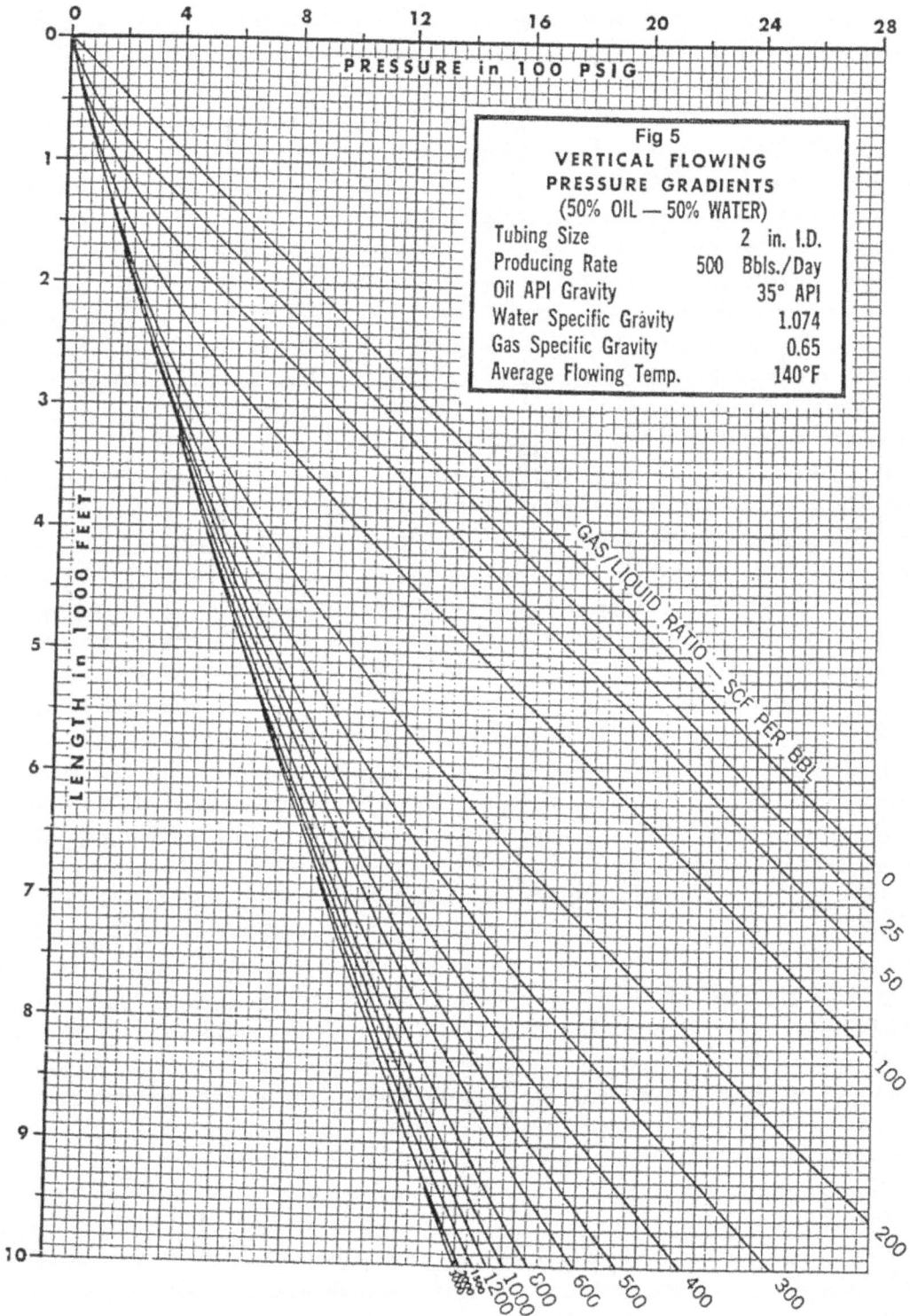

Fig 5
VERTICAL FLOWING
PRESSURE GRADIENTS
(50% OIL — 50% WATER)

Tubing Size	2 in. I.D.
Producing Rate	500 Bbls./Day
Oil API Gravity	35° API
Water Specific Gravity	1.074
Gas Specific Gravity	0.65
Average Flowing Temp.	140°F

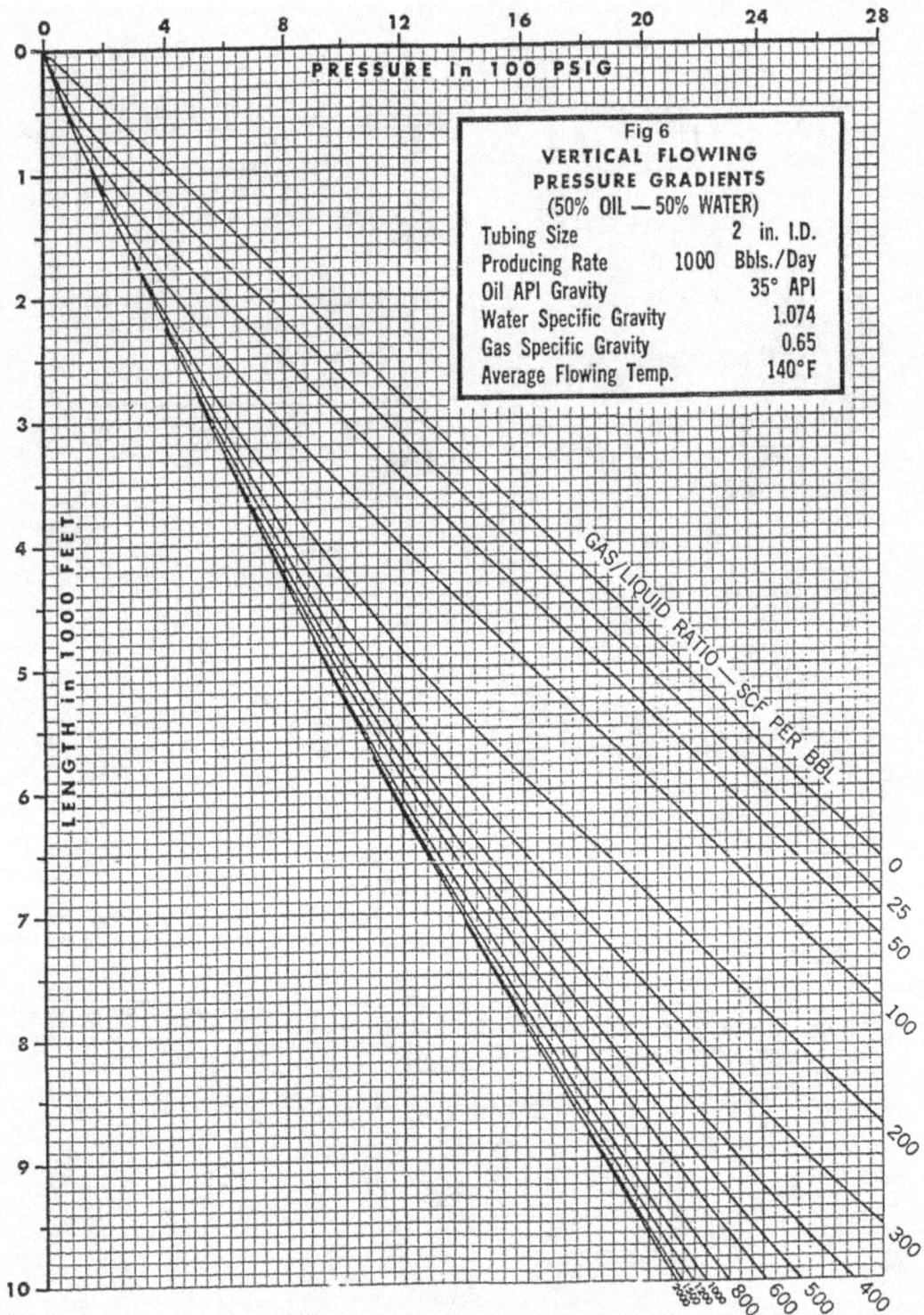

Fig 6
VERTICAL FLOWING
PRESSURE GRADIENTS
(50% OIL — 50% WATER)

Tubing Size	2 in. I.D.
Producing Rate	1000 Bbls./Day
Oil API Gravity	35° API
Water Specific Gravity	1.074
Gas Specific Gravity	0.65
Average Flowing Temp.	140°F

PRESSURE in 100 PSIG

LENGTH in 1000 FEET

GAS/LIQUID RATIO — SCF PER BBL

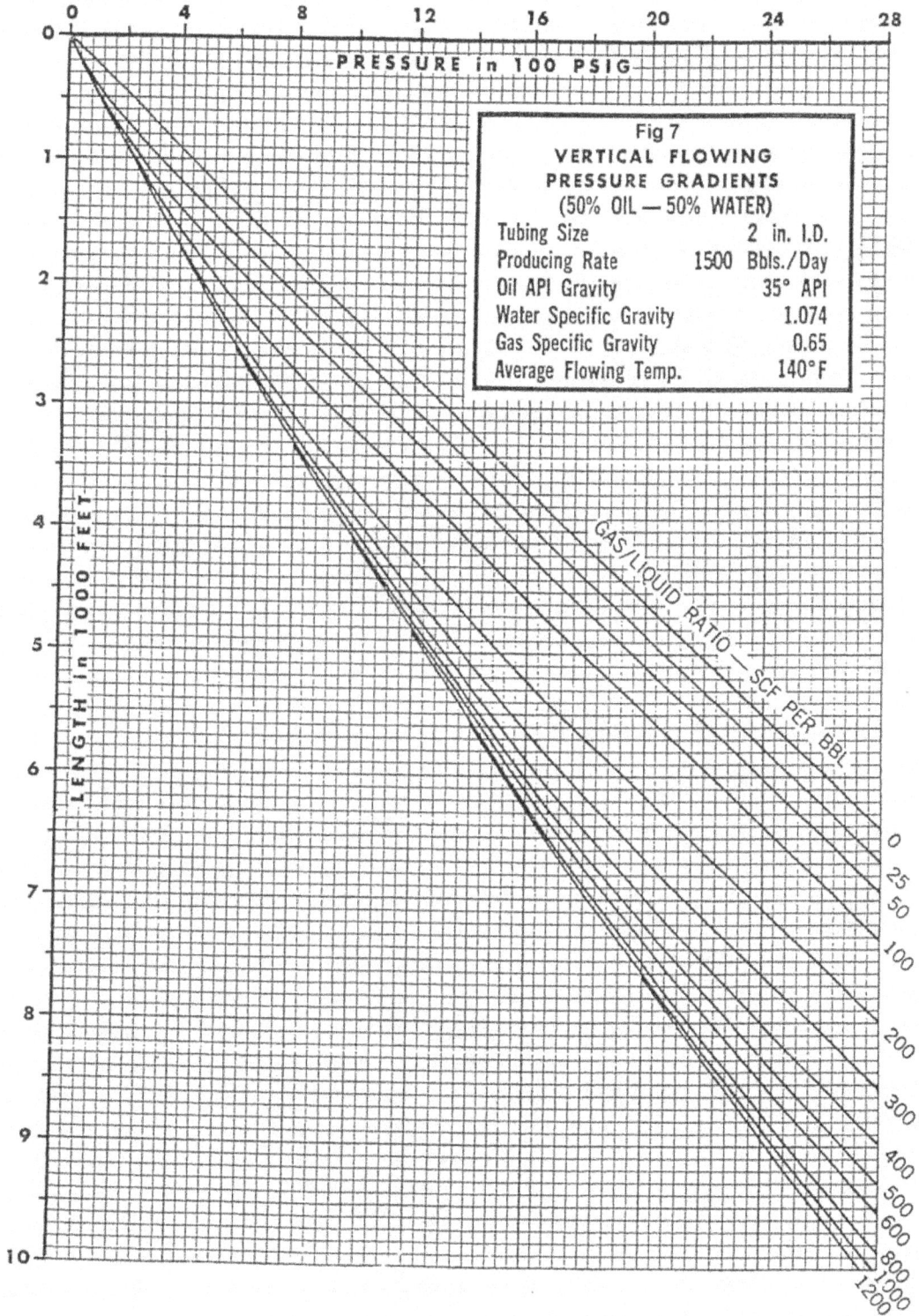

Fig 7
VERTICAL FLOWING
PRESSURE GRADIENTS
(50% OIL — 50% WATER)

Tubing Size	2 in. I.D.
Producing Rate	1500 Bbls./Day
Oil API Gravity	35° API
Water Specific Gravity	1.074
Gas Specific Gravity	0.65
Average Flowing Temp.	140°F

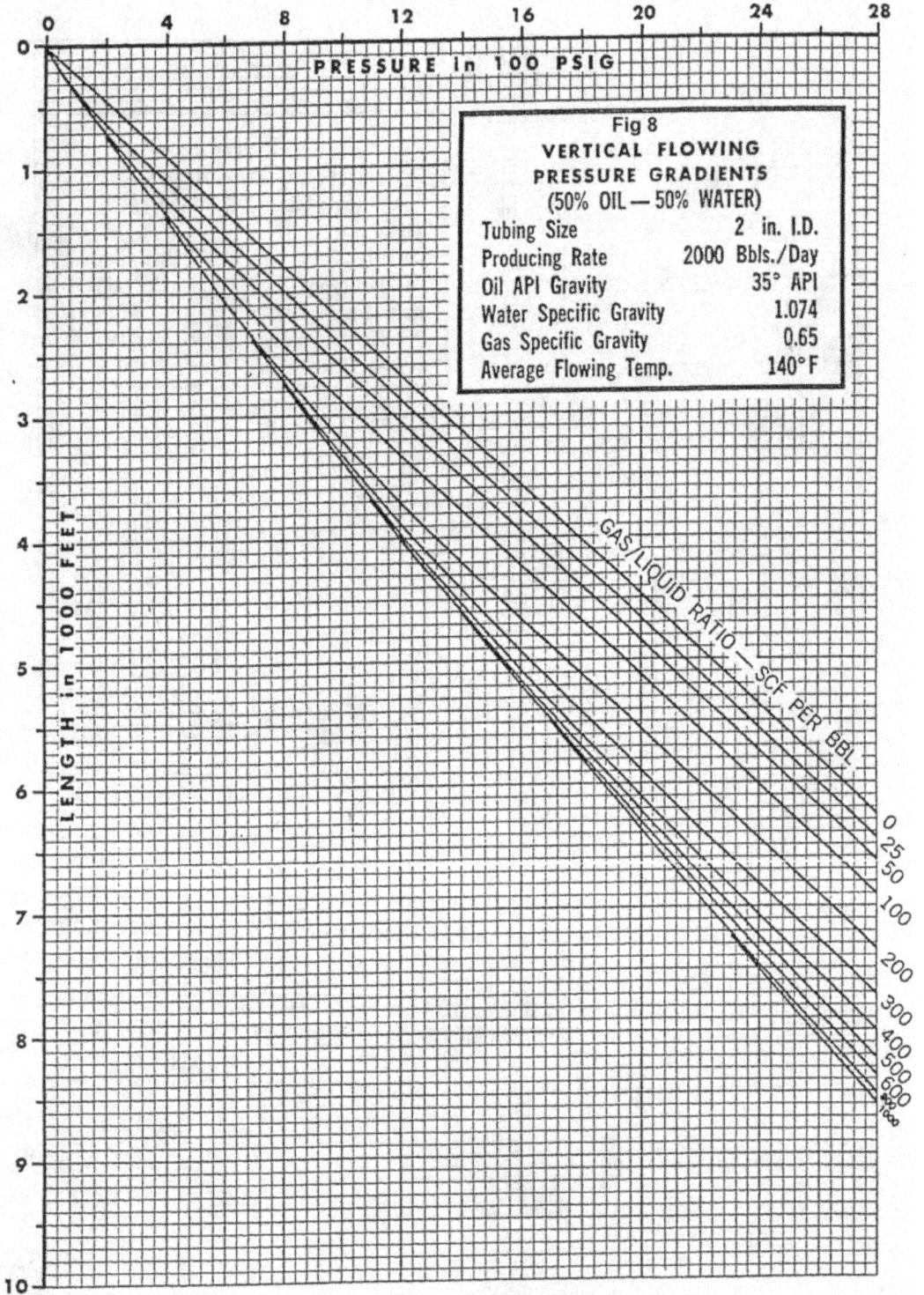

Fig 8
VERTICAL FLOWING
PRESSURE GRADIENTS
(50% OIL — 50% WATER)

Tubing Size	2 in. I.D.
Producing Rate	2000 Bbls./Day
Oil API Gravity	35° API
Water Specific Gravity	1.074
Gas Specific Gravity	0.65
Average Flowing Temp.	140° F

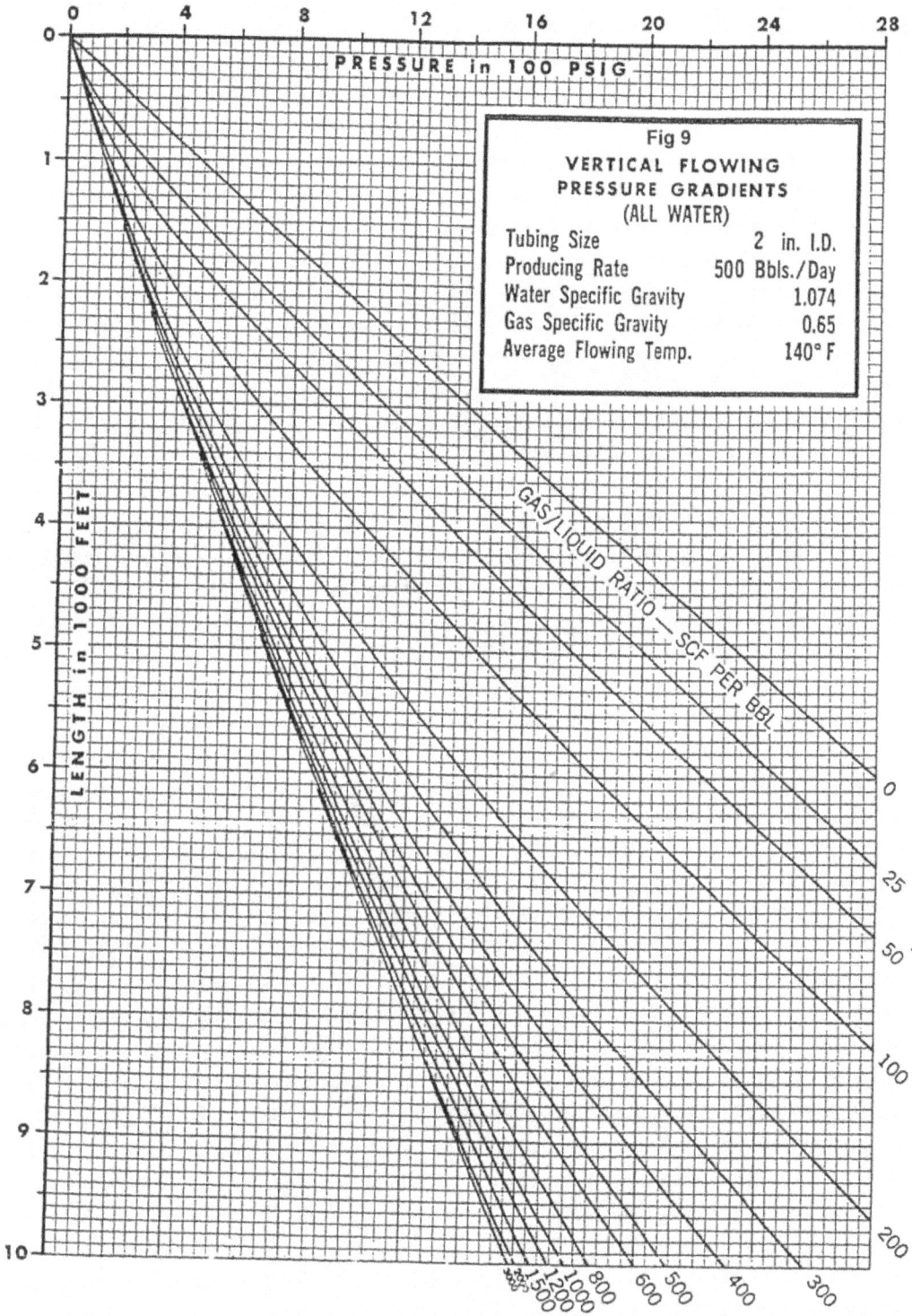

Fig 9
VERTICAL FLOWING
PRESSURE GRADIENTS
(ALL WATER)

Tubing Size	2 in. I.D.
Producing Rate	500 Bbls./Day
Water Specific Gravity	1.074
Gas Specific Gravity	0.65
Average Flowing Temp.	140° F

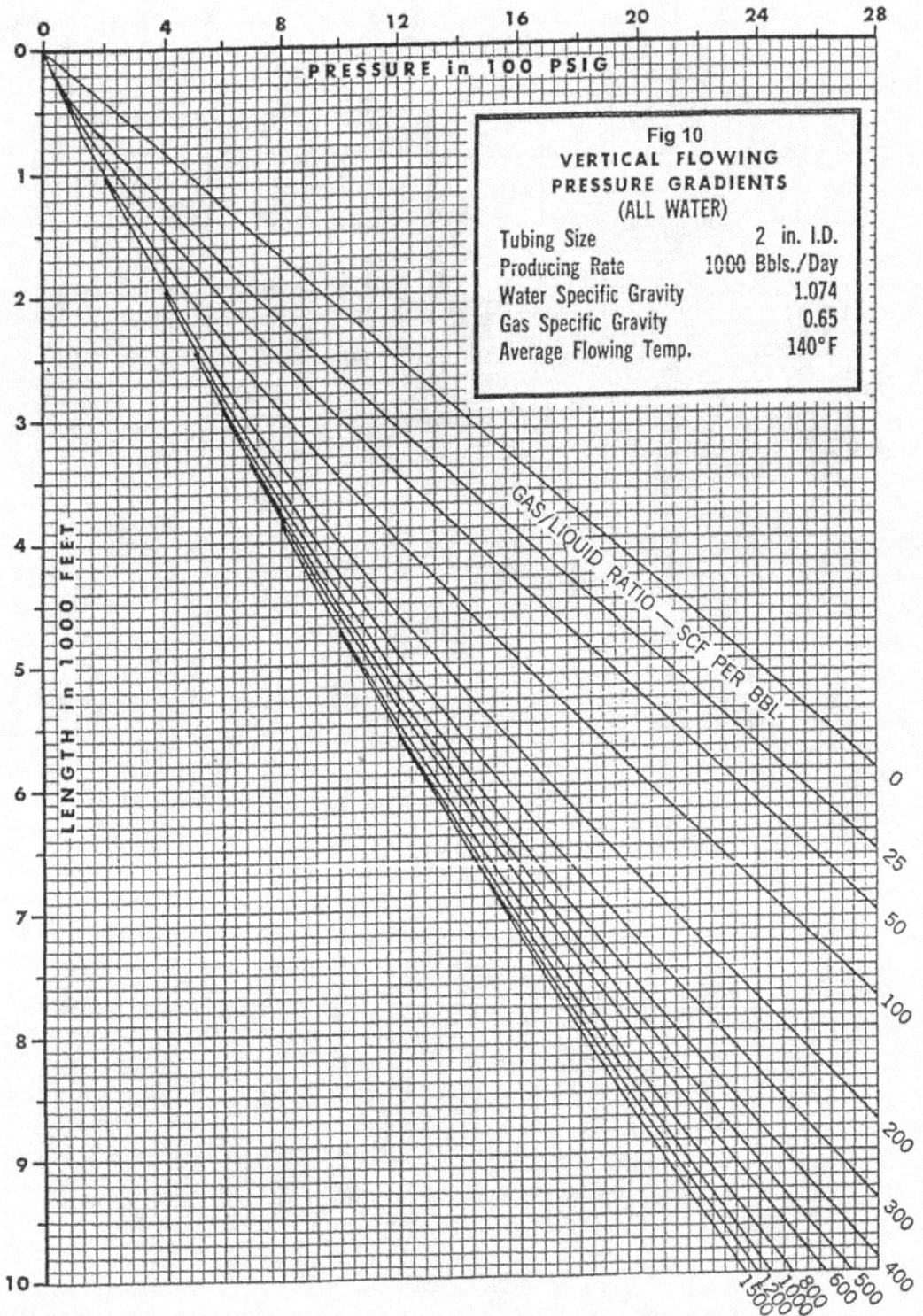

Fig 10
VERTICAL FLOWING
PRESSURE GRADIENTS
(ALL WATER)

Tubing Size	2 in. I.D.
Producing Rate	1000 Bbls./Day
Water Specific Gravity	1.074
Gas Specific Gravity	0.65
Average Flowing Temp.	140°F

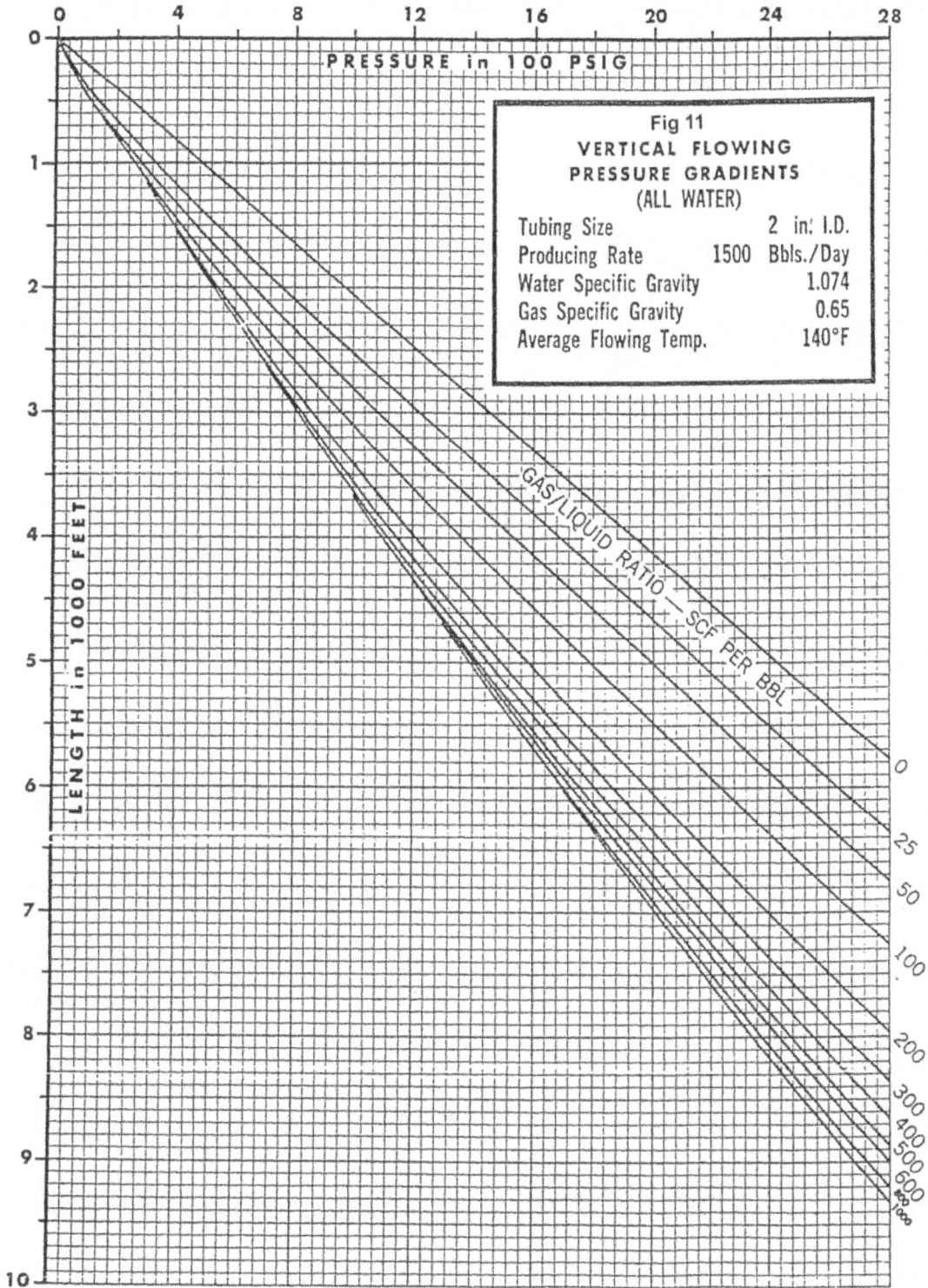

Fig 11
VERTICAL FLOWING
PRESSURE GRADIENTS
(ALL WATER)

Tubing Size	2 in. I.D.
Producing Rate	1500 Bbls./Day
Water Specific Gravity	1.074
Gas Specific Gravity	0.65
Average Flowing Temp.	140°F

PRESSURE in 100 PSIG

LENGTH in 1000 FEET

GAS/LIQUID RATIO — SCF PER BBL

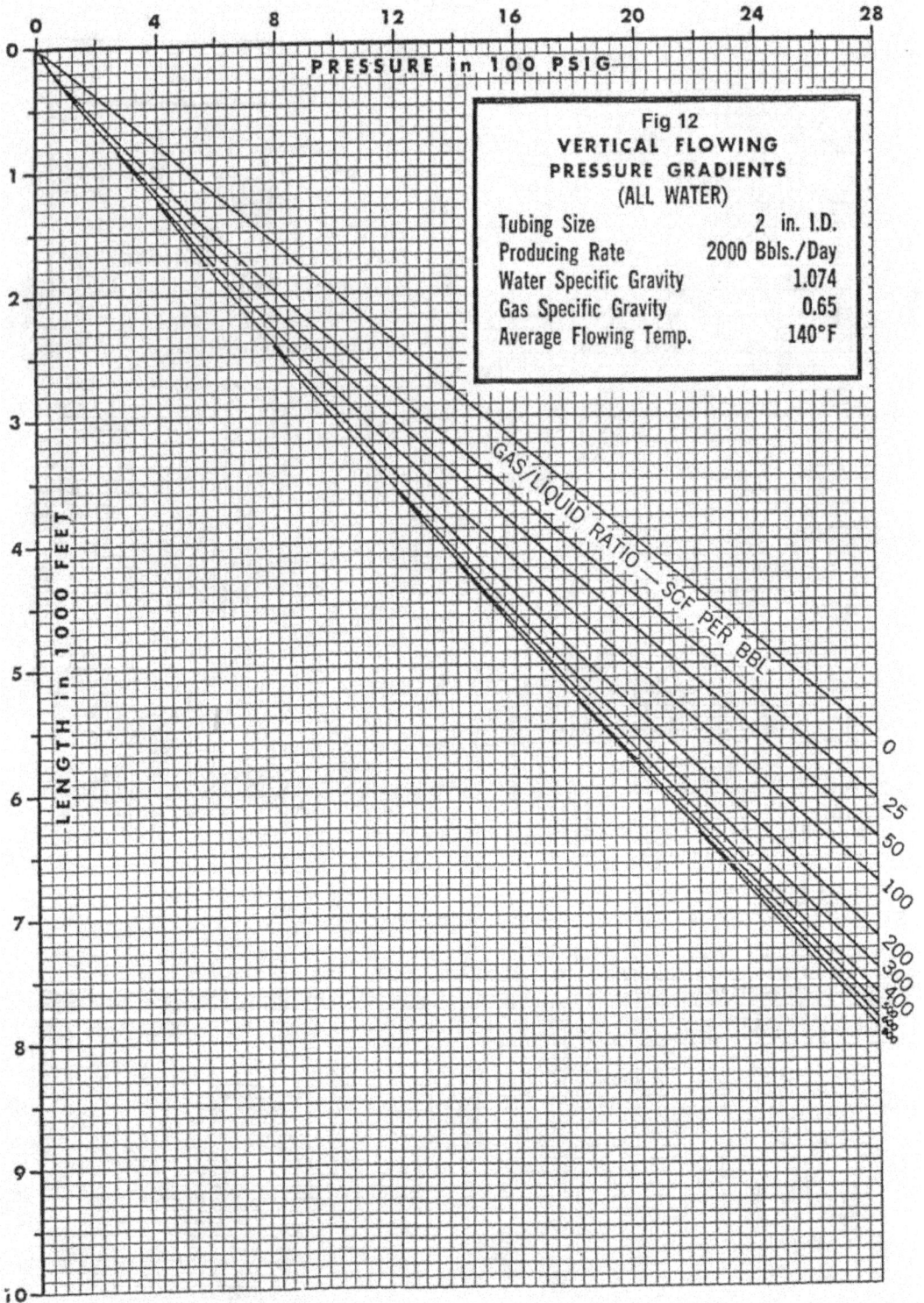

Fig 12
VERTICAL FLOWING
PRESSURE GRADIENTS
(ALL WATER)

Tubing Size	2 in. I.D.
Producing Rate	2000 Bbls./Day
Water Specific Gravity	1.074
Gas Specific Gravity	0.65
Average Flowing Temp.	140°F

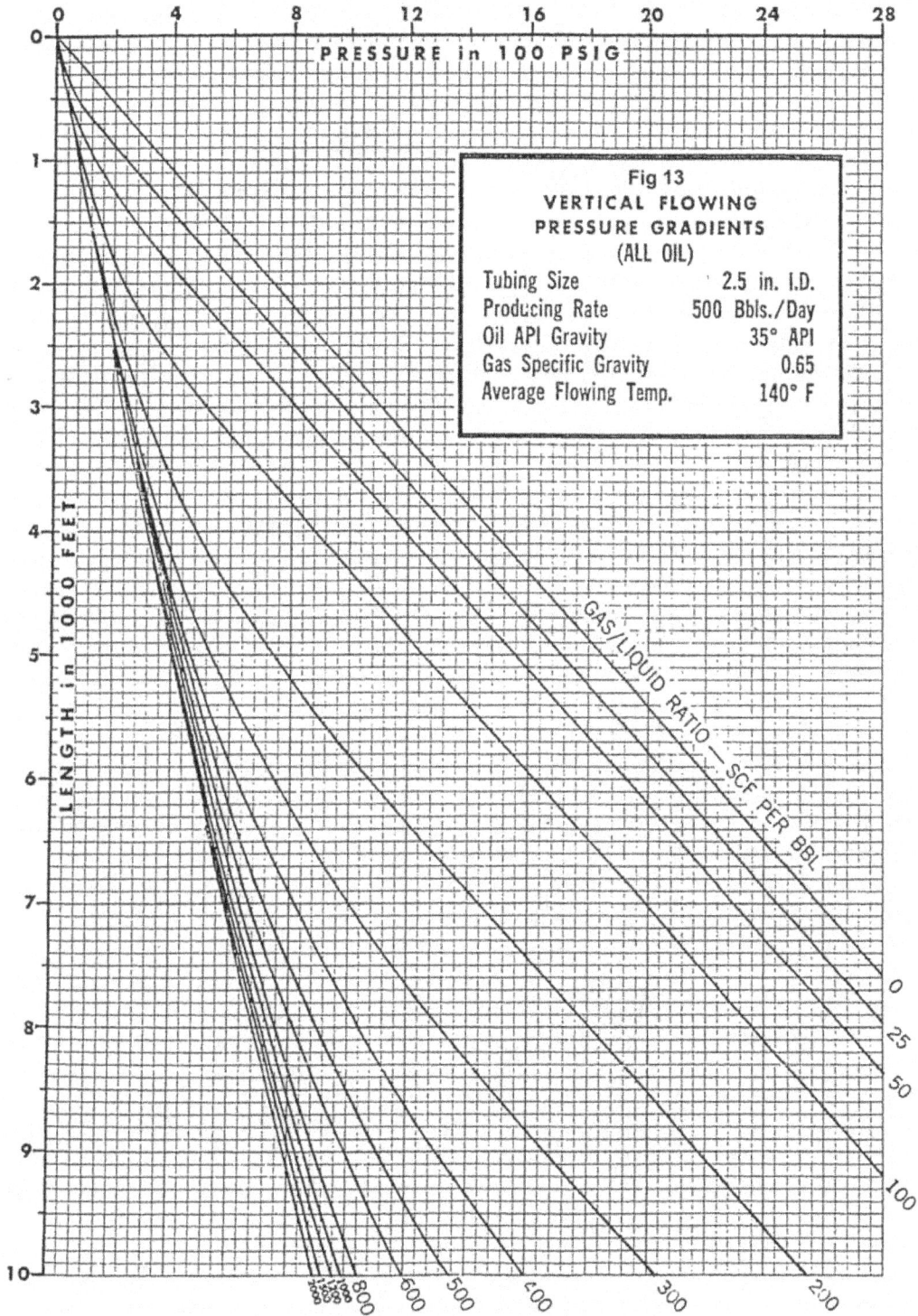

Fig 13
VERTICAL FLOWING
PRESSURE GRADIENTS
(ALL OIL)

Tubing Size	2.5 in. I.D.
Producing Rate	500 Bbls./Day
Oil API Gravity	35° API
Gas Specific Gravity	0.65
Average Flowing Temp.	140° F

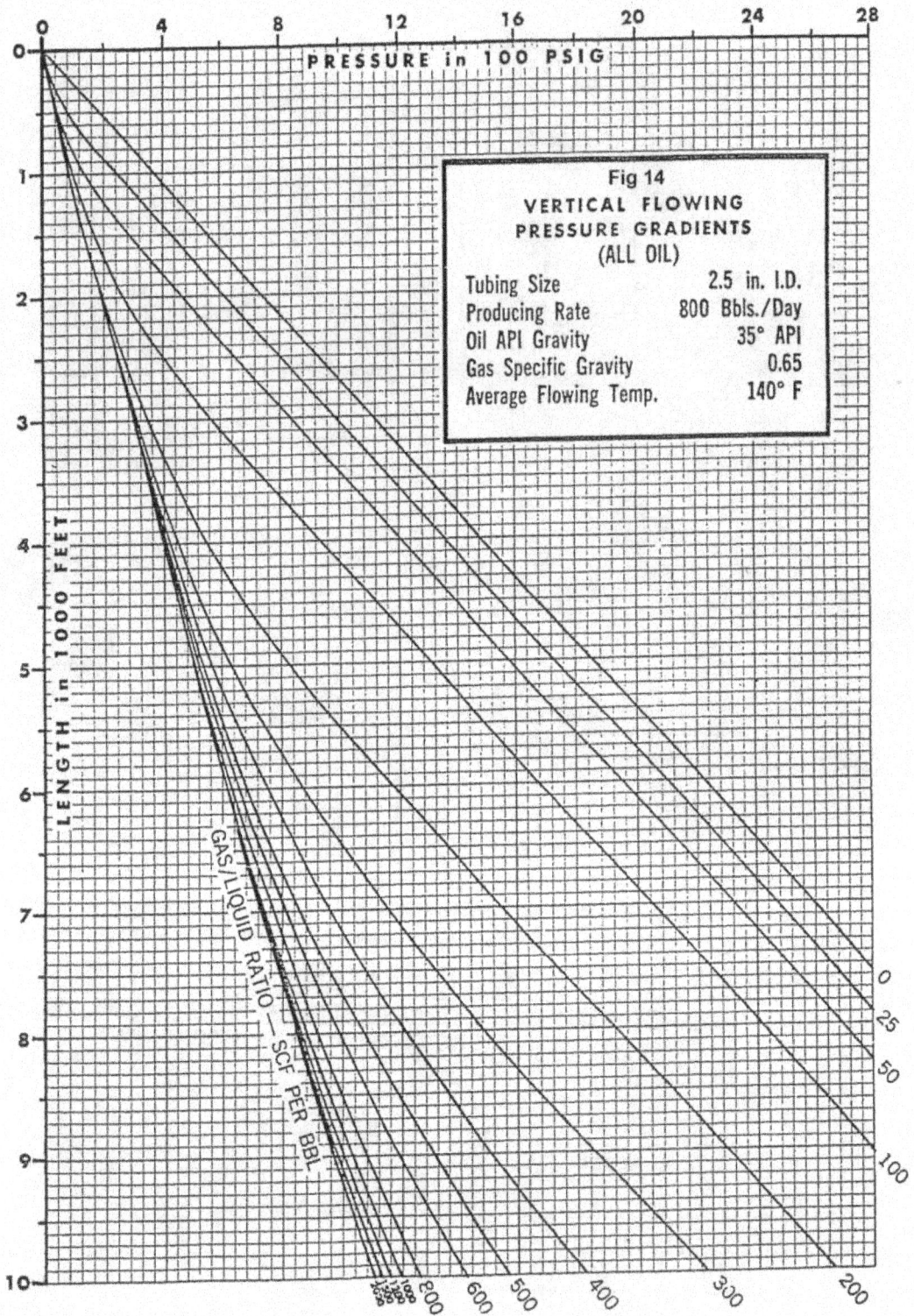

Fig 14
VERTICAL FLOWING
PRESSURE GRADIENTS
(ALL OIL)

Tubing Size	2.5 in. I.D.
Producing Rate	800 Bbls./Day
Oil API Gravity	35° API
Gas Specific Gravity	0.65
Average Flowing Temp.	140° F

PRESSURE in 100 PSIG

LENGTH in 1000 FEET

GAS/LIQUID RATIO — SCF PER BBL

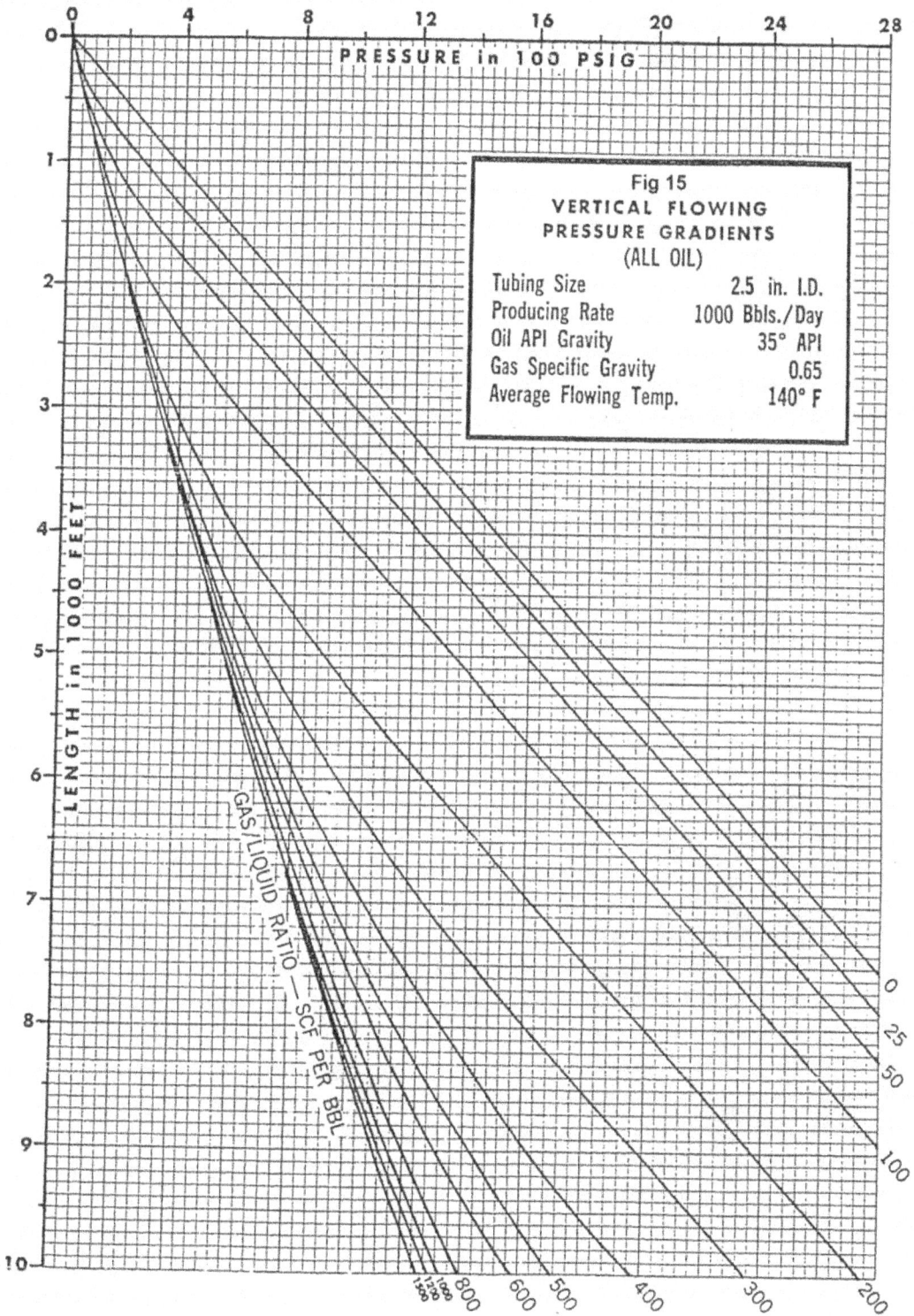

Fig 15
VERTICAL FLOWING
PRESSURE GRADIENTS
(ALL OIL)

Tubing Size	2.5 in. I.D.
Producing Rate	1000 Bbls./Day
Oil API Gravity	35° API
Gas Specific Gravity	0.65
Average Flowing Temp.	140° F

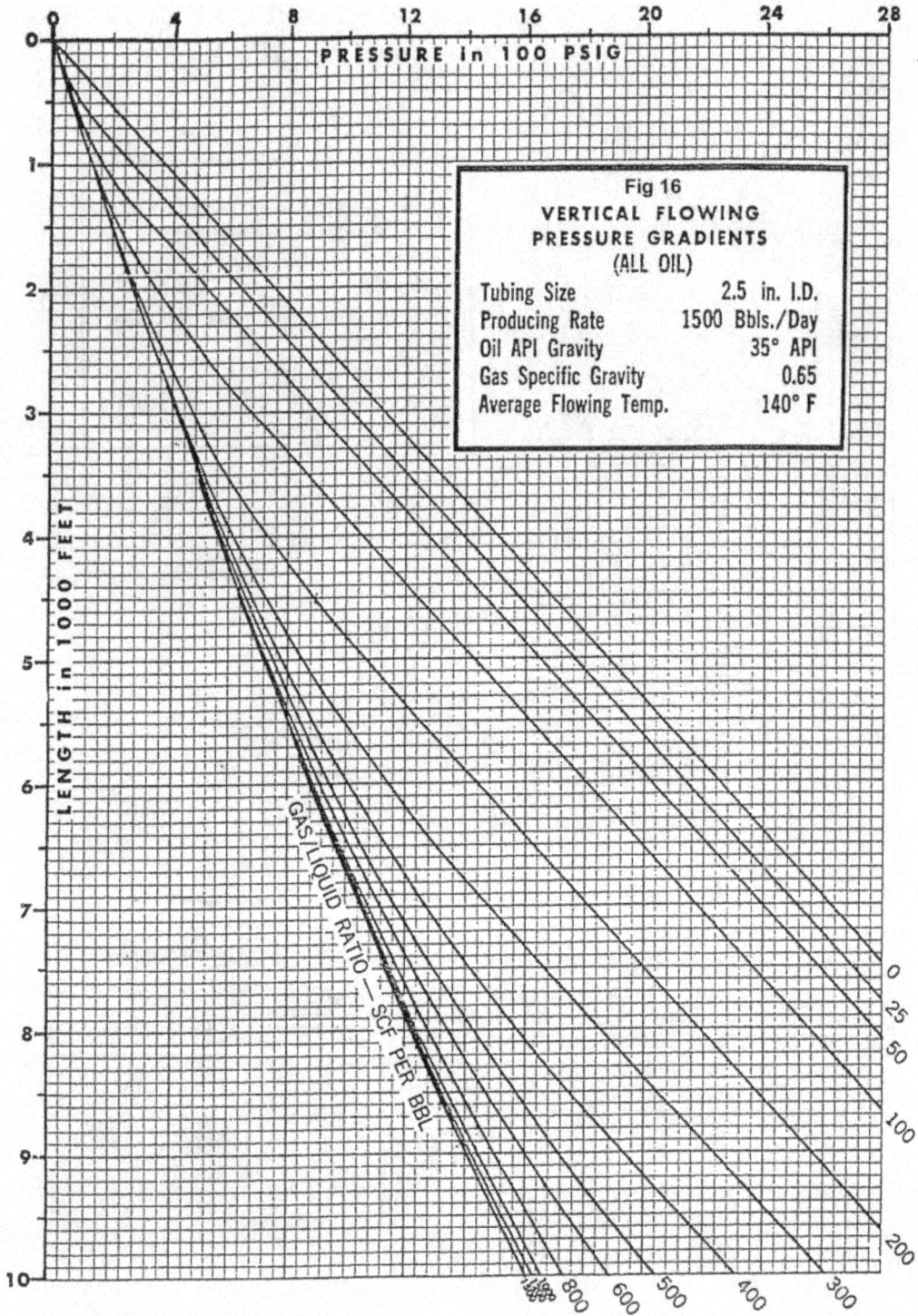

Fig 16
VERTICAL FLOWING
PRESSURE GRADIENTS
(ALL OIL)

Tubing Size	2.5 in. I.D.
Producing Rate	1500 Bbls./Day
Oil API Gravity	35° API
Gas Specific Gravity	0.65
Average Flowing Temp.	140° F

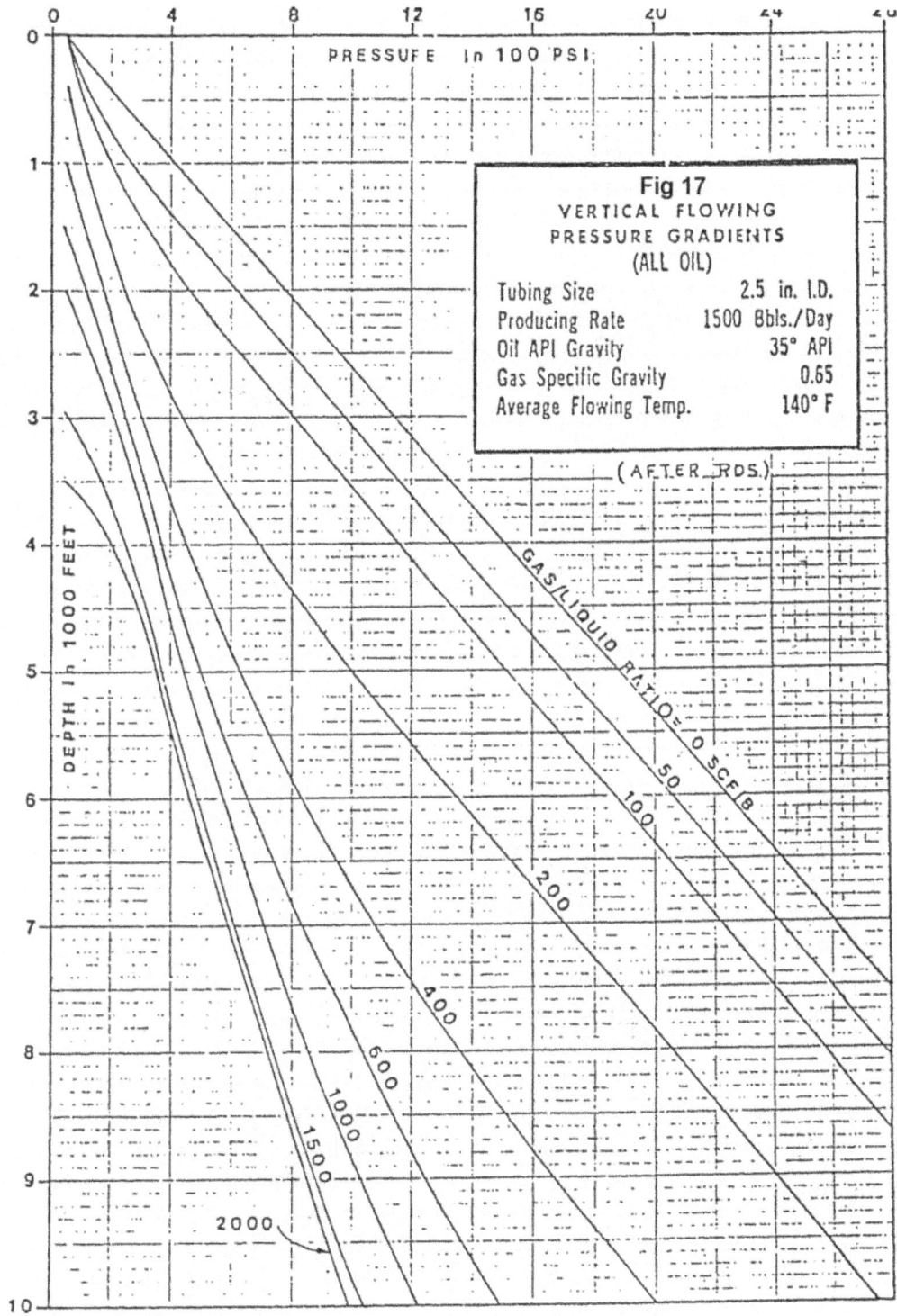

Fig 17
VERTICAL FLOWING
PRESSURE GRADIENTS
(ALL OIL)

Tubing Size	2.5 in. I.D.
Producing Rate	1500 Bbls./Day
Oil API Gravity	35° API
Gas Specific Gravity	0.65
Average Flowing Temp.	140° F

(AFTER RDS.)

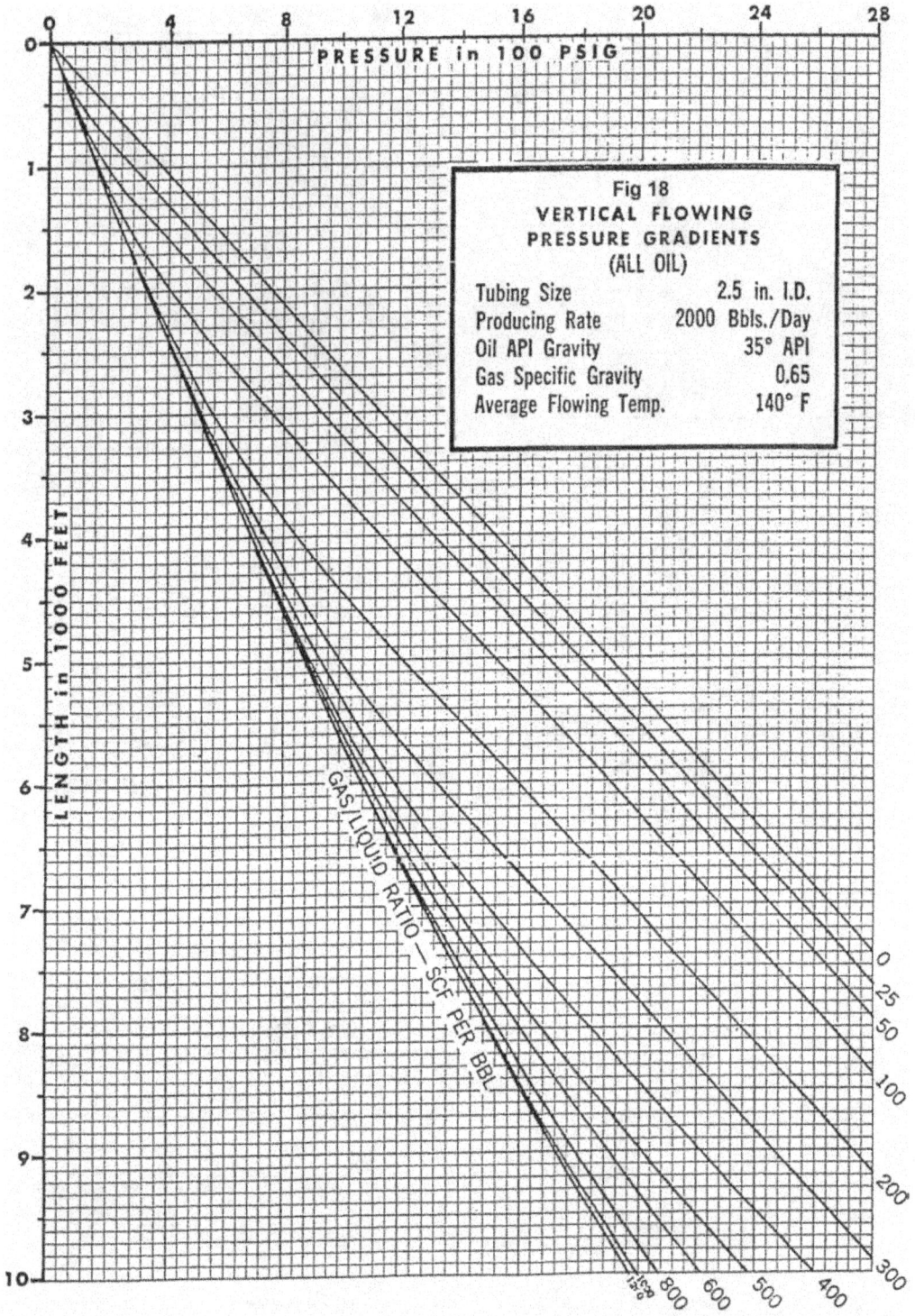

Fig 18
VERTICAL FLOWING
PRESSURE GRADIENTS
(ALL OIL)

Tubing Size	2.5 in. I.D.
Producing Rate	2000 Bbls./Day
Oil API Gravity	35° API
Gas Specific Gravity	0.65
Average Flowing Temp.	140° F

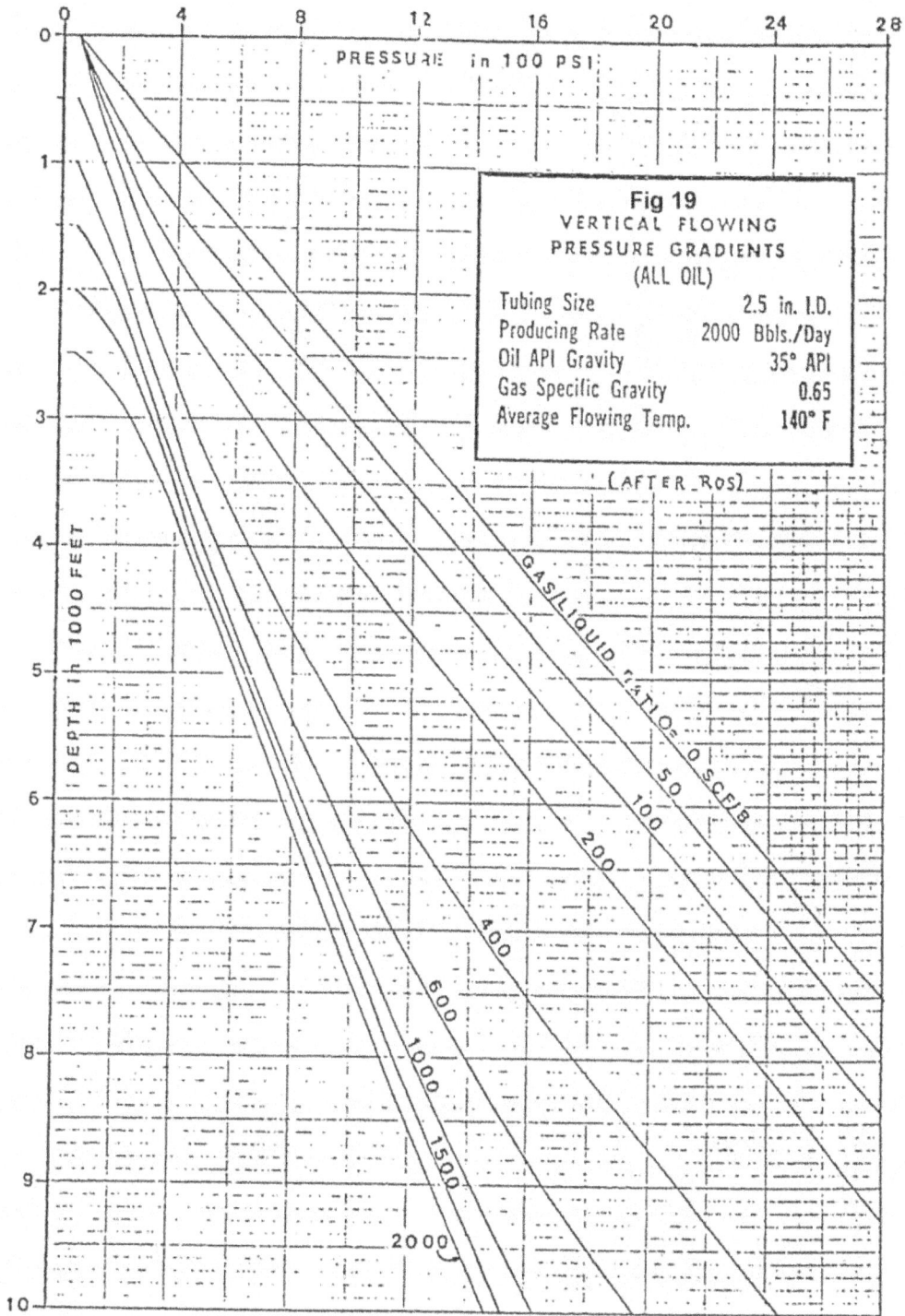

Fig 19
VERTICAL FLOWING
PRESSURE GRADIENTS
(ALL OIL)

Tubing Size	2.5 in. I.D.
Producing Rate	2000 Bbls./Day
Oil API Gravity	35° API
Gas Specific Gravity	0.65
Average Flowing Temp.	140° F

(AFTER ROS)

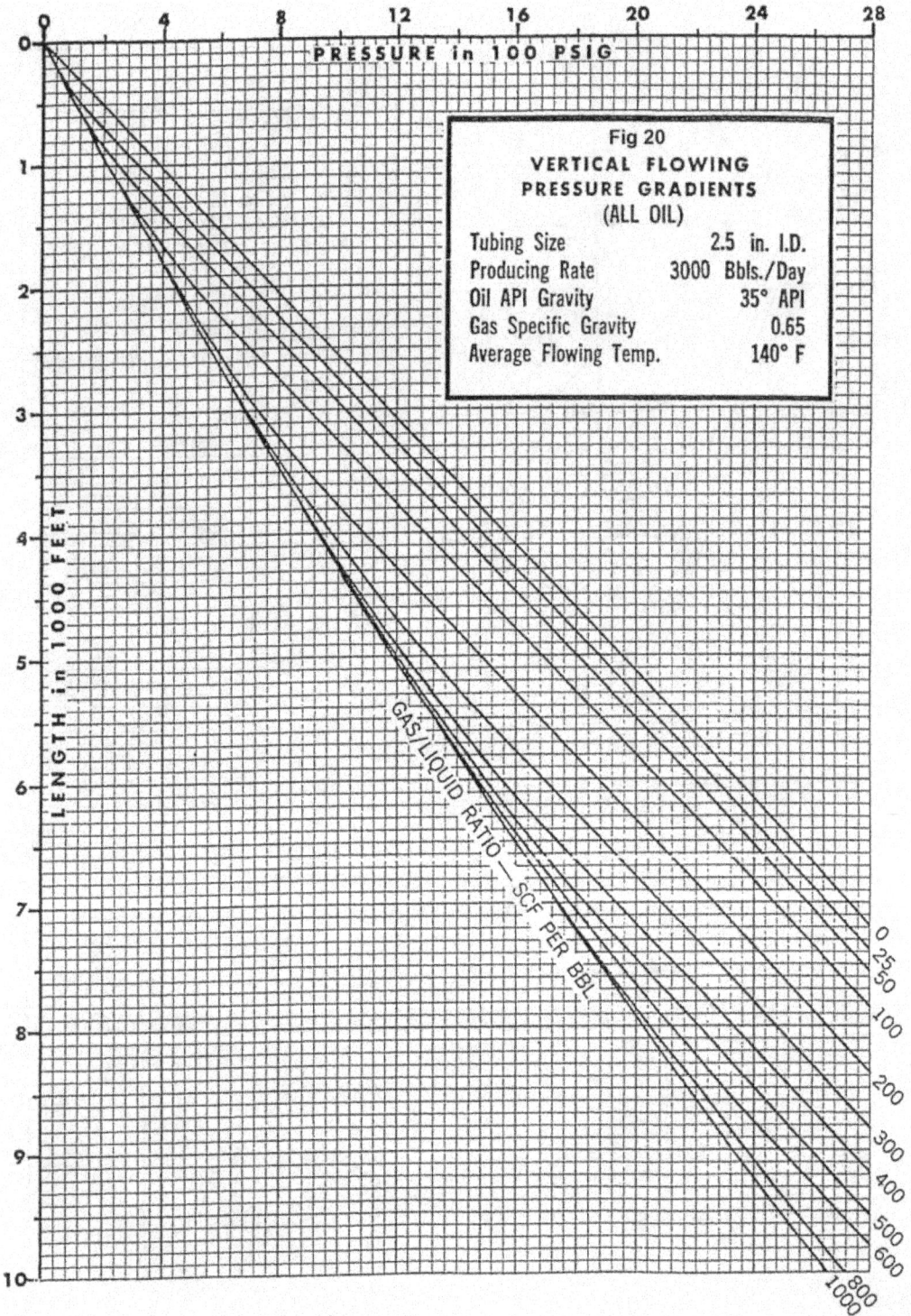

Fig 20
VERTICAL FLOWING
PRESSURE GRADIENTS
(ALL OIL)

Tubing Size	2.5 in. I.D.
Producing Rate	3000 Bbls./Day
Oil API Gravity	35° API
Gas Specific Gravity	0.65
Average Flowing Temp.	140° F

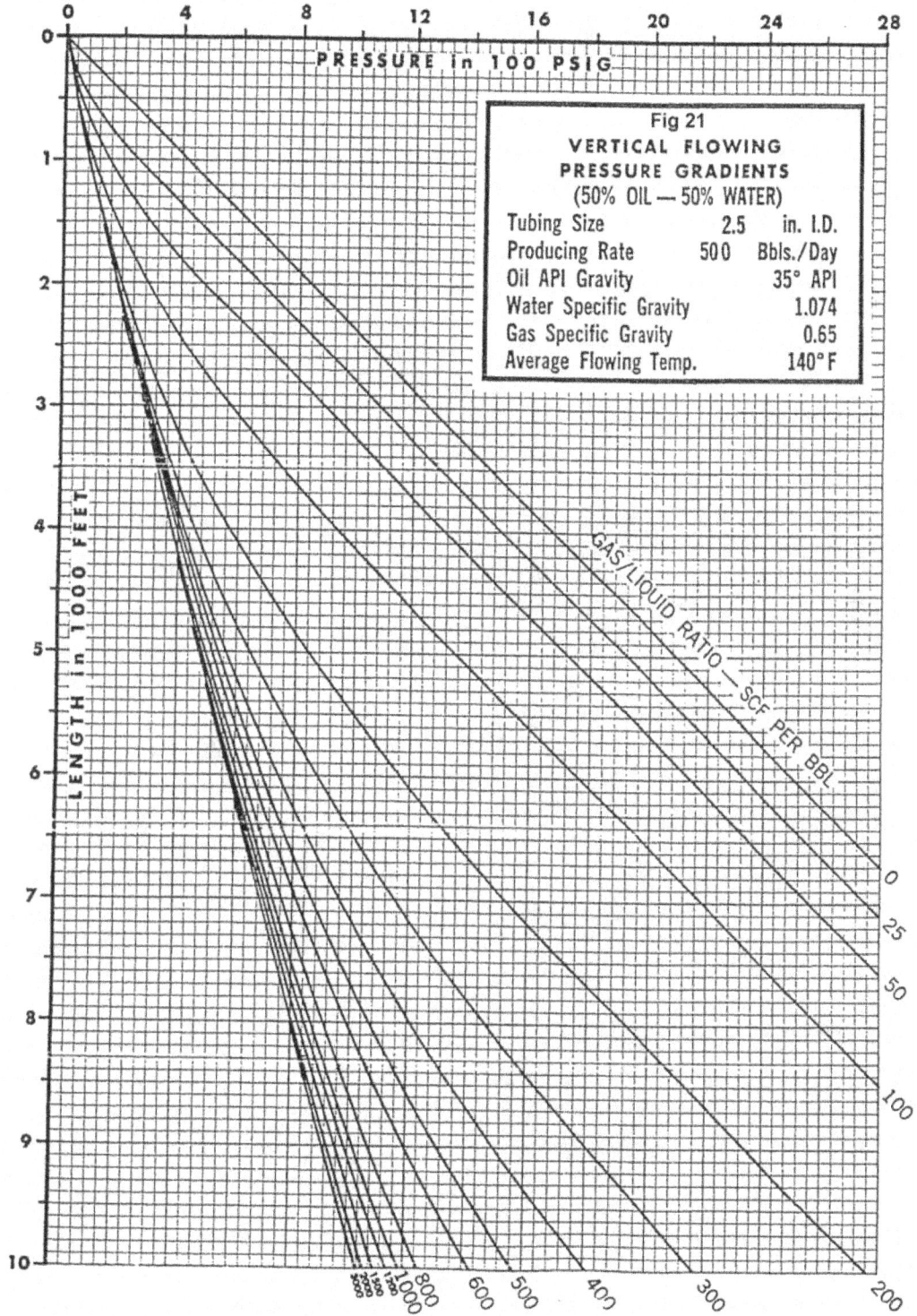

Fig 21
VERTICAL FLOWING
PRESSURE GRADIENTS
(50% OIL — 50% WATER)

Tubing Size	2.5	in. I.D.
Producing Rate	500	Bbls./Day
Oil API Gravity		35° API
Water Specific Gravity		1.074
Gas Specific Gravity		0.65
Average Flowing Temp.		140°F

PRESSURE in 100 PSIG

LENGTH in 1000 FEET

GAS/LIQUID RATIO — SCF PER BBL

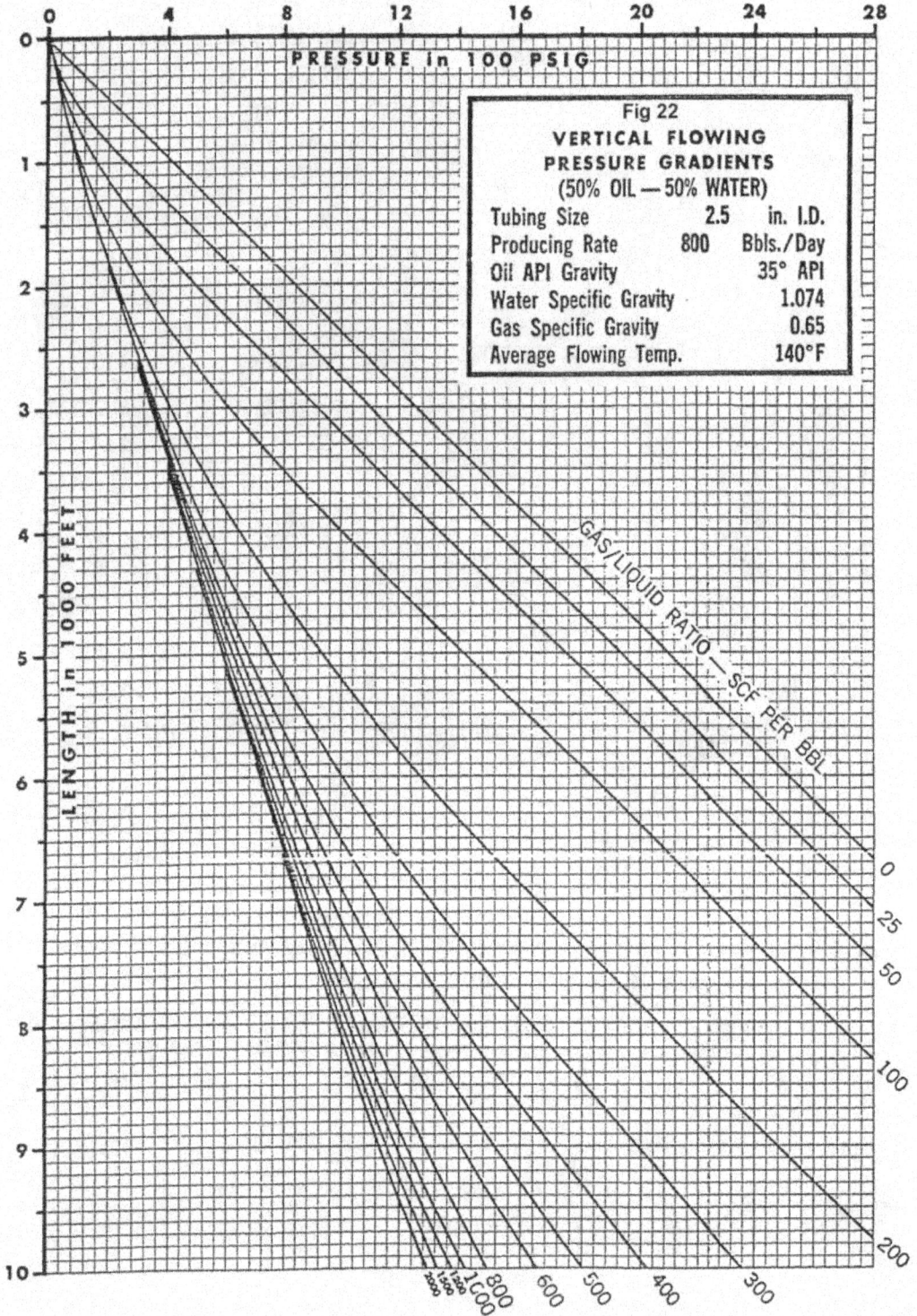

Fig 22
VERTICAL FLOWING
PRESSURE GRADIENTS
(50% OIL — 50% WATER)

Tubing Size	2.5	in. I.D.
Producing Rate	800	Bbls./Day
Oil API Gravity		35° API
Water Specific Gravity		1.074
Gas Specific Gravity		0.65
Average Flowing Temp.		140°F

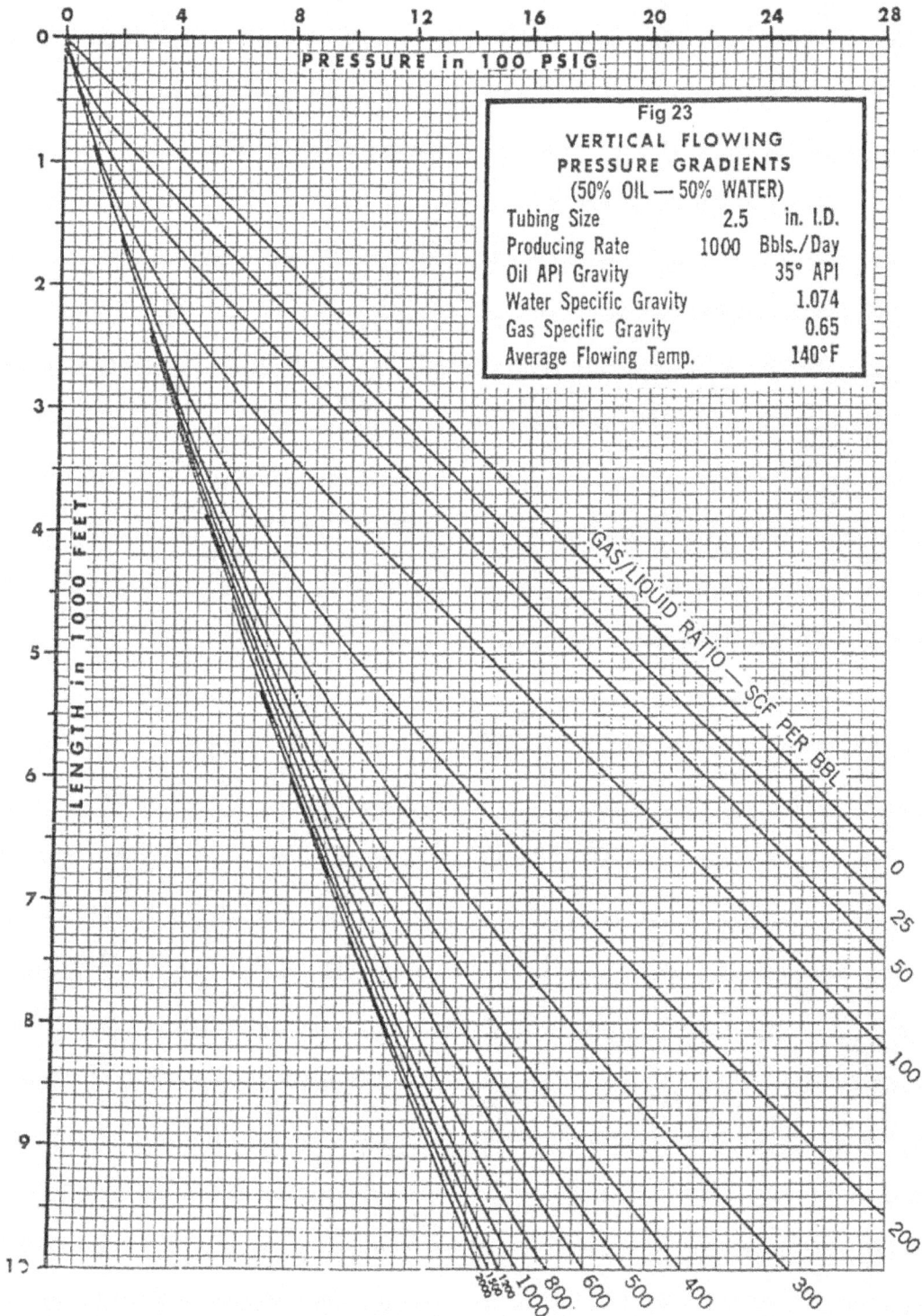

PRESSURE in 100 PSIG

LENGTH in 1000 FEET

Fig 23
VERTICAL FLOWING
PRESSURE GRADIENTS
(50% OIL — 50% WATER)

Tubing Size	2.5	in. I.D.
Producing Rate	1000	Bbls./Day
Oil API Gravity		35° API
Water Specific Gravity		1.074
Gas Specific Gravity		0.65
Average Flowing Temp.		140°F

GAS/LIQUID RATIO — SCF PER BBL

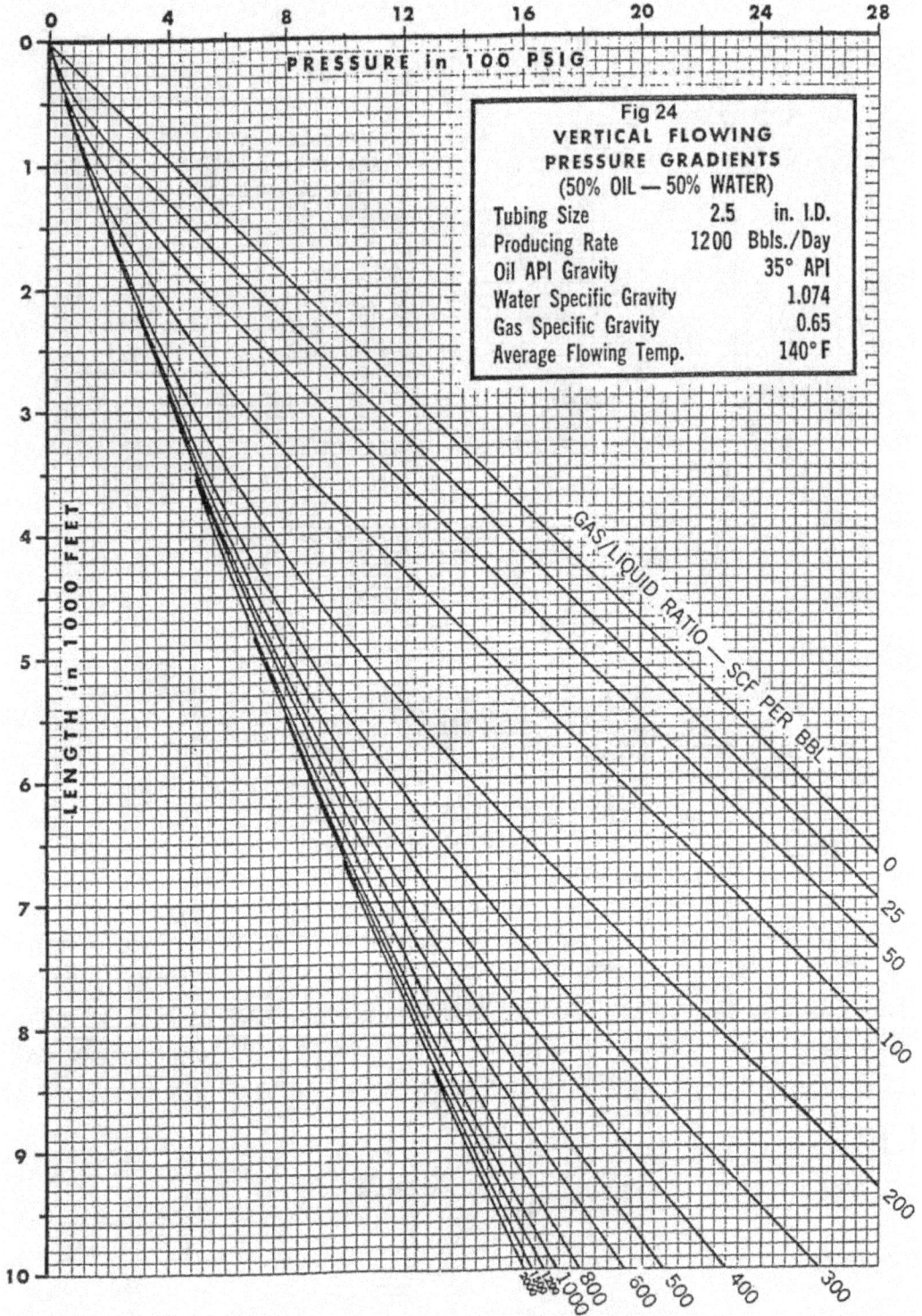

Fig 24
VERTICAL FLOWING
PRESSURE GRADIENTS
(50% OIL — 50% WATER)

Tubing Size	2.5	in. I.D.
Producing Rate	1200	Bbls./Day
Oil API Gravity		35° API
Water Specific Gravity		1.074
Gas Specific Gravity		0.65
Average Flowing Temp.		140° F

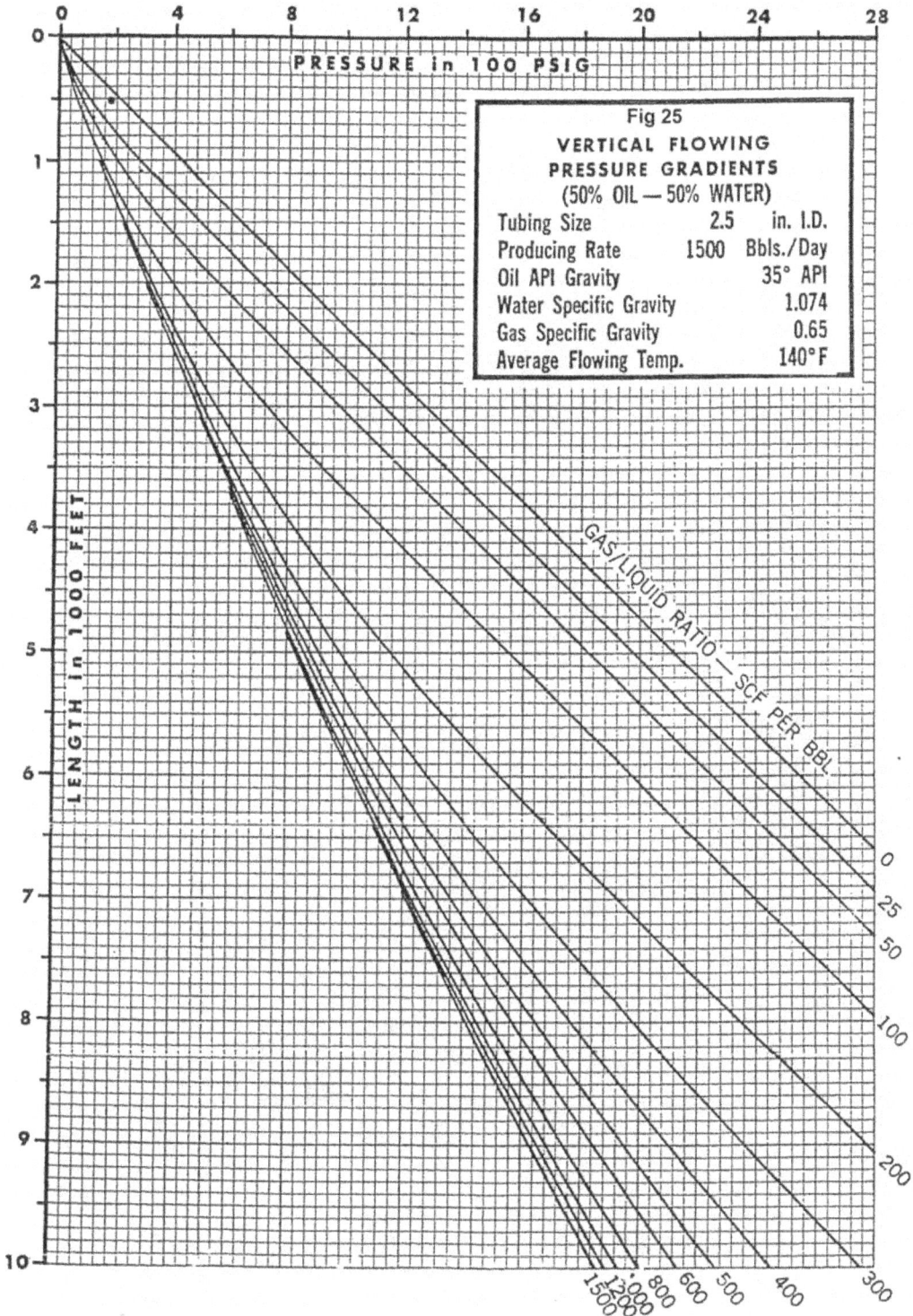

Fig 25

VERTICAL FLOWING
PRESSURE GRADIENTS
(50% OIL — 50% WATER)

Tubing Size	2.5	in. I.D.
Producing Rate	1500	Bbls./Day
Oil API Gravity		35° API
Water Specific Gravity		1.074
Gas Specific Gravity		0.65
Average Flowing Temp.		140°F

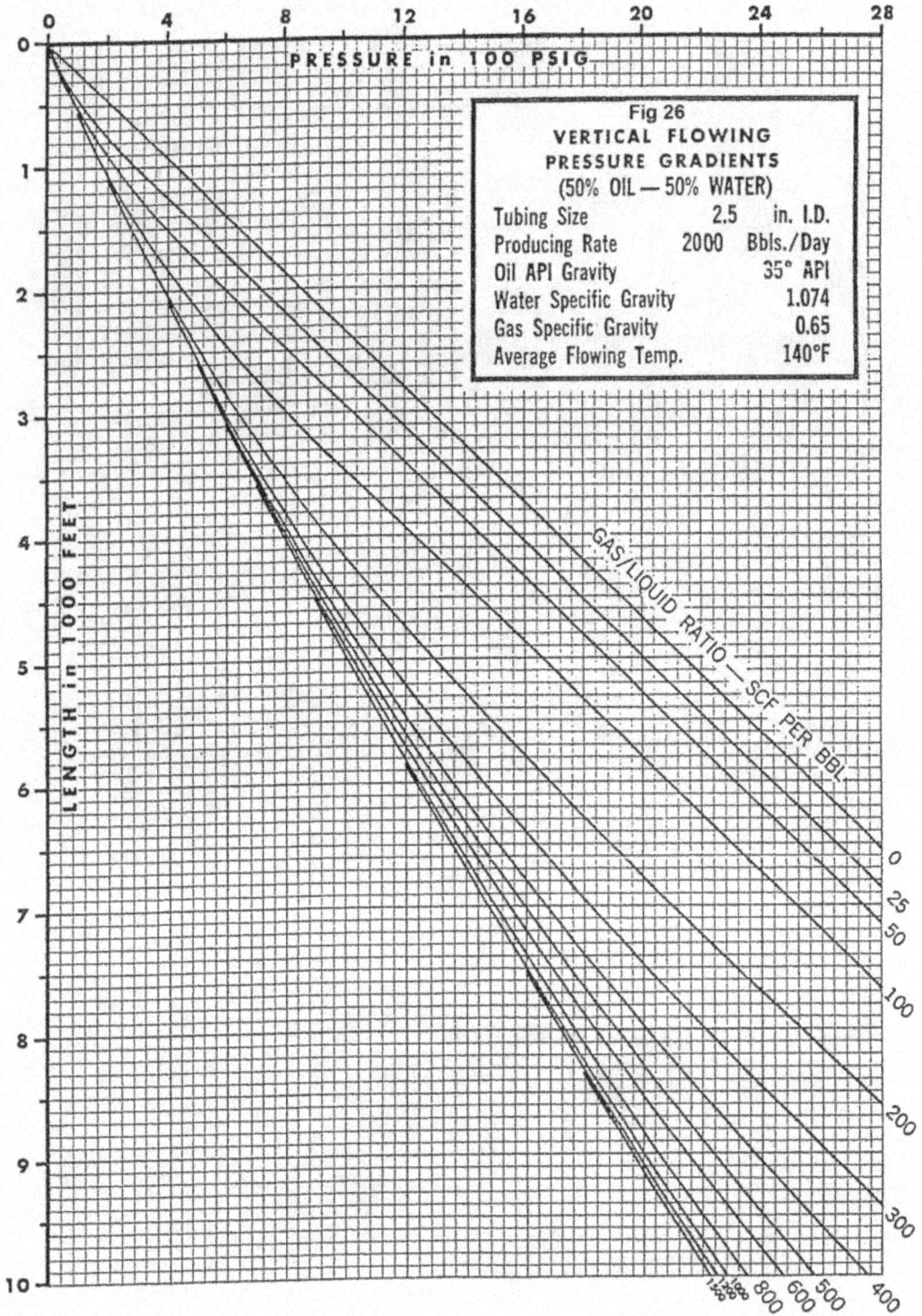

Fig 26
VERTICAL FLOWING
PRESSURE GRADIENTS
(50% OIL — 50% WATER)

Tubing Size	2.5	in. I.D.
Producing Rate	2000	Bbls./Day
Oil API Gravity		35° API
Water Specific Gravity		1.074
Gas Specific Gravity		0.65
Average Flowing Temp.		140°F

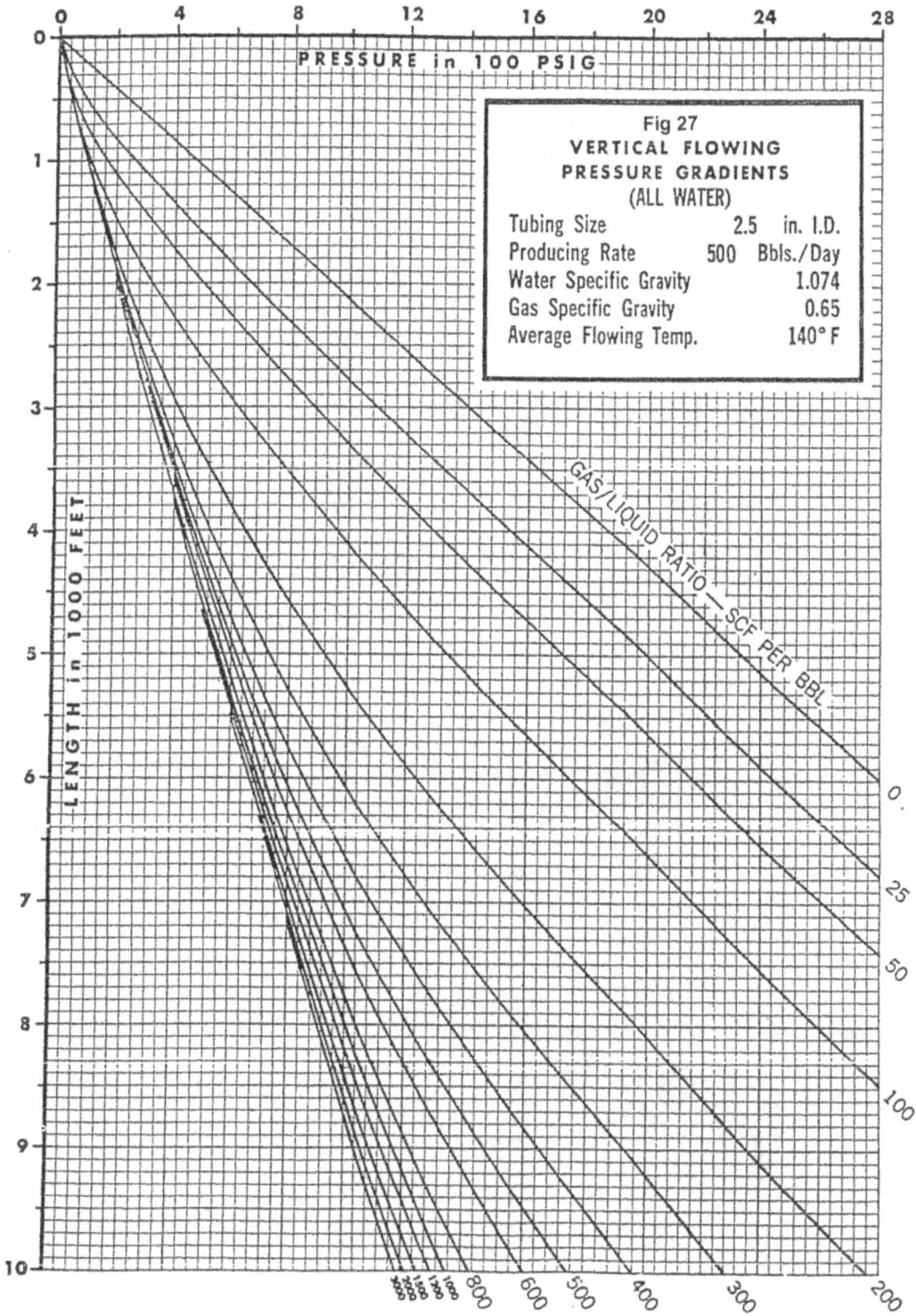

Fig 27
VERTICAL FLOWING
PRESSURE GRADIENTS
(ALL WATER)

Tubing Size	2.5 in. I.D.
Producing Rate	500 Bbls./Day
Water Specific Gravity	1.074
Gas Specific Gravity	0.65
Average Flowing Temp.	140° F

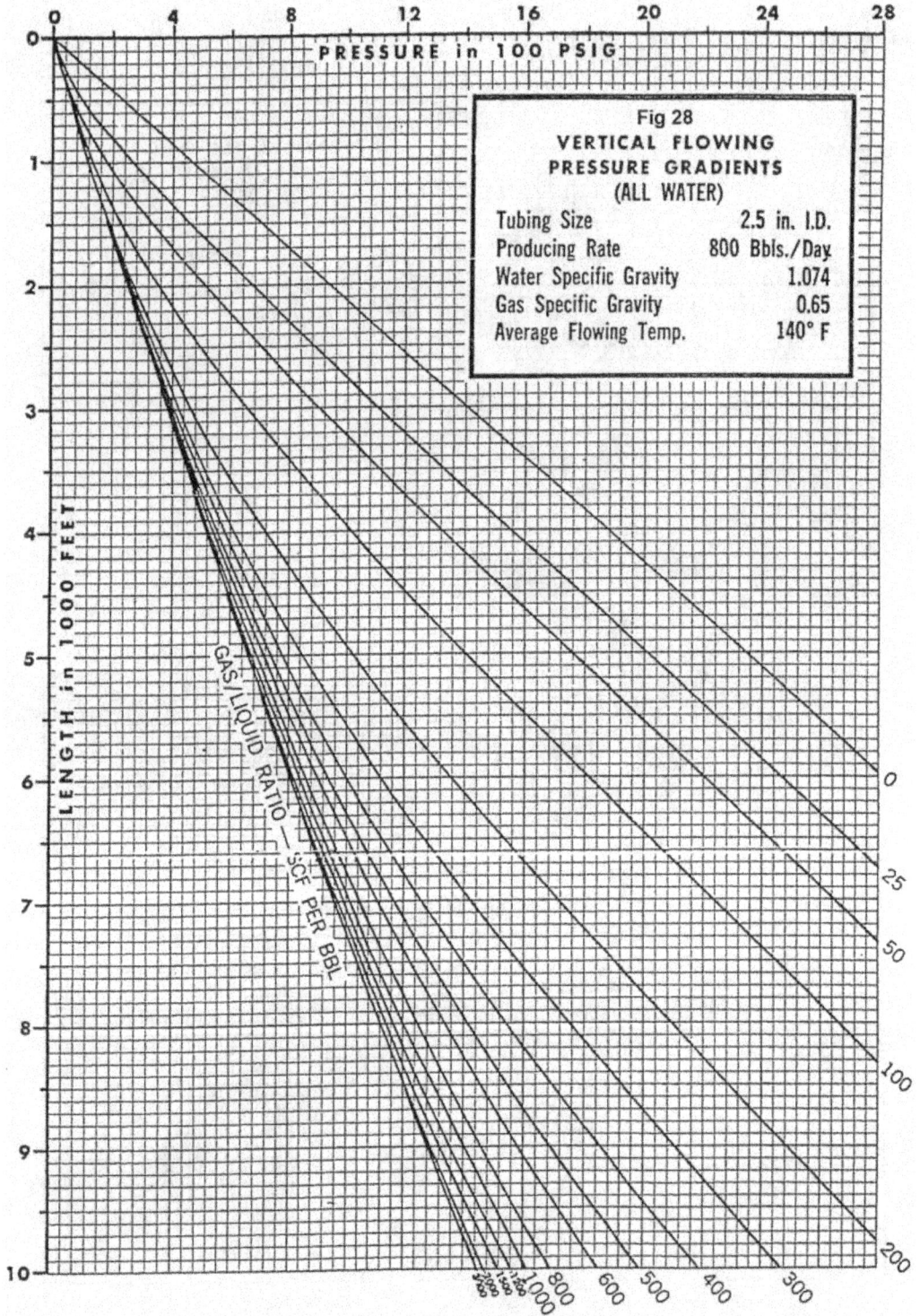

Fig 28
VERTICAL FLOWING
PRESSURE GRADIENTS
(ALL WATER)

Tubing Size	2.5 in. I.D.
Producing Rate	800 Bbls./Day.
Water Specific Gravity	1.074
Gas Specific Gravity	0.65
Average Flowing Temp.	140° F

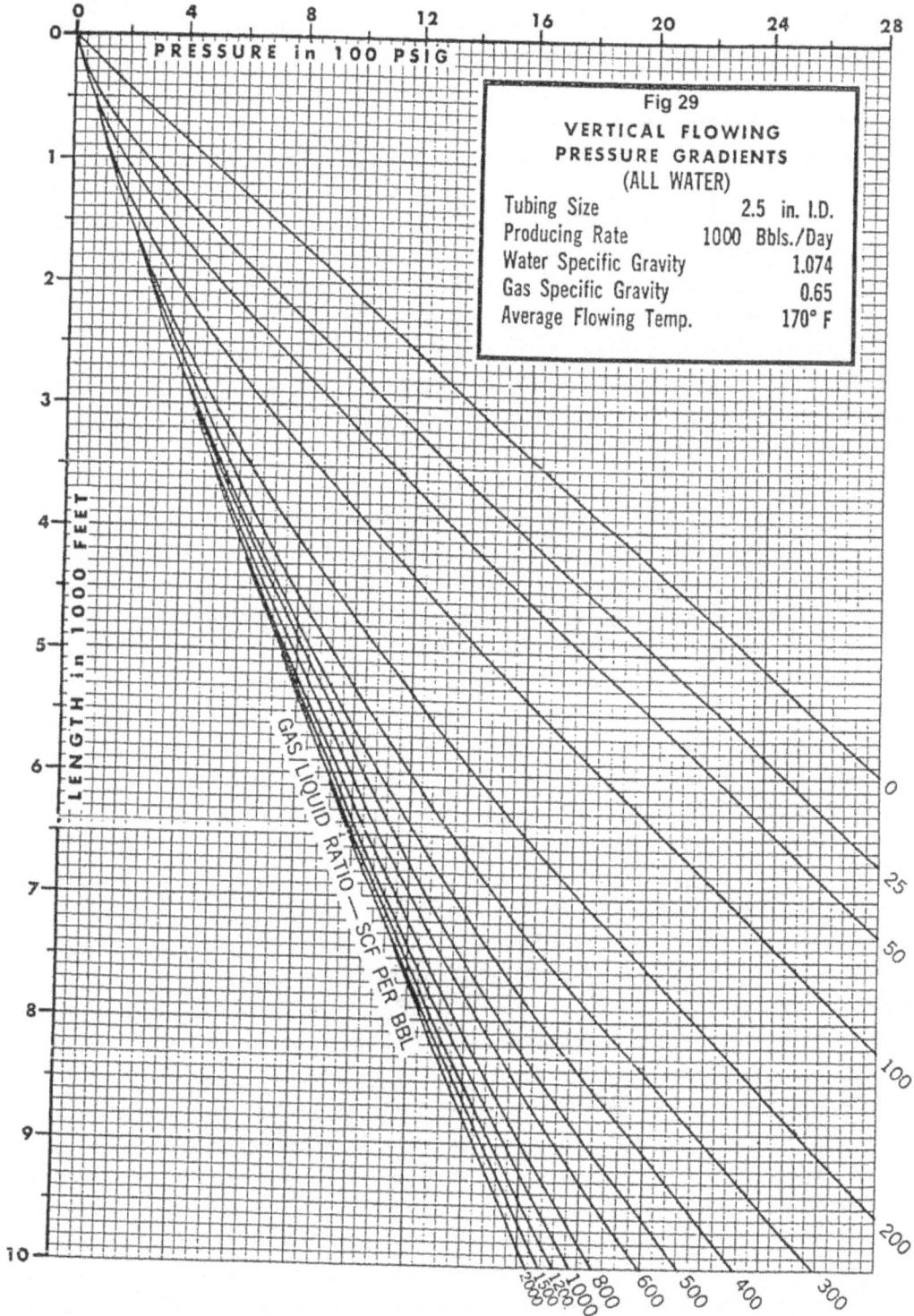

Fig 29
VERTICAL FLOWING
PRESSURE GRADIENTS
(ALL WATER)

Tubing Size	2.5 in. I.D.
Producing Rate	1000 Bbls./Day
Water Specific Gravity	1.074
Gas Specific Gravity	0.65
Average Flowing Temp.	170° F

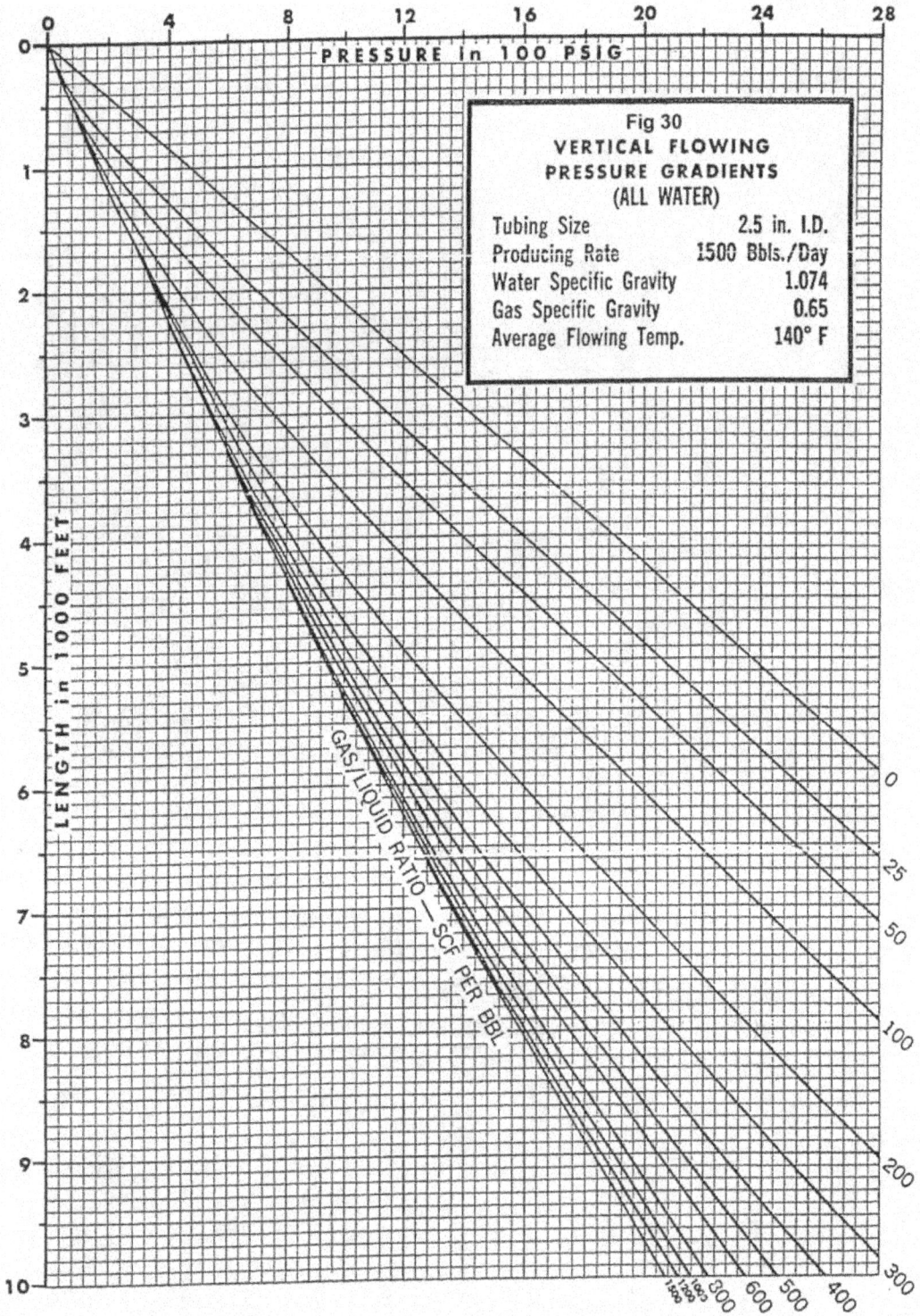

Fig 30
VERTICAL FLOWING
PRESSURE GRADIENTS
(ALL WATER)

Tubing Size	2.5 in. I.D.
Producing Rate	1500 Bbls./Day
Water Specific Gravity	1.074
Gas Specific Gravity	0.65
Average Flowing Temp.	140° F

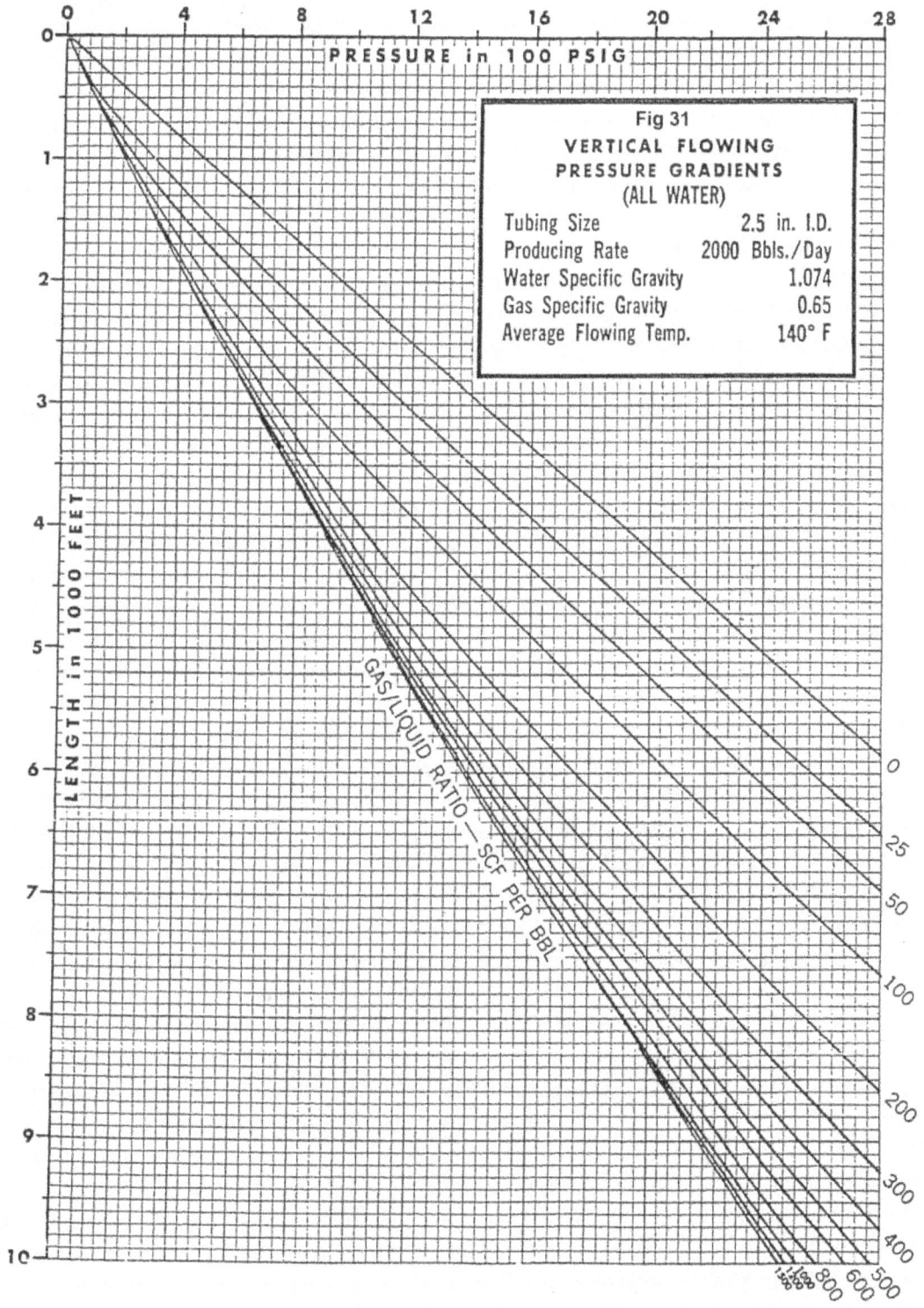

Fig 31
VERTICAL FLOWING
PRESSURE GRADIENTS
(ALL WATER)

Tubing Size	2.5 in. I.D.
Producing Rate	2000 Bbls./Day
Water Specific Gravity	1.074
Gas Specific Gravity	0.65
Average Flowing Temp.	140° F

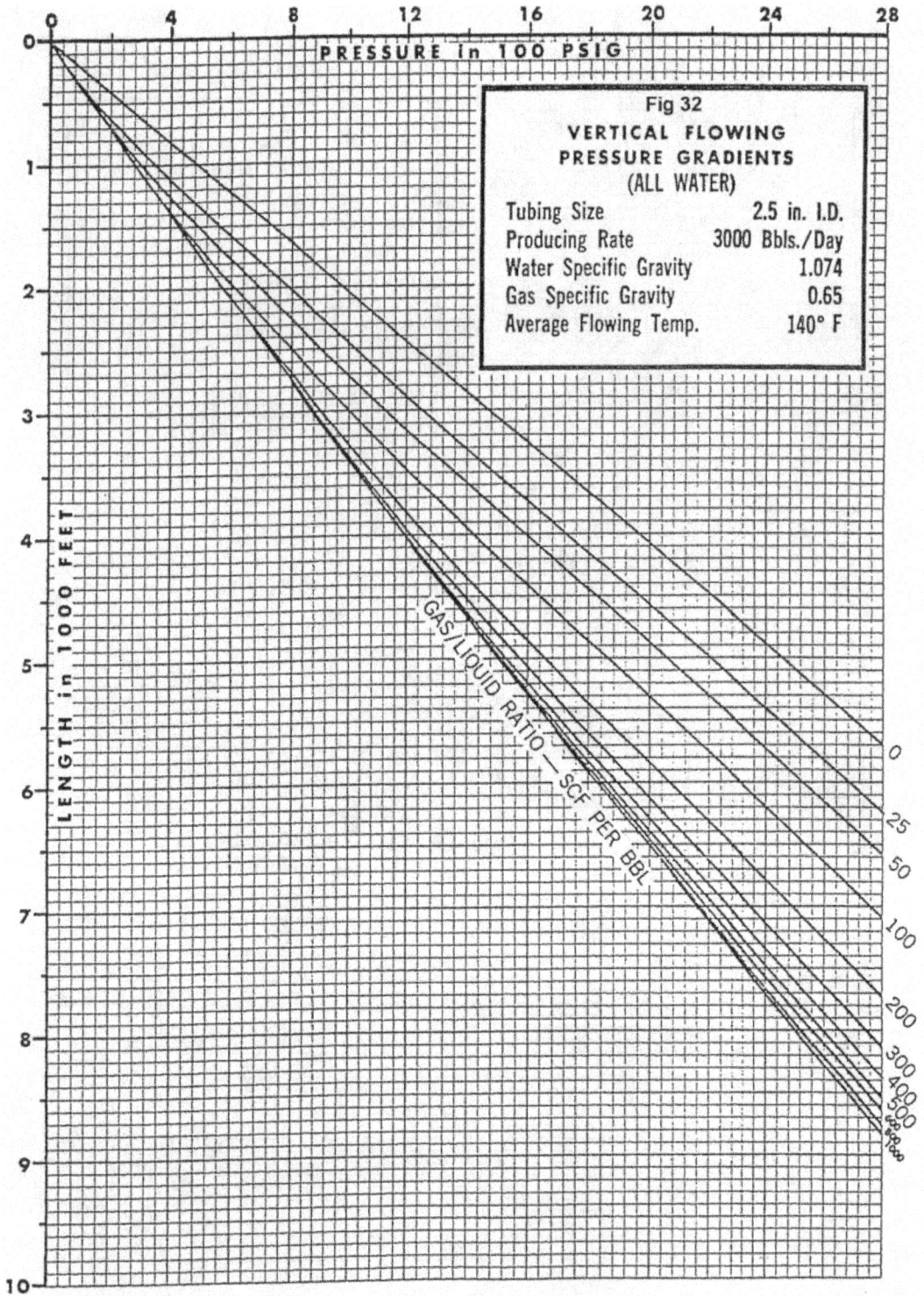

Fig 32
VERTICAL FLOWING
PRESSURE GRADIENTS
(ALL WATER)

Tubing Size	2.5 in. I.D.
Producing Rate	3000 Bbls./Day
Water Specific Gravity	1.074
Gas Specific Gravity	0.65
Average Flowing Temp.	140° F

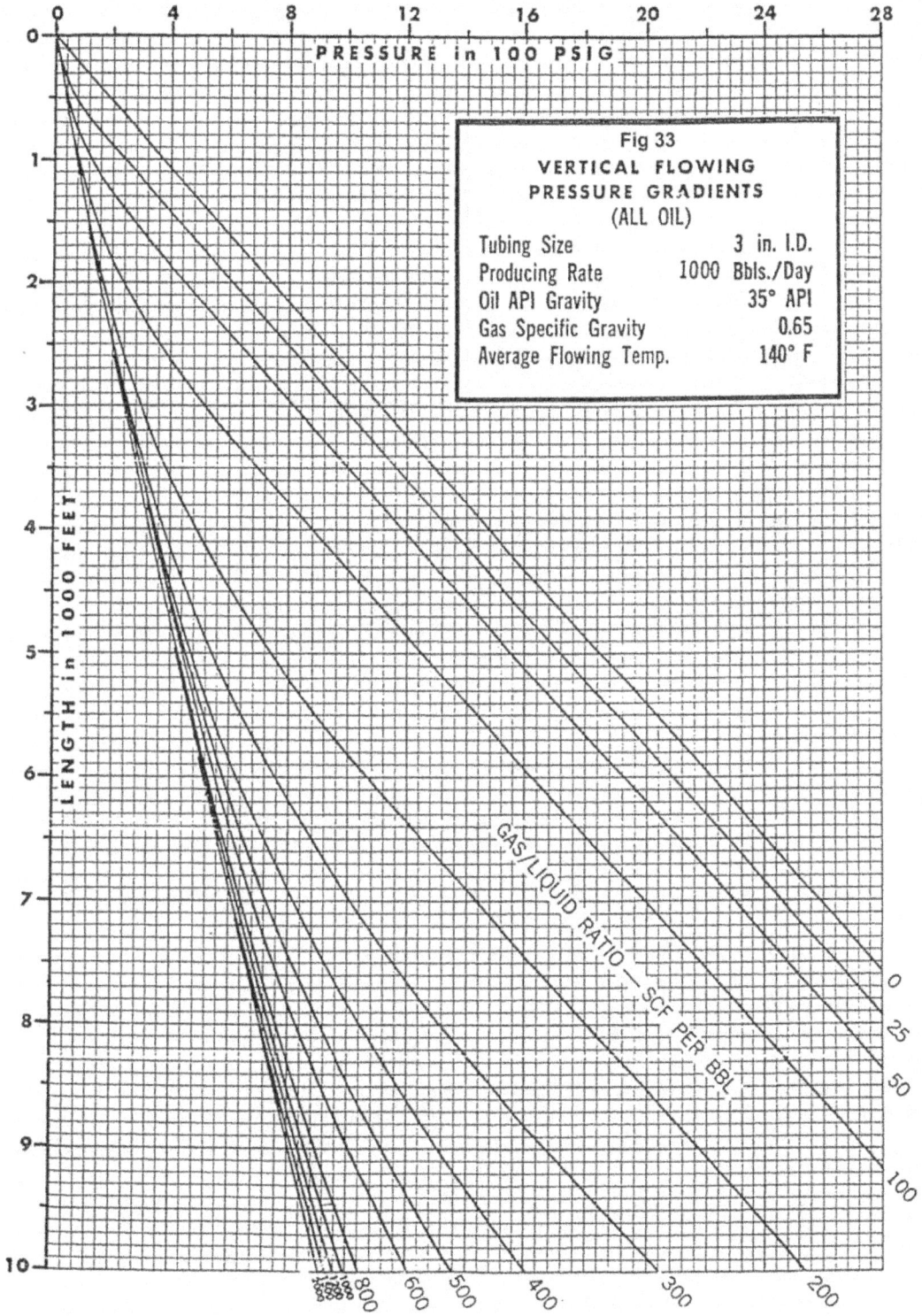

Fig 33
VERTICAL FLOWING
PRESSURE GRADIENTS
(ALL OIL)

Tubing Size	3 in. I.D.
Producing Rate	1000 Bbls./Day
Oil API Gravity	35° API
Gas Specific Gravity	0.65
Average Flowing Temp.	140° F

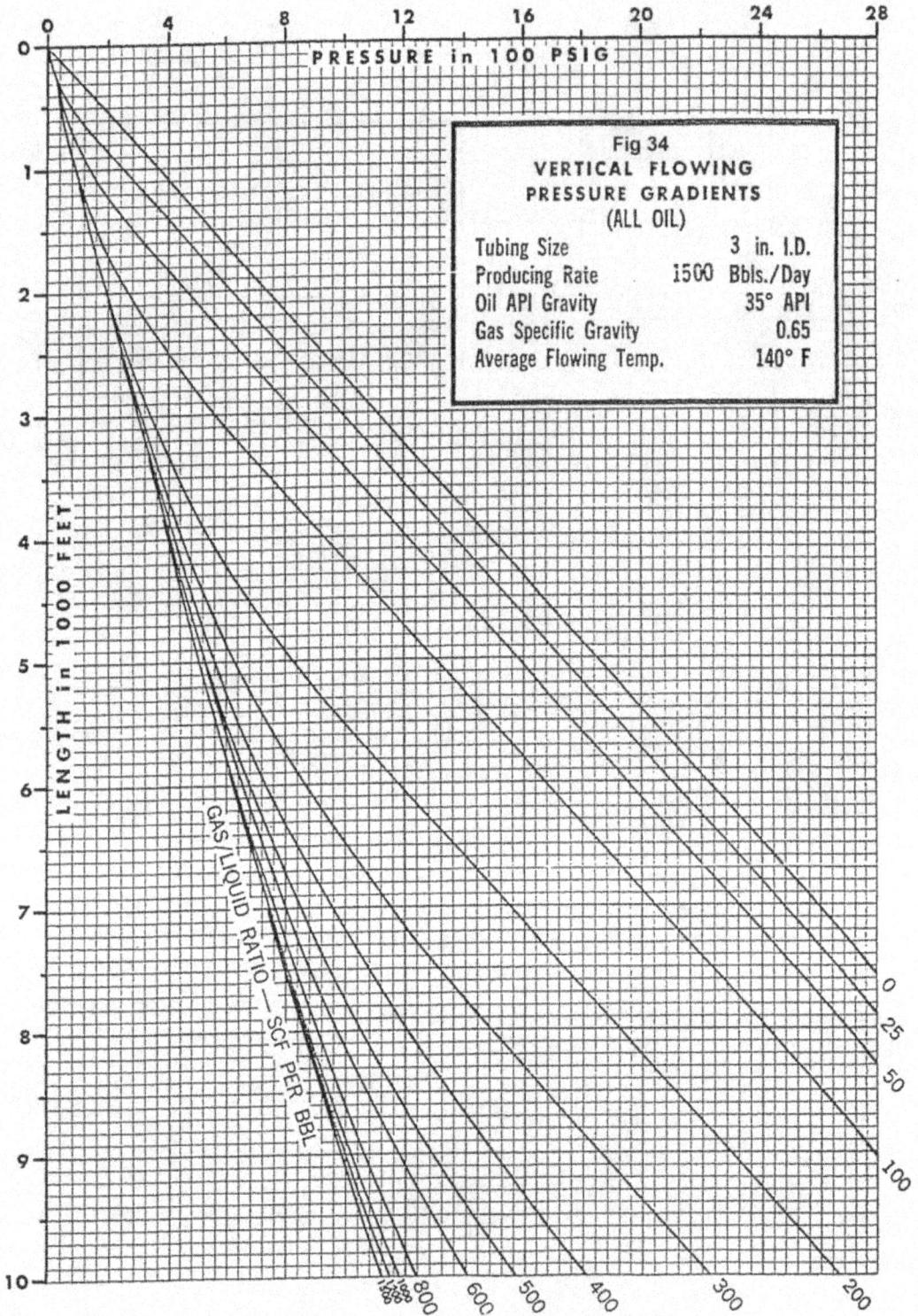

Fig 34
VERTICAL FLOWING
PRESSURE GRADIENTS
(ALL OIL)

Tubing Size	3 in. I.D.
Producing Rate	1500 Bbls./Day
Oil API Gravity	35° API
Gas Specific Gravity	0.65
Average Flowing Temp.	140° F

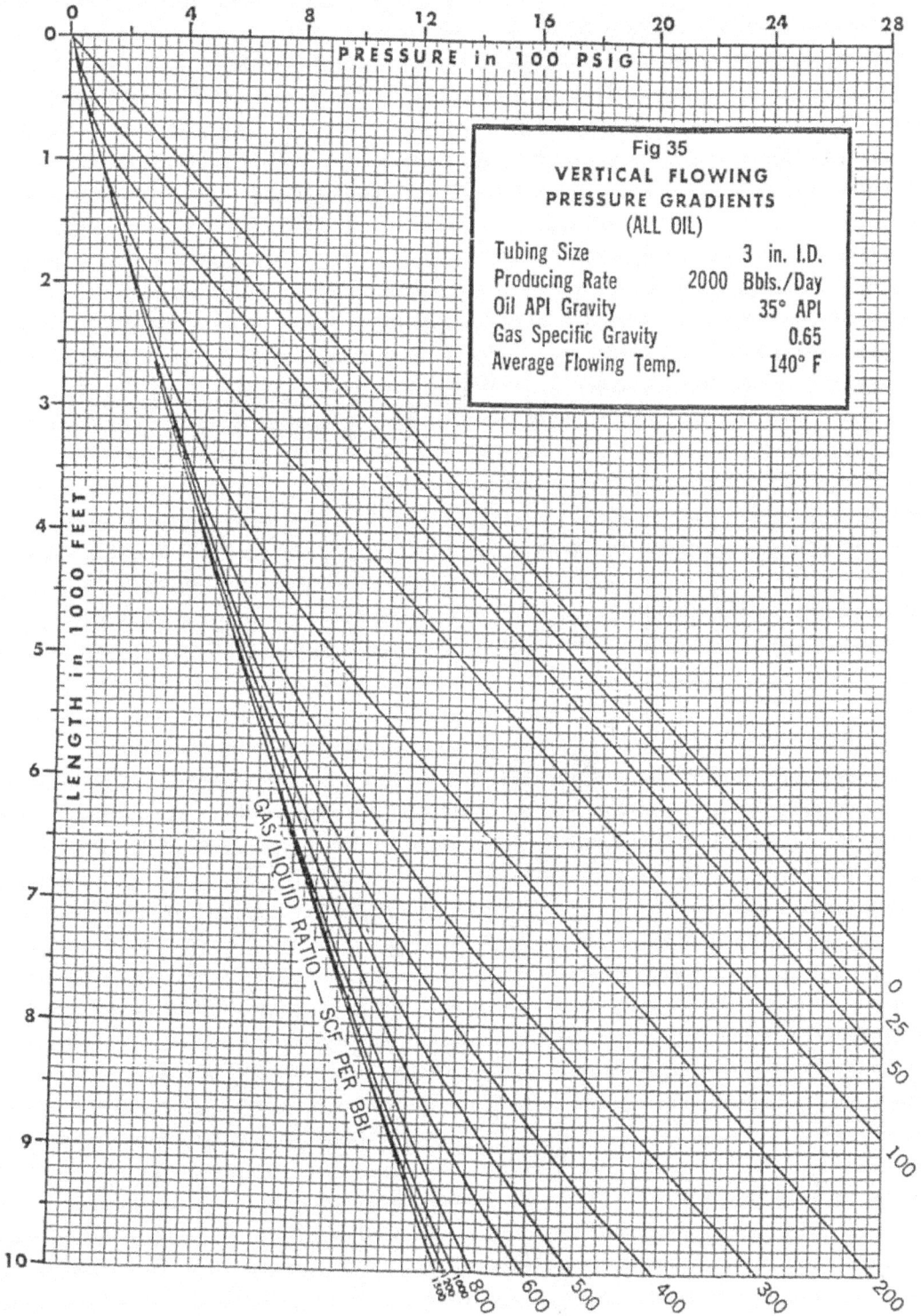

Fig 35
VERTICAL FLOWING
PRESSURE GRADIENTS
(ALL OIL)

Tubing Size	3 in. I.D.
Producing Rate	2000 Bbls./Day
Oil API Gravity	35° API
Gas Specific Gravity	0.65
Average Flowing Temp.	140° F

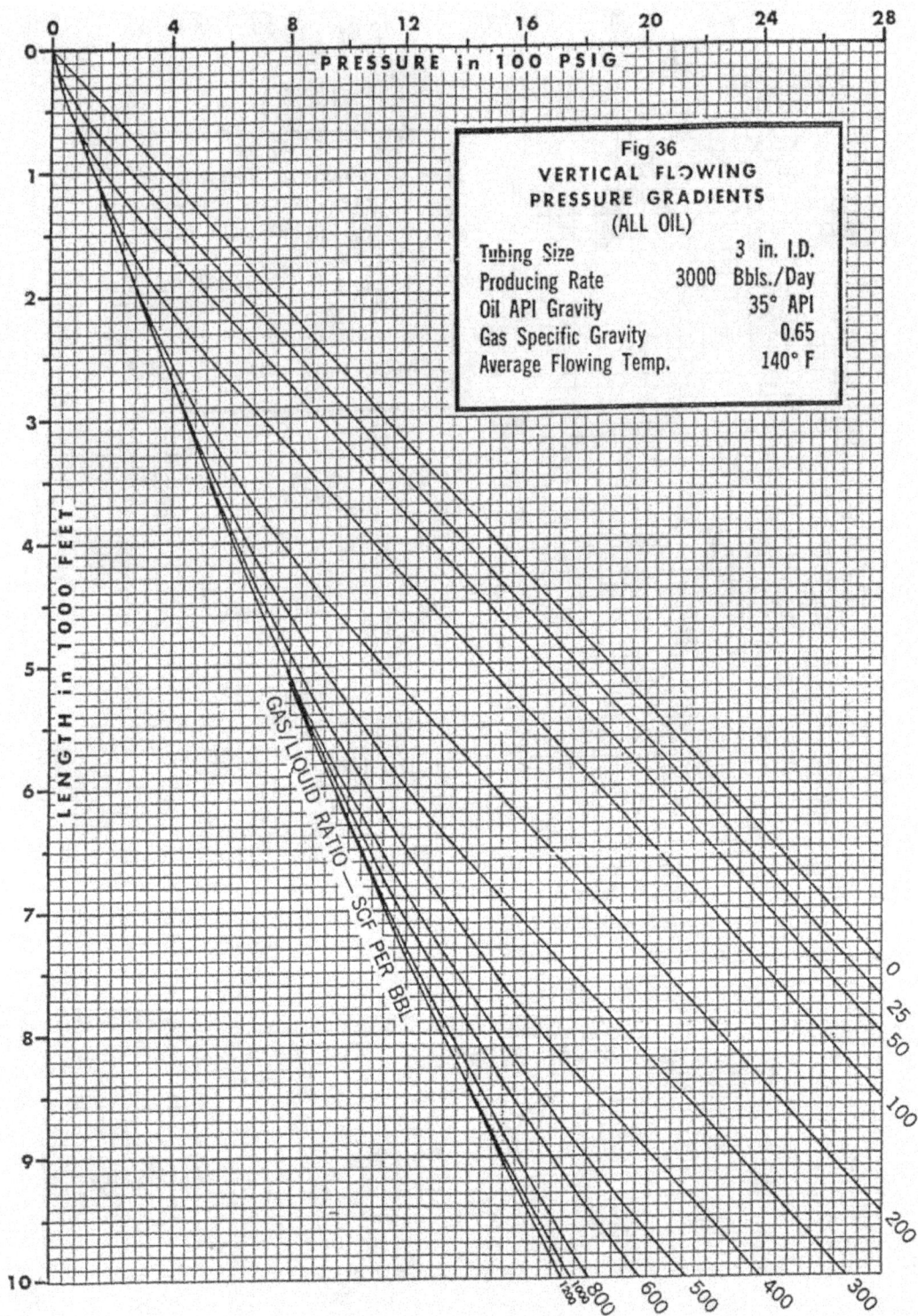

Fig 36
VERTICAL FLOWING
PRESSURE GRADIENTS
(ALL OIL)

Tubing Size	3 in. I.D.
Producing Rate	3000 Bbls./Day
Oil API Gravity	35° API
Gas Specific Gravity	0.65
Average Flowing Temp.	140° F

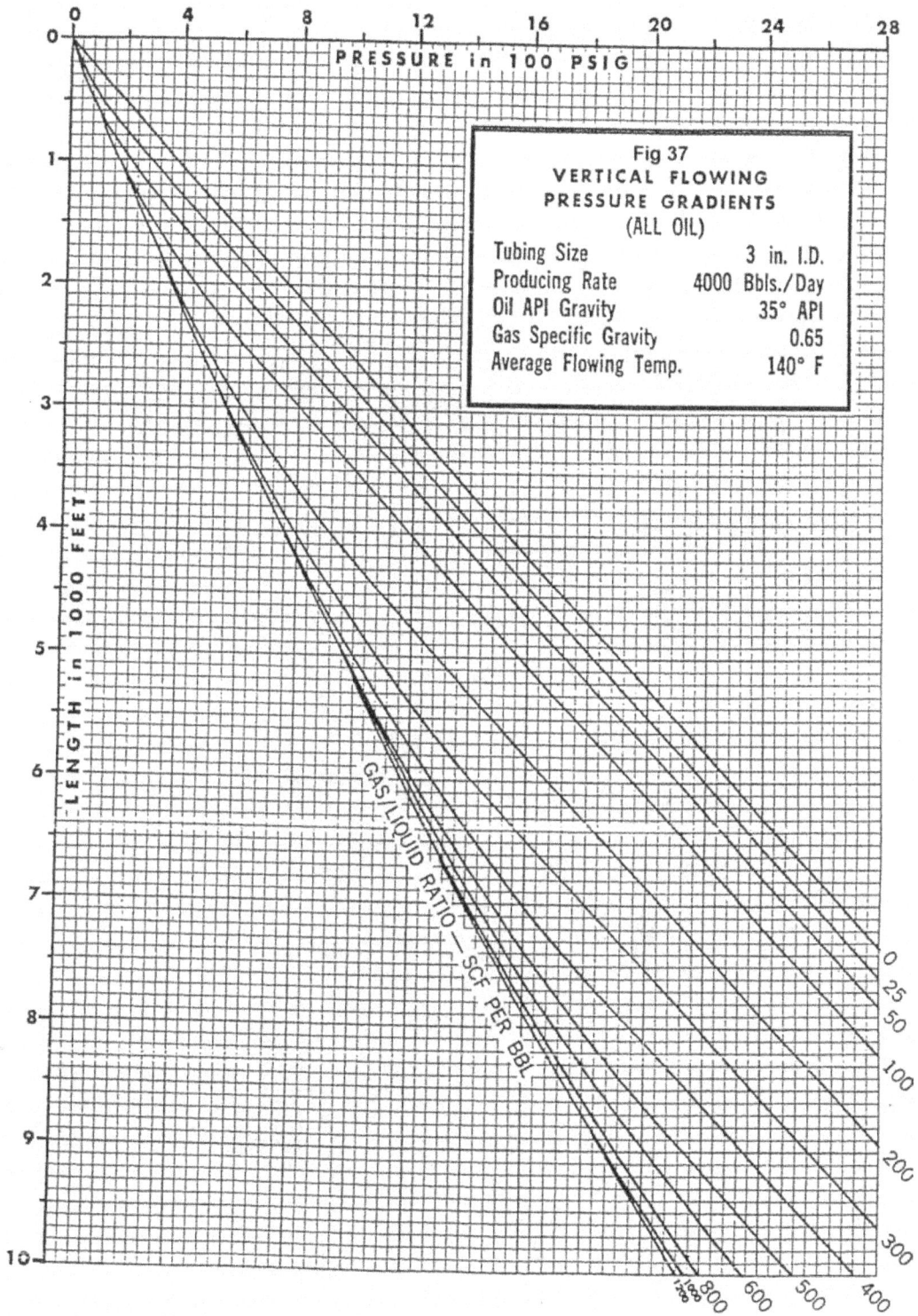

Fig 37
VERTICAL FLOWING
PRESSURE GRADIENTS
(ALL OIL)

Tubing Size	3 in. I.D.
Producing Rate	4000 Bbls./Day
Oil API Gravity	35° API
Gas Specific Gravity	0.65
Average Flowing Temp.	140° F

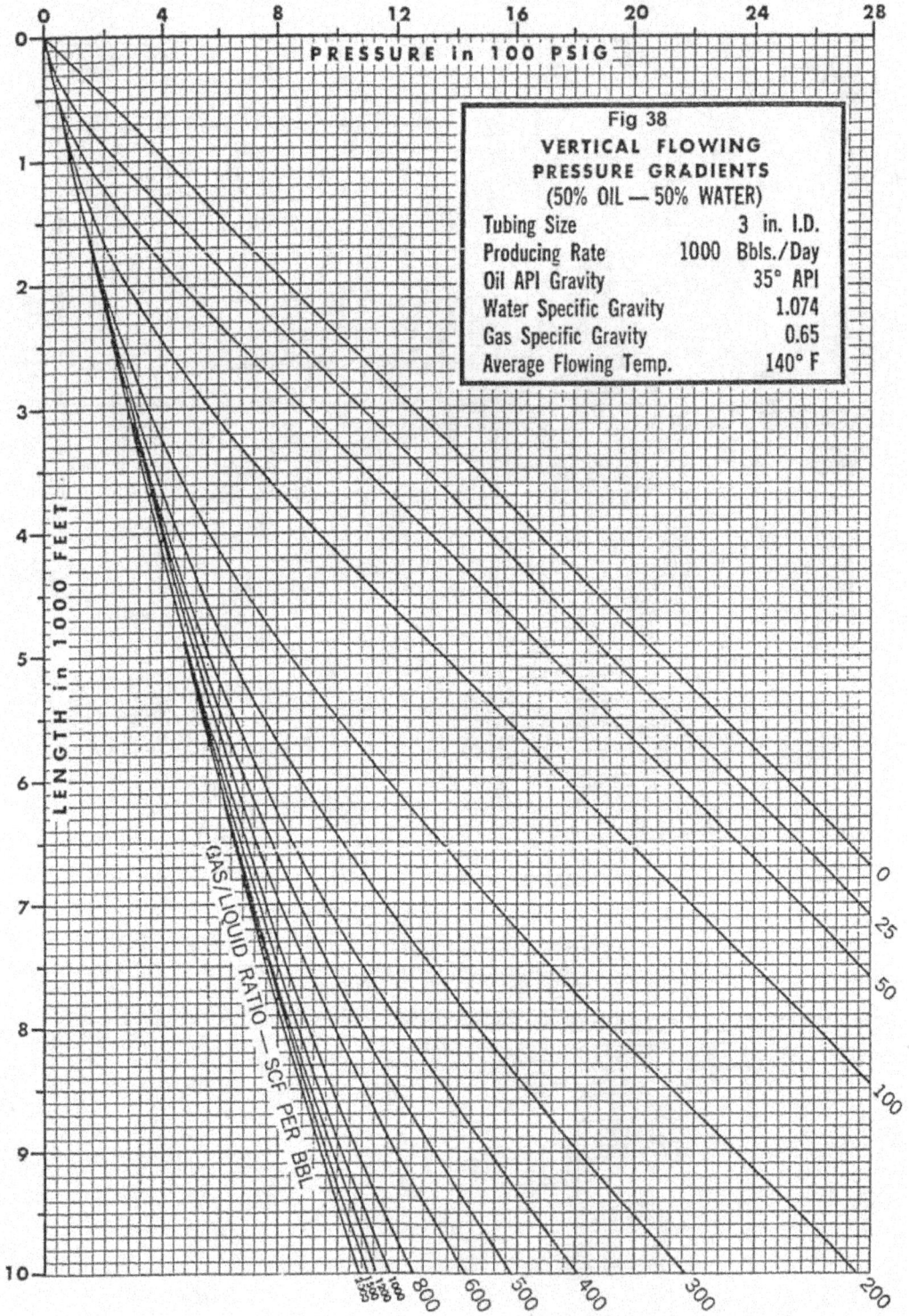

Fig 38
VERTICAL FLOWING
PRESSURE GRADIENTS
(50% OIL — 50% WATER)

Tubing Size	3 in. I.D.
Producing Rate	1000 Bbls./Day
Oil API Gravity	35° API
Water Specific Gravity	1.074
Gas Specific Gravity	0.65
Average Flowing Temp.	140° F

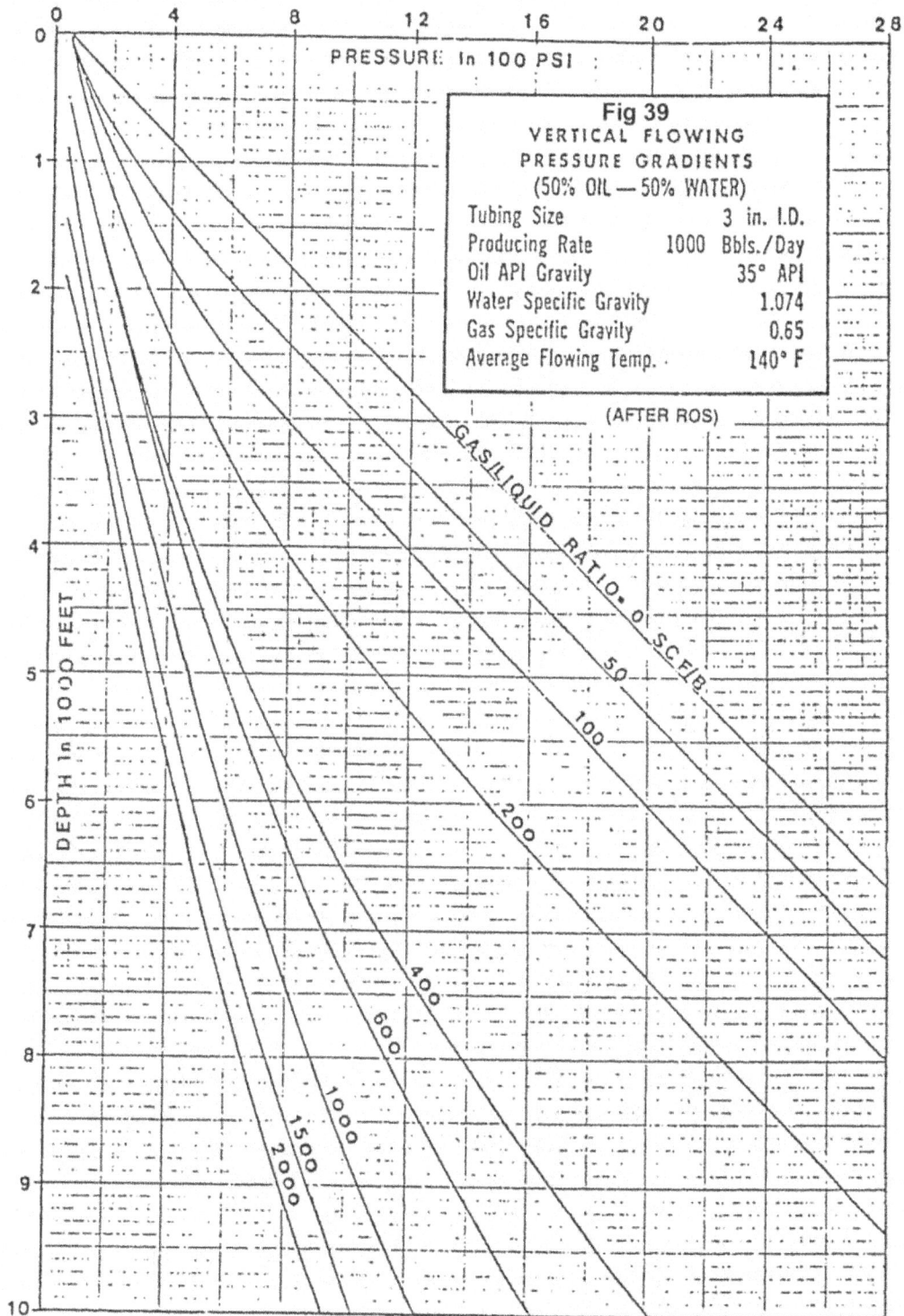

Fig 39
VERTICAL FLOWING
PRESSURE GRADIENTS
(50% OIL — 50% WATER)

Tubing Size	3 in. I.D.
Producing Rate	1000 Bbls./Day
Oil API Gravity	35° API
Water Specific Gravity	1.074
Gas Specific Gravity	0.65
Average Flowing Temp.	140° F

(AFTER ROS)

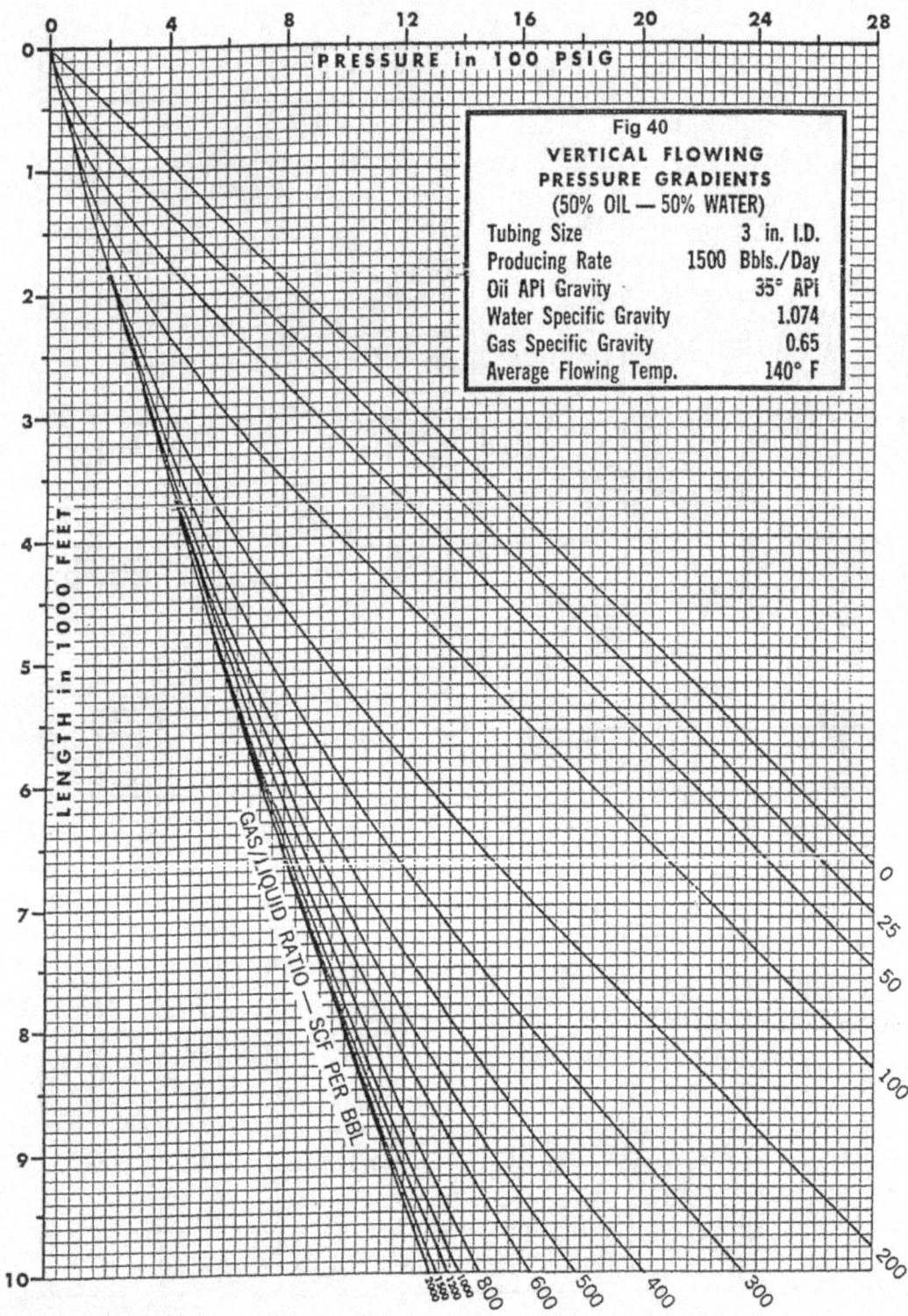

Fig 40
VERTICAL FLOWING
PRESSURE GRADIENTS
(50% OIL — 50% WATER)

Tubing Size	3 in. I.D.
Producing Rate	1500 Bbls./Day
Oil API Gravity	35° API
Water Specific Gravity	1.074
Gas Specific Gravity	0.65
Average Flowing Temp.	140° F

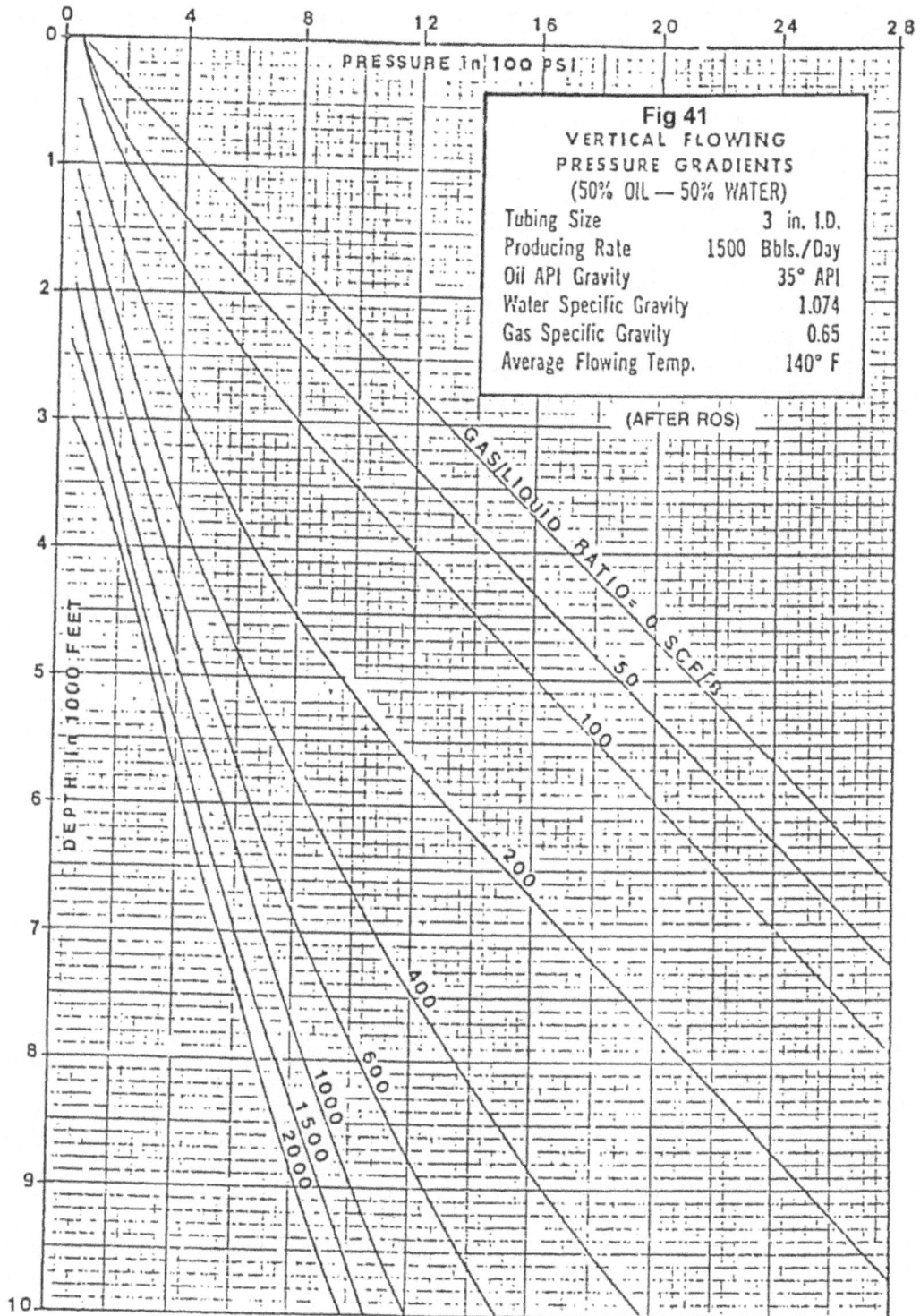

Fig 41
VERTICAL FLOWING
PRESSURE GRADIENTS
(50% OIL — 50% WATER)

Tubing Size	3 in. I.D.
Producing Rate	1500 Bbls./Day
Oil API Gravity	35° API
Water Specific Gravity	1.074
Gas Specific Gravity	0.65
Average Flowing Temp.	140° F

(AFTER ROS)

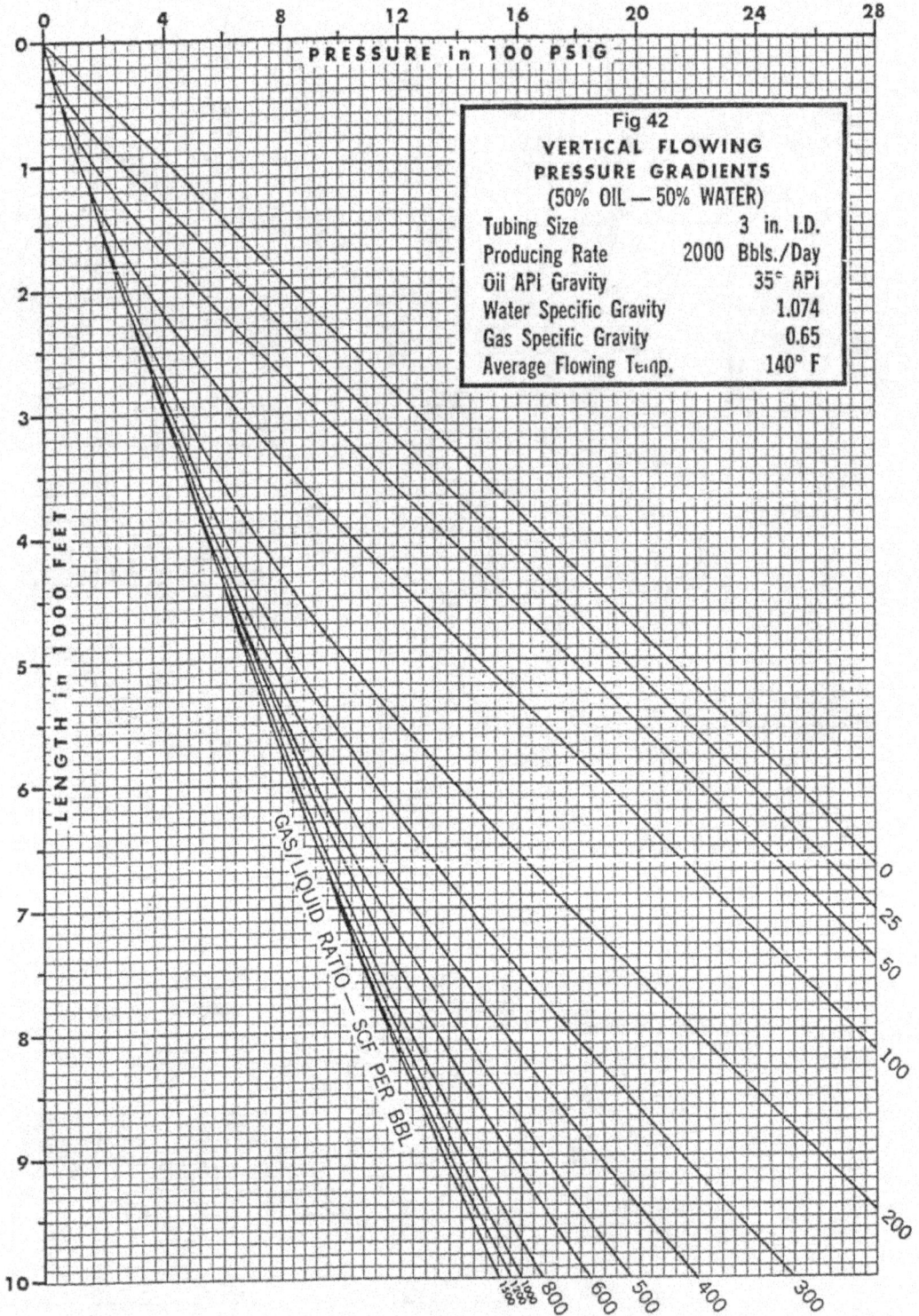

Fig 42
VERTICAL FLOWING PRESSURE GRADIENTS
(50% OIL — 50% WATER)

Tubing Size	3 in. I.D.
Producing Rate	2000 Bbls./Day
Oil API Gravity	35° API
Water Specific Gravity	1.074
Gas Specific Gravity	0.65
Average Flowing Temp.	140° F

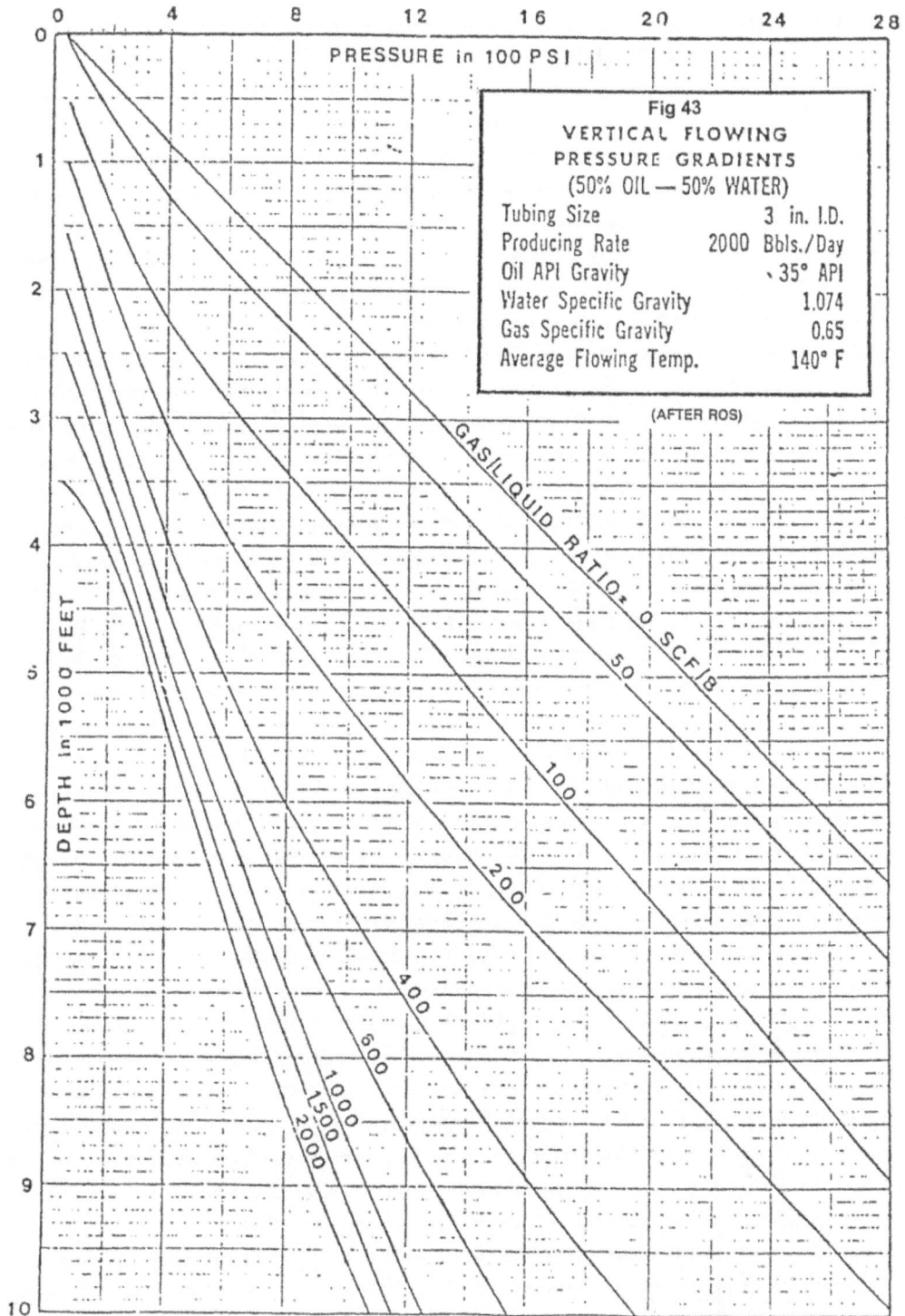

Fig 43
VERTICAL FLOWING
PRESSURE GRADIENTS
(50% OIL — 50% WATER)

Tubing Size	3 in. I.D.
Producing Rate	2000 Bbls./Day
Oil API Gravity	⟍35° API
Water Specific Gravity	1.074
Gas Specific Gravity	0.65
Average Flowing Temp.	140° F

(AFTER ROS)

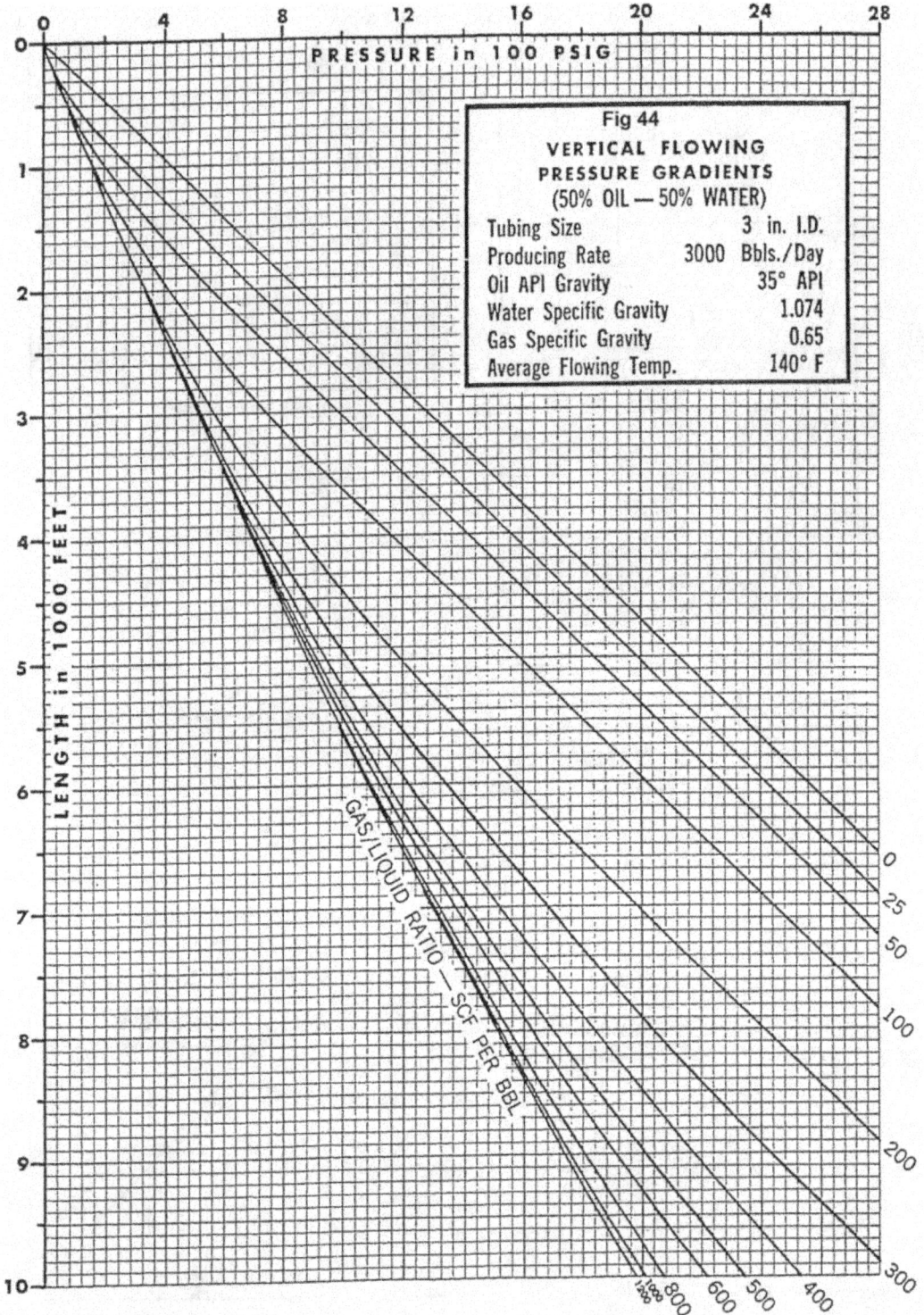

Fig 44
VERTICAL FLOWING
PRESSURE GRADIENTS
(50% OIL — 50% WATER)

Tubing Size	3 in. I.D.
Producing Rate	3000 Bbls./Day
Oil API Gravity	35° API
Water Specific Gravity	1.074
Gas Specific Gravity	0.65
Average Flowing Temp.	140° F

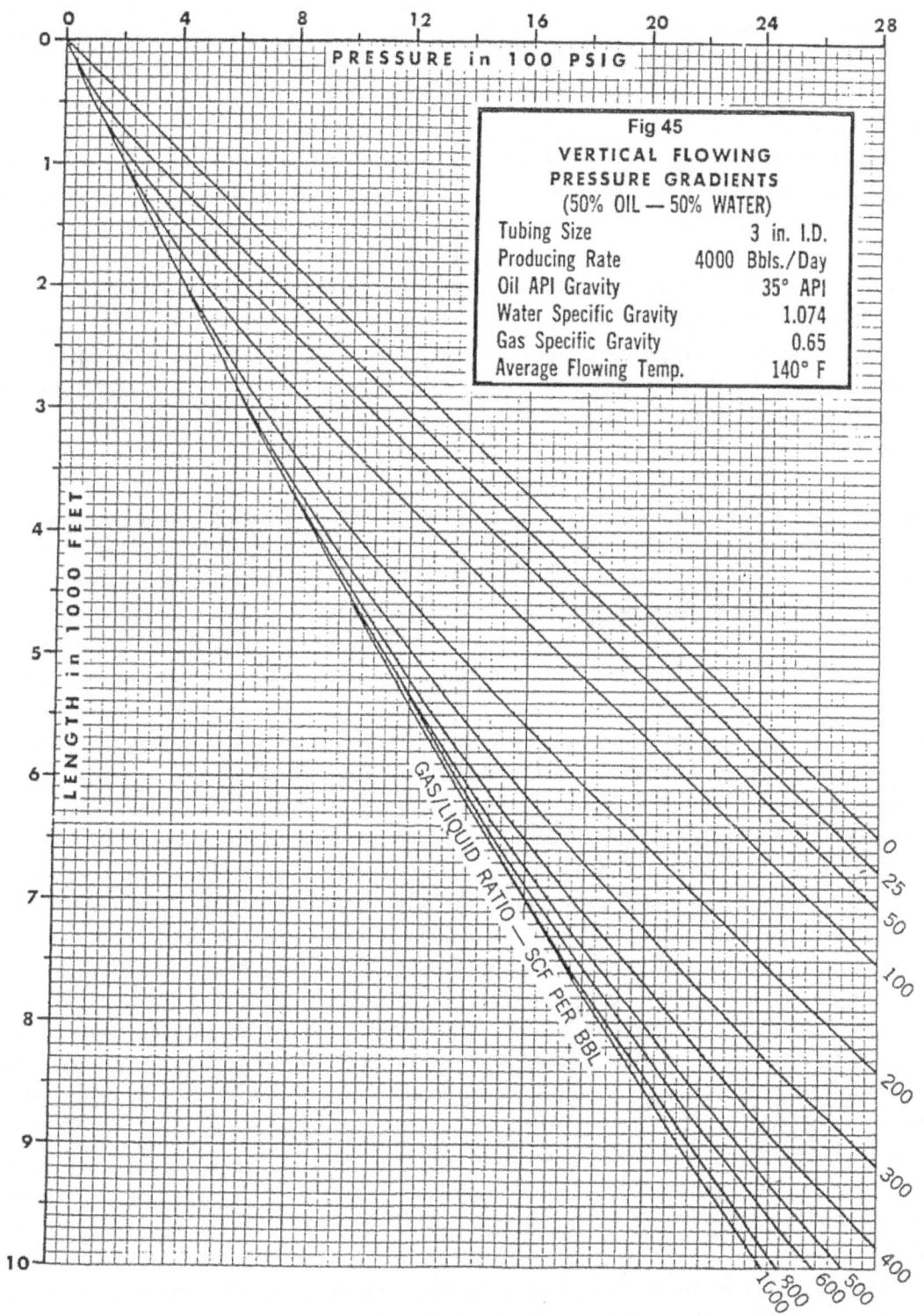

Fig 45

VERTICAL FLOWING
PRESSURE GRADIENTS
(50% OIL — 50% WATER)

Tubing Size	3 in. I.D.
Producing Rate	4000 Bbls./Day
Oil API Gravity	35° API
Water Specific Gravity	1.074
Gas Specific Gravity	0.65
Average Flowing Temp.	140° F

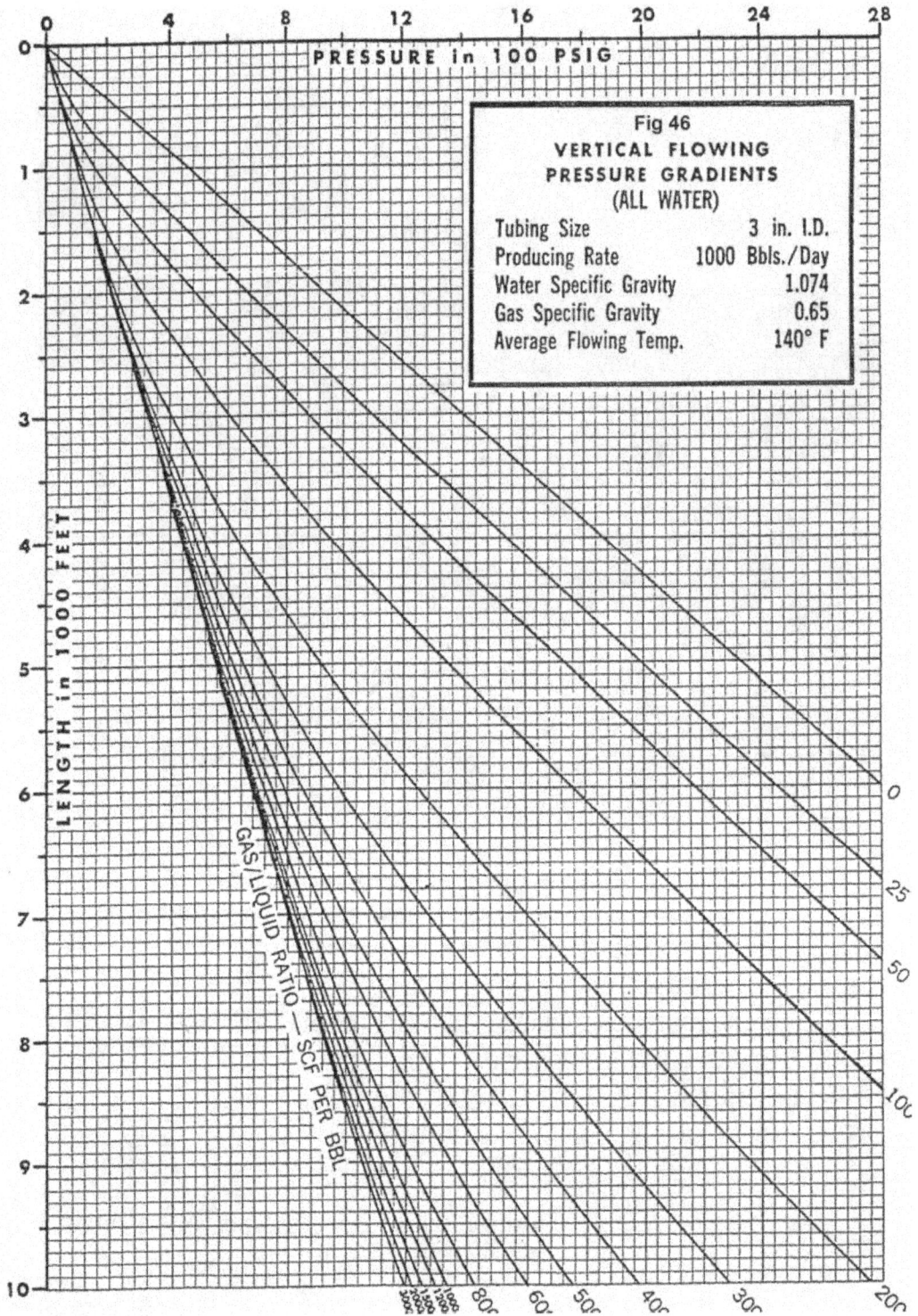

Fig 46
VERTICAL FLOWING
PRESSURE GRADIENTS
(ALL WATER)

Tubing Size	3 in. I.D.
Producing Rate	1000 Bbls./Day
Water Specific Gravity	1.074
Gas Specific Gravity	0.65
Average Flowing Temp.	140° F

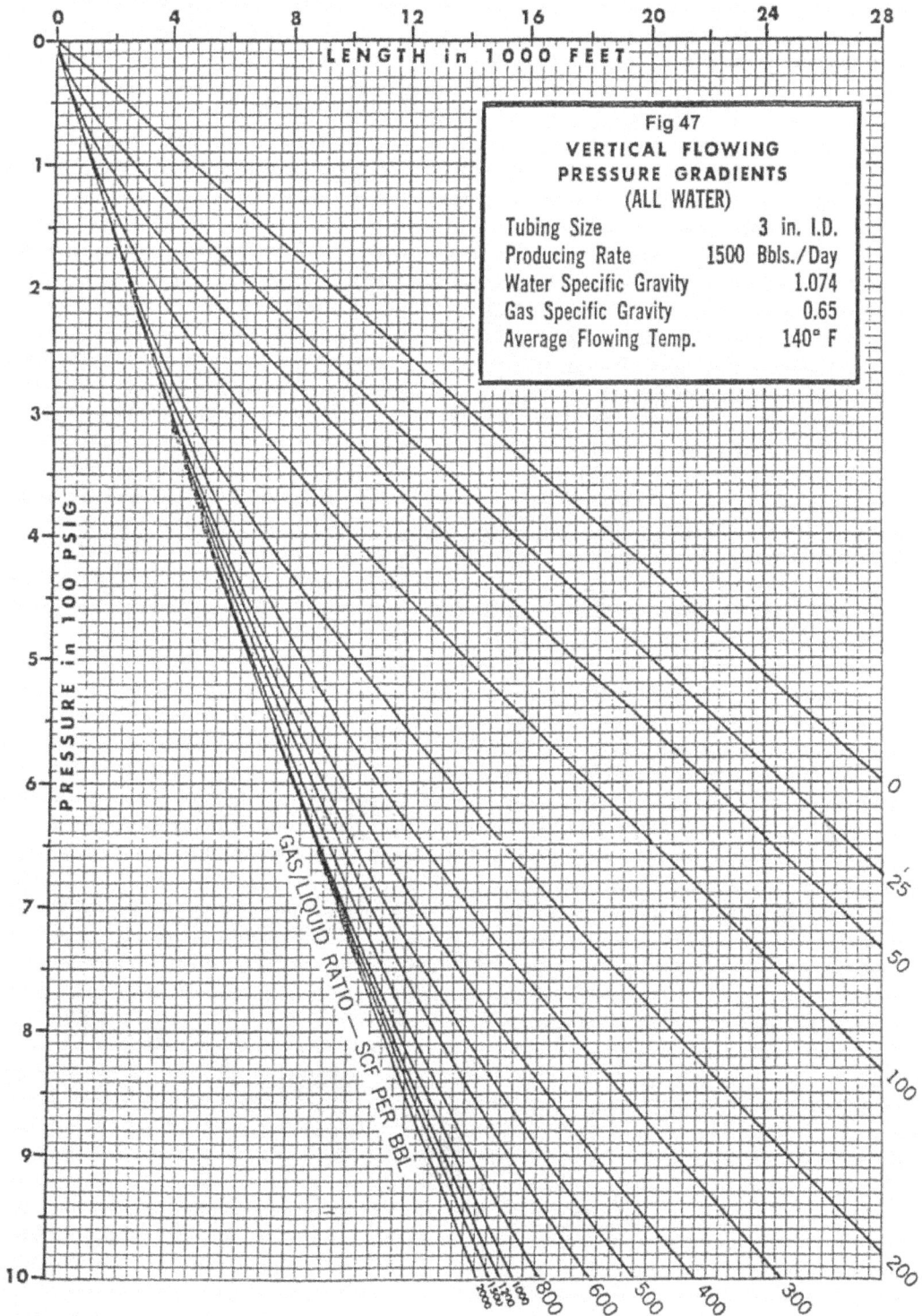

Fig 47
VERTICAL FLOWING
PRESSURE GRADIENTS
(ALL WATER)

Tubing Size	3 in. I.D.
Producing Rate	1500 Bbls./Day
Water Specific Gravity	1.074
Gas Specific Gravity	0.65
Average Flowing Temp.	140° F

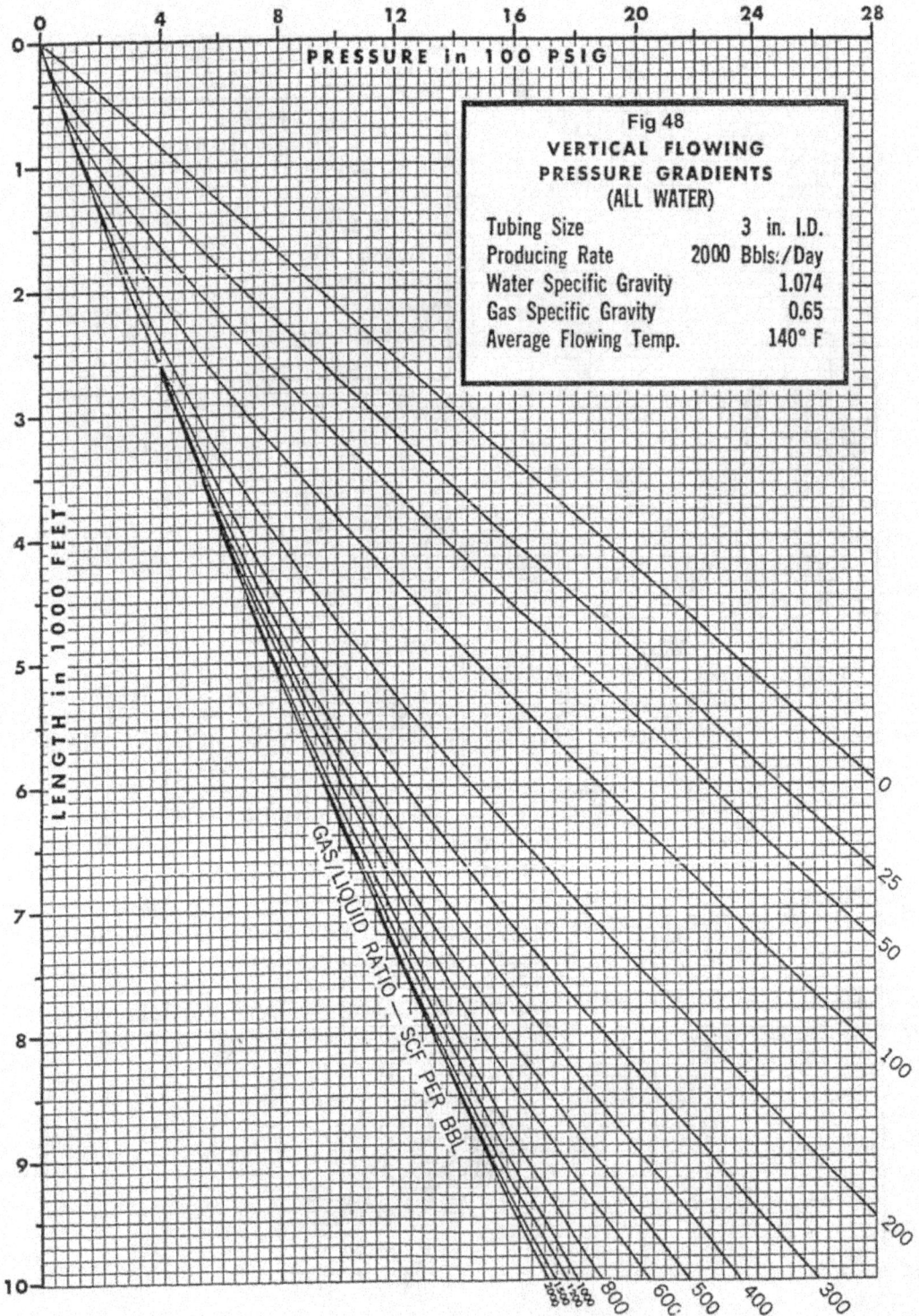

Fig 48
VERTICAL FLOWING
PRESSURE GRADIENTS
(ALL WATER)

Tubing Size	3 in. I.D.
Producing Rate	2000 Bbls:/Day
Water Specific Gravity	1.074
Gas Specific Gravity	0.65
Average Flowing Temp.	140° F

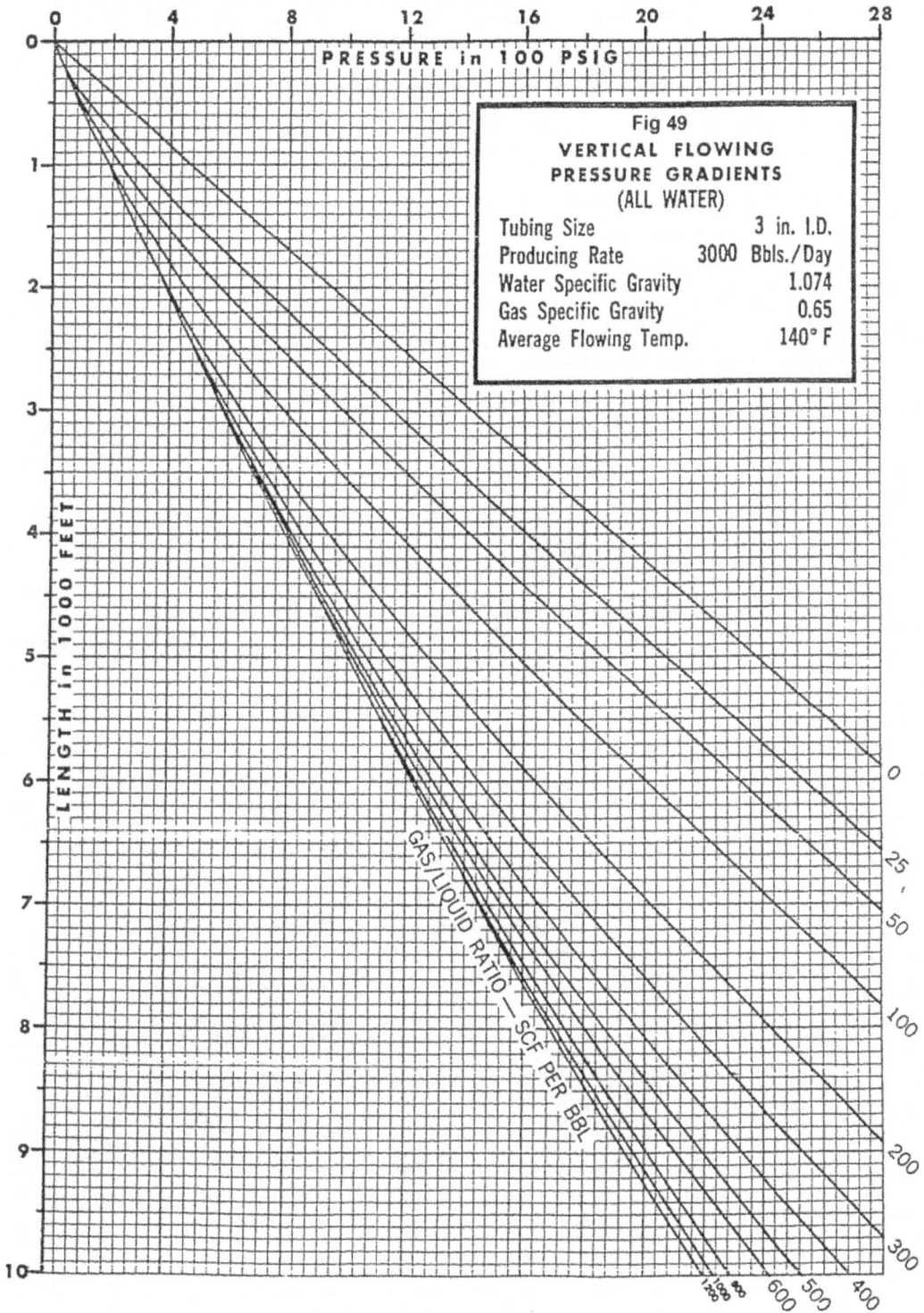

Fig 49
VERTICAL FLOWING
PRESSURE GRADIENTS
(ALL WATER)

Tubing Size	3 in. I.D.
Producing Rate	3000 Bbls./Day
Water Specific Gravity	1.074
Gas Specific Gravity	0.65
Average Flowing Temp.	140° F

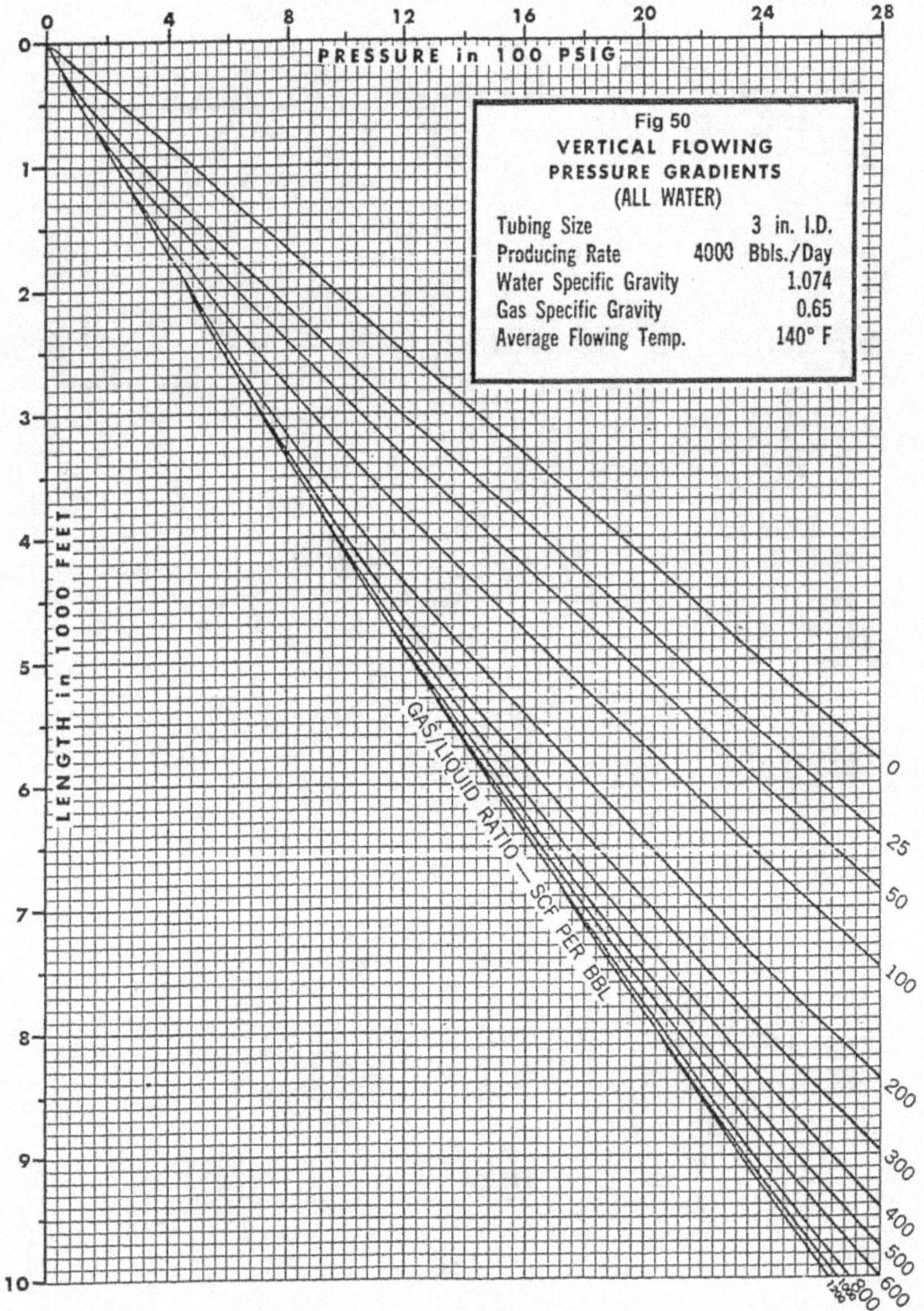

Fig 50
VERTICAL FLOWING
PRESSURE GRADIENTS
(ALL WATER)

Tubing Size	3 in. I.D.
Producing Rate	4000 Bbls./Day
Water Specific Gravity	1.074
Gas Specific Gravity	0.65
Average Flowing Temp.	140° F

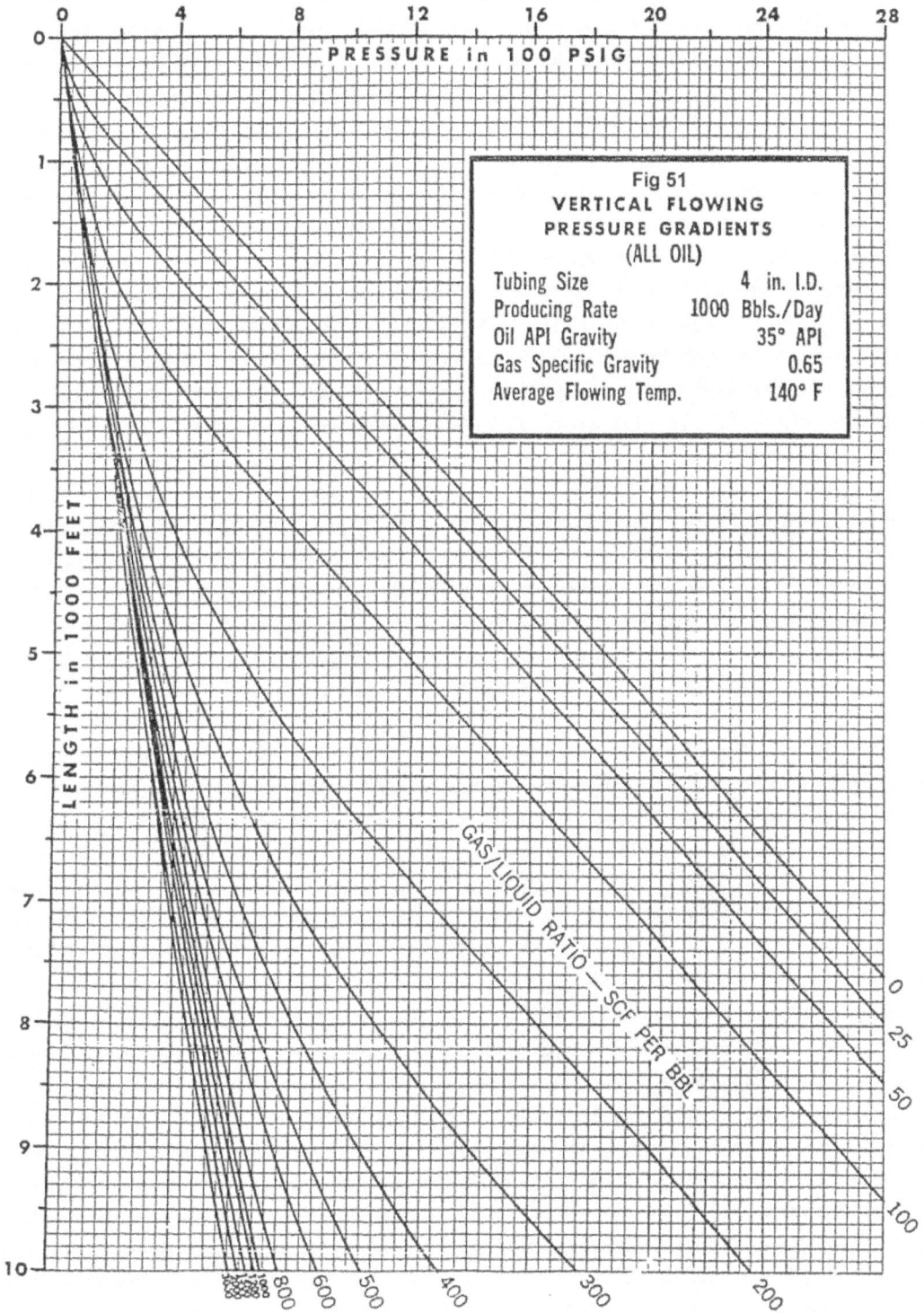

Fig 51
VERTICAL FLOWING
PRESSURE GRADIENTS
(ALL OIL)

Tubing Size	4 in. I.D.
Producing Rate	1000 Bbls./Day
Oil API Gravity	35° API
Gas Specific Gravity	0.65
Average Flowing Temp.	140° F

PRESSURE in 100 PSIG

LENGTH in 1000 FEET

GAS/LIQUID RATIO — SCF PER BBL

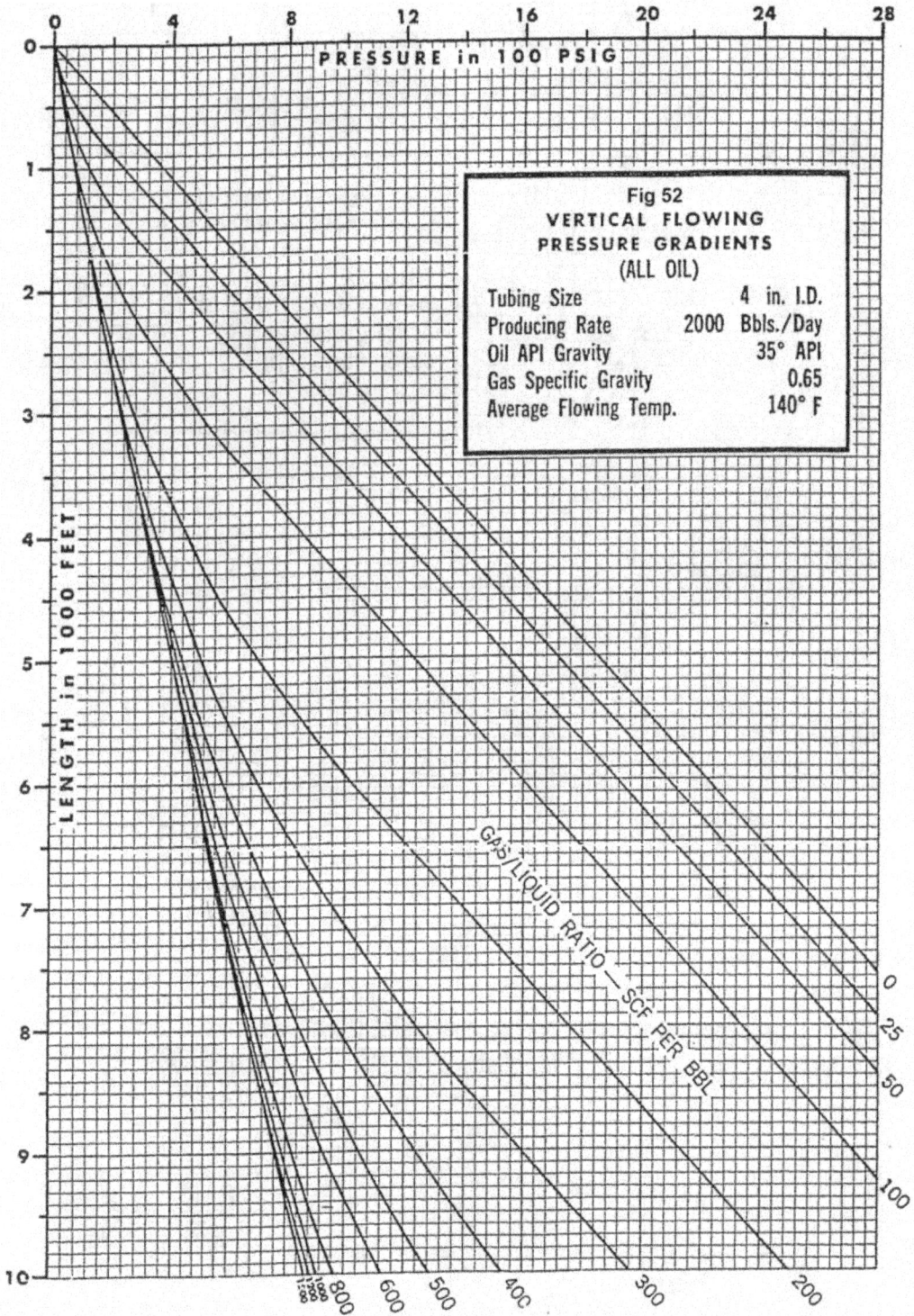

Fig 52
VERTICAL FLOWING
PRESSURE GRADIENTS
(ALL OIL)

Tubing Size	4 in. I.D.
Producing Rate	2000 Bbls./Day
Oil API Gravity	35° API
Gas Specific Gravity	0.65
Average Flowing Temp.	140° F

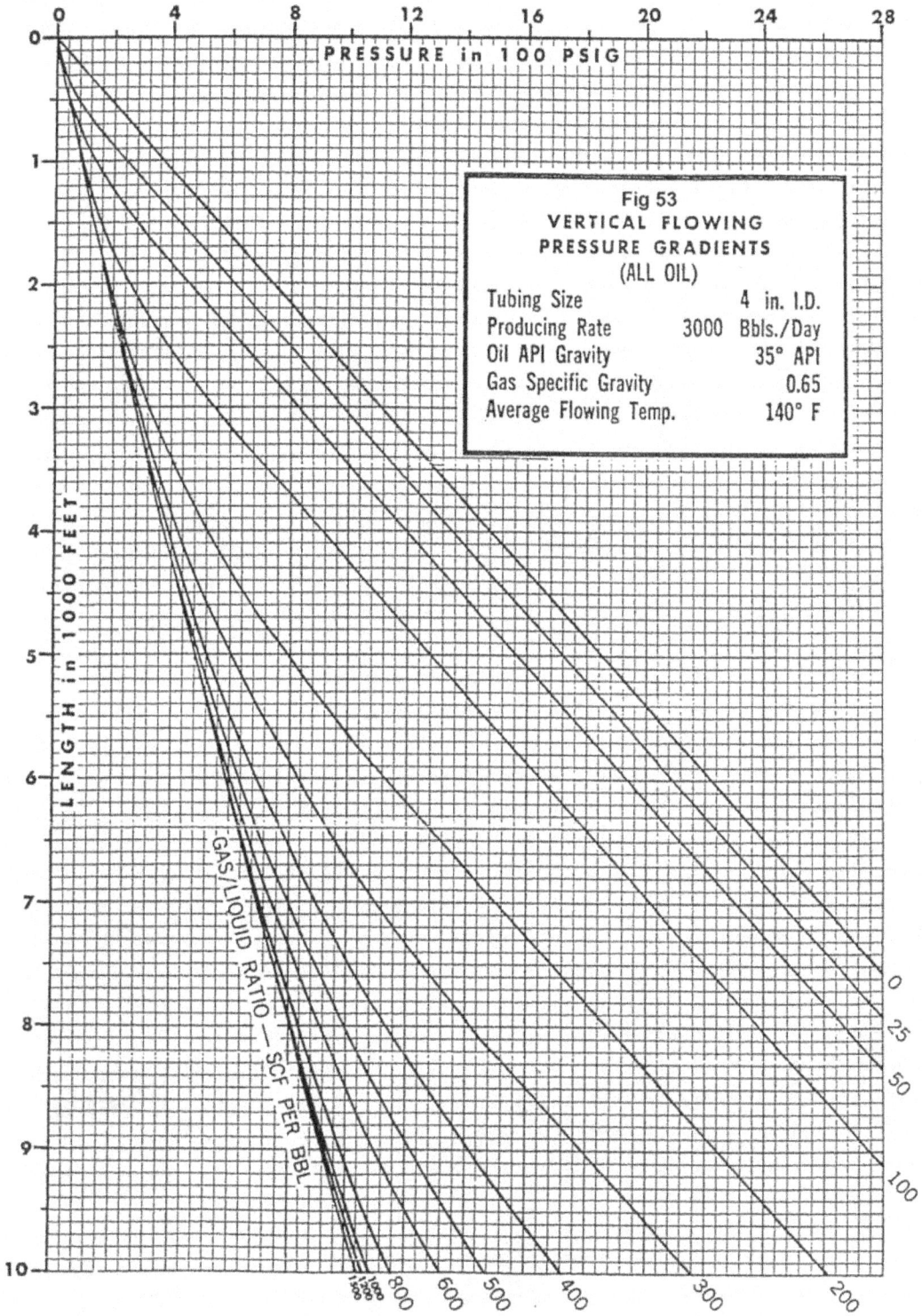

Fig 53
VERTICAL FLOWING
PRESSURE GRADIENTS
(ALL OIL)

Tubing Size	4 in. I.D.
Producing Rate	3000 Bbls./Day
Oil API Gravity	35° API
Gas Specific Gravity	0.65
Average Flowing Temp.	140° F

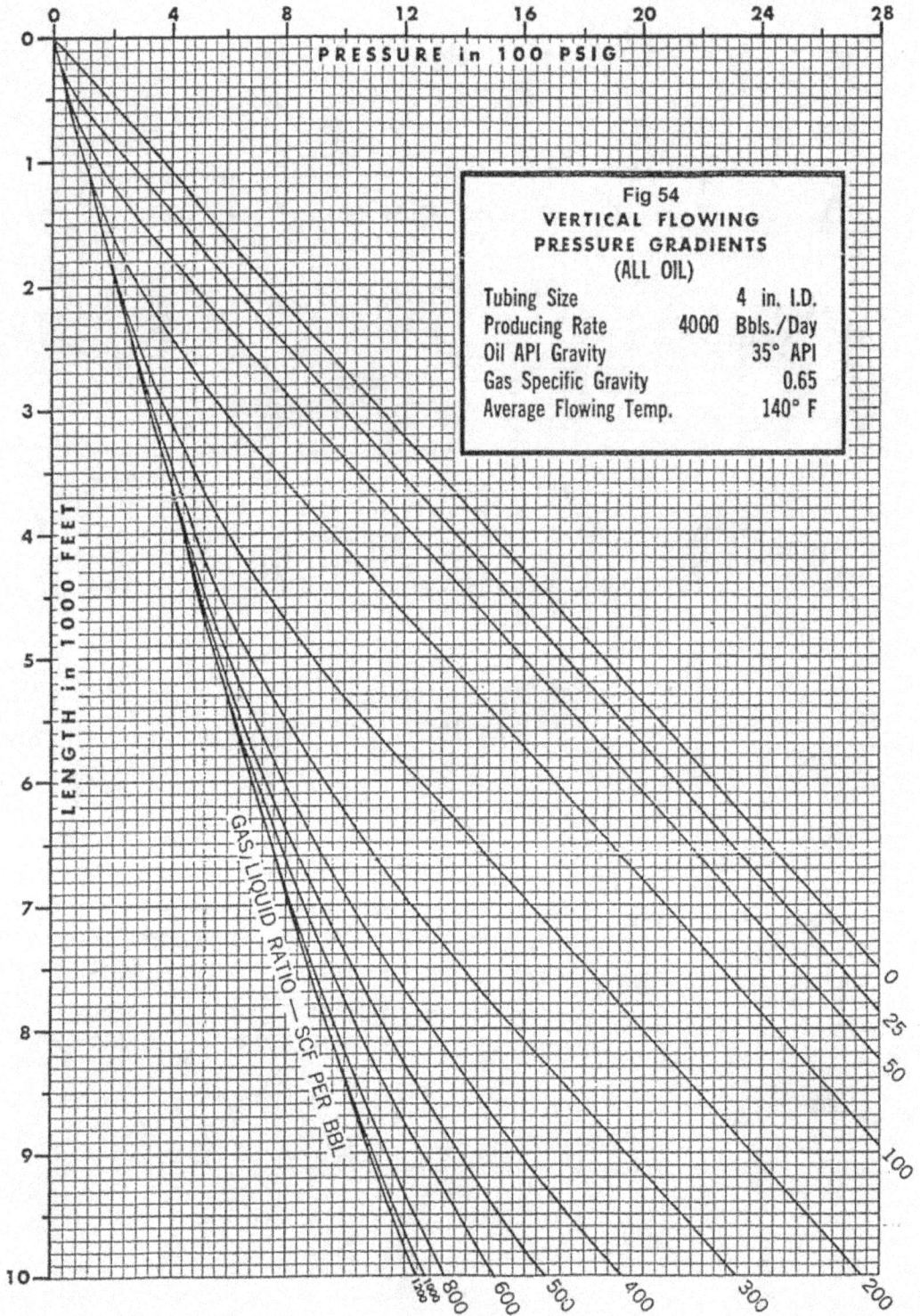

Fig 54
VERTICAL FLOWING
PRESSURE GRADIENTS
(ALL OIL)

Tubing Size	4 in. I.D.
Producing Rate	4000 Bbls./Day
Oil API Gravity	35° API
Gas Specific Gravity	0.65
Average Flowing Temp.	140° F

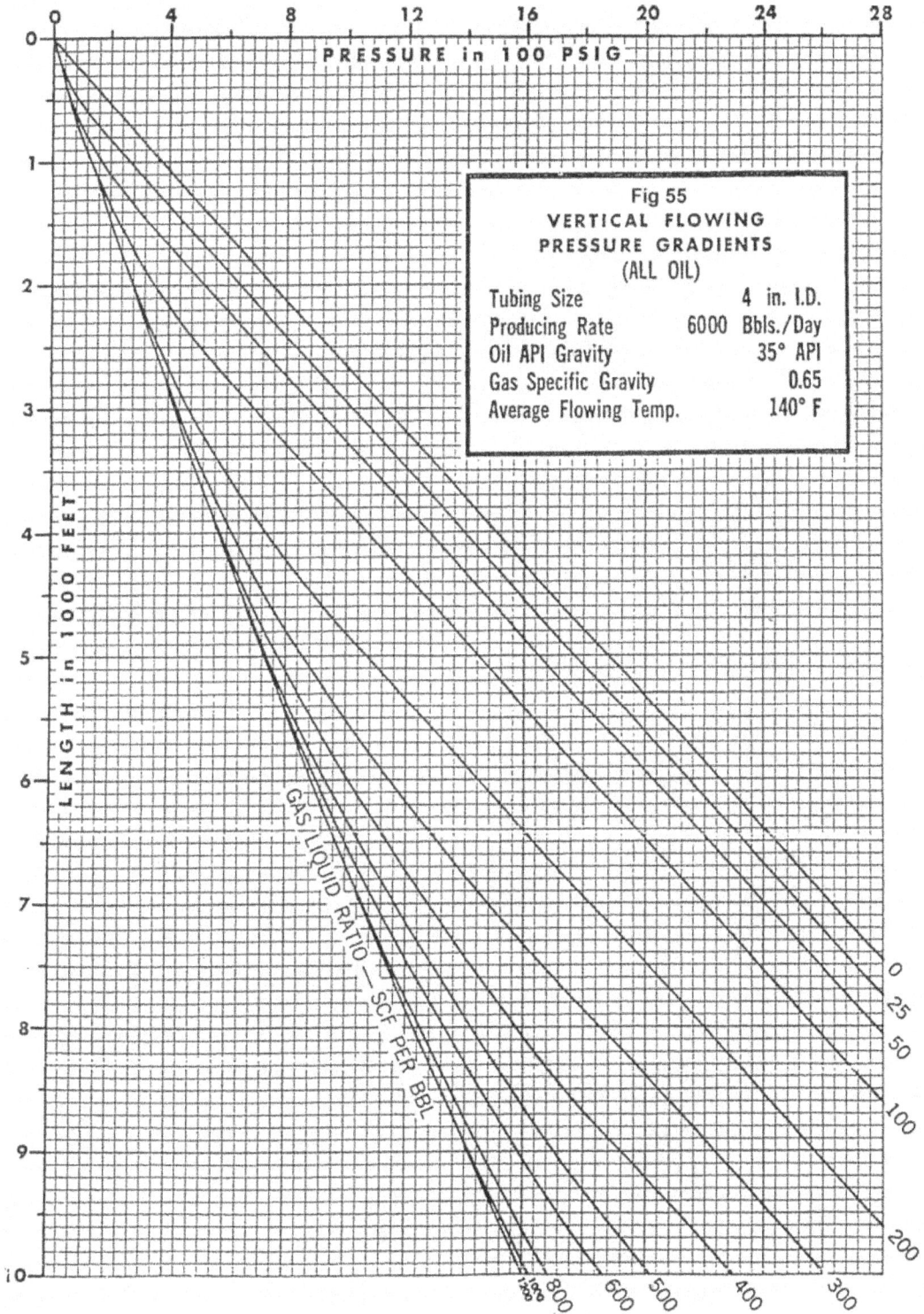

Fig 55
VERTICAL FLOWING
PRESSURE GRADIENTS
(ALL OIL)

Tubing Size	4 in. I.D.
Producing Rate	6000 Bbls./Day
Oil API Gravity	35° API
Gas Specific Gravity	0.65
Average Flowing Temp.	140° F

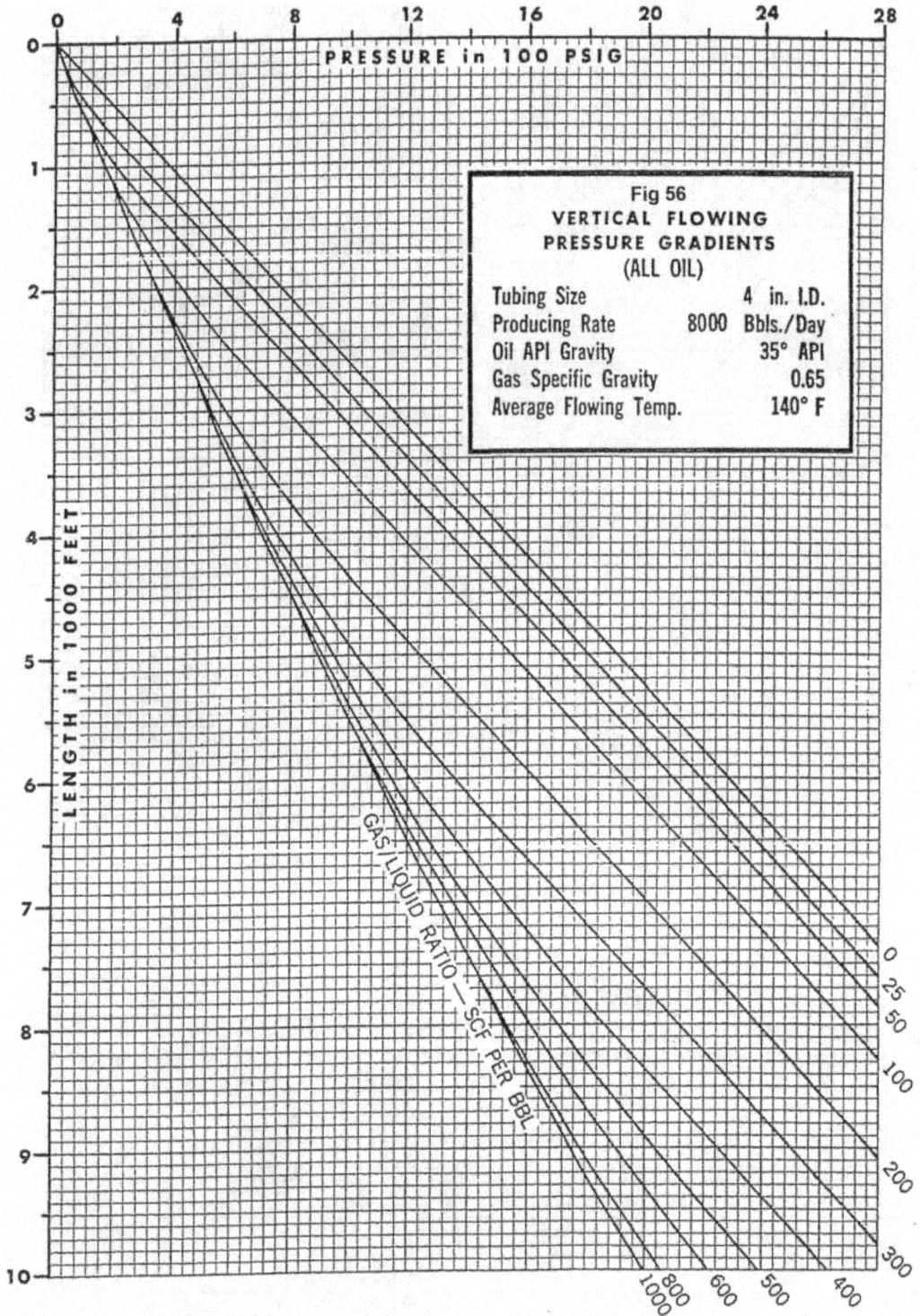

Fig 56
VERTICAL FLOWING
PRESSURE GRADIENTS
(ALL OIL)

Tubing Size	4 in. I.D.
Producing Rate	8000 Bbls./Day
Oil API Gravity	35° API
Gas Specific Gravity	0.65
Average Flowing Temp.	140° F

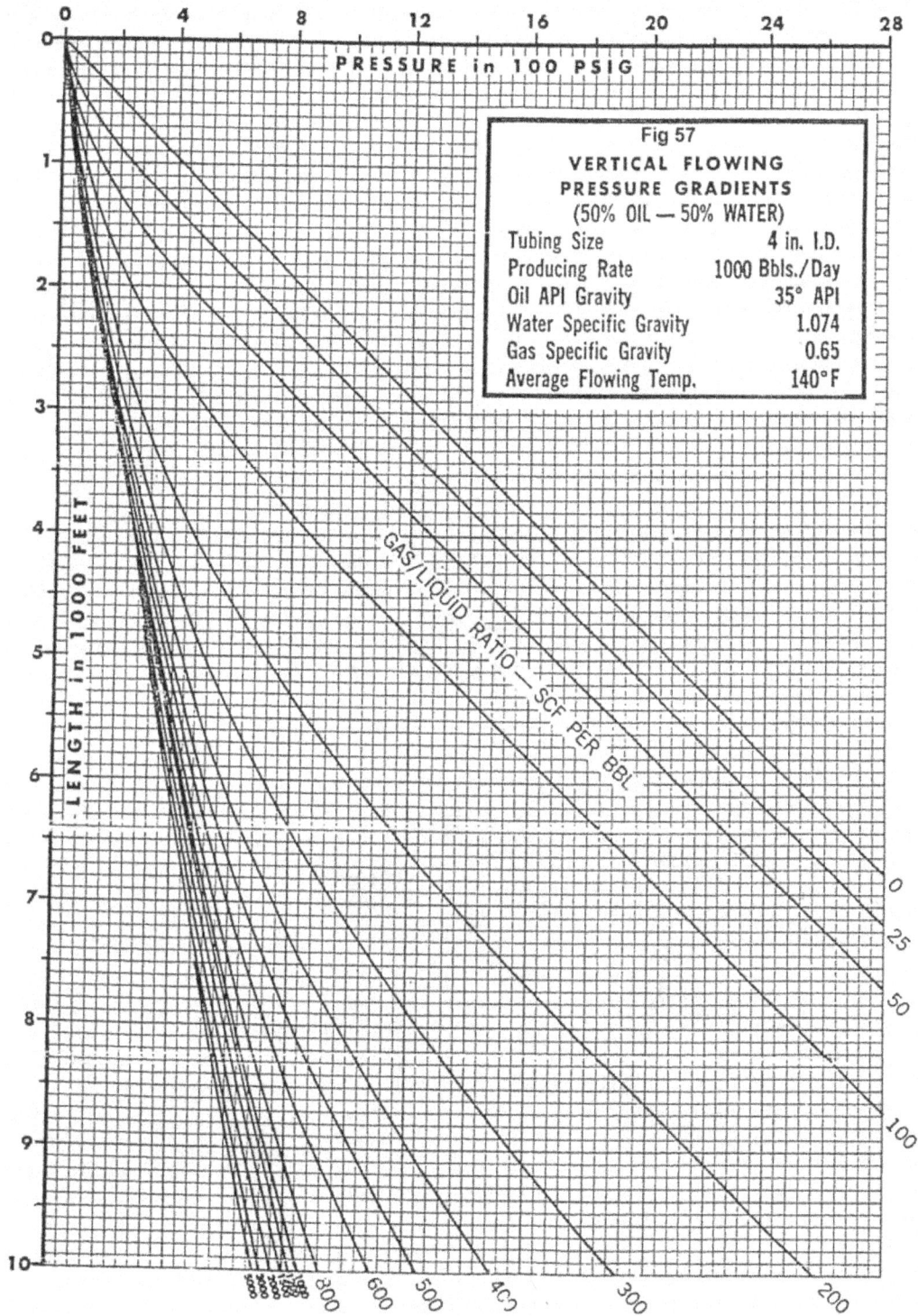

Fig 57

VERTICAL FLOWING
PRESSURE GRADIENTS
(50% OIL — 50% WATER)

Tubing Size	4 in. I.D.
Producing Rate	1000 Bbls./Day
Oil API Gravity	35° API
Water Specific Gravity	1.074
Gas Specific Gravity	0.65
Average Flowing Temp.	140°F

PRESSURE in 100 PSIG

LENGTH in 1000 FEET

GAS/LIQUID RATIO — SCF PER BBL

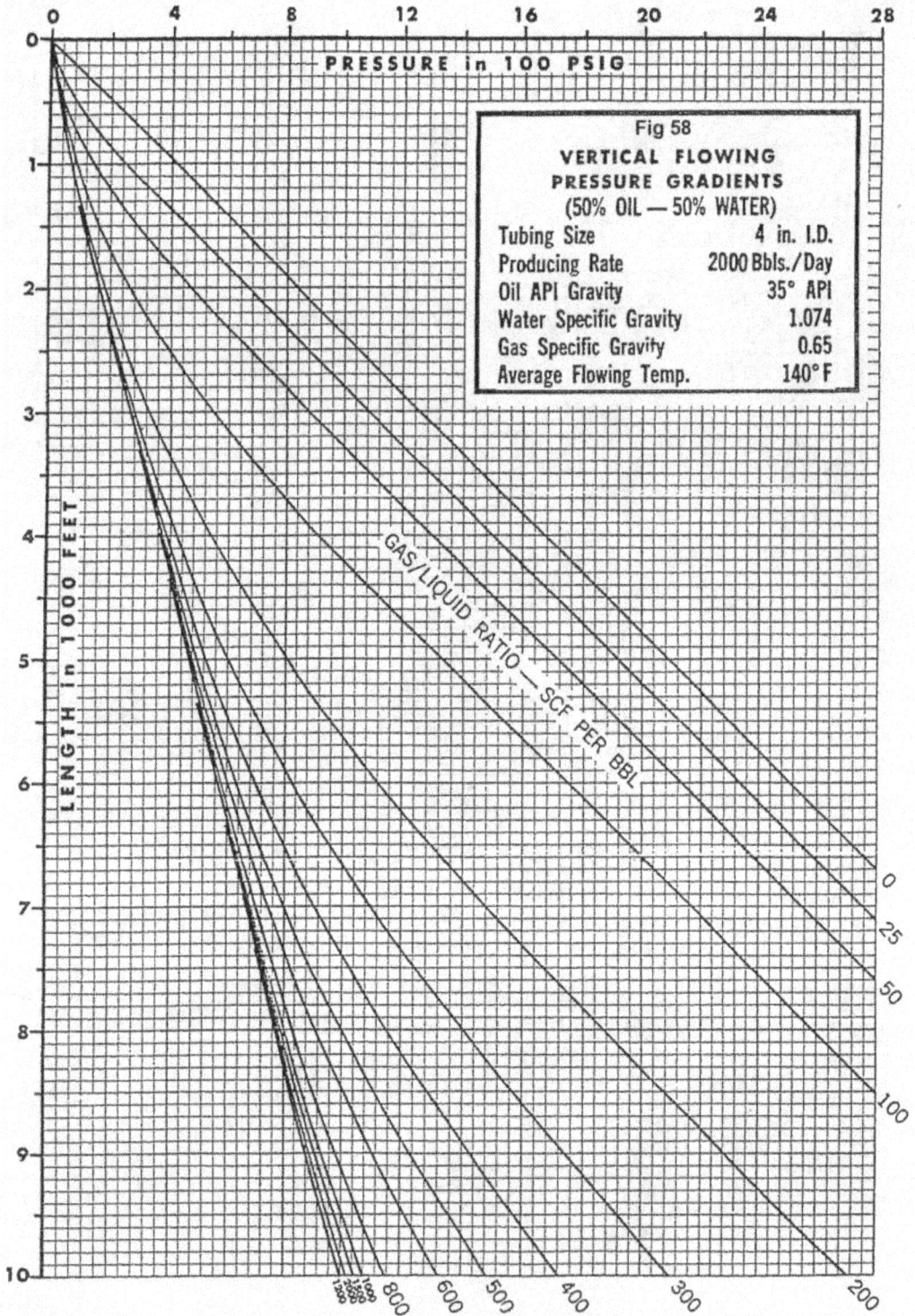

Fig 58
VERTICAL FLOWING
PRESSURE GRADIENTS
(50% OIL — 50% WATER)

Tubing Size	4 in. I.D.
Producing Rate	2000 Bbls./Day
Oil API Gravity	35° API
Water Specific Gravity	1.074
Gas Specific Gravity	0.65
Average Flowing Temp.	140°F

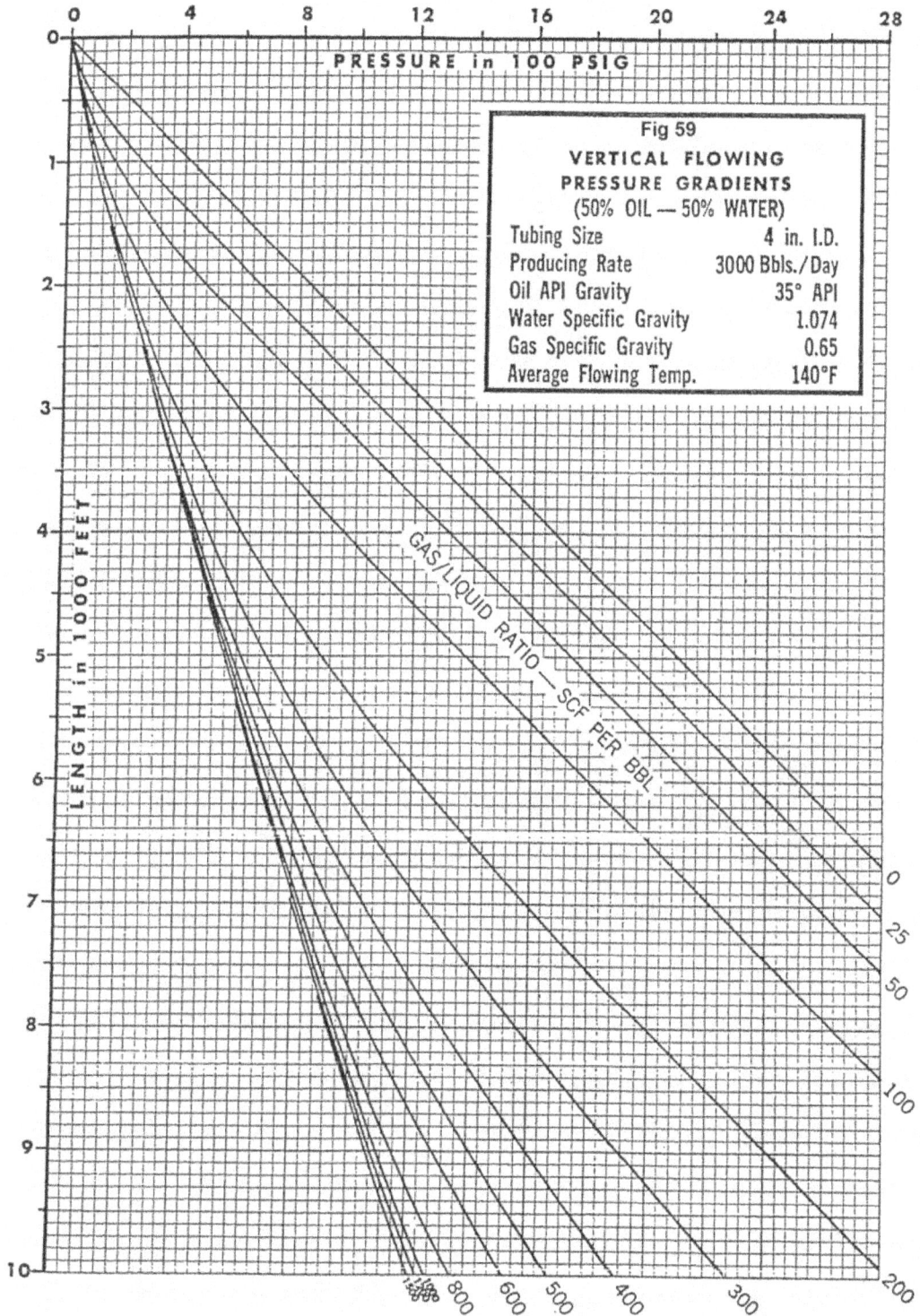

Fig 59

VERTICAL FLOWING
PRESSURE GRADIENTS
(50% OIL — 50% WATER)

Tubing Size	4 in. I.D.
Producing Rate	3000 Bbls./Day
Oil API Gravity	35° API
Water Specific Gravity	1.074
Gas Specific Gravity	0.65
Average Flowing Temp.	140°F

PRESSURE in 100 PSIG

LENGTH in 1000 FEET

GAS/LIQUID RATIO — SCF PER BBL

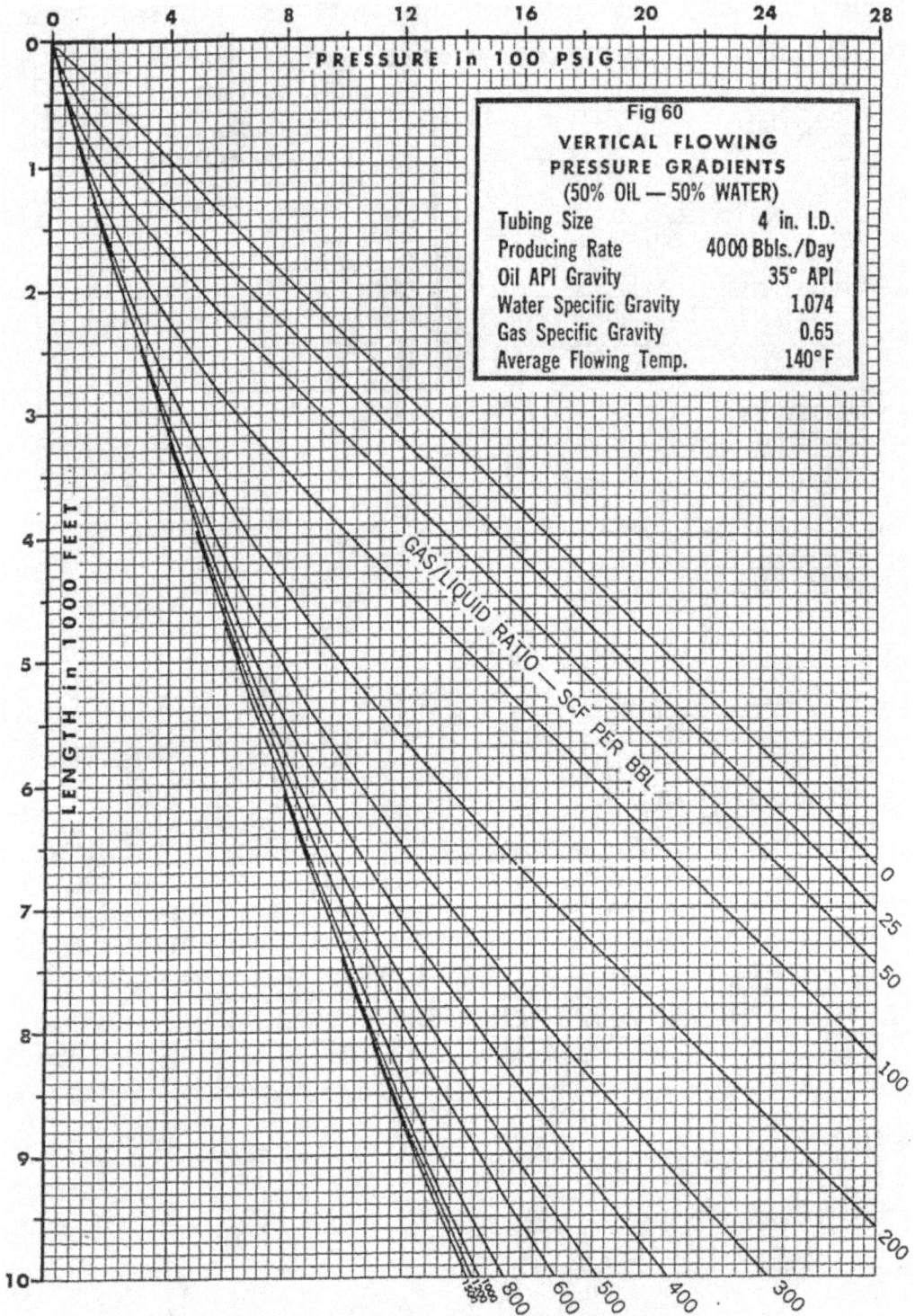

Fig 60

VERTICAL FLOWING
PRESSURE GRADIENTS
(50% OIL — 50% WATER)

Tubing Size	4 in. I.D.
Producing Rate	4000 Bbls./Day
Oil API Gravity	35° API
Water Specific Gravity	1.074
Gas Specific Gravity	0.65
Average Flowing Temp.	140°F

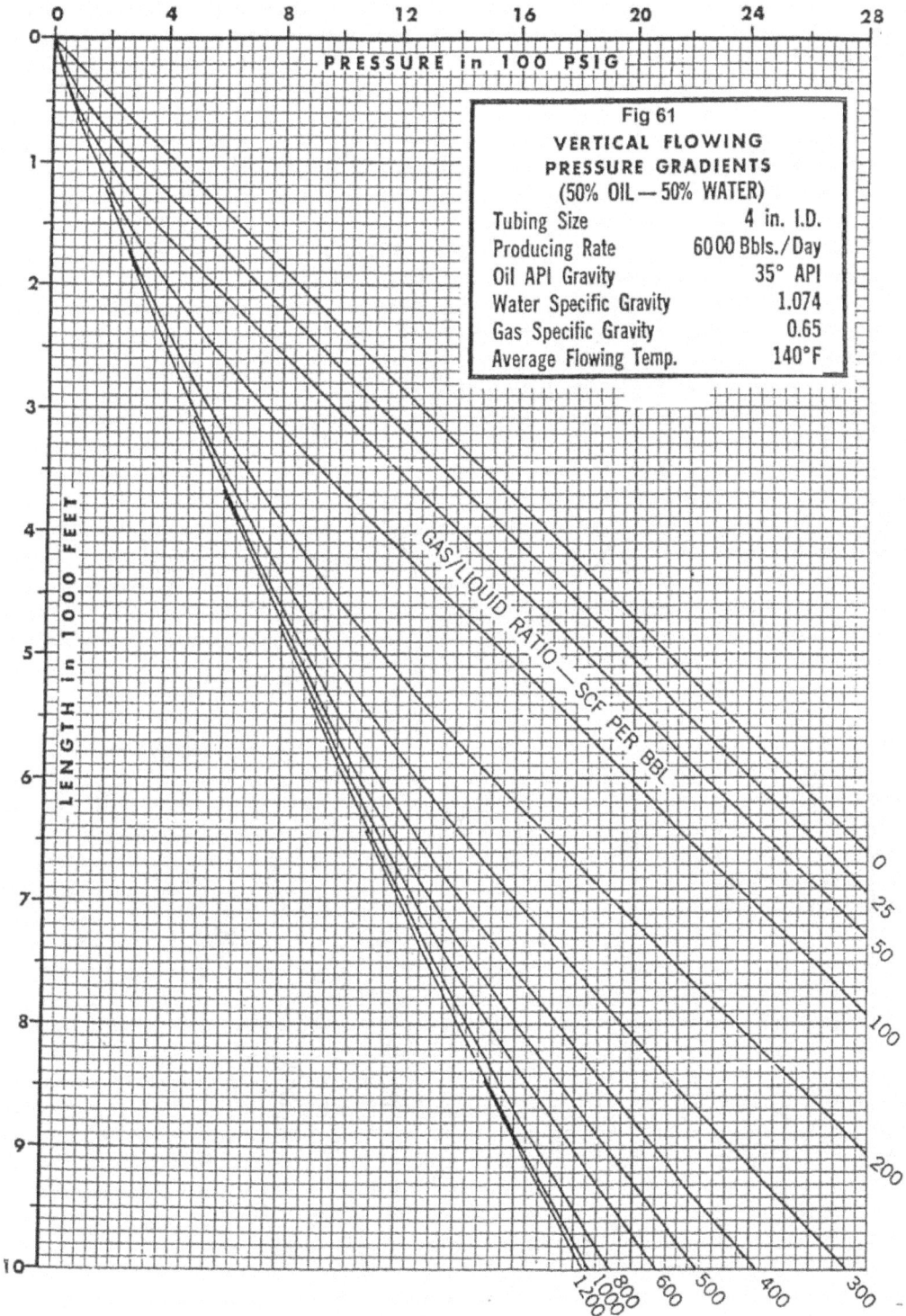

Fig 61
VERTICAL FLOWING
PRESSURE GRADIENTS
(50% OIL — 50% WATER)

Tubing Size	4 in. I.D.
Producing Rate	6000 Bbls./Day
Oil API Gravity	35° API
Water Specific Gravity	1.074
Gas Specific Gravity	0.65
Average Flowing Temp.	140°F

PRESSURE in 100 PSIG

LENGTH in 1000 FEET

GAS/LIQUID RATIO — SCF PER BBL

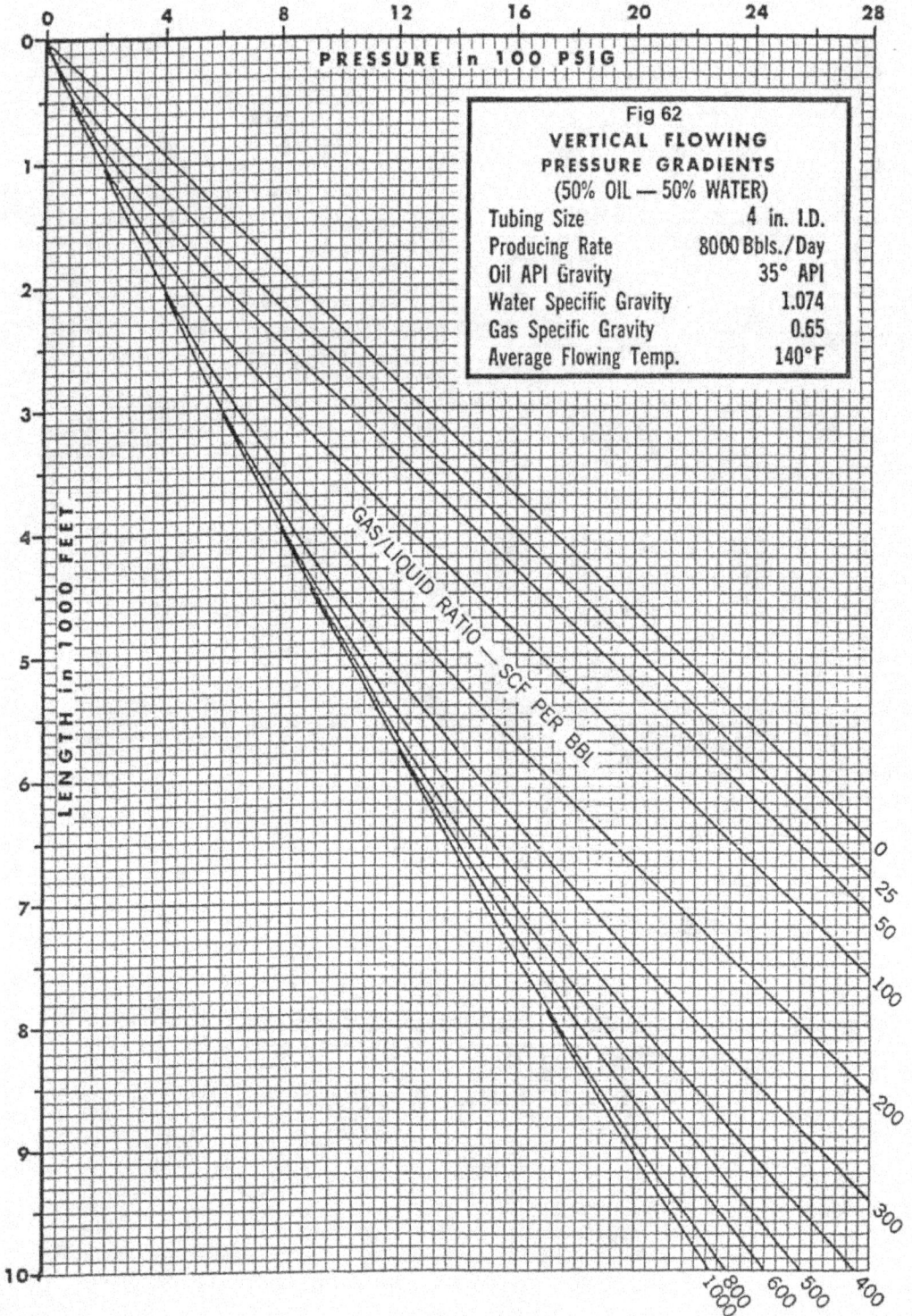

Fig 62
VERTICAL FLOWING
PRESSURE GRADIENTS
(50% OIL — 50% WATER)

Tubing Size	4 in. I.D.
Producing Rate	8000 Bbls./Day
Oil API Gravity	35° API
Water Specific Gravity	1.074
Gas Specific Gravity	0.65
Average Flowing Temp.	140°F

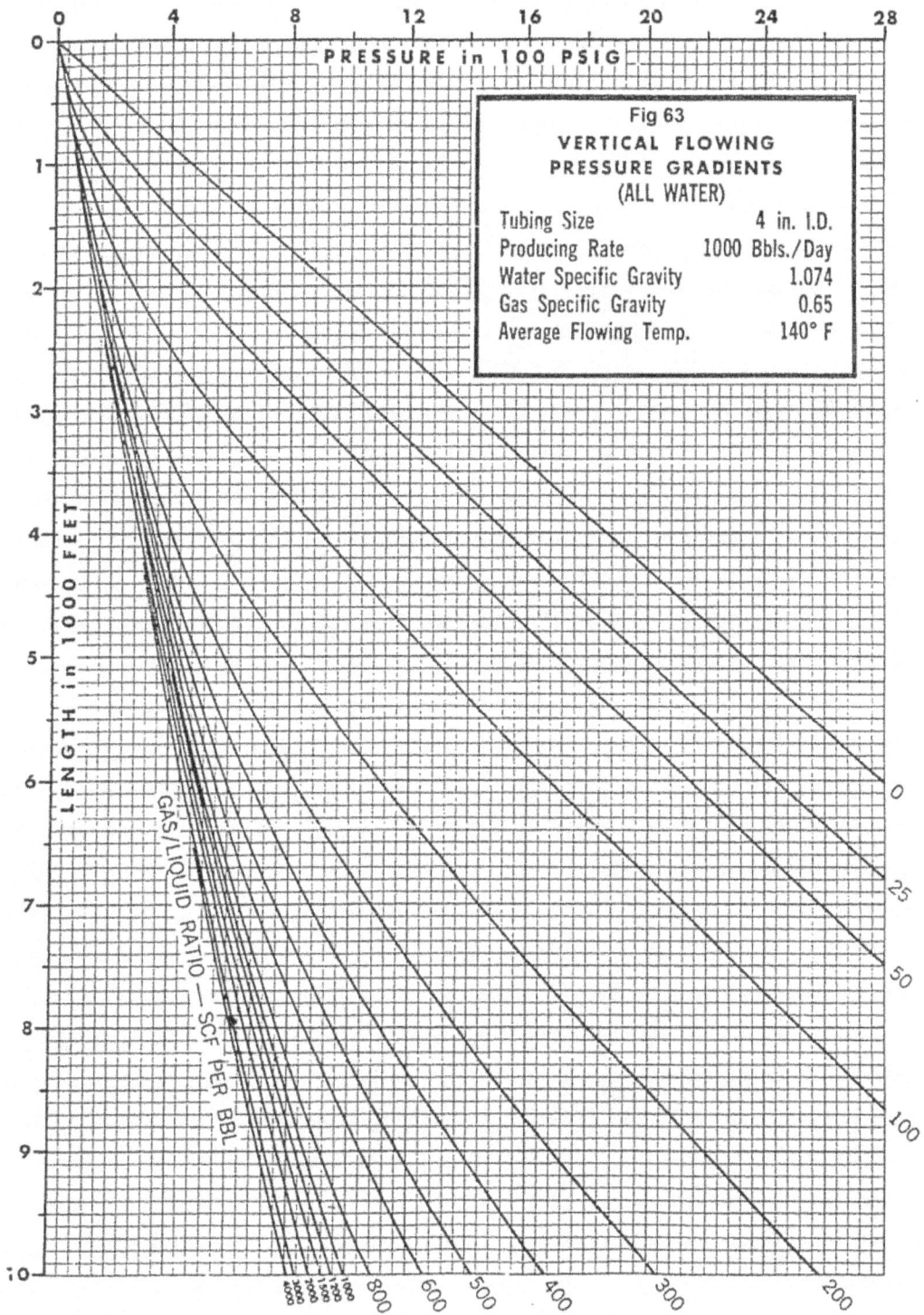

Fig 63
VERTICAL FLOWING
PRESSURE GRADIENTS
(ALL WATER)

Tubing Size	4 in. I.D.
Producing Rate	1000 Bbls./Day
Water Specific Gravity	1.074
Gas Specific Gravity	0.65
Average Flowing Temp.	140° F

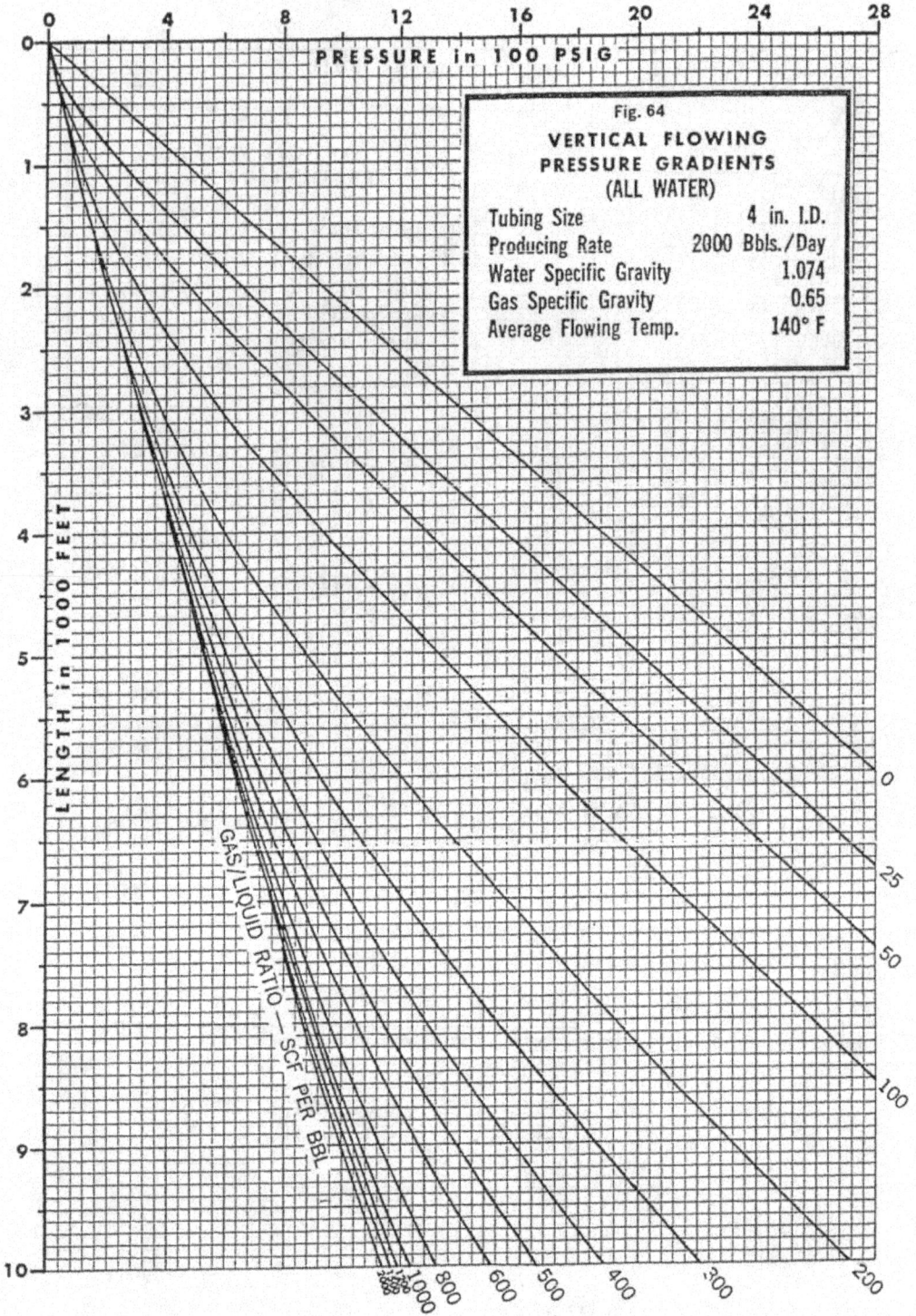

Fig. 64
VERTICAL FLOWING
PRESSURE GRADIENTS
(ALL WATER)

Tubing Size	4 in. I.D.
Producing Rate	2000 Bbls./Day
Water Specific Gravity	1.074
Gas Specific Gravity	0.65
Average Flowing Temp.	140° F

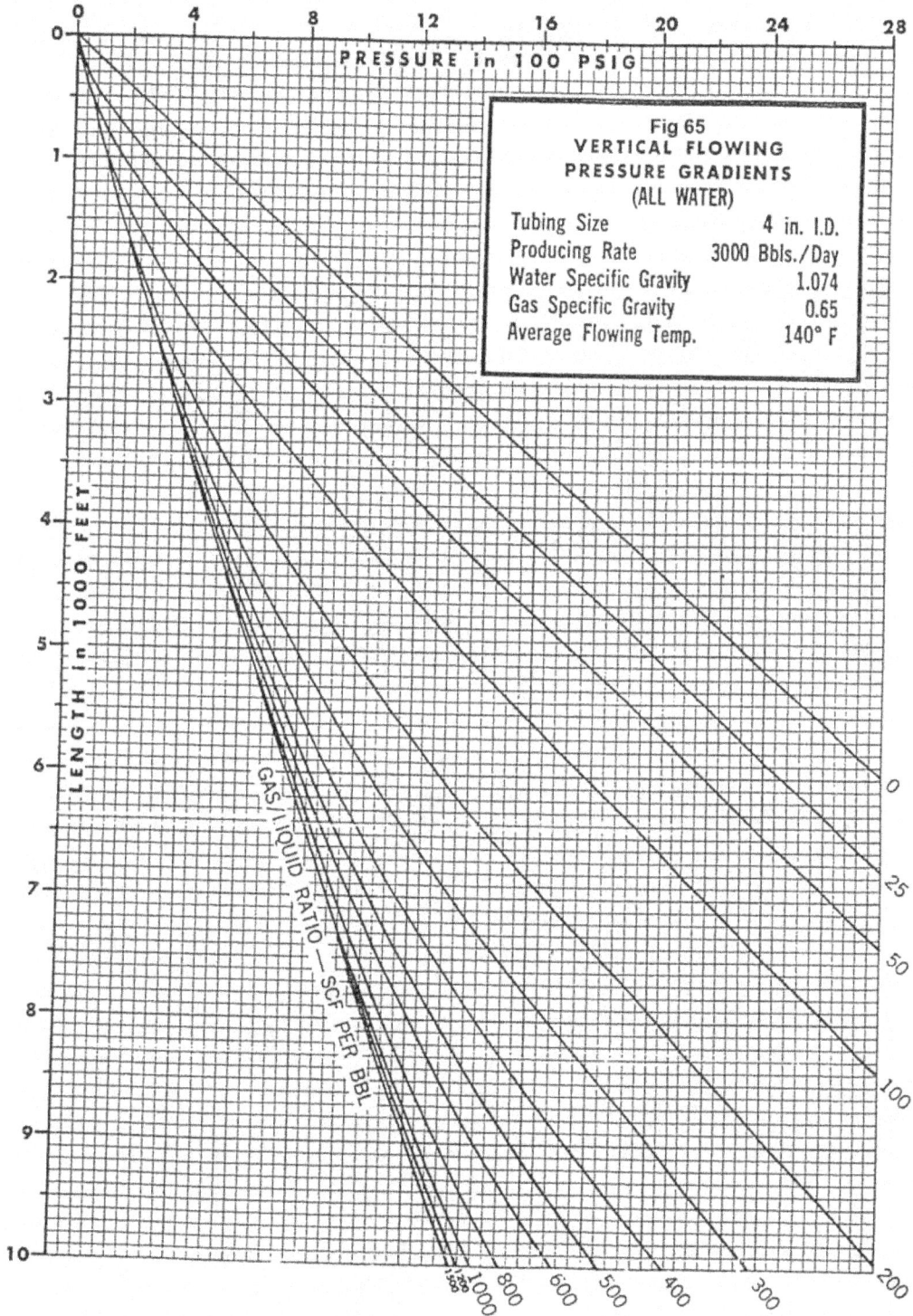

Fig 65
VERTICAL FLOWING
PRESSURE GRADIENTS
(ALL WATER)

Tubing Size	4 in. I.D.
Producing Rate	3000 Bbls./Day
Water Specific Gravity	1.074
Gas Specific Gravity	0.65
Average Flowing Temp.	140° F

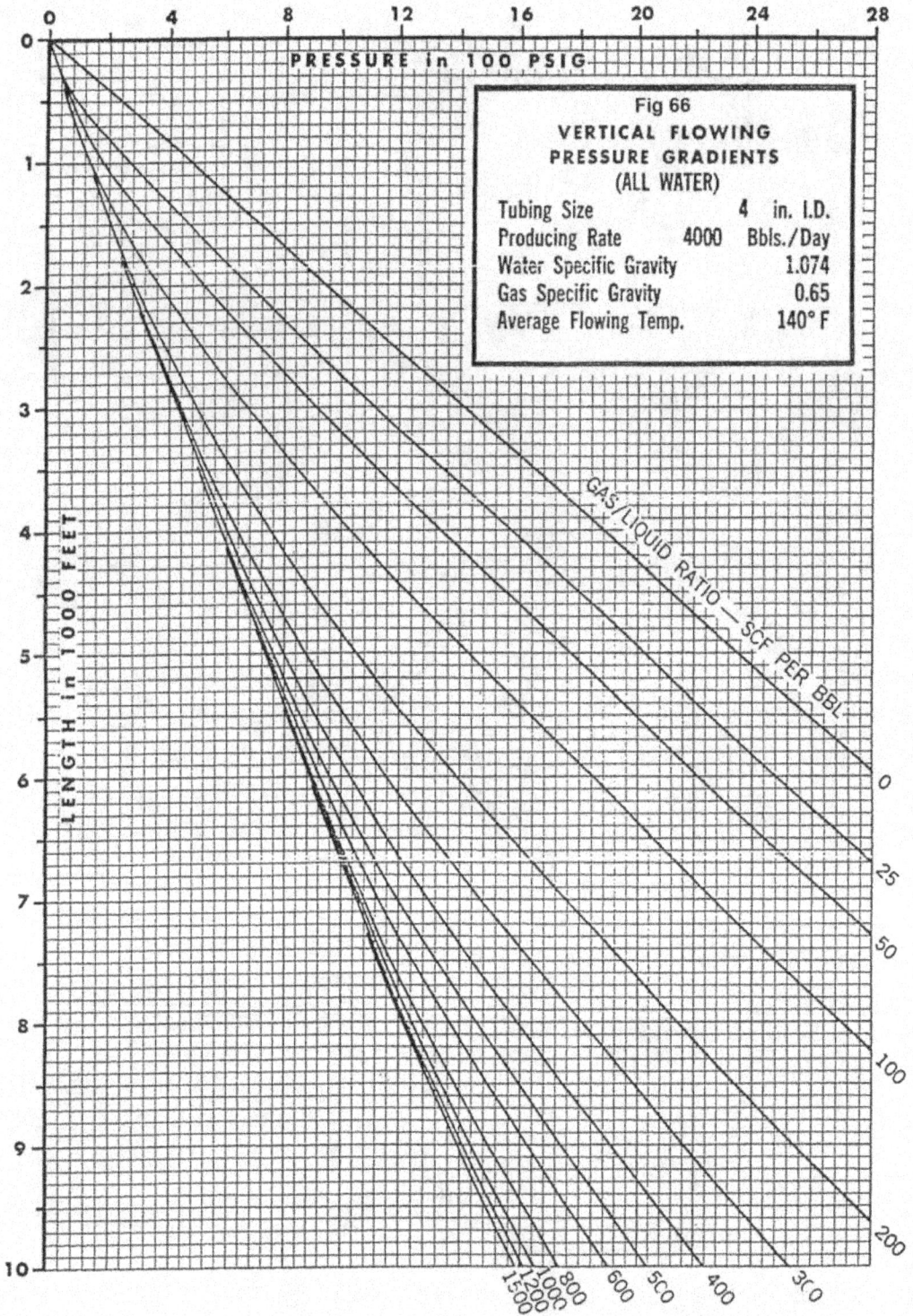

Fig 66
VERTICAL FLOWING
PRESSURE GRADIENTS
(ALL WATER)

Tubing Size	4	in. I.D.
Producing Rate	4000	Bbls./Day
Water Specific Gravity		1.074
Gas Specific Gravity		0.65
Average Flowing Temp.		140° F

PRESSURE in 100 PSIG

LENGTH in 1000 FEET

GAS/LIQUID RATIO — SCF PER BBL

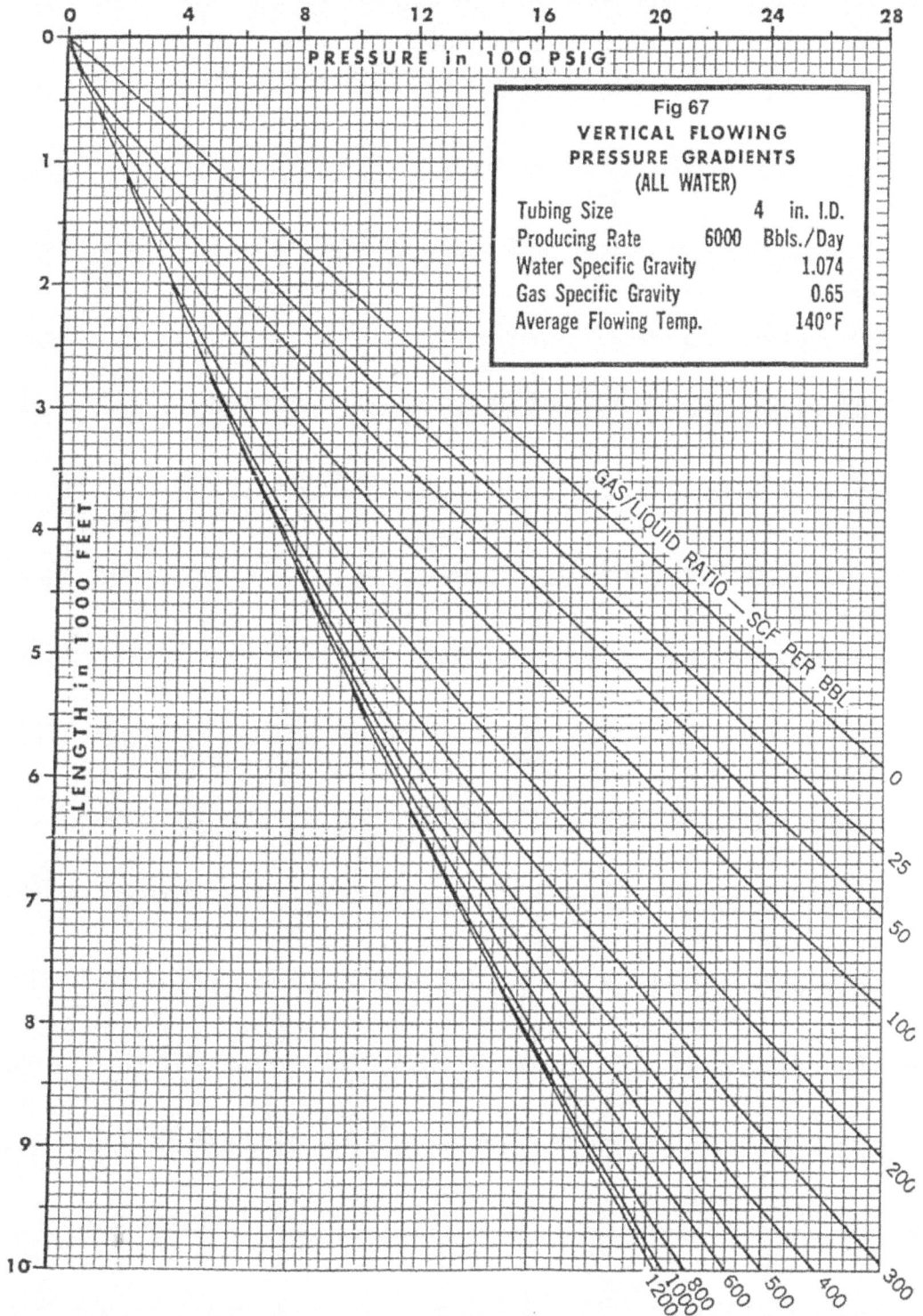

Fig 67
VERTICAL FLOWING
PRESSURE GRADIENTS
(ALL WATER)

Tubing Size	4	in. I.D.
Producing Rate	6000	Bbls./Day
Water Specific Gravity		1.074
Gas Specific Gravity		0.65
Average Flowing Temp.		140°F

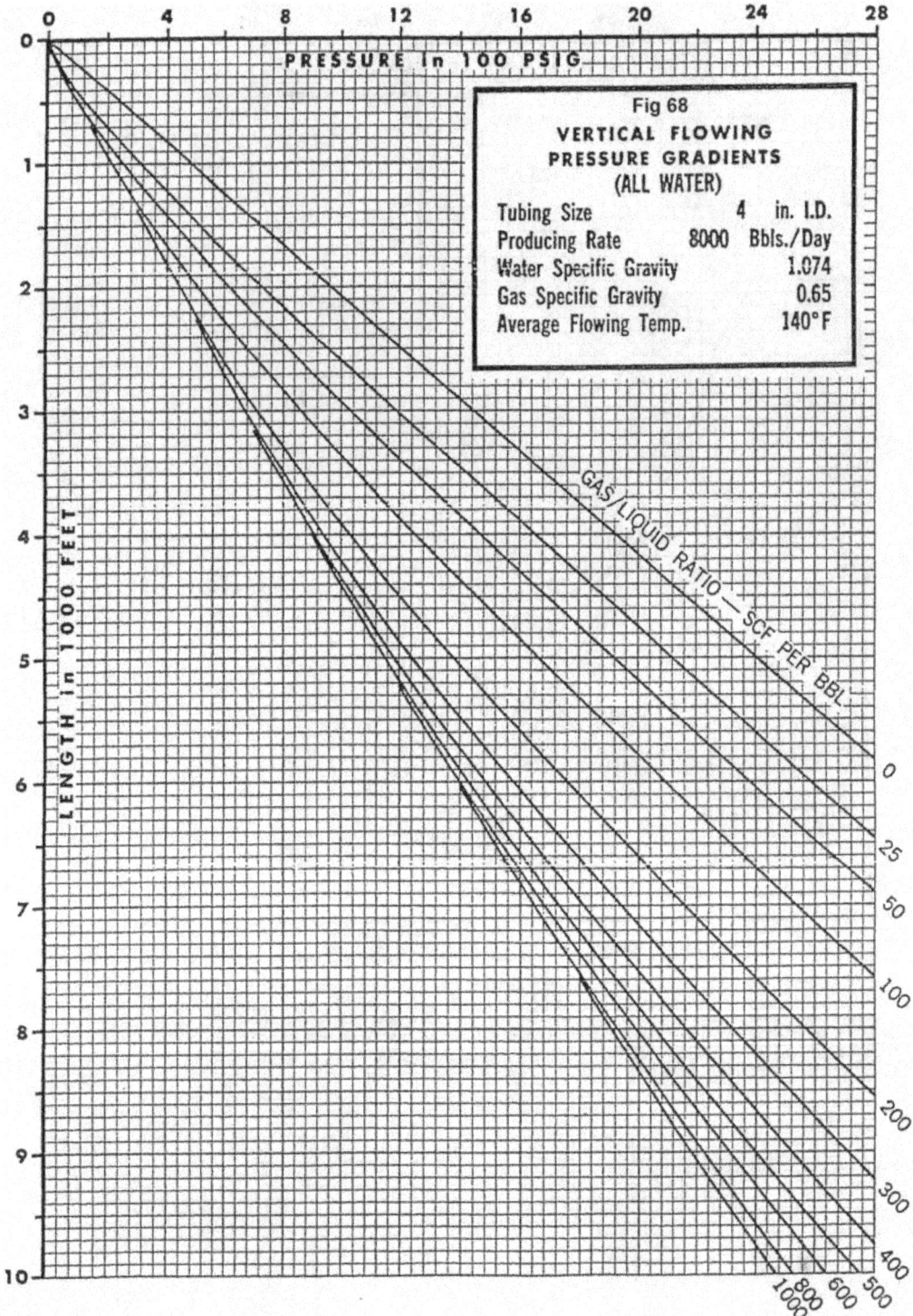

Fig 68
VERTICAL FLOWING
PRESSURE GRADIENTS
(ALL WATER)

Tubing Size	4 in. I.D.
Producing Rate	8000 Bbls./Day
Water Specific Gravity	1.074
Gas Specific Gravity	0.65
Average Flowing Temp.	140°F

Appendix B

PROBLEMS

Gas Lift

1. 1 MM SCFD gas is available for a gas lift well. 1,000 psia is kick off pressure of the well and then gas injection pressure of 800 psia a wellhead pressure is 130 psia. Calculate the casing pressure margin is 50 psi.

Solution: The hydrostatic pressure of well fluid (26° API oil) is (0.39 psi/ft.) (5,200 ft.), or 2,028 psig, which is greater than the given reservoir pressure of 2,000 psia. Therefore, the well does not flow naturally. The static liquid level depth,

or, $5,200 - (2,000 - 14.7)/(0.39) = 110$ ft.

Depth of the top valve is calculated:

or, $D_1 = \dfrac{1,000 - 50 - 130}{0.39 - \dfrac{1,000 - 50}{40,000}} = 2,245$ ft. > 110 ft.

Tubing pressure margin at surface is (0.25) (800), or 200 psi. Depth of the second valve is:

or, $D_2 = \dfrac{1,000 - 50 - 130 + (0.39 - 0.052)(2,245)}{0.39 - \dfrac{1,000 - 50}{40,000}} = 3,004$ ft.

or, $D_3 = \dfrac{1,000 - 50 - 330 + (0.39 - 0.052)(3,004)}{0.39 - \dfrac{1,000 - 50}{40,000}} = 3,676$ ft.

or, $D_4 = \dfrac{1,000 - 50 - 330 + (0.39 - 0.052)(3,676)}{0.39 - \dfrac{1,000 - 50}{40,000}} = 4,269$ ft.

or, $D_5 = \dfrac{1,000 - 50 - 330 + (0.39 - 0.052)(4,269)}{0.39 - \dfrac{1,000 - 50}{40,000}} = 4,792$ ft.

This is the depth of the operating valve.

2. Calculate the design of gas lifts valves based on the following data:

Pay Zone depth	:	6,500 ft.
Casing size and weight	:	7 in. (PPF 23 lb)
Tubing	:	2 3/8 in., 4.7 lb. (1.995 in ID)

Liquid level surface

Kill fluid gradient	:	0.4 psi/ft
Gas gravity	:	0.75
Bottom-hole-temperature	:	170°F
Temperature surface flowing	:	100°F
Injection depth	:	6,300 ft.

Minimum tubing pressure at injection

Point	:	600 psi
Pressure kickoff	:	1,000 psi
Pressure surface operating	:	900 psi
Pressure of wellhead	:	140 psi
Tubing pressure margin at surface	:	180 psi
Casing pressure margin	:	0 psi

Solution: Design tubing pressure at surface (phf, d)

	:	140 + 180 = 320 psia
Design tubing pressure gradient (G_{fd})	:	(600 – 320)/6,300 = 0.044 psi/ft.
Temperature gradient (G_t)	:	(170 –100)/6,300 = 0.011 F/ft.
		1 – R 1.0 – 0.2562 = 0.7438
		T.E.F. = R/(1 – R) 0.2562/0.7438
		= 0.3444

Depth of the top valve is,

$$D_1 = \frac{1,000 - 0 - 120}{0.40 - \dfrac{1,000 - 0}{40,000}} = 2,347 \text{ ft.}$$

Temperature at the top valve	:	100 + (0.011) (2,347) = 126°F
Design tubing pressure at the top valve	:	320 + (0.044) (2,347) = 424 psia

For constant surface opening pressure of 900 psia the valve opening pressure is:

$$P_{vo1} = (900) \left(1 + \frac{2,347}{40,000}\right) = 953 \text{ psia}$$

The dome pressure at the valve depth is:

$$P_{vc} = 817 + (0)(0.2526)(424) = 817 \text{ psia}$$

The dome pressure at 60°F can be calculated with trial-and-error method. The first estimate is given by ideal gas law:

$$P_d \text{ at } 60°\text{ F} = \frac{520 P_d}{T_d} = \frac{(520)(817)}{(126+460)} = 725 \text{ psia}$$

Spread sheet programs give $Z_{60F} = 0.80$ at 725 psia and 60°F. The same spread sheet gives $Z_d = 0.85$ at 817 psia and 126°F.

$$P_d \text{ at } 60°\text{F} = \frac{(520)[0.80]P_d}{(126+460)(0.85)} (817) = 683 \text{ psia}$$

Test rack opening pressure is,

$$P_{tro} = \frac{683}{0.7438} + 0 = 918 \text{ psia}$$

Summary of Results for Problem 2

Valve No.	Valve Depth (ft.)	Tempe-rature (°F)	Design Tubing Pressure (psia)	Surface Opening Pressure (psia)	Dome Pressure at Depth (psia)	Valve Closing Pressure (psia)	Valve Closing Pressure (psia)	Dome Pressure at 60°F (psia)	Test Rack Opening (psia)
1	2,347	126	424	900	953	817	817	683	918
2	3,747	142	487	900	984	857	857	707	950
3	5,065	156	545	900	1,014	894	894	702	944
4	6,300	170	600	900	1,042	929	929	708	952

3. Calculate the depth of the operating Valve and the minimum GLR ratio for the following well data:

$$\text{Depth} = 8,000 \text{ ft.}$$
$$P_{SO} = 800 \text{ psig}$$
$$2 \text{ } 3/8 \text{ in. tubing} = 1.995 \text{ in. ID}$$
$$5 \text{ } 1/2 \text{ in., 20 IB/ft. casing}$$

No water production
$$Y_0 = 0.8762$$
$$\text{BHP (SI)} = 2,000 \text{ psig}$$
$$\text{PI} = 0.10 \text{ bbl/day/psi}$$
$$P_{tf} = 50 \text{ psig}$$
$$T_{av} = 127°\text{F}$$

Cycle time : 45 minutes

Desired production : 100 bbl/day

$$Y_g = 0.80$$

Solution: The static gradient is,

$G_s = 0.8762(0.433) = 0.379$ psi/ft.

Thus, the average flowing BHP is,

$P_{bhfave} = 2,000 - 1,000 = 1,000$ psig.

The depth to the static fluid level with the $p_{tf} = 50$ psig, is

$$D_s = 8,000 - \left(\frac{2,000 - 50}{0.379} \right) = 2855 \text{ ft.}$$

The hydrostatic head after a 1,000 psi drawdown is,

$$D_{dds} = \frac{1,000}{0.379} = 2,639 \text{ ft.}$$

Thus, the depth to the working fluid level is,

$WFL = D_s + D_{dds} = 2,855 + 2,639 = 5,494$ ft.

The number of cycles per day will be,

$$\frac{24(60)}{45} = 32 \text{ cycles/day}$$

The number of bbls per cycle is $\frac{100}{32} \approx 3$ bbls/cycle.

In intermittent – gas lift, operation depending on depth, 30–60% of the total liquid slug is lost due to slippage or fallback.

If a 40% loss of starting slug is assumed, the volume of the starting slug is $\frac{3}{0.60} \approx 5.0$ bbl/cycle.

Because the capacity of this tubing is 0.00387 bb/ft, the length of the starting slug is $\frac{5.0}{0.00387} \approx 1,292$ ft.

This means that the operating valve should be located $\frac{1,292}{2} = 646$ ft. below the working fluid level. Therefore, the depth to the operating valve is 5,494 + 646 = 6,140 ft.

The pressure in the tubing opposite the operating valve with the 50 psig surface back-pressure is,

$P_1 = 50 + (1,292)(0.379) = 540$ psig.

For minimum slippage and fallback, a minimum velocity of the slug up the tubing should be 100 ft./min. This is accomplished by having the pressure in the casing opposite the operating valve opens to be at least 50% greater than the tubing pressure with a minimum differential of 200 psi. Therefore, for a tubing pressure at the valve depth is 540 psig, at the instant that valve opens, the minimum casing pressure at 6,140 ft. is,

$$P_{minc} = 540 + 540/2 = 810 \text{ psig,}$$
$$P_{so} = 707 \text{ psig.}$$

The minimum volume of gas required to lift the slug to the surface will be that required to fill the tubing from injection to surface, less the volume the slug. Thus this volume is (6,140 + 1,29218.8 bbls, which converts to 105.5 ft.^{3}).

The approximate pressure in the tubing is under a liquid slug at the instant the slug equal to the pressure due to the slug length plu back pressure. This is,

$$P_{ts} = \left(\frac{3.0}{0.00387} \right) (0.379) = 344 \text{ ps}$$

Thus, the average pressure in the tubing is,

$$P_{tave} = \frac{810 + 344}{2} = 577 \text{ psig} = 591.7 \text{ p}$$

The average temperature in the tubing is 127°

This gives z = 0.886. The volume of gas at standard condition (API 60°F, 14.695 psia) is,

$$V_{sc} \ 105.5 \left(\frac{591.7}{14.695} \right)\left(\frac{520}{587} \right)\frac{1}{0.886} = 4,246.$$

4. Design of gas lift valves based on the following well data:

Pay zone depth	: 6,500 ft.
Casing size and weight	: 7 in. (PPF 23 lb)
Tubing	: 2 3/8 in., 4.7 lb. (1.995 in ID)

Liquid level surface

Kill fluid gradient	: 0.4 psi/ft.
Gas gravity	: 0.75
Bottom-hole-temperature	: 170°F
Temperature surface flowing	: 100°F
Injection depth	: 6,300 ft.

Minimum tubing pressure at injection

Point	:	600 psi
Pressure kickoff	:	1,000 psi
Pressure surface operating	:	900 psi
Pressure of wellhead	:	120 psi
Tubing pressure margin at surface	:	200 psi
Casing pressure margin	:	0 psi

Valve specifications given in problem 1.

Solution: Design tubing pressure at surface (phf, d)

: $120 + 200 = 320$ psia

Design tubing pressure gradient (G_{fd}) : $(600 - 320)/6{,}300 = 0.044$ psi/ft.

Temperature gradient (G_t) : $(170 - 100)/6{,}300 = 0.011$ F/ft.

$1 - R\ 1.0 - 0.2562 = 0.7438$

$T.E.F. = R/(1 - R)\ 0.2562/0.7438$

$= 0.3444$

Depth of the top valve is,

$$D_1 = \frac{1{,}000 - 0 - 120}{0.40 - \dfrac{1{,}000 - 0}{40{,}000}} = 2{,}347 \text{ ft.}$$

Temperature at the top valve   : $100 + (0.011)(2{,}347) = 126\,^\circ F$

Design tubing pressure at the top valve : $320 + (0.044)(2{,}347) = 424$ psia

For constant surface opening pressure of 900 psia the valve opening pressure is,

$$P_{vo1} = (900)\left(1 + \frac{2{,}347}{40{,}000}\right) = 953 \text{ psia}$$

The dome pressure at the valve depth is calculated with Eq. (13.43):

$$P_{vc} = 817 + (0)(0.2526)(424) = 817 \text{ psia}$$

The dome pressure at 60°F can be calculated with trial-and-error method. The first estimate is given by idea gas law:

$$P_d \text{ at } 60\,^\circ F = \frac{520\ P_d}{T_d} = \frac{(520)(817)}{(126 + 460)} = 725 \text{ psia}$$

Spread sheet programs give $Z_{60F} = 0.80$ at 725 psia and 60°F. The same spread sheet gives $Z_d = 0.85$ at 817 psia and 126°F.

$$P_d \text{ at } 60°F = \frac{(520) [0.80] P_d}{(126+460) (0.85)} (817) = 683 \text{ psia}$$

Test rack opening pressure is,

$$P_{tro} = \frac{683}{0.7438} + 0 = 918 \text{ psia.}$$

5. An oil well has a pay zone around the mid-perf depth of 5,200 ft. The formation oil has a gravity of 30°API and GLR of 500 scf/stb. Water cut remains 10%.

Solution: The IPR of the well is expressed as,

$$q = J[P - P_{wf}]$$

where,

    J = 0.5 stb/day/psi
    P = 2,000 psia

A 2 in. tubing (1.995 in. ID) can be set with a packer at 200 ft. above the mid-perf. What is the maximum expected oil production rate from the well with continuous gas lift at a wellhead pressure of 200 psia if:

(a) Unlimited amount of lift gas is available for the well?
(b) Only 1.2 MMscf/day of lift gas is available for the well?

An oil has a pay zone around the mid-perf depth of 6,200 ft. The formation oil has a gravity of 30°API and GLR of 500 scf/stb. Water cut remains 10%. The IPR of the well is expressed as,

$$q = q_{max} \left[1 - 0.2\frac{P_{wf}}{P} - 0.80.2\left(\frac{P_{wf}}{P} \right) \right]$$

where,

    J = 0.5 stb/day/psi
    P = 2,000 psia

A 2 1/2 in. tubing (1.995 in. ID) can be set with a packer at 200 ft. above the mid-perf. What is the maximum expected oil production rate from the well with continuous gas lift at a wellhead pressure of 200 psia if:

(a) Unlimited amount of lift gas is available for the well?
(b) Only 1.2 MMscf/day of lift gas is available for the well?

6. An oil field has 30 oil wells. The gas lift at the central compressor station is first pumped to four injection manifolds with 4 in. ID, 1.5 mile lines and then distributed to the wellheads with 4 in. ID, 0.4 mile lines. Given the following data, calculate the required output pressure of compression station:

Gas-specific gravity (Y_g)	:	0.70
Base temperature (T_b)	:	60°F
Base pressure (P_b)	:	14.7 psia

7. Calculate the port size from following data:

Upstream Pressure	:	950 psia
Downstream pressure		
For subsonic flow	:	550 psia
Tubing ID	:	1.995 in.
Gas specific gravity	:	0.70 (1 for air)
Gas specific heat ratio	:	1.3
Upstream temperature	:	80°F
Gas viscosity	:	0.03 cp
Choke discharge coefficient	:	0.6

Use Otis Spread master Valve.

8. Calculate the design of gas lift valves based following data:

Pay Zone depth	:	5,400 ft.
Casing size and weight	:	7 in. (PPF 23 lb)
Tubing	:	2 3/8 in., 4.7 lb. (1.995 in ID)

Liquid level surface

Kill fluid gradient	:	0.4 psi/ft.
Gas gravity	:	0.65
Bottom-hole-temperature	:	150° F
Temperature surface flowing	:	80°F
Injection depth	:	5,300 ft.

Minimum tubing pressure at injection

Point	:	550 psi
Pressure kickoff	:	950 psi
Pressure surface operating	:	900 psi
Pressure of wellhead	:	150 psi
Tubing pressure margin at surface	:	200 psi
Casing pressure margin	:	0 psi

Otis 1 1/2 in. OD valve with 1/2 in. diameter seat R = 0.2562

9. Calculate the design of gas lift valves based on following data:

Pay zone depth	: 7,200 ft.
Casing size and weight	: 7 in., 23 lb
Tubing	: 2 3/8 in., 4.7 lb (1.995 in ID)

Liquid level surface

Kill fluid gradient	: 0.4 psi/ft.
Gas gravity	: 0.70
Bottom-hole-temperature	: 160° F
Temperature surface flowing	: 90°F
Injection depth	: 7,300 ft.

Minimum tubing pressure at injection

Point	: 650 psi
Pressure kickoff	: 1,050 psi
Pressure surface operating	: 950 psi
Pressure of wellhead	: 150 psi
Tubing pressure margin at surface	: 200 psi
Casing pressure margin	: 10 psi
Otis $1^1/_2$ in. OD valve with ½ in. diameter seat	
	: R = 0.1942

Sucker Rod Pump

10. A well is pumped off (fluid level is the pump depth) with a sucker rod pump. A 3 in. tubing string (3.5 in. OD, 2,995 in ID) in the well is not anchored. Calculate (a) expected liquid production rate (use pump volumetric efficiency 0.8), and (b) required prime mover power (use safety factor 1.35).

Input data

Pump setting depth (D)	: 4,000 ft.
Depth to the liquid level in annulus (H)	: 4,000 ft.
Flowing tubing head pressure (p_{tf})	: 100 ft.
Tubing outer diameter (d_{to})	: 3.5 in.
Tubing inner diameter (d_{ti})	: 2.995 in.
Tubing anchor (1 = yes; 0 = no)	: 0
Plunger diameter (d_p)	: 2.5 in.
Rod section 1, diameter (d_{r1})	: 1 in.
Length (L_1)	: 0 ft.
Rod section 2, diameter (d_{r2})	: 0.875 in.

Length (L$_2$)   : 0 ft.
Rod section 3, diameter (d$_{r3}$)   : 0.75 in.
Length (L$_3$)   : 4,000 ft.
Rod section 4, diameter (d$_{r4}$)   : 0.5 in.
Length (L$_4$)   : 0 ft.
Type of pumping unit (1 = conventional; –1 = Mark II or Air-balanced)
                                                       : 1
Polished rod stroke length (S)   : 86 in.
Pumping speed (N)   : 22 spm
Crank to pitman ratio (c/h)   : 0.33°
Oil gravity (API)   : 25° API
Fluid formation volume factor (B°)   : 1.2 rb/stb
Pump volumetric efficiency (E$_y$)   : 0.8
Safety factor to prime mover power (F$_s$): 1.35

Solution:

$$A_t = \frac{\pi d^2}{4} = 2.58 \text{ in.}^2$$

$$A_p = \frac{\pi d^2}{4} = 4.91 \text{ in.}^2$$

$$A_r = \frac{\pi d^2}{4} = 0.44 \text{ in.}$$

$$W_f = S_f(62.4)\frac{DA_p}{144} = 7,693 \text{ lb}$$

$$W_r = \frac{Y_r DA_r}{144} = 6,013 \text{ lb}$$

$$M = 1\pm\frac{c}{h} = 1.33$$

$$S_p = S - \frac{12D}{E}\left[WF\left(\frac{1}{A_r}+\frac{1}{A_r}\right) - \frac{SN^2M}{70471.2}\frac{W_r}{A_r}\right] = 70 \text{ in.}$$

$$Q = 0.1484\frac{A_p NS_p E_v}{B_o} = 753 \text{ stb/day}$$

$$L_N = H + \frac{P_{ft}}{0.433\,yl} = 4,255 \text{ ft.}$$

$$P_h = 7.36 \times 10^{-6}{}_{QYl}L_N = 25.58 \text{ hp}$$
$$P_f = 6.31 \times 10^{-7}W_r SN = 7.2 \text{ hp}$$
$$P_{pm} = F_s(P_h + P_f) = 44.2 \text{ hp.}$$

11. The following geometric dimensions are for the pumping unit C-320D – 213-86:

$$D_1 = 96.05 \text{ in.}$$
$$D_2 = 111 \text{ in.}$$
$$c = 37 \text{ in.}$$
$$c/h = 0.33$$

If this unit is used with a 2 1/2 in. plunger and 7/8 in. rods to lift 25°API gravity crude (formation volume factor 1.2 rb/stb) at depth of 3,000 ft., answer the following questions:

(a) What is the maximum allowable pumping speed if L = 0.4 is used?
(b) What is the expected maximum polished rod load?
(c) What is the expected peak torque?
(d) What is the desired counterbalance weight to be placed at the maximum position on the crank?

Solution: The pumping unit C-320D – 213-86 has a peak torque of gear box rating of 320,000 in. lb, a polished rod rating of 21,300 lb, and a maximum polished rod stroke of 86 in.

(a) The polished rod stroke length can be estimated as,

$$S = 2c\frac{d_2}{d_1} = (2)(37)\frac{111}{96.05} = 85.52 \text{ in.}$$

The maximum allowable pumping speed is,

$$N = \sqrt{\frac{70,47.2\,L}{S\left(1-\dfrac{c}{d}\right)}} = \sqrt{\frac{(70,471.2)(0.4)}{(85.52)(1-0.33)}} = 22 \text{ SPM.}$$

(b) The 25°API gravity has an $S_f = 0.9042$. The area of the $2^{1}/_{2}$-in. plunger is $A_2{}^p = 4.91 \text{ in.}^2$ The area of the 7/8 in. rod is $A_r = 0.60 \text{ in.}$ Then,

$$W_f = S_f(62.4)\frac{DA_p}{144}$$

$$= (0.9042)(62.4)\frac{(3,000)(4.91)}{144} = 5,770 \text{ lbs}$$

$$W_r = \frac{Y_S DA_r}{144} = \frac{(490)(3,000)(0.60)}{144} = 6,138 \text{ lbs}$$

$$F_1 = \frac{SN^2\left(1+\dfrac{c}{h}\right)}{70,471.2} = \frac{(85.52)(22^2)(1+0.33)}{70,471.2} = 0.7940.$$

Then the expected maximum PRL is,

$$PRL_{max} = W_f - S_f\,(62.4)\,\frac{W_r}{Y_s} + W_r + W_rW_1$$
$$= 5,770 - (0.9042)\,(62.4)\,(6,138)/(490) + 6,138 + (6,138)\,(0.794)$$
$$= 16,076 \text{ lbs} < 21,300 \text{ lb.}$$

(c) The peak torque is calculated:

$$T = \frac{1}{4}S\left(W_f + \frac{2SN^2Wr}{70471.2}\right)$$
$$= \frac{1}{4}(85.52)\left(5,770 + \frac{2(85.52)\,(22)^2(6,138)}{70,471.2}\right)$$
$$= 280,056 \text{ lb in.} < 320,000 \text{ lb in.}$$

(d) Accurate calculation of counterbalance load requires the minimum PRL:

$$F_2 = \frac{SN^2\left(1 + \dfrac{c}{h}\right)}{70,471.2} = \frac{(85.52)\,(22^2)\,(1-033)}{70,471.2} = 0.4$$

$$PRL_{min} = -S_f\,(62.4)\,\frac{W_r}{Y_s} + W_r - W_r\,F_2$$

$$= -(0.9042)\,(62.4)\,\frac{6,138}{490} + 6,138 - (6,138)\,(0.4)$$

$$= 2,976 \text{ lb.}$$

$$C = \frac{1}{2}\,(PRL_{max} + PRL_{min})$$

$$= \frac{1}{2}\,(16,076 + 2,976) = 9,526 \text{ lb.}$$

A product catalog of LUFKIN Industries indicates that the structure unbalanced is 450 lb and 4 No. 5ARO counterweights placed at the maximum position (c in this case) on the crank will produce an effective counterbalance load of 10,160 lb, that is,

$$W_c = \frac{(37)\,(96.05)}{(37)\,(111)} + 450 = 10,160$$

Which gives $W_c = 11,221$ lb. To generate the ideal counter-balance load of $C = 9,526$ lb, the counterweights should be placed on the crank at,

$$R = \frac{(9,526)\,(111)}{(11,221)\,(96.02)} + (37) = 36.30 \text{ inch.}$$

12. Solution given by computer program sucker rod pumping load .xls.

Sucker Rod Pumping Load .Xls

Description: This spread sheet calculates the maximum allowable pumping PRL, the minimum PRL. Peak torque, and counterbalance load.

Instruction: (1) update parameter values in the Input section; and (2) view result in the solution section.

Input data

Pump setting depth (D)	:	3,000 ft.
Plunger diameter (d_p)	:	2.5 in.
Rod section 1, diameter (d_{r1})	:	1 in.
Length (L_1)	:	0 ft.
Rod section 2, diameter (d_{r2})	:	0.875 in.
Length (L_2)	:	0 ft.
Rod section 3, diameter (d_{r3})	:	0.75 in.
Length (L_3)	:	4,000 ft.
Rod section 4, diameter (d_{r4})	:	0.5 in.
Length (L_4)	:	0 ft.

Type of pumping unit (1 = conventional; –1 = Mark II or Air-balanced)

	:	1
Polished rod stroke length (S)	:	86 in.
Pumping speed (N)	:	22 spm
Crank to pitman ratio (c/h)	:	0.33°
Oil gravity (API)	:	25°API

Maximum allowable acceleration factor (L)

	:	0.4

Solution:

$$S = 2c\frac{d_2}{d_1} = 85.52$$

$$N = \sqrt{\frac{70471.2\,L}{S\left(1-\dfrac{c}{h}\right)}} = 22\text{ SPM}$$

$$A_p = \frac{\pi d^2}{4} = 4.91 \text{ in.}^2$$

$$A_r = \frac{\pi d^2}{4} = 0.60 \text{ in.}$$

$$W_f = S_f(62.4)\,\frac{DA_p}{144} = 5{,}770 \text{ lb.}$$

$$W_r = \frac{Y_S DA_r}{144} = 6{,}138 \text{ lb}$$

$$F_1 = \frac{SN^2\left(1+\dfrac{c}{h}\right)}{70{,}471.2} = 0.7940^\circ$$

$$\text{PRL}_{max} = W_f - S_f\,(62.4)\,\frac{W_r}{Y_S} + W_r + W_rF_1 = 16{,}076 \text{ lb}$$

$$T = \frac{1}{4}S\left(W_f + \frac{2SN^2 Wr}{70{,}471.2}\right) = 280{,}056 \text{ lb}$$

$$F_2 = \frac{SN^2\left(1+\dfrac{c}{h}\right)}{70{,}471.2} = 0.40$$

$$\text{PRL}_{min} = -S_f\,(62.4)\,\frac{W_r}{Y_S} + W_r - W_r\,F_2 = 2{,}976 \text{ lb}$$

$$C = \frac{1}{2}\,(\text{PRL}_{max} + \text{PRL}_{min}) = 9{,}526 \text{ lb.}$$

13. A well is pumped off with a rod pump. A 2 1/2 in. tubing string (2.875 in. OD, 2,441 ID) in the well is not anchored. Calculate (a) expected liquid production rate (use pump volumetric efficiency 0.80) and (b) required prime mover power (use safety factor 1.3).

14. A well is pumped with a rod pump to a liquid level of 2,800 ft. A 3 in. tubing string (3 1/2 in. OD, 2,995 in. ID) in the well is anchored. Calculate (a) expected liquid production rate (use pump volumetric efficiency 0.85) and (b) required prime mover power (use safety factor 1.4).

15. A well is to be put on a sucker rod pump. The proposed pump setting depth is 4,500 ft. The anticipated production rate is 500 bbl/day of oil of 40°API gravity against wellhead pressure 150 psig. It is assumed that the working liquid level is low, and a sucker rod string having a working stress of 30,000 psi is to be used. Select surface and subsurface equipment for the installation. Use a safety factor of 1.40 for prime mover power.

16. A well is to be put on a sucker rod pump. The proposed pump setting depth is 4,000 ft. The anticipated production rate is 550 bbl/day oil of 35° API gravity against wellhead pressure 120 psig. It is assumed that working liquid level will be

about of 30,000 psi is to be used. Select surface and subsurface equipment for the installation. Use a safety factor of 1.40 for prime mover power.

17. A pumping well is production 230 bbl/day of oil and 85 bbl/day of water with a tubing GOR of 140 scf/bbl. The pump displacement rate is estimated to be 385 bbl/day, the PVT data of the oil are given in chart. Estimate the pressure at the pump intake.

Ans.: 625 psig.

18. SV and TV checks are carried out on a well completed with 27000 ft of 3/4 in., rods and 2 in. diameter plunger. A load of 3810 Ib is recorded during the former and a load of 4820 IB during the latter test. Estimate the pressure exerted by the formation at the SV while the well is pumping, if the THP is 100 psig.

Ans.: 333 psig.

Chamber Lift

19. What would be the effect if a chamber-lift installation were made in well 7 of Block A, the installation involving 2000 ft. of 4 1/2 in. tubing (internal cross-sectional area 0.0850 sq ft.) run on 9000 ft. if 2 7/8 in. tubing? (In this problem it is necessary to assume a value for the THP during production of the slug; it is suggested 300 psig as a reasonable figure. The explanation for this point of difference is the gas pressure required to move the slug up the 2 7/8 in. tubing is considerably greater than the pressure needed to lift the liquid out of the 4 1/2 in. chamber because of the length of the slug in the smaller tubing).

Ans.: Optimum number of cycles per day, 31; production rate, 720 bbl/day ($\pm$ 50); volume of injected gas per cycle, 53.5 mcf ($\pm$ 3.5); volume of injected gas per day, 1660 mcf ($\pm$ 110); maximum gas-injection pressure, 1880($\pm$ 115).

Note: If a figure of 400 psig were assumed for the THP during production of the slug, the answer for the optimum number of cycle per day and for the production rate would not be changed. However, the predicted volume of injected gas per cycle would rise to 56.7 mcf ($\pm$ 3.5) and the maximum gas-injection pressure to 1980 psig ($\pm$ 115).

20. A well completed with 4492 ft of 2 3/8 in. tubing has static pressure of 1100 psig at 4500 ft. It is production on plunger lift with an average CHP of 220 psig. What are the production rate and efficiency of the operation, if it is assumed that the trap pressure may be neglected?

Ans.: 518 bbl/day; 78 percent.

Plunger Lift

21. A well completed with 3400 ft. 2 3/8 in tubing has static pressure of 800 psig at 3450 ft. It is production on plunger lift with an average CHP – 160 psig. What are the production rate and efficiency of the operation, if it is assumed that the trap pressure may be neglected?

References

Alhanati, F.J.S., Schmidt, Z, Doty, D.R. *et al.* (1993). Condition Gas-Lift Instability: Diagnosis. *Criteria and Solutions,* SPE 26554.

Allan, J.C., Moore, P.C. and Adair, P. (1989). Design and Application of an Integral Jet Pump/ Safety Valve in a North Sea Oilfield. SPE 19279.

Anderson, J., Freeman, R. and Pugh, T. (2005). Hydraulic Jet Pumps Prove Ideally Suited for Remote Canadian Oil Field, SPE 94263.

Baklid, A., Apeland, O.J. and Teigen, A.S. (1998). CT ESP for Yme, Converting the Yme Field Offshore Norway from a conventional Rig-Operated Field to CT-Operated for Work over and Drilling Applications, SPE 46018.

Bayh III, R.I. and Neuroth, D.H. (1989). Enhanced Production from Cable-Deployed Electrical Pumping Systems, SPE 19707.

Beauquin, J.-L., Boireau, C., Lemay, L., *et al.* (2005). Development Status of a Metal Progressing Cavity Pump for Heavy Oil and Hot Production Wells, SPE/S-CIM/CHOA 97796.

Blanksby, J., Hicking, S. and Milne, W. (2005). Development of High-Horsepower ESPs to Extend Brent Field Life, SPE 96797.

Boothby, L.K. Garred, M.A. and Woods, J.P. (1988). Application of Hydraulic Jet Pump Technology on an Offshore Production Facility, SPE 18236.

Boyun Guo, Willam C. Lyons and Ali Ghalam, bor-Petroleum Production Engineering.

Breit, S., Sikora, K. and Akerson, J. (2003). Overcoming the Previous Limitations of Variable Speed Drives on Submersible Pump Applications. SPE 81131.

Brinkhorst, J.W. (1998). Successful Application of High GOR ESPs in the Lekhwair Field, SPE 49466.

Brown, K.E. (1982). Artificial Lift Volume 2.

Brown, K.E. (1982). Overview of Artificial Lift Systems, SPE 9979.

Brown, K.E., The Technology of Artificial Lift Methods Vol. 2a Tulsa, OK Petroleum Publish Co, 1980.

Butlin, D.M. (1991). The Effect of Motor Slip on Submersible Pump Performance, SPE 23529.

Chacin, J.E. (1994). Selection of Optimum Intermittent Lift Scheme for Gas Lift Wells, SPE 27986.

Chen, A., Li, H., Zhang, Q., *et al.* (2007). Circulating Usage of Partial Produced Fluid as Power Fluid for Jet Pump in Deep Heavy-Oil Production, SPE 97511.

Christ, F.C. and Zubin, J.A. (1983). The Application of High Volume Jet Pump in North Slope Water Source Wells, SPE 11748.

Corteville, J.C., Ferschneider, G., Hoffmann, F.C., *et al.* (1987). Research on Jet Pumps for Single and Multiphase Pumping of Crudes, SPE 16923.

De Ghetto, Riva, M. and Giunta, P. (1994). Jet Pump Testing in Indian Heavy Oils, SPE 27595.

Evans, R.D. and Weaver, P. (1985). Performance Analysis and Field Testing of a Compact Dual-Piston, Hydraulic Sucker Rod Pumping Unit, SPE 13807.

Fairuzov, Y.V., Geurrero-Sarabia, I., Calva-Morales, C., *et al*. (2004). Stability Maps for Continuous Gas Lift Wells: A New Approach to Solving an Old Problem, SPE 90644.

Fairuzov, Y.V., Geurrero-Sarabia, I., Calva-Morales, C., *et al*. (2004). Stability Maps for Continuous Gas-Lift Wells, SPE 97275.

Faustinelli, J., Briceno, W. and Padron, A. (1998). Gas Lift Jet Applications Offshore Lake Maracaibo, SPE 48840.

Ferguson, S.E. and Moyes, P.B. (1997). Preventing Fluid Losses in ESP Well Completions: Avoid Formation Damage and Improve Pump Life. SPE 38041.

Filho, C.O.C. and Boradlo, S.N. (2005). Assessment of Intermittent Gas Lift Performance Through Simultaneous and Coupled Dynamic Simulation, SPE 94946.

Gadbrashitov, I.F. and Sudeyev, I.V. (2006). Generation of Curves of Effective Gas Separation at the ESP Intake on the Basis of Processed Real Measurements Collected in the Priobskoye Oil Field, SPE 102272.

Gamboa, J., Olivet, A. and Espin, S. (2003). New Approach for Modeling Progressive Cavity Pumps Performance, SPE 84137.

Ghareeb, M.M., Shedid, S.A. and Ibrahim, M. (2007). Simulation Investigations for Enhanced Performance of Beam Pumping System for Deep, High-volume Wells, SPE 108284.

Gibbs, S.G. (1982). A Review of Methods for Design and Analysis of Rod Pumping Installations, SPE 9980.

Gibbs, S.G. (1991). Application of Fiberglass Sucker Rods, SPE 20151.

Giuggioli, A. and De Ghetto, G. (1995). Innovative Technologies Improve the Profitability of Offshore Heavy.

Grupping, A.W., Coppers, J.L.R. and Groot, J.G. (1998). Fundamentals of Oilwell Jet Pumping, SPE 15670.

Harun, A.F., Prado, M.G. and Doty, D.R. (2003). Design Optimization of Rotary Gas Separator in ESP Systems, SPE 80890.

Hatzlavramidis, D.T. (1991). Modeling and Design of Jet Pumps, SPE 19713.

Heuman, W.R., Moore, E.R.B., Yeu, Y., *et al*. (1995). ESP Run Life Maximisation for the Xijiang Field Development, SPE 29970.

Hongen, D. (1995). Technologies Used in the Production of High Pour Point Crude Oil in Shenyang Oilfield, SPE 29953.

James F. Lee, Gas Well Deli quification.

Jariwala, H., Davies, J. and Hepburn, Y. (1996). Advances in the Completion of 8 km Extended Reach ESP Wells, SPE 36579.

Jennings, J.W. (1989). Design of Sucker—Rod Pump System, SPE 20152.

Jonathan Bellarly, Well Completion Design.

Klein, S.T. (2002). Development of Composite Progressive Cavity Pumps, SPE 78705.

Kulyuan, L. (1995). The Application Experience of Electrical Submersible Pump (ESP) in Offshore Oilfields, Bohai Bay, China, SPE 29952.

Leismer, D. (1993). A System Approach to Annular Control for Total Well Safety, SPE 26740.

Macary, S., Mohamed, I., Rashad, R., *et al*., (2003). DownholePermanent Monitoring Tackles Problematic Electrical Submersible Pumping Wells, SPE. 84138.

Mack, J. and Donnell, J. (2007). Coil Tubing Deployed ESP in 5½ inch Casing: Challenges in Designing Down, SPE 106875.

Mahgoub, I S., Shahat, M.M. and Fattah, S.A. (2005). Overview of ESP Application in Western Desert of Egypt—Strategy for Extending Lifetime, IPTC 10142.

Manson, D.M. (1986). Artificial Lift by Hydraulic Turbine—Driven Downhole Pump: Its Development/Application, and Selection, SPE 14134.

Martinez, J., "Downhole Gas Lift and the Facility". Paper Presented at the ASME/API Gas Lift Workshop, Held in Houston, TX, February 4–5, 2003.

MCAfee, R.V. "The Evaluation of Vertical Lift performance in Producing wells" JPT, April 1961, 3908.

McCoy, J.N., Becker, D.J., Rowlan, O.L., *et al.* (2002). Minimizing Energy Cost by Maintaining High Volumetric Pump Efficiency, SPE 78709.

Mills, R.A.R. and Gaymard, R. (1996). New Applications for Wellbore Progressing Cavity Pumps, SPE 35541.

Mitra, N.K. and Singh, Y.K . (2007). Increased Oil Recovery from Mumbai High through ESP Campaign, OTC 18748.

Miwa, M., Yamada, Y. and Kobayashi, O. (2000). ESP Performance in Mubarraz Field, SPE 87257.

Moore, P.C. and Adair, P. (1991). Dual Concentric Gas-Lift Completion Design Changes for High GLR and High Sand Production; Apache Stag Project SPE 77801.

Neely, A.B. Montgomery, J.W. and Vogel, J. (1974). A Field Test and Analytical Study of Intermittent Gas Lift, SPE Jour, 502–12; Trans, AIME, 257.

Neely, B., "Section of Artificial Lift Methods". Paper SPE 10337 Presented at the 56th Annual Fall Technical Conference and Exhibition held in San Antonio, TX, October 5–7, 1981.

Noonan, S.G., Decker, K.L. and Mathisen, C.E. (2000). Subsea Gas Lift Design for the Angola Kuito Development, OTC 11874.

Noronha, F.A.F., Franca, F.A. and Alhanati, F.J.S. (1998). Improved Two-Phase Model for Hydraulic Jet Pumps, SPE 50940.

Norris, C. and A1-Hinai, S.H. (1996). Operating Experience of ESP's in South Oman, SPE 36183.

NORSOK Standard D010 (2004). Well Integrity in Drilling and Well Operations, Standards Norway.

Novillo, G. and Cedeno, H. (2001). ESP's Application in Oritupano-Leona block, East Venezuela, SPE 69434.

Ogunsina, O.O. and Wiggins, M.L. (2005). A Review of Downhole Separation Technology, SPE 94276.

Pankratz, R.E. and Wilson, B.L. (1998). Predicting Power Cost and Its Role in ESP Economics, SPE 17522.

Patterson, M. M. (1996). On The Efficiency of Electrical Submersible Pumps Equipped with Variable Frequency Drives: A Field Study, SPE 25445.

Perrin, D. (1999). Well completion and Servicing. Institut Francais Du Petrole Publications, ISBN 2-7108-0765-3.

Pessoa, R. and Prado, M. (2003). Two-Phase Flow Performance for Electrical Submersible Pump Stages, SPE 81910.

Pickford, K.H. and Morris, B.J. (1989). Hydraulic Rod-Pumping Units in Offshore Artificial-Lift Applications, SPE 16922.

Poblano, E., Camacho, R. and Fairuzov, Y.V. (2005). Stability Analysis of Continuous—Flow Gas Lift Wells, SPE 77732.

Powers, M.L. (1998). Economic Considerations for Sizing Tubing and Power Cable for Electric Submersible Pumps, 15423.

Ramirez, M., Zdenkovic, N. and Medina, E. (2000). Technical/Economical Evaluation of Artificial Life Systems for Eight Offshore Reservoirs, SPE 59026.

Sawaryn, S.J. (2003). The Dynamics of Electrical-Submersible –Pump Populations and the Implication for Dual-ESP Systems, SPE 87232.

Sawaryn, S.J., Grames, K.N. and Whelehan, O.P. (2002). The Analysis and Prediction of Electric Submersible Pump Failure in the Milne Point Field, Alaska, SPE 74685.

Sawaryn, S.J., Norrell, K.S. and Whelehan, O.P. (1999). The Analysis and Prediction of Electric Submersible—Pump Failures in the Milne Point Field, Alaska, SPE 56663.

Schmidt, Z., Dotyu, D.R., Lukong, P.B., *et al.* (1984). Hydrodynamic Model for Intermittent Gas Lifting of Viscous Oil, SPE 10940.

Shaw, S.F., Gas Lift Principles and practice Houston, TX: Gulf Publishing Co, 1939.

"Recommend Practice for Operation, Maintenance and Trouble Shooting of Gas Lift Installation" API RP-11 - V$_5$, 2nd Edition, American Petroleum Institute, 1999.

"Recommend Practice for Design of Continuous flow Gas Lift Installations Using Injection Pressure Operated Valves." API RP-11 – V$_6$, 2nd Edition, American Petroleum Institute.

Stephens, R.K., Loveland, K.R., Whitlow, R.R., *et al.* (1996). Lessons Learned on Coiled Tubing Completions SPE 35590.

Stewart, D.R. and Holland, B. (1997). Innovative ESP Completions for Liverpool Bay Development, SPE 36936.

Sun, D. and Prado, M. (2006). Single-Phase Model for Electric Submersible Pump (ESP) Head Performance, SPE 80925.

T.R.W., Nind, Principle of Oil Well Production.

Taufan, M., Adriansyah, R. and Satriana, D. (2005). Electrical Submersible Progressive Cavity Pump (ESPCP) Application in Kulin Horizontal Wells, SPE 93594.

Thoma, O., Allen, Alam, P., Robert, Production Operations volume.

Tischler, A., Woodward, T.A. and Becker, B.G. (2005). Coiled-Tubing Gas Lift Reclaims 2000 BOPD of Lost Crude, SPE 95682.

Treadway, R.B. and Focazio, K.R. (1981). Fiberglass Sucker Rods—A Futuristic Solution to Today's Problem Wells, SPE 10251.

Tripp, H.A. (1988). Mechanical Performance of Fiberglass Sucker—Rod Strings, SPE 14346.

Williams, C.R., Bayh III, R.I., O'Dell, P.M., *et al.* (1992). A Subsurface Safety Valve Specifically Designed for Jet Pump Applications, SPE 24066.

Wilson, B.L., Mack J. and Foster, D. (1998). Operating Electrical Submersible Pumps below the Perforations, SPE 37451.

Winkler, H.W. (1994). Misunderstood or Overlooked Gas-Lift Design and Equipment Considerations, SPE 27991.

Winkler, H.W. and S.S. Smith. Gas Lift Manual Camco Ine, 1962.

Xu, Z.G. and Golan, M. (1989). Criteria for Operation Stability of Gas-Lift Wells, SPE 19362.

Zuvanich, P.L. (1959). High Volume Lift with Hydraulic Long Stroke Pumping Units, SPE 1223.

www.ingramcontent.com/pod-product-compliance
Lightning Source LLC
Chambersburg PA
CBHW080131220326
41598CB00032B/5023